The British Dam Society

Long-term benefits and performance of dams

Proceedings of the 13th Conference of the British Dam Society and the ICOLD European Club meeting held at the University of Kent, Canterbury, UK from 22 to 26 June 2004.

Edited by Henry Hewlett

Conference organised by the British Dam Society. www.britishdams.org

Organising Committee: Andy Hughes (Chairman), Keith Gardiner, Jon Green, Henry Hewlett, David Littlemore, Mark Morris and Neil Sandilands

In addition to members of the organising committee, the following personnel kindly assisted with the review of papers: Chris Binnie, Rod Bridle, Alan Brown, Wendy Daniell, Kenny Dempster, Michael Kennard, Howard Lovenbury, Mark Noble and Paul Tedd

Published by Thomas Telford Publishing, Thomas Telford Ltd, 1 Heron Quay, London E14 4JD. www.thomastelford.com

Distributors for Thomas Telford books are
USA: ASCE Press, 1801 Alexander Bell Drive, Reston, VA 20191-4400, USA
Japan: Maruzen Co. Ltd, Book Department, 3–10 Nihonbashi 2-chome, Chuo-ku, Tokyo 103
Australia: DA Books and Journals, 648 Whitehorse Road, Mitcham 3132, Victoria

First published 2004

The cover photographs show Bewl Water, courtesy of Kent County Council and Leigh Barrier, Tonbridge, and the Thames Barrier, both courtesy of the Environment Agency

Also available from Thomas Telford Books
Interim guide to quantitative risk assessment for UK reservoirs. DEFRA. ISBN 0 7277 3267 6
Reservoirs in a changing world, British Dam Society. ISBN 0 7277 3139 4
The reservoir as an asset. British Dam Society. ISBN 0 7277 2528 9
Reservoir safety and the environment. British Dam Society. ISBN 0 7277 2010 4
Water resources and reservoir engineering. British Dam Society. ISBN 0 7277 1692 1
The prospect for reservoirs in the 21st century. British Dam Society. ISBN 0 7277 2704 4
A guide to the Reservoirs Act 1975. DETR and ICE. ISBN 0 7277 2851 2
Land drainage and flood defence responsibilities: 3rd edition. ICE. ISBN 0 7277 2508 4
Dams 2000. British Dam Society. ISBN 0 7277 2870 9
Reservoir engineering. Guidelines for practice. Edward Gosschalk. ISBN 0 7277 3099 1
Practical dam analysis. M A M Herzog. ISBN 0 7277 2725 7

A catalogue record for this book is available from the British Library

ISBN: 0 7277 3268 4

© The authors and Thomas Telford Limited 2004

Printed and bound in Great Britain by MPG Books, Bodmin

Preface

This book contains the proceedings of the 13th Conference of the British Dam Society, *Long-term benefits and performance of dams,* held at the University of Kent, Canterbury, in June 2004.

The 55 papers cover a wide variety of issues. There are papers covering the benefits that reservoirs can provide in terms of water supply and recreation and also on the environmental impacts they can have. The use of new materials in reservoir construction is discussed, in particular the use of geomembranes to provide watertightness. Several papers describe portfolio risk assessments undertaken both in the UK and overseas, and various other issues related to reservoir management are covered, including changes in the enforcement of UK reservoir safety legislation. The issue of climate change and its effect on reservoir safety is considered and there is an update on European research into dambreak analysis. A number of flood alleviation schemes involving flood storage reservoirs are described, with particular emphasis on hydraulic controls and environmental aspects. The performance of dams in the UK, Norway, Iran and Kenya is analysed from long-term records of instrumentation and monitoring. Case histories of dam failures and 'near misses' are described with lessons that can be learnt from them. Various studies and works to rehabilitate a number of dams of differing types, sizes and age are also described.

The conference included the presentation of the biennial Geoffrey Binnie Lecture by Roy Coxon. The 2004 Lecture, entitled *'Matters related to adverse incidents associated with dams and the role of Review Boards'* is published in the Society's journal *Dams and Reservoirs.*

Contents

6. Incidents and rehabilitation case histories

1. Benefits and social impacts of reservoirs

The benefit of dams to society

C.J.A. BINNIE, Independent Consultant

SYNOPSIS.   Dams have been constructed from historical times to provide the needs of many civilisations. Focussing primarily on the UK, the paper sets out the benefits of improved health and life from the provision of a clean water supply from reservoirs, protection from drowning and damage from floods, the provision of power from hydro schemes, water for irrigation, as well as the recreation and environmental benefits of the reservoirs.

INTRODUCTION

At the launch of the World Commission on Dams Report, Nelson Mandela said that for all the problems around some dams; "the problem is not the dams. It is the hunger, It is the thirst. It is the darkness of a township. It is the townships and rural huts without running water, lights, or sanitation." (Bridle 2003). How true. This paper looks at the benefits of dams, and the problems that society would face without them, concentrating primarily on the UK but with illustrations from other countries.

Dams provide water for society to drink and use, protection from both river and marine floods, hydro electric power, irrigation water to grow food, a pleasant recreation area, and enhanced environment. Dams have been constructed during different periods depending on the needs of society at the time.

EARLY DAMS

Dams have been reported from earliest historical times such as the Maan Dam which provided water for irrigation and water supply for the Queen of Sheba's people.

Some of the oldest small reservoirs in UK were constructed by the medieval monasteries to provide supplies of fish, generally carp. The provision of fresh food over a longer season must have been of nutritional benefit in those times.

MILL DAMS

The Doomsday Book, compiled in 1086, included some 7,000 mills in Britain,(Binnie G.M.1987) Many of these would have used a low dam to control water flow in the mill leat and the stream. During medieval times these were used to generate power for flour milling and later for fulling wool. Few mill dams survive today.

The Wealden iron industry, boomed between about 1540 and 1640 using water power to drive bellows to generate heat, to drive the hammer mills that were used to form the iron and to bore cannon (Binnie G.M. 1987). A few hammer ponds survive today.

From about 1750 blast furnaces powered by coal along with water powered hammer mills were developed in Shropshire. Water power was also used to power the spinning and textile mills. Because of the high rainfall these were located on each side of the Pennines (Binnie G.M.1987).

Thus water power from dams and rivers provided the beginnings of manufacturing that led to the industrial revolution and Britain becoming a major exporting nation and the ensuing wealth.

ORNAMENTAL LAKES

The industrial revolution resulted in uncontrolled development, often with unsanitary housing conditions, so the wealthy classes sought separation by constructing large houses and elaborate gardens, often with ornamental lakes. The leading exponent of this was Lancelot "Capability" Brown. Examples today include Stowe, Sheffield Park in Sussex and Stourhead created by Henry Hoare (Binnie G.M. 1987). Many of these are now run by the National Trust and give pleasure to hundreds of thousands of visitors each year.

CANAL DAMS

With the start of the industrial era, based initially on water power, and the opening of the coal mines, a means of transport for coal, iron ore and heavy goods was required. The roads were frequently poorly maintained, rutted tracks and not suitable for transporting heavy loads, particularly coal. Between 1770 and 1830 over 2,000 km of canals were constructed. Water is required for locking and so reservoirs were constructed to provide water to the summit pounds of almost all canals.

As an example of the benefit that canals and their reservoirs can bring, Birmingham, while near to coal and iron mines, was too far from them to be served by the then roads. A ring of canals was constructed both to bring in raw materials and also to carry manufactured products to London and other ports. This enabled Birmingham industry to flourish.

Without reservoirs canals could only have been built in the lower reaches of a valley where the natural flow in the river was sufficiently in excess of that needed by the mill owners and other users to allow enough for canal locking. Without reservoirs the canal network would have been inadequate for more than local transport, there would have been no link between Yorkshire industry and the important port of Liverpool, Birmingham would have been virtually land locked, and there would have been no inland route to deliver coal to London: see Figure 1 (Dutton 2003). This would have seriously restricted and delayed the industrial revolution on which the wealth of our country was based.

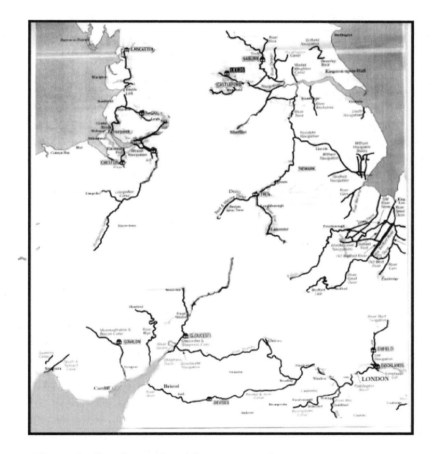

Figure 1: Canal system with no reservoirs.

The canals are now used mainly for recreation with many canal barges being used for sedate leisure, and canal banks providing solitude for anglers often close to urban environments.

In addition water side property in urban areas now provides desirable views and features so these canal areas have attracted redevelopment in such places as Birmingham, Manchester, the Little Venice area of London, and in Leeds.

Thus, over 200 years after their construction, the reservoirs that enabled the canals to be viable continue to serve society.

Dams and reservoirs also support canals in other countries. Probably of most note are the dams at the south end of the Panama Canal which stored water and raised the canal to the extent that its construction could eventually be completed. Without them it would not have been, certainly for many decades later, thus restricting the development and naval defence of the Unites States.

DOMESTIC WATER SUPPLY
As the industrial revolution developed it resulted in much overcrowding and squalor in the expanding industrial cities. Reformers, in particular Edwin Chadwick, realised that conditions, and therefore the health of the people, would be improved by the provision of a clean water supply and the disposal of sewage (Binnie GM 1981). Following Chadwick's report in 1842 (Chadwick 1842), the Public Health Act of 1848 provided, through the Central Board of Health, the means to support towns and cities in providing water supply and sewerage. The health benefits of the wholesome, generally upland, water supplies are illustrated by cholera statistics. In 1832 there were 30,000 deaths from cholera and in 1849 60,000. Deaths continued in the large cities. In 1857 John Snow published his paper on the Broad Street pump episode, demonstrating that infection occurred not from the supposed miasma in the air but from sewage contaminated well water. It was then realised that almost all the rivers in and downstream of urban areas were also polluted both from the sewers and from the filth from the generally unpaved streets.

Steam pumping was expensive so most new water supplies were provided by gravity from reservoirs constructed in upland areas. Because the need for clean water was understood but the methods of water treatment were known to have little effect, most reservoirs had any sources of pollution, such as people and cattle, removed from the catchment area (Binnie C 1995).

The City fathers were not entirely altruistic in improving the health of the industrial workers. Their output increased as well.

The large towns then started to construct upstream reservoirs, Manchester in 1848, Liverpool in 1852 and London in about 1870. The benefit of clean water supplies can be seen in the graph of Enteric deaths in Figure 2, (Binnie C 1995).

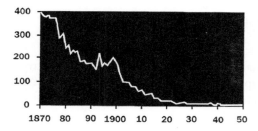

England & Wales - Enteric Fever
Standardised Death Rates at all ages
for each year from 1870 to 1945 (after Bradley)

Figure 2: Enteric Fever standardised death rates

There were factories driven by water power downstream of many of the new reservoirs. The mill owners were powerful and demanded that a steady release of water be made, generally amounting to one third of the yield of the reservoir. Today the mills are long closed but this compensation water flow continues providing the benefit of a healthy river environment all year.

It is interesting to consider what the sustainable population would have been without reservoirs. We today use much more water per person than people did in the late Victorian era when reservoirs were first being built. However steam pumping then was both inefficient and costly so long distance transfer of water then would have been impractical.

Taking the instance of Greater London the population in 1870 when reservoirs started to be built was about 4 million. Today the water supplied to Greater London is about 2,200 Mld (Arkell 2003). The river Thames already goes below its minimum environmental flow and the amount available from groundwater is about 200 Mld. Allowing conjunctive use to double this amount would mean, without reservoirs, a population limitation of about 20 percent of the current, or about one and a half million. My great grandfather submitted a Bill to Parliament to bring water by gravity pipeline from Llangorst in Wales to London. However without the benefit of storage the sustainable dry weather flow of the Welsh uplands would have been low. Until the development of desalination plants during the 1960's London's population would have been severely constrained.

Looking at England and Wales as a whole, the total water supplied today is about 15,000 Mld (Water Facts 2000). The total groundwater abstraction licensed is 8,476 Mld. The Environment Agency consider that about 1,000 Mld of this is over licensed and unsustainable. On the other hand it considers that about 1,000 Mld of river flow could be abstracted during dry weather (Watts 2003). Conjunctive use could increase the amount of water available. However, it can be seen that, without reservoirs, the total population of the country would have been appreciably constrained until the first economic electric pumps became available for long distance transfer, and then the advent of desalination systems.

HYDRO AND TIDAL POWER
After the 1939 – 1945 War the nationalisation of the power industry facilitated a major initiative to develop the hydropower potential offered by the terrain and water in the Highlands of Scotland (Bridle and Sims 1999). Governments throughout the world have used hydropower development to create employment, not only on the project itself, but through a Keynesian multiplier affecting other industries attracted by the energy. The British Government is no exception and the development of hydropower in Scotland was motivated to some extent in this way. By 1980 the hydropower installed in the North of Scotland was 1756MW with an annual output of over 3,000 GWh.

In the North of Scotland over 2,400 km of transmission circuits were constructed. The development of hydropower opened up the Highlands. The construction of a wide transmission system enabled industries to prosper and provide skilled jobs, thereby retaining young people in the Highlands and sustaining a society there with a complete cross section of jobs and income levels. Hydro-production funded the spread of transmission capacity into the glens and farms started to be connected to electricity for the first time. This brought them electric lighting, a fundamental improvement in a region where the winter nights are long, and once the farmers became familiar with the benefits of electricity, they started to use it for milking and to develop their output in other ways. By 1980, 94% of all farms in the neighbourhood were connected and were using a total of 241 MWh. The construction of the dams and power stations also required the construction of new high quality access roads which in turn provided much improved access and in turn brought in tourism.

The problem with nuclear and coal fired power stations was that they were unable to respond to rapid fluctuations of power demand such as when a system tripped as occurred in 2003 in the eastern United States or when popular sporting events had half time or finished such as the 2003 Rugby World Cup. To respond to this pumped storage schemes were constructed in Wales with a high and low reservoir. The first, the 360MW Ffestiniog scheme was completed in 1964 and the second, the 1800 MW Dinorwig scheme, in 1981. Dinorwig can be brought to full generating load in 12 seconds and is also used to control the frequency of the national grid system. Constructing dams, particularly the 70m high Marchlyn Dam within the Snowdonia National Park, was a challenge but the schemes are now major tourist attractions.

Dams can also be formed in the sea where the tidal range is high and thus generate tidal power. There are several tide mills dating from medieval times and the Carew mill in Pembrokeshire is still in operation today. The 230MW La Rance scheme in Brittany was constructed in the mid 1960s and is the largest in operation.

The English Stones Barrage near the Severn Bridge could develop about 970 MW. (Binnie C.J.A. and Roe 1986) The Severn Barrage lower down the Bristol Channel between Lavenock Point and Brean Down could have an installed power of 7,2000MW, and annual energy output of 14.4 TWh. (STPG 1986). Taylor (1998) estimated this could provide up to 7% of the demand of England and Wales without the emission of polluting gases or the generation of toxic waste products.

Hydropower could contribute much to the UK's efforts in meeting the objectives of the Rio and Kyoto Conferences in reducing green house gas emissions to minimise the impact of climate change. Each kilowatt-hour generated by hydropower saves about 900 grams of carbon dioxide when compared to coal generated power. The hydropower generated between 1947 and 1980 therefore saved a total of 62 million tonnes of carbon dioxide in the atmosphere. (Bridle and Sims 1999)

Internationally hydropower is the world's main source of renewable energy providing about 20% of the world's energy generation. (British Hydropower Association 2003.) Installed capacity is 674,000MW with a further 103,000MW under construction. Dams are required to provide almost all of this.

FLOOD PROTECTION FROM RIVER FLOODING
Dams provide the benefit of protection from flooding from rivers in two ways, either by direct protection or by routing the flood through a reservoir provided for other means thus reducing the peak flow in the river downstream of it.

Examples of the former include many of the dykes through Holland. A good example of the latter in this country is the Leigh Barrier which protects Tonbridge. Without the benefit of the barrier, Tonbridge would have suffered severely in the Autumn 2000 floods.

Nowadays the Environment Agency insist that new development does not increase flooding downstream and that storage be provided. This can either be achieved by excavation of compensation storage but more often by the construction of a dam and empty reservoir. Flood defence reservoirs are often used as amenity areas or used for grazing or other agricultural purposes. Thus dams provide the benefit of being able to carry out development without the risk of increased flooding downstream.

The experience of Dublin described by Mangan (1996) is typical of the contribution by dams to flood relief in the British Isles. Huarricane Charlie produced intense rainfall and flooding on 25th and 26th August 1986. Twenty four hour rainfall in excess of 200 mm was recorded in the Dublin Mountains. The peak inflow to the Pollaphuca Reservoir, at the top of the cascade of the dams in the Liffey valley was 445m^3/s. No flooding was experienced in Dublin. A hydrological model simulating the flow in the Liffey at Dublin without the retention provided by the reservoirs suggests a flow there of 380m^3/s, which would have caused considerable damage in the city.

Severn Trent Water have formalised its agreements with the Environment Agency to hold its Derwent Valley reservoirs at 80% of capacity from October to the end of January better to provide flood reduction downstream. This is typical of arrangements made by other owners of large reservoirs (Bridle and Sims 1999).

Similar features occur overseas. The Yangtze River in China has drowned about 300,000 people in the last century, displaced several million and in 1954 alone inundated 3 million hectares. The Three Gorges Dam will provide flood protection to the 15 million people who now live in the flood plain, converting what used to be a flood every 10 years into one in one hundred years, and to 1 in 1000 years when the Dongting Lakes downstream have been rehabilitated to store flood waters.

SEA DEFENCE

Whilst almost all of Britain is above sea level, there are areas along the coast which have been reclaimed from the sea to provide agricultural or development land. This has been achieved by constructing dams, called sea defences, to keep out the sea. A good example is the sea defences in the Wash to protect the highly productive Fens from being inundated. These sea defences have resulted in a significant increase in national agricultural output. Some of these are now several metres below high water. A good example of the protection of development is Canvey Island. These sea defences have provided extra land for housing and industrial development particularly for installations needing connection to the sea such as refineries.

Several of the estuaries were developed as ports and centres of commerce. With the south east of England falling relative to sea levels, several of these estuaries here are at risk of higher relative tidal levels. In 1953 a surge tide came down the North Sea and breached the sea defences. This caused 300 deaths in East Anglia but 3000 deaths in Holland. In England this resulted in the raising of the tidal defences and the construction of the Thames Barrier (Gilbert and Horner 1985). Figure 3 shows the area of London provided with protection by these dams (NCE 2003). About one and a half million people work in this area. The 1953 event was lucky for central London in one aspect, at the last moment the extreme meteorological condition curved away and struck Holland instead. Had it not parts of London would have flooded. The benefit of raised river walls and a new barrage provides protection for London against a one in 1,000 year marine flooding event. In 2002 the barrier was shut for 30 tides to prevent either marine or fluvial flooding demonstrating the increasing benefit obtained by this dam system.

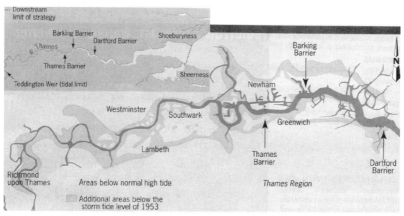

Figure 3: Thames flood zone

IRRIGATION

Most supermarkets have strict requirements for quality and size of vegetables and generally require any producer from whom they buy to have irrigation to ensure uniform quality and security of supply.

Since the Environment Agency no longer allow new summer water abstraction in most of the south, east and Midlands of England, farmers have had to construct on farm storage reservoirs. Thus much of the commercial vegetable crops produced in Britain rely on irrigation water from farm reservoirs.

Overseas weather requirements are often more extreme. Irrigation water is often required to cover an entire dry season. In 1877 when a low Nile flood failed to irrigate adequately, there was famine and death among the six and a half million Egyptian population. In 1902 the Aswan Dam was constructed to provide two crops a year and the Aswan High Dam constructed in the 1960s extended this to provide perennial irrigation for a much larger area and a much larger population as well as 12,000 MW of hydropower. Now Egypt supports a population of over 70 million. Without these dams Egypt's population would have been much smaller than it is today.

RECREATION, CONSERVATION AND ENVIRONMENT
When Victorians built reservoirs they knew they could not treat the water so access to the reservoir, and often to the whole catchment, was often severely restricted. Now many reservoirs are recognised for their recreation, conservation and environmental benefits

On the environment, almost all reservoirs release water downstream thus ensuring the downstream environment is maintained even in a drought. At Roadford freshets are released to mimic the natural river and bring salmon up to the spawning beds. That facility would have been most welcome on many rivers during the drought of 2003.

Many reservoirs constructed on ordinary farmland are now Sites of Special Scientific Interest. Nine reservoirs are now internationally registered under the Ramsar Convention (1971) as "Wetlands of International Importance especially as Waterfowl Habitats (Ramsar Convention Bureau 1999). One of them (Abberton) is cited as *"...a roost for the local estuarine population of wildfowl. It is outstandingly important as an autumn arrival point, moulting and wintering locality for wildfowl. Thirteen species of waterfowl occur in nationally important numbers, including Widgeon, whose winter numbers are of international significance, Mute Swan, Gadwall, Shoveler, Pochard, Tufted Duck, Goldeneye, Goosander and Coot"*.

All new reservoirs are landscaped. This includes forming fillets and adjusting the slopes of the dams to minimise its apparent height, forming artificial islands so that birds can nest free from the predation of foxes, and forming lagoons along the foreshore to maintain shallow wetlands for wildfowl even during drawdown. Many now have woodland plantations near the margin. Extensive planting often screens car parks and facilities.

Rutland Water, one of the largest reservoirs in the Britain, is, in the words of Sir David Attenborough, *"one of the finest examples of creative conservation in Great Britain"* (Anglian Water, 1995).

Like most reservoirs it is now stocked with fish. As a result otters and ospreys have been encouraged to breed there, increasing the bio-diversity.

Reservoirs are now extensively utilised for recreation. Most have fishing. Many have sailing clubs. Several have peripheral paths for walkers, bicyclists, and sometimes horse riders. Many have quiet environmental areas where bird watching hides allow visitors to watch many species of birds. Rutland Water attracts 50,000 birdwatchers a year.

Some reservoirs such as Kielder and Carsington commercialise these features with large carparking areas, a large visitor centre, and even caravan parks and chalets. Visitors to Carsington each year are about 1.2 million, to Kielder 1 million, and to Rutland between ¾ and 1 million.

Thus our reservoirs now provide the benefit of good, albeit changed, environment, and extensive recreation facilities.

SUMMARY
In summary, dams and the reservoirs they form, have provided considerable benefit to society from early times providing water for drinking, growing food, and power when it would not otherwise be available. They also provide an enhanced environment and recreation for many. Without dams and reservoirs the industrial revolution on which our wealth was based would have been much delayed. The population of our major towns would have been curtailed. Without hydropower green house gas emissions would have been greater, and hence climate change would have increased. Without reservoirs providing irrigation water more of our food would be imported. However society will only support more reservoirs provided the benefits they can bring are both provided to the full and publicised.

ACKNOWLEDGEMENTS
I would like to thank Rod Bridle for providing me with information for this paper as well as all those others who have contributed.

REFERENCES
Anglian Water, 1995. *Managing our Future Naturally. Rutland Water.* 1995 European Prize for Tourism and the Environment, A Great Britain Entry from Anglian Water.
Arkell, B. 2003 *personal communication*

Binnie GM, 1981. *Early Victorian Water Engineers.* Thomas Telford, London.

Binnie GM, 1987. *Early Dam Builders in Britain.* Thomas Telford, London.

Binnie CJA 1995 Centenary Address of the Chartered Institution of Water and Environmental Management 1895-2045

Binnie CJA and Roe DE, 1986. *Civil Engineering Aspects of an English Stones barrage in Tidal Power.* Thomas Telford.

Bridle R. 2003. *Dams for life.* Water and Environment Manager, July/August 2003.

Bridle R. and Sims G. The benefits of dams to British Society. Dams and Reservoirs, December, 1999.

British Hydropower Association 2003 personal communication, October 2003.

Chadwick Report on the sanitary conditions of the labouring population of Gt Britian, 1842.

Dutton D. Personal communication 2003.

Gilbert and Horner R. 1985. The Thames Barrier. Thomas Telford.

Mangan BJ and Hayes TA, 1996. River Liffey Reservoirs: 50 years of protecting and supplying Dublin City. Proceedings of the 9th Conference of the British Dam Society. The Reservoir as an Asset. Thomas Telford, London.

NCE 2003 New Civil Engineer October Flood Risk Management

Ramsar Convention Bureau, 1999. Lists of Wetlands of International Importance. Ramsar Convention Bureau, Switzerland, Web: http://ramsar.org/

Smith 2002. The Centenary of the Aswan High Dam 1902-2002 ICE Thomas Telford.

STPG 1986. Tidal Power from the Severn. Report by the Severn Tidal Power Group 1986.

Taylor SJ, 1998. Sustainable Development in the Use of Energy for Electricity Generation. Proceedings of the Institution of Civil Engineers; Civil Engineering, Volume 126, Issue 3, August 1998, pp126-132, Thomas Telford, London.

Water Facts 2000 Water UK

Watts G 2003 personal communication

Lake Hood - Creating Waves in the Community

G. A. LOVELL, Tonkin & Taylor Ltd, Christchurch, New Zealand

SYNOPSIS. Lake Hood is the largest artificial recreational lake in New Zealand. Located in the South Island of New Zealand, 100km south of Christchurch it services the Ashburton district, which has a population of 30,000 people. The lake area is just over 70 hectares with approximately 7000 m of shoreline and was developed principally for water sport activities. It provides for an international length-rowing course (2km), as well as water skiing, sailing, dive training, swimming and sunbathing. As part of the development of the lake a new residential subdivision on its shores has been planned. This includes a staged construction of 150 sections with lake or canal frontages.

From its initial conception the social impact of the lake's construction on both the township and its surrounding population was considered. Throughout this innovative project the close liaison with the local community, through public meetings, public open days and transparent media coverage has meant that support has grown in parallel to this community spirit. The community resource has impacted, both socially and commercially, on the lives of those living in and around the district.

INTRODUCTION

In 1987 Ken Kingsbury, who had seen the creation of man-made lakes in Britain, decided that such a project was feasible and desirable for the keen water sport enthusiasts in Ashburton. He called a public meeting and a sufficient number of people attended the initial meeting to encourage those present to form a working party to investigate suitable sites.

A number of sites were considered and in 1989 a site was chosen within 6 kilometres of the main road and adjacent to the banks of the Ashburton River. The initial committee was enlarged and the committee formed an Incorporated Society with the aim of negotiation and purchase of land.

In 1990 the site became available to purchase with a price tag of NZD$120,000. The Society decided on a funding scheme of $100 joining fee and a $20 per year annual subscription. The local paper ran free advertisements and within seven weeks the society had purchased the land.

The Society, over a period of three years, obtained limited technical assistance, using local civil contractors and volunteers to prepare and apply for water resource consents. After three hearings, 29 resource consents were obtained relating to diversion and use of water to construct a dam to form a recreational lake. The majority of these consents related to the takes and discharges of water and sediments from/to Ashburton River and a number of minor streams.

A local contractor developed the idea of a staged construction sequence involving progressive impoundment with comprehensive monitoring of seepage piezometric gradient. The aim was to take flood flows from the Ashburton River and use the flood sediment to line the lake floor.

The Society had limited funds so a separate entity was created to control the development and construction of the recreational lake giving more protection to the Society and the new Trustees of the Ashburton Aquatic Park Charitable Trust (Trust)

The Trust was now responsible for management of construction and operation of the lake. The Society was responsible for fund raising to meet requests by the Trust.

DEVELOPMENT OF LAKE CONCEPT
Tonkin & Taylor Ltd (T&T), Environmental and Engineering Consultants became involved during the last resource consent hearing and provided detailed technical support. This led on to the development of the lake layout and development of a construction sequence for the Trust.

T&T suggested an assessment of all the risks to the project. A 'risk management' workshop was held to help give clear focus and direction to the Trust. T&T then developed a staged programme to address/manage each risk, involving and reporting to the Trust with up to date cost estimates.

Each of the project risks was broken down into separate packages for the Trust to consider. Each risk and mitigation measure had to be seen as practical and affordable.

The approach became "which is the current highest risk to the Trust". T&T spent considerable time and energy breaking down the risks and the steps needed to resolve and react if needed.

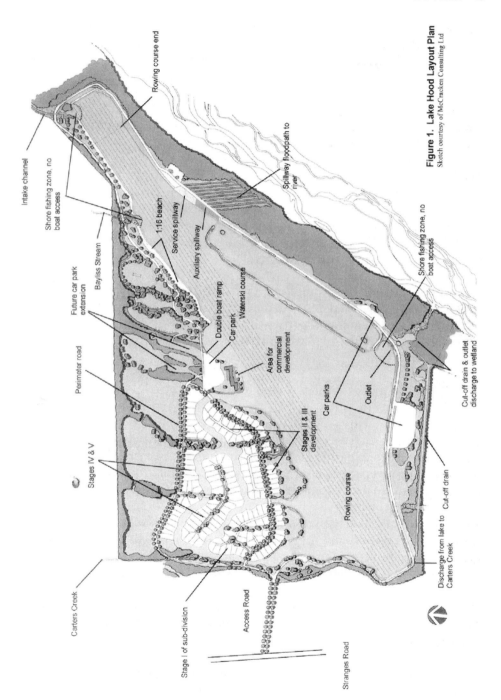

Intake channel

Shore fishing zone, no boat access

Rowing course end

Bayliss Stream

Future car park extension

1:16 beach

Service spillway

Auxiliary spillway

Spillway floodpath to river

Waterski course

Double boat ramp

Car park

Perimeter road

Area for commercial development

Shore fishing zone, no boat access

Stages IV & V

Stages II & III development

Car parks

Outlet

Carters Creek

Cut-off drain & outlet discharge to wetland

Stage I of sub-division

Rowing course

Discharge from lake to Carters Creek

Cut-off drain

Access Road

Stranges Road

Figure 1. Lake Hood Layout Plan
Sketch courtesy of McCracken Consulting Ltd

The Trust kept the Society and community informed of each risk item that was being addressed. This helped when it came to the fund raising for each item. The community became increasingly committed and enthusiastic about the project as it developed to fruition.

Field trials
One of the major risks to the project was the source and installation of an economical liner material.

The construction of the lake above the ground water table, over an existing floodplain with highly pervious cobbles, resulted in the crucial design task of preventing excessive seepage losses. Construction of an adequate lake floor liner that ensured water would be retained was critical to the success or failure of the project. The deepest section of the lake is about 6.5 metres. This was the largest risk to the project.

A modified silt liner was proposed and, with detailed computer modelling combined with field trials, was decided upon as the best way forward. A farmer from an adjoining property, who supported the project, indicated that silt on his property could be used. The silt was from 1m bgl (below ground level) to 2.5 m bgl.

A MODFLOW model was developed for the 2.5 ha trial pond with in excess of 20 peizometres installed.

The 2.5ha trial pond was constructed to determine the depth of silt to be placed over the existing soils to meet several important conditions:
   a) Reduce seepage to hold the lake above natural ground water level
   b) To ensure lake seepage was less than 500l/s as required by the resource consent
   c) To ensure that groundwater rise at the downstream boundary was less than 150mm.
Silt was spread over the ground to predetermined depths and cultivated into the existing soils to a depth of 300mm, using typical farming equipment.

The new soil mix was then compacted to form the lake floor liner. The trial showed that an average of 150 to 200 mm of silt was necessary to provide a suitable liner material. It was difficult to confirm the risk and options available should the liner not meet the Resource Consent conditions.

A local source of natural bentonite clay was found (250 km round trip). T&T investigated the material and decided that is was suitable if it could be made into slurry and dispersed. Local transport firms were informed and several came forward and transported, at no cost, several loads to the site for a trial.

A team of 20 local Society members came to help break up the clumps of natural bentonite before it was put into a grout pump and pumped to the trial lake floor. After half a day it was found that this was not going to be practical on a large scale. The material was too "plastic".

Bentonite was placed at 10m centres around the lake edge, chopped up as much as possible by the Society members using shovels and spades, and then thrown into the trial pond using an excavator. To disburse the bentonite the Society members used two jet boats and one outboard powered boat for a period of five hours.

The piezometre readings over the next couple of weeks showed quite a step in reduced permeability of the liner as the bentonite moved to areas of high seepage. The trial pond was drained and on visual inspection a thin film of bentonite was found on most of the trial pond floor. The MODFLOW model was now calibrated ready for the main lake. Seepage was estimated at 200 to 250 l/s, half of the consent requirements.

Construction
Major fund raising began in 1999 and lake construction was tendered and prices confirmed. Major grants were sought to raise the required NZD $3.95 M including 10% contingency and comprised the following:

i)	New Zealand lotteries board	$1,200,000
ii)	Community Trust	$750,000
iii)	Ashburton District Council	$650,000
iv)	Loan from Ashburton District Council	$1,000,000
v)	Ashburton Trust	$200,000
vi)	Public donations	$150,000

Construction started December 2000. Public viewing platforms were built with controlled access to areas for the public to view construction progress.

Public open days were held every three months on site with buses taking the public around the site explaining where the status of construction was at and what was to happen next. This Public Relations exercise was considered necessary as the Trust depended on local support.

During construction another trial lake was developed (15ha. Sited as part of the final lake) and it was used to check that the assumptions made in the trial pond and MODFLOW model were correct.

The other purposes of the trial lake were
  a) To determine the response of the water table to a known recharge
  b) To locate areas of floor liner with high leakage by identifying local groundwater increases
  c) To establish the need for a groundwater cut-off drain along the southern boundary (mitigation measure to stop ground water rise being greater than 150 mm)

The test was to give certainty and assurance to the Trust in several areas.
  a) That the liner was working
  b) Would the contingency allowed for bentonite be required? If not, the budget surplus would be used to redesign the lake to eight rowing lanes not six
  c) Could all the resource consents be met, in particular, the groundwater rise at the boundary?

The MODFLOW model predicted the groundwater rise at the boundary, would be in excess of the Resource Consent requirement, however the consent conditions could be met with the installation of the cut-off drain.

	Field Results	MODFLOW Model Results
Predicted Seepage	77 to 140 l/s	93 to 151 l/s

Several meetings were held with the Trust to explain the 15 ha trial lake results and make recommendations from these results. The Trust decided to install the cut-off drain and go back to the public to raise money for the additional rowing lanes.

The lake was completed on 15th December 2001 and during Christmas and New Year 2001 the mean annual flood in the Ashburton River occurred. The lake was quickly filled by the floodwater. It transpired that this was the best thing that could happen for the lake, for a week floodwaters were taken, which successfully helped seal the lake floor with natural flood sediments.

Intake during normal flow

Intake during flood flow Jan 02

The lake was monitored for nearly a month and showed the average seepage to ground was between 101 l/s and 115 l/s with 97.5 % confidence.

The cut off drain was installed along the southern property boundary between 28/1/02 and 8/2/02.   Ground water level readings dropped and became stable and the resource consent conditions were met.

The lake was officially opened on 28 April 2002. The high level of attendance reflected the support from the community.

Lake Hood – Typical weekend

CURRENT USE OF THE LAKE

Service clubs
As with any rural district and community, Ashburton has a multitude of active service clubs. These clubs have become increasingly supportive in several areas that in time will see an increase in the use of the lake and any ongoing fund raising. These clubs have attended to landscape plantings on the site and developed walking paths and mountain bike tracks. Ashburton Jaycees, who have run a triathlon for the last 17 years, had a new venue for the event almost purpose built.

Ashburton College
Ashburton High School has a role of 1150 pupils and accommodates year 9 through to year 13 students. Currently the school is encouraging students to join the Rowing Club and gives students leave to attend training and events.

In time, Ashburton College would like to have training courses in place for yachting and canoeing. Unfortunately New Zealand Government legislation under the Health and Safety Act, combined with the personal responsibility that teachers/instructors now take for school field trips, has had a negative impact on outdoor school activities. The New Zealand legislation has made it so much of a burden on schools that on many occasions schools do not contemplate activities off the school grounds.

Ashburton College's Principal has already seen the 35 students involved in rowing become more focused and willing to accept challenges. The school is looking at ways of managing the risk of programmes involving outdoor water events.   Once this is remedied Lake Hood will become a great resource to Ashburton College.

Lake community
At the end of November 2003, two families live permanently at the lake sub-division with a further four houses currently being built. A total of 31 sections out of 35 Stage 1 sections have been sold.   Stage 2 of the sub-division is currently being designed for construction in 2004.

New residential houses under construction

The completion of the first houses has resulted in a dramatic increase in the sale of remaining sections.   Now that families are residing at Lake Hood there is already the feel of a community.   In time these local residents will enjoy a rural lifestyle with a water front aspect.

Sports clubs
With Ashburton previously being approximately 1½ hours away from facilities suitable for water sports activity (other than jet boating), Lake Hood provides an ideal venue at their back doorstep. Consequently the level of activity in leisure water sports in the Ashburton area has risen.

Listed below are some of the new clubs recently established in the Ashburton area:

Rowing   75 members
Sailing   32 boats
Water skiing Club   28 members

Sports clubs an hour away in Christchurch travel to Ashburton for training and 'day out' events.

The Lake's effect on sports clubs has already shown signs of being of significant benefit to those other than water sports. There has been an increase in general support for other clubs e.g. cricket, tennis etc. It was found that parents of children playing cricket or tennis on a Saturday now became more involved in the sport. Where previously parents would drop the children off, go home and pack up the boat to go away 'up country' for water-skiing etc., this was now not necessary.

The resulting effect on these clubs is viewed by locals as having a very positive influence on community spirit and on the sporting clubs themselves.

Ashburton sailing - Club day

Ecology
The new lake has had an impact on the local ecology. Transforming what was grass farmland into a lake and wetland hinterland. Already there are signs of wildlife taking up residence. Trout have been released for recreational fishing. Careful plantings of native and appropriately introduced species have initially had positive results both aesthetically and practically on the lake environ.

Commercial
Local businesses have invested in the lake during feasibility investigations, construction and by way of sponsorship of clubs and events on the lake. They are already seeing results from their investments in terms of increased sales, new developments and new industries.

New businesses have emerged catering for water sports selling new and used powerboats, used sailing yachts, water ski equipment, canoes, kayaks and other boating accessories. Local motorcycle shops have expanded to cater for jet skis and mechanical servicing of boats.

The community is affected each time there is a significant event held on the lake. Events such as the New Zealand Long Distance Canoe Meet or the New Zealand Powerboat Racing National Championships impact right throughout the community. Such businesses as petrol stations, hotels, motels, company groups, restaurants, and supermarkets are all positively affected.

New Zealand Power Boat National Championships April 2003

The hotels have noticed increased use of their facilities, conference rooms for meetings and after match functions. The closest Tavern to the lake is doing major redevelopment, increasing meeting room and restaurant capacity and installing a drive through bottle store.

The local hotels are part of a District Licensing Trust. The trust is proactive at giving support at sponsoring events or with capital support for equipment for water sport clubs. They have become the 'anchor' sponsor for the annual 'Aquafest.'

Local attraction
A passive use of Lake Hood has been use of the lake as a local point of interest. Local residents and tourists use the lake as a quiet place for picnicking, walking and as spectators of water activities. As facilities grow this type of use will only increase.

Lessons learnt when dealing with the community
- A community-based project invariably starts with a few keen individuals who volunteer their time.
- Keep the community involved and informed from inception to completion
- Keep development transparent – so that everyone knows what is happening
- Ask for help
- Where possible use local suppliers and businesses.

CONCLUSIONS
Ashburton is fortunate to have long twilights in the summer and a warm climate. With the lake so close to the township locals comment it is noticeable over just one summer the changes in family use of the lake. Whether involved in water sports or not, families appear at the lake edge to have a barbecue in the evenings. On the weekends the lake abounds with water craft of all shapes and sizes and the continuing development of the lakeside subdivision is offering a choice of lifestyle opportunities.

Over time and generations the culture of the community will adapt and embrace the lake as part of its fabric.

The lake has had a ripple effect throughout all aspects of the community.
The dreams of a small but determined group of people have been realised to benefit the individuals and community as a whole, not just now but in the years to come.

ACKNOWLEDGEMENTS
The Author wishes to thank McCracken Consulting for supply of the concept sketch and the efforts and support of the Chairman (David West), Trustees and committee members of the Ashburton Aquatic Park Charitable Trust and Society.

Balancing the costs and benefits of dams: an environmental perspective

Dr. U COLLIER, Living Waters Programme, WWF International

SYNOPSIS While dams have brought many benefits, they have also caused major environmental impacts, especially on freshwater ecosystems which are suffering a serious decline. This paper explores how the benefits promised by dam schemes can be gained without excessive, unacceptable environmental costs, with a particular focus on Europe. The Spanish National Hydrological Plan is used as an example of an ill-conceived, unbalanced scheme. The paper then looks at examples from Zambia and Switzerland to show how mitigation measures can reduce the impacts of some dams, while still maintaining economic benefits. The paper promotes the decision-making framework of the World Commission on Dams as the way forward.

INTRODUCTION

Dams have played an important role in development for centuries, if not millennia, and have created a range of socio-economic benefits (WWF, 2003a). However, the World Commission on Dams (WCD) found that these benefits often come at an unacceptable and unnecessary environmental and social cost (WCD, 2000). Perceptions as to what is acceptable or not vary between different sides of the dams debate. While economic cost and benefits are relatively easy to calculate in financial terms, environmental costs are often less quantifiable, thus making it more difficult to arrive at a balanced assessment of all costs and benefits. To some extent, the same applies to social costs, although when it comes to the displacement of people or loss of agricultural lands, such costs are easier to calculate.

The environmental impacts of dam projects can be wide-ranging and diverse. Some impacts are directly related to the construction phase and flooding through the reservoir. Downstream impacts from the operation of dams can be significant. Major impacts can also be caused by civil works such as access roads and power lines. In many cases, some of the worst effects can be avoided through mitigation measures, yet sadly such measures are not applied universally. Unnecessary costs can also be caused by the failure to carry out a comprehensive options assessment (as proposed by the WCD), resulting in the construction of dams where there may have been suitable alternatives, such as demand side management.

Not all impacts can be mitigated and in the worst case, they can result in the destruction of unique habitats or even species extinction. Such cases are likely to be considered 'unacceptable costs' - not only by environmental organisations but also by key decision-makers. For example, the World Bank uses the loss of endangered species as a key criterion for evaluating dam projects.

VALUING ENDANGERED SPECIES
According to WWF's Living Planet Report, the world is currently undergoing a very rapid loss of biodiversity comparable with the great mass extinction events that have previously occurred only five or six times in the Earth's history (WWF, 2002). In the last 30 years, freshwater species have seen a particular serious decline, with 54% of 195 indicator species showing a population decline. Dams are one of the factors in this decline, in particular through their effects on fish migration and impacts on downstream wetlands.

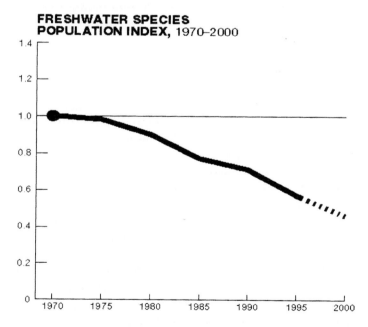

FRESHWATER SPECIES POPULATION INDEX, 1970–2000

Figure 1: Freshwater species decline (from WWF, 2002)

Some dam sites are particularly threatening as far as endangered species are concerned. One extreme case is that of the Kihansi spray toad (*Asperginus nectophrynoides*) which lives only in the fine mist created by the cascading waters of the Kihansi Falls in the Southern Udzungwa Mountains of Tanzania. Diversion of the Kihansi River for power production resulted in the threat of global extinction for the spray toad and possibly other species in the Gorge. The original EIA failed to look at the downstream impacts of the dam, which was obviously a major omission. Mitigation measures (i.e. spraying the toads' habitat artificially) have resulted in a loss of 15 MW of capacity at the 180 MW plant. Saving the toad has thus come at a considerable economic cost.

But how do you value the survival of a unique species of toad? Clearly, one cannot put a monetary value on such a species. The Convention on Biological Diversity, ratified by 188 countries, recognises the intrinsic value of biodiversity and requires the promotion of viable species populations. In Europe, habitats and species are also protected by a various national and European Union (EU) legislative measures, as discussed below.

EU ENVIRONMENTAL PROTECTION
In Europe, there are currently around 5500 large dams in operation. Few European rivers (or stretches of them) are unregulated and there has already been a major freshwater habitat loss. At the same time, environmental protection is high on the agenda in many European countries, hence new dam proposals are often met with considerable opposition.

With the expansion of the EU in 2004, the majority of European countries (including those who aspire to future membership, such as Turkey) will have to conform to EU environmental legislation. In principle, this legislation will provide clear guidance as to where dam projects might be acceptable. Key legislative measures in this context are the Birds (79/409/EEC) and Habitats Directives (92/43/EEC), as well as the Water Framework Directive and the Environmental Impact Assessment Directive. In the future, the Strategic Environmental Impact Assessment Directive (2001/42/EC) will also play a role for programmes and plans.

Under the Water Framework Directive (2000/60/EC) member states are expected to prevent the deterioration of surface waters. This should in principle protect sites that achieve good or very good ecological status. The Habitats Directive requires member states to prevent the deterioration of natural habitats and the disturbance of species in designated areas (so-called Natura 2000 sites), which should mean protection from developments such as dams. However, 'overriding public interest' can be used by member states as a way out. Despite these ambiguities in European legislation, habitats and species protection are a fundamental requirement that needs to be taken into account in siting decisions for dams in the region. Yet, as the following example shows, not all member states take this seriously.

SNHP – UNACCEPTABLE AND UNNECESSARY

One project that stands out both in terms of its sheer scale and its infringement of European legislation is the Spanish National Hydrological Plan (SNHP). The SNHP, approved in Spanish law in 2001, consists of two parts:

- A water transfer from the Ebro River, impacting the Pyrenees, Lower Ebro basin and Ebro Delta;
- An investment programme to build more than 100 dams and associated reservoirs and canal networks throughout the rest of the country, re-routing another 35 rivers and tributaries.

© WWF-Canon / WWF-Spain

Figure 2: Spanish National Hydrological Plan (SNHP). Map of the planned use of Ebro waters in the SNHP. Spain

The SNHP is likely to have major negative environmental impacts. Following an initial assessment, WWF found that 47 of the planned dams are likely to have a significant impact on at least 46 of the official Sites of Community Importance (SCI) proposed by the Spanish Environment Ministry for the Natura 2000 network. There are 35 dams that are situated completely or partly in Natura 2000 sites. At least 126 Important Birds Areas and 86 Special Protected Areas (designated under the Birds Directive) will be affected, including at least 14 habitat types and 18 species. While the exact impact is not known in the absence, at this stage, of individual EIAs, the plan undoubtedly puts huge development pressures onto sites that should be protected under European law.

Furthermore, the Ebro Delta is currently undergoing serious erosion due to lack of sediments (held back by existing dams in the Ebro river basin). This condition will deteriorate with the SNHP because the Plan does not acknowledge the need for a minimum flow of solids.

The Plan fails to give proper recognition to alternatives such as water demand management and makes assumptions about future demand (especially from agriculture) that are unlikely to materialize. According to an independent assessment by the Third World Centre for Water Management, the SNHP in its present form cannot be justified for economic and environmental reasons and it would be a very expensive 'white elephant' (Biswas and Tortajada, 2002).

THE WAY FORWARD – TOWARDS GREENER DAMS?

Implementing the WCD in Europe
With 5500 large dams in operation, Europe has already heavily dammed most of its major rivers. However, there are still numerous new dam projects, especially in Spain and Turkey. Obviously, the development pressures (as well as water stress) in these countries are greater than elsewhere in Europe, while their storage and hydropower potential is much less developed. At the same time, those countries also have some of Europe's most valuable ecosystems.

So how can some of this potential be developed, without causing large scale destruction? Obviously, there are various requirements under European legislation, as mentioned above. However, the WCD provides additional guidance which needs to be implemented in Europe to avoid further large-scale damage.

Out of the strategic priorities of the WCD, 'comprehensive options assessment' and 'sustaining rivers and livelihoods' are particularly critical for protecting vulnerable environments.

Firstly, options assessment will ensure that alternatives to dams are given due consideration. As the example of the SNHP shows, demand-side management (energy or water conservation) can in many cases reduce the need for new supply through dams. Not every proposed dam can be replaced by a demand-side programme but there can be little doubt that the current supply-side mentality in water and energy supply needs to be redressed. This makes both economic and environmental sense.

Secondly, 'sustaining rivers and livelihoods' recognised the importance of rivers, watersheds and aquatic ecosystems as the basis for life and livelihoods of local communities.   A basin-wide understanding of the impacts of development options such as dams is crucial.   WWF promotes Integrated River Basin Management (IRBM) which aims to maximize the economic and social benefits derived from water resources in an equitable manner while preserving and, where necessary, restoring freshwater ecosystems (WWF, 2003b).   The WFD also requires river basin management, although its implementation schedule is slow.

Under this strategic priority, the WCD also suggested the development of national policies for maintaining selected rivers with high ecosystem functions and values in their natural state. WWF has long been campaigning for the designation of free flowing rivers.   For example, in Iceland, where the Kárahnjúkar hydropower plant will cause considerable damage to two glacial rivers, WWF is urging the Icelandic government to afford protection to a third glacial river, Jökulsá á Fjöllum, including its designation as a Ramsar site.

Addressing existing dams
The WCD also stressed the need to address the environmental and social problems caused by existing dams.   Considering Europe has already 5500 dams in operation, a key challenge is to ensure that they operate in an environmentally acceptable way.   In the past, many dams where built without EIAs and without mitigation measures.   Some mitigation measures can be introduced at a later stage, as the following two examples show.

Environmental flows in Zambia
The reduction of the downstream low of a river is one of the key ecosystems impacts of dams.   Maximising the output of a dam can have serious consequences both for ecosystems and other users downstream.   However, in many cases it is possible to adjust the operational regime of a dam to better meet a variety of needs.   So-called 'environmental flows' provide critical contributions to river health, economic development and poverty alleviation (IUCN, 2003).

To demonstrate that environmental flows are not just the 'luxury' for rich developed nations, WWF is working with the Zambian Ministry of Water and Energy Development and the Zambian Electricity Supply Company to introduce environmental flows at the Itezhi Tezhi dam, upstream of the Kafue flats wetland. The restored flow regime will have benefits not just for wildlife but also for fisheries and cattle grazing downstream of the dam.

Naturemade hydropower in Switzerland
The naturemade green electricity label in Switzerland supported by WWF (www.naturemade.ch), accredits new and existing hydropower plants under certain conditions. To achieve the highest standard, the 'naturemade star' label, hydro plants have to meet strict environmental conditions. These include environmental flows, sediment flushing, fish ladders and protection of wetland habitats. Additionally, operators have to pay a percentage of their income into a fund for environmental improvement measures, including habitat recreation. 14 Swiss electricity suppliers have gained certification under this label.

EUROPE'S POOR RECORD
Despite some good examples, a recent WWF report on water management in Europe has shown key gaps in national water policies as far as dams and environmental protection are concerned (WWF, 2003c). In particular, the report identified the lack of strategies to maintain free-flowing rivers and too few regulations to monitor and reduce the impact of existing dams. For example less than 40% of the surveyed countries have obligations to maintain ecologically acceptable flow regimes downstream of dams and fewer than 30% require fish ladders or passes specifically tailored to the site and species where the dam is located. Even where these requirements exist (e.g. Switzerland, Poland, Hungary, Slovakia and Turkey), their practical implementation and effectiveness is poor and there little or no monitoring to check that measures have been put into place. So far, there is little evidence of the implementation of the WCD's recommendations.

CONCLUSIONS
Freshwater ecosystems are of crucial importance to human survival – they serve as spawning grounds for fisheries, as cleansing systems for pollution, and as sources for our fresh water. Nevertheless, the loss of freshwater biodiversity continues at a rapid pace. Dams are a major culprit in this process – yet the destruction caused is quite often unnecessary. While dams bring benefits in terms of water supply, electricity generation or flood control, often alternatives are available to provide the same services, sometimes even at lower cost. Where they are not available, careful siting and balanced operation can significantly reduce the impacts of dams. There can be little doubt that we need to find a better balance between costs and benefits. WWF believes that the decision-making framework proposed by the WCD points the way forward. Even in Europe, where the planning of dam projects is subject to various environmental directives, the WCD framework provides additional guidance that if adhered to, should enhance decision-making and help protect precious ecosystems.

REFERENCES

Biswas, A.K. and Tortajada, C. (2002). *Assessment of Spanish National Hydrological Plan.* Third World Centre for Water Management, Mexico.

IUCN (2003). *Flow – the essentials of environmental flows.* IUCN, Gland.

World Commission on Dams (2000). *Dams and Development: a new framework for decision-making.* Earthscan, London.

WWF (2002a). *Living Planet Report 2002.* WWF, Gland. http://www.panda.org/downloads/general/LPR_2002.pdf

WWF (2002a). *An Investor's Guide to Dams.* WWF, Gland.

WWF (2003b). *Managing water wisely.* WWF, Gland. http://www.panda.org/downloads/freshwater/managingriversintroeng.pdf

WWF (2003c). *WWF's water and wetland index: critical issues in water policy across Europe.* WWF, Gland. http://www.panda.org/downloads/europe/wwireport.pdf

Follow up to the WCD Report - where has it gone?

J BIRD, Independent Consultant, formerly Coordinator of UNEP Dams and Development Project[1]

SYNOPSIS.   Despite the wide range of responses to the report of the World Commission on Dams, there has been an increasing realization of the need to address its recommendations through appropriate national and institutional processes. Neither rejection of the report nor full endorsement hold the answer. This paper outlines some of the momentum being built by national follow-up processes and the actions taken by an increasing number of inter-governmental, bilateral and private sector organizations. Minimizing the financial, environmental, social and reputational risks associated with dam projects is at the centre of these initiatives and key concepts such as options assessment, public acceptance, benefit sharing and environmental flows are beginning to enter the mainstream of planning processes.

REACTIONS TO WCD: FROM REJECTION TO ENDORSMENT

A full spectrum of responses
It is hard to conceive a wider range of reactions to the World Commission on Dams Report (WCD, 2000) than those received, but maybe that is not so surprising given the intensity and polarity of the debate itself (DDP, 2003). There are those that reject the report outright and those that call for its immediate implementation as if it were law. What is interesting about the reactions is that they do not fit as neatly into pigeon-holes as our characterizations of stakeholder type would suggest. There is considerable diversity of reaction both between and within organizations, whether they be government agencies, professional associations, financing agencies, NGOs or affected peoples' groups.

Extreme headline reactions are there for those who wish to continue the polarization of the debate. At one end of the spectrum there is outright rejection of the Report by the Ministry of Water Resources of India (MWR, 2001) and a former President of the International Commission on Large Dams who stated that the Report *'made dams look like villains, to be avoided unless there is no other way out'* (van Robbroeck, 2002). At the other end, passionate endorsement. For example on the day of the Report's launch, the International Rivers Network commented that it *'vindicates much of what dam critics have long argued'* and if applied, *'the era of destructive dams should come to an end'* (IRN, 2000).

Some critics' responses were influenced as much by their perspective on the composition of the Commission or the process it adopted as by its content. For example, the case studies in India and Thailand significantly influenced Government's subsequent positions on the WCD Report. Similarly, a number of agencies from developing country governments felt their views were not adequately represented (Dubash et al, 2001, p43). Other reactions were strongly influenced by concerns that the Report's recommendations could further burden the project appraisal process through incorporation, in their raw form, into safeguard policies of the multi-lateral organizations.[2] The World Bank explains that this will not be the case (World Bank, 2001).

Criticism was not limited to those involved with dam building. Amidst their support for the Report, some NGOs felt that it fell short of calling for a moratorium on dams. They proposed to test commitment to a new approach by requiring the legacy of past projects to be addressed before initiating new projects. Some had wanted more of a challenge to the prevailing development model and condemnation of vested private sector interests. Reactions voiced by a range of stakeholder groups after the Report's launch are recorded in the proceedings of the Third WCD Forum meeting (WCD, 2001).

Yet, between these extremes more than one hundred responses have been formally recorded and analyzed. It is evident from the follow-up around the world, that many more responses and comments are not available in the public domain. As is often the case, there exists a large middle ground, the silent majority, who neither reject nor endorse the Report. An analysis of reactions received provides an important reflection on the WCD Global Review and its three-tier recommendations, the 5 core values, 7 strategic priorities and associated policy principles, and the 26 guidelines (DDP, 2003).

Derailed or on track?

So, has the report fuelled or calmed the debate? Conflict has not mysteriously vanished. However, the process itself has built a culture and atmosphere wherein advocates both for and against dams can enter into a civilized and constructive discussion. There has been an opening up of space for dialogue. Follow-on discussions have started at a range of appropriate levels – global, regional, national, sub-national and community levels using the framework from the Report. It is a framework that considerably narrows the areas of controversy, allowing areas of agreement to be acknowledged and areas for more intense analysis to be flagged.

Where the controversy lies is the more detailed recommendations for implementation – the guidelines. What is important now is to look beyond the extreme reactions that continue to occupy the public limelight, and examine the extent and way in which the Report is influencing planning processes and implementation procedures.

In its independent analysis of the WCD process, the World Resources Institute outlined its view on how the Report will be taken up, '*Over the long term, the bridge back to formal government and intergovernmental processes will likely be built incrementally, by incorporating practice into formal laws, in part through continued pressure by non-governmental actors*' (Dubash et al, 2001, p127). But added to these actors are the large number of people occupying the middle ground who also recognize that change is needed.

INITIATIVES FOR CHANGE

What are the driving forces behind the various follow-on processes, given that the WCD Report has no legal status internationally? Clearly it is not the Commission. That disbanded on the date of the Report's publication. The initiative of its Forum members taken in February 2001, to continue with dissemination and promote dialogue on its findings, certainly has played an important role. But even then, there need to be catalysts to sustain any process within countries or organizations.

Three primary drivers come to mind. Most prominent is campaigning by international and national NGOs at both project level and targeted towards specific individual stakeholder groups. They have kept the WCD report and the issues it addresses firmly on the global agenda. WWF also has a campaign to engage with financing organizations to promote the WCD recommendations (WWF, 2003) and at the same time has used hard-hitting advertisements in high profile magazines to deliver its message (for example The Economist, 2003). In this case, globalization, at least in respect of information exchange, is something fully embraced by NGOs (Gyawali, 2001).

Secondly, a number of governments from developed countries have indicated their broad support for the WCD recommendations. There is considerable synergy with their domestic policies and these positions are reflected in their influence on the multi-lateral development banks and in their own development assistance programs.

Thirdly, the trend over the past decade towards corporate social responsibility and triple bottom line accounting on financial, social and environmental aspects of operations in the private sector has led to companies voluntarily subscribing to international initiatives such as the UN Global Compact, Global Reporting Initiative, UNEPs Finance Initiative and environmental management procedures under ISO 14001. Due diligence procedures have been strengthened accordingly in order to reduce reputational risk and caution association with potentially problematic projects. The example of the Brent Spar platform from the oil and gas industry demonstrates the adverse impact that negative publicity can generate and also highlights the lessons learnt and benefits of dialogue.[3]

In less developed countries and emerging economies, the drivers for change reflect a combination of the above sources, the influence of each depending upon the prevailing development paradigm, the institutional and governance structures and inevitably, the extent that the country is dependant on external financing for project development. Reformers within some government agents have initiated dialogue processes aimed at introducing appropriate reforms.

Facilitating follow up internationally
Both 'godparents' of the WCD process, the World Bank and IUCN-The World Conservation Union, have published detailed responses to the WCD Report outlining the subsequent actions they would take as follow-up.

After consultations with a number of agencies in its member countries, the World Bank's Board of Director's endorsed a statement that '......*shares the core values and concurs with the need to promote the seven strategic priorities..*' while outlining where World Bank policy differs from the guidelines. As a practical element of its response, the World Bank promoted a 'Dams Planning and Management Action Plan' to promote good practice and support innovations in projects involving water resources, energy and dams. The Plan uses the seven strategic priorities as a framework to look at projects in the pipeline and intends to provide operational support services for critical elements identified by the Commission. A first output of the Plan is the development of a Sourcebook on Options Assessment (World Bank, 2003b).

There has been a considerable polarisation over the Bank's response, with a number of government agencies in developing countries encouraged by the decision not to amend its safeguard policies, while critics pointed to a lack of commitment to the outcome of a process that it helped to initiate.

IUCN's response was more supportive. It recognized that work needs to be done to operationalize the WCD recommendations and encourage multi-stakeholder groups to progress further (IUCN, 2001). Three priority areas identified were regional strategies for engagement and supporting multi-stakeholder process; work on global policy processes related to sustainable development and links with the Ramsar Convention (Ramsar, 2002), the Convention on Biodiversity (CBD) and the private sector; and work on strategic analysis and tools related to dam development and operation, including a toolkit of environmental flows and improved economic valuation of ecosystem services. Many of the principles in the WCD report also feature in IUCNs Water and Nature Initiative.

Both the World Bank and IUCN were key players in establishing a global follow on initiative to WCD in the form of UNEPs Dams and Development Project (DDP, 2001). Together with representatives from a government basin agency, affected peoples' groups, the private sector and advocacy NGOs, they worked within the mandate provided by the Third WCD Forum meeting (WCD, 2001) to craft a multi-stakeholder process with a goal *'To promote a dialogue on improving decision-making, planning and management of dams and their alternatives based on the World Commission on Dams core values and strategic priorities'*. As part of the formulation process, the six member liaison group was expanded to a 14 member Steering Committee, adding two other government representatives, indigenous peoples' groups, utilities, inter-governmental organizations, professional associations, organizations working on options, and research groups.

In selecting this route, the global multi-stakeholder follow-on process was brought into the UN inter-governmental system, thereby providing confidence among some agencies critical of the WCD Report that the follow-up process would take account of their views and provide an environment within which they could participate in the project through the Forum. Taking over what he described as a 'hot potato', UNEP Executive Director Klaus Toepfer captured the challenge of the DDP, *'I believe that we have no choice but to find ways of crossing traditional divides, to act together and find solutions to what has often been a conflict ridden way of working..'* (DDP, 2002). Responding to this challenge, the membership of the DD Forum has increased to include the Brazilian National Water Agency (ANA), the Chinese Ministry of Water Resources, Turkey's General Directorate of State Hydraulic Works (DSI), India's Planning

Commission, Nepal's Ministry of Water Resources and Uganda's Ministry of Energy among others. The Forum of stakeholders forming part of the global dialogue process has expanded to 120 organizations.

Also at the global scale, the response of the World Water Council provides an insight into some of the challenges in taking the dialogue on dams and development further. Pointing to both positive and negative feedback from its members, the Council's official response acknowledges the important contribution of the WCD, supports the core values and strategic priorities, and recognizes that they have relevance to other infrastructure (WWC, 2001). In practice however, members of the Task Force on Dams established by the Council actively campaigned against acknowledgement of the WCD. This was evident at the Third World Water Forum, where they objected to direct reference to the WCD in the theme summary on dams, but was able to broadly accept its recommendations through a reference to '*A framework for planning and implementation based on values of equity, efficiency, participatory decision-making, sustainability and accountability*' and a series of principles that reflect many of the WCD strategic priorities (WWF3, 2003).

Beyond the perspectives of international organizations, there has been action at regional and national levels.

Regional initiative in Southern Africa
In response to a call from its Ministers, the Southern African Development Community (SADC) is adopting a two-fold strategy (SADC, 2003). It will comprise of a formal statement providing SADCs position on the WCD Report and a policy document on dams and development to guide future SADC involvement with dams related activities. Supported by the German agency GTZ and the DDP, initial drafts of the position paper and policy document are being prepared for review by a multi-stakeholder workshop in early 2004. They will be submitted for discussion in the formal committee processes of SADC and ultimately reviewed by the Committee of Ministers and approved at a SADC Summit.

National dialogues
A wide range of multi-stakeholder national processes have emerged since the launch of the WCD Report, many of which have been encouraged and supported by the DDP. A number are outlined below. Common characteristics include participation of all key stakeholders, government, endorsement by the responsible government agency, and a preliminary scoping stage leading ultimately towards recommendations on policy and procedures relevant to the local context. In some cases, translation of the WCD Overview and Report have been a pre-requisite to wider discussion.

Experience of these national dialogues outline in this paper is based on the writer's involvement (Bird and Wallace, 2002) and updates provided by the DDP.[4]

South Africa

Probably the most advanced of all the national follow-up processes, the South African initiative on WCD started life as a proposal to hold a meeting among two groups - the professional organization, SANCOLD, and the Department of Water Affairs. However, based on discussions with local NGOs and the transition WCD Secretariat, the process took on a more multi-stakeholder character with a Symposium organized for 23-24 July 2001. The overall consensus of the Symposium was reflected in the resolution that *"declares itself to be broadly supportive of the strategic priorities outlined in the WCD report, but believes that the guidelines need to be contextualized in the South African situation"*.

Since then the elected Coordinating Committee, representative of diverse stakeholder groups, has met approximately at two monthly intervals. There have been two further multi-stakeholder forums to review a draft Scoping Report and assess recommendations on policy reform measures for the first three of the WCDs seven strategic priorities. The process is scheduled to be completed by October 2004 when the Committee's recommendations will be submitted to Government for consideration.

Vietnam

A multi-stakeholder consultation on the report of the WCD was held in Hanoi in October 2002 organized by the Ministry of Agriculture and Rural Development (MARD) with financial support from the Asian Development Bank (ADB). In advance of the workshop, MARD arranged the translation of the WCD Report and Overview into Vietnamese with assistance from DDP. Based on the outcome of the consultation, a proposal emerged for a two phase follow-up. Phase 1 prepared a scoping paper to examine the WCD recommendations in the context of Vietnam and identified areas of agreement, disagreement, opportunities and constraints. Workshop discussions on the draft scoping paper will then define the second phase to analyze key outstanding issues and make specific recommendations on policy and procedures to Government decision-makers.

Nepal

Presentations and discussions on the WCD Report were organized in Nepal in the two years since its launch, both by professional associations and NGOs. Although there was strong interest to build on these meetings, the lack of involvement of government agencies was a major constraint. A change of Government and a facilitation role from DDP saw the establishment of a broad-based task force on dams and development and, in January 2003, the launch of a multi-stakeholder dialogue. Its aim is, *"To carry out national consultations on dams and development to consider the relevance of the recommendations of the WCD and other bodies in the Nepalese context with the ultimate aim of recommending the development and adoption of a national guideline for improved decision-making, planning and management of dams and alternatives for Nepal"*. By September 2003, a scoping report had been prepared comparing the legal and regulatory framework in Nepal with the WCD recommendations, and identifying where reforms were considered appropriate in the local context. Discussions on a second phase started in November 2003.

Thailand

Translation of the WCD Overview into the Thai language formed the basis for a national multi-stakeholder meeting organized by the National Water Resources Committee in March 2003. The two-day meeting concluded with general support to the framework of core values and strategic priorities and agreed to establish a national task force on dams and development to take the process further and develop locally appropriate recommendations for government. In July 2003, the Ministry of Natural Resources and the Environment formally constituted the task force comprising government agencies responsible for and related with dam projects, river basin water user associations, NGOs and academic institutions. By examining the issues in a local context, the process has broadened its participation and included agencies initially reluctant to consider the Commission's recommendations.

Pakistan

In 2001, IUCN was requested by the Ministry of Environment to facilitate discussions about the WCD final report and develop locally appropriate recommendations. The process, supported by the Royal Netherlands Embassy in Islamabad, was delayed while institutional arrangements were worked out to ensure involvement of key government agencies responsible for dam projects in the water and energy sector. During this period, advocacy NGOs voiced concerns about being alienated from the dialogue. Subsequently, the WCD consultative process re-started with a series of workshops scheduled for September to December 2003. Other provincial consultations were initiated by the Pakistan Water Partnership, an affiliate of the Global Water Partnership.

Other national processes

Similar consultative processes are beginning to emerge in other countries. In Asia, an initial multi-stakeholder meeting was held in the Philippines in August 2001 sponsored by ADB and preparations are now underway to hold a second meeting in early 2004 with a view to setting up a national follow-on activity. In Sri Lanka, a workshop was held in December 2003 initiated jointly by a Government agency and an NGO. In Latin America, a core group of stakeholders has met in Argentina to plan for a multi-stakeholder consultation on the report tentatively scheduled for March 2004 and in Brazil, an international meeting on dams and reservoirs that will also have a focus on domestic dams and development issues is being convened. In Africa, national consultations linked to the SADC process at various stages of preparation in Lesotho, Malawi, Mozambique, Namibia, and Zambia.

In Europe, a number of countries have developed a response to the WCD, with some convening multi-stakeholder meetings to consider both their domestic situation and their influence on international activities. These include Germany, UK, and most recently the Netherlands (Both Ends, 2003).

Interaction between the DDP and government agencies in China is opening a channel of dialogue on dams and development despite the clear reservations of the Chinese Ministry of Water Resources on certain aspects of the Report. DDP's entry into the UN system, coupled with the World Bank's response not to add any additional layers of safeguard policy, encouraged this engagement. The Chinese Ministry of Water Resources joined the DD Forum as an opportunity to both participate in the global arena and make known their experience and perspectives. The WCD Report is now being translated into Chinese.

In contrast, there has not been a similar relationship developing with the Water Resources Ministry in India that took a position of non-engagement on the recommendations of the WCD Report. As water resources is predominately a State matter, the opportunities for dialogue may be more promising at a decentralized level.

Private sector financing and export credit
On 4 June 2003, a group of four private banks signed up to the Equator Principles', in which they require an Environmental Assessment for sensitive projects and subscribe to the safeguard policies of the International Finance Corporation of the World Bank Group.[5] The number of banks endorsing the Principles has increased to eighteen as of November 2003. This initiative demonstrates an unprecedented realization in the financing sector of the need to address social and environmental issues to minimize risk to business, both financial and reputational risk. In parallel and leading on from this, an increasing number of organisations are addressing the WCD Report. Swiss Re, the reinsurance group, prepared a Focus Report on Dams[6] stating its support for the WCD's five core values and seven strategic priorities concluding that, *'It is Swiss Re's conviction that in the future,*

large projects should be handled in accordance with these principles and practices'. The banking group, HSBC, is working in conjunction with WWF to develop a freshwater policy that is expected to address many of the issues in the Report. Henderson Global Investors have used the Report in assessing whether companies are eligible to be included in their investment funds.

Some Export Credit Agencies have referenced the WCD recommendations as an influence on their new environmental policies (Neumann-Silkow, 2003). The Swiss export credit agency, ERG, has explicitly referenced the WCD recommendations in its EIA guidelines and requires an EIA Report to outline how the seven strategic priorities will be addressed in the context of a proposed project. New environmental guidelines of the Japan Bank for International Cooperation also drew on the WCD Report and include a number of the elements of the strategic priorities including the importance of environmental and social considerations in assessing alternatives, priority to the prevention rather than mitigation of impacts, early disclosure of information, recognition of the rights of indigenous peoples, agreement with affected people on mitigation measures and an emphasis on improving livelihoods.[7]

In June 2003, the Overseas Private Investment Corporation of the United States (OPIC) released a consultation draft revision to its Environmental Handbook to accommodate new polices on large dams and forestry.[8] OPIC announced *'it believes it is important to show leadership in adopting and implementing those elements of the WCDs recommendations that inform good development policy and that are within OPICs capacity to implement'*. The draft revision includes extensive references to specific strategic priorities and guidelines. Recent guidelines of the French ECA, Coface, also refer to the Report and incorporate some of its recommendations including benefit-sharing and environmental flows.[9]

With many of these processes, NGOs have expressed concerns that the organizations have been too selective and not gone far enough in endorsing the principles contained in the WCD Report. There are also many commercial financial agencies whose policies are not disclosed and have not yet addressed the Report. Whatever one's perspective on this, in comparison with the situation of five years ago, it is evident that a process has started to substantially address social and environmental issues in a more comprehensive manner and that it is likely to gain further momentum and evolve over time.

The regional development banks have generally responded by promising reviews of their existing policies. The Asian Development Bank, for example, published the preliminary results of its review in January 2002.[10]

Professional associations

Of the professional associations, the International Hydropower Association (IHA) has taken the most pro-active role in following up on the WCD Report. In contrast to the position of ICOLD and ICID, it engaged with the DDP as a Steering Committee member and Forum member in order that the position of its constituency on the future potential and direction of the industry is well represented. In parallel, IHA prepared Sustainability Guidelines that have embraced some of the WCD principles within a framework of promoting hydropower as a clean, renewable and sustainable technology.[11] They include, the concept of options assessment, informing and involving local communities in the decision-making process, benefit sharing and environmental flows.

Although not supportive as an international organization, individual national committees of ICOLD have been proactive in the DDP process, notably the British Dams Society that made a financial contribution and the South African National Committee on Large Dams that is a founding member of the SA Initiative on WCD.

WHERE TO FROM HERE?

The above responses and follow-up actions can be viewed in the light of the Commissioners own expectations. In the final chapter, Commissioners suggested that 'Nobody can of course simply pick up the report and implement it in full. It is not a blueprint'(WCD, 2000, p311). Instead they proposed a series of entry points for different stakeholder groups among which are to include reviews of existing national procedures and regulations, encourage multi-stakeholder partnerships, address the legacy of past social and environmental problems, refer to the WCD principles in corporate policy documents, use the guidelines for screening and evaluating potential projects, and refine the tools proposed. Considerable progress is being made in these fields, but there are many other aspects still to be addressed.

Assessing the extent to which people have benefited as a result of the WCD Report is a long-term process and will gradually be informed by case by case experiences. The factors and influences are many and such a discussion will no doubt be as diverse as the debate on dams itself. However, there are signs that several of the principles espoused in the Report are beginning to enter into common usage. Many indeed entered the arena prior to the Commission as indicated in its broad knowledge base, albeit in limited cases. The endorsement of such innovations within the comprehensive framework of the Report has raised awareness and provided examples of good practice with an added impetus. But no doubt, as with a dam project, the true benefits and costs of the WCD Report will not be known for many years after its 'commissioning'.

In the meantime, where will the dialogue go? Business as usual seems increasingly to be an option of the past. In addressing the issues and recommendations in the WCD Report, government agencies, utilities, developers, financiers and others proposing dam projects require more certainty that their proposals are both effective and sustainable, minimizing the financial, social, environmental, technical and reputational risk. They question though whether advocacy NGOs will continue to insist on full endorsement of the WCD strategic priorities and guidelines as a pre-requisite. In practice, the national dialogues based on the framework provided by the Commission, demonstrate that polarized positions can be set aside and progress made towards a more common understanding of what is appropriate within the local context.

The examples of the Equator Principles and OECD harmonization process for ECAs point to the advantages in taking a common approach to policy development among finance agencies. This could be extended to the arena of dams. But bearing in mind on government responses to the WCD report, such policy statements should incorporate sufficient flexibility to reflect differing contexts and the results of the relevant national multi-stakeholder dialogue on dams and development.

The national dialogues have indicated a way forward. The synergy with broader processes that encourage sustainable development, greater accountability and corporate social responsibility all provide an enabling environment for these reforms to emerge. But the process is not an easy one. For those with an engineer's training like me, used to traveling a path from A to B in a direct line, the uncertainties, deviations and delays associated with what are essentially political dialogue processes takes some adjustment. The ongoing processes show considerable promise and there are signs that some groups vehemently opposed to the WCD report are prepared to enter into dialogue under the new institutional arrangements. These are encouraging signals given the inevitable increase in calls for dam projects that will come in a response to the UN Millennium Development Goals for water supply, renewable energy and food production. However, despite this momentum, there remains a considerable challenge ahead to translate the outcomes of national level dialogues into firm commitments in the legal and policy framework.

A comment from the risk-averse private sector provides a fitting conclusion. In its Focus Report, Swiss Re makes a point about dams that is fundamental to all developers – private or public, *'For projects of this magnitude and complexity, risk mitigation and limitation must be a top priority'*. Failing to acknowledge and address the recommendations of the WCD Report is a strategy unlikely to minimize those risks.

REFERENCES
Bird, J and Wallace, P, 2002. *Progress of 'Dams and Development' dialogue at national and global levels*. Presentation to Second South Asia Water Forum, 14-16 December 2002, Islamabad
Both Ends, 2003. WCD in the Netherlands at http://www.bothends.org/project/project_info.php?id=12&scr=tp
DDP – Dams and Development Project, 2001. Objectives and Work Programme at http://www.unep-dams.org/document.php?cat_id=40
DDP, 2002. Confluence Newsletter No 1, May at http://www.unep-dams.org/document.php?cat_id=27
DDP, 2003. Analysis of reactions to the WCD Report at http://www.unep-dams.org/files/Interim_Report_on_Analysis_of_Reactions.doc
Dubash, N, Dupar, M, Kothari, S, and Lissu, T, 2001. *'A Watershed in Global Governance? An Independent Assessment of the World Commission on Dams'*, World Resources Institute, Washington available at www.wcdassessment.org
Gyawali, D, 2001. *Water in Nepal*, Himal Books, Kathmandu.
IRN - International Rivers Network, 2000, at http://www.unep-dams.org/document.php?doc_id=79
IUCN – The World Conservation Union, 2001 at http://www.unep-dams.org/document.php?doc_id=176
MWR – Ministry of Water Resources of India, 2001 at http://www.unep-dams.org/document.php?doc_id=160
Neumann-Silkow, F, 2003. *The Use of Environmental and Social Criteria in Export Credit Agencies' Practices*. Report prepared for GTZ, available at http://www.ecologic.de/modules.php?name=News&file=article&sid=829
Ramsar, 2002, Declaration on Resolution VIII. 2 on the World Commission on Dams, 8th Conference of the Contracting Parties, at http://www.ramsar.org/key_res_viii_02_e.htm
SADC – Southern African Development Community, 2003. Summary of the regional initiative on dams and development at http://www.unep-dams.org/document.php?cat_id=60
The Economist, 8 November 2003, p69.

van Robbroeck, 2002. Back to Our Roots? Presentation to the 70th Annual Meeting of ICOLD, September at http://www.icold-cigb.net/Back_Roots.htm

World Bank, 2001, at http://lnweb18.worldbank.org/ESSD/ardext.nsf/18ByDocName/Official WorldBankResponsetotheWCDReport/$FILE/TheWBPositionontheReportoftheWCD.pdf

World Bank, 2003a. Water Resources Sector Strategy at http://lnweb18.worldbank.org/ESSD/ardext.nsf/18ByDocName/WaterResourcesManagement

World Bank, 2003b. Stakeholder Assessment in Options Assessment at http://wbln0018.worldbank.org/esmap/site.nsf/files/wb+dam+booklet+10.9.03.pdf/$FILE/wb+dam+booklet+10.9.03.pdf

WCD - World Commission on Dams, 2000, *Dams and Development: A New Framework for Decision-Making*, Earthscan, London.

WCD, 2001. *Proceedings of the Third WCD Forum Meeting*, Spier Village 25-27 February, Cape Town at http://www.unep-dams.org/document.php?cat_id=13.

WWC - World Water Council, 2001 at http://www.unep-dams.org/document.php?doc_id=238

WWF, 2003. *Dam Right! An Investor's Guide to Dams*, available at www.panda.org/dams

WWF3 – Third World Water Forum, 2003. Theme Statement on Dams and Development at http://www.world.water-forum3.com/wwf/DAMS1_dams.doc

---

ENDNOTES

[1] The writer would like to acknowledge the assistance provided by the Dams and Development Project in compiling this paper and providing information on the various follow-up activities described.

[2] For example, the position of the Chinese delegation attending a regional workshop on the WCD report held in at ADB in Manila on 19-20 February 2001, see http://www.adb.org/Documents/Events/2001/Dams_Devt/Dams_devt.asp

[3] For a reflection on the Brent Spar experience see http://archive.greenpeace.org/pressreleases/oceandumping/1998nov25.htm l. There are parallels with the dams debate and interesting lessons learnt by Shell *"Dialogue should start as early as possible in decision-making 'Dialogue-Decide-Deliver' is better and less costly than 'Decide-Announce-Defend'"*.

[4] See http://www.unep-dams.org/document.php?cat_id=16

[5] See *'An approach for Financial Institutions in Determining, Assessing and Managing Environmental and Social Risk in Project Financing'*
http://www.equator-principles.com/index.html

[6] See http://www.swissre.ch/INTERNET/pwswpspr.nsf/fmBookMarkFrame Set?Read Form&BM=../vwAllbyIDKeyLu/BMER-5HNHW9?OpenDocument

[7] See http://www.jbic.go.jp/english/environ/guide/finance/index.php and http://www.jbic.go.jp/english/environ/guide/finance/check/list02.php

[8] See http://www.opic.gov/EnvironASP/envbook_revisions.htm

[9] See *Environmental Guidelines on Hydroelectric Power Stations and Large Dams*
http://66.102.11.104/search?q=cache:s8KkRk2nL9QJ:www.coface.com/_docs/barragesgb.pdf+coface+dams&hl=en&ie=UTF-8

[10] See http://www.adb.org/NGOs/adb_responses.asp

[11] See http://www.hydropower.org/1_5.htm

Political ecology of dams in Teesdale.

C.S.MCCULLOCH, University of Oxford, UK.

SYNOPSIS. Between 1894 and 1970, six dams were built in the beautiful Pennine landscape of Upper Teesdale in North East England to supply industrial consumers on Teesside. Political influences on the decisions to build these impounding reservoirs are explored to discover the reasons for ignoring alternatives, some of them much less intrusive on the rural environment. Was the concept of a sequence of dams in upland dales overtaken by a megadam with consequent major transfers of water between catchments? With hindsight, should preference have been given to provision of domestic and industrial water storage by the "Metropolitan" solution of pumped storage off-river reservoirs close to the point of use? By asking who benefits and who pays, economically, socially and environmentally, this historical analysis presents a wide perspective on the social and environmental impacts of dams and reservoirs with implications for future choices.

INTRODUCTION

From the 19th Century, Pennine dams were regarded as a "natural" solution for water supply for growing industrial cities in the valleys and nearby lowlands. Over 200 were built between 1840 and 1970. The physical advantages of altitude allowing gravity flow from upland sources to lowland consumers, high rainfall and low evaporation, rivers transporting soft water in valleys topographically-suited for impoundment, gave the impression that this solution to water supply was pre-determined, a right and proper use of natural resources.

Industrial Teesside with its thirsty iron and steel works and heavy chemical factories sited around the estuary of the Tees, in a rain shadow area, followed this pattern of looking to the hills for water for a century. But a closer look at the history of the six dams built in Teesdale shows that the choices were strongly influenced by politics. Increasing wealth of urban industrialists on Teesside bargained with an almost feudal society of aristocratic Pennine landowners, threatened by new taxes, and their small tenant farmers, who had few resources and little power. Rights to build dams were easily negotiated with the gentry but post-Second World War opposition grew from middle-class defenders of the countryside.

The argument is proposed that engineering solutions to water supply to Teesside have been influenced strongly by politics. Historical vignettes, illustrate the role of engineers exercising power, in varying contexts, over development of water resources. Attention will be given to neglected alternatives. Once the interplay between technology and politics is recognised, "what if?" games may be played to assess how different political priorities might have led to outcomes more in tune with 21st Century ambitions in Europe for a water environment with a high degree of biological health.

JAMES MANSERGH AND JULIUS KENNARD: ENGINEERS AND POLITICAL ACTORS

James Mansergh and the first phase of dam building.
The first water undertaker for Middlesbrough and Stockton was a private company set up by the local industrialists, who organised direct abstraction from the Tees at Broken Scar (Figure 1), where a steam pump was installed in 1860, designed by Messrs J & C Hawkesley (Mansergh 1882). Later, the local Corporations claimed that the water supplied was sometimes unfit to drink and that the Tees was being ruined by abstraction. The Mayor of Middlesbrough had ambition to bring purer water from Pennine reservoirs in the manner of Manchester Corporation who, in 1847, took the whole of the Longendale valley to construct a series of stepped reservoirs (Walters 1936). He needed the help of an engineer who was a skilled politician as well as an expert in dam building to help him take over the private company by compulsory purchase.

He hired James Mansergh, who had designed a series of six dams in the Elan and Claerwen valleys in 1870-71 for the water supply of Birmingham. Mansergh held that it was "incontestable" that "the purveying of water to the public should be one of the distinctive functions of the responsible sanitary authority of any district" (Anon. 1905). His political beliefs suited and his advocacy skills won the day; the Stockton and Middlesbrough Corporations Act of 1876 was passed after a struggle lasting 42 days in committee in both Houses of Parliament. The Act authorised a new body, later to be called the Tees Valley Water Board, to abstract 39,096m^3/d from the Tees at Broken Scar and to construct, in the tributary Lune and Balder valleys, six reservoirs starting with Hury and Blackton.

Far from leading to an instant improvement of water supply with increased investment, taking the company into public ownership paralysed activity for years. Compulsory purchase did not come cheap: the legalities of the Act cost Middlesbrough and Stockton Corporations each £12,403 (£0.56M) (Note 1), whilst the cost of purchasing the company amounted to £845,986 (£38.3m) (MRO 1898a). This financial burden was so substantial that progress with the proposed upland reservoirs, then estimated to cost a further £700,000 (£31.7M), was seriously delayed. Until 1882, the new Water Board ran at a loss (MRO 1898b). Without the backing of the

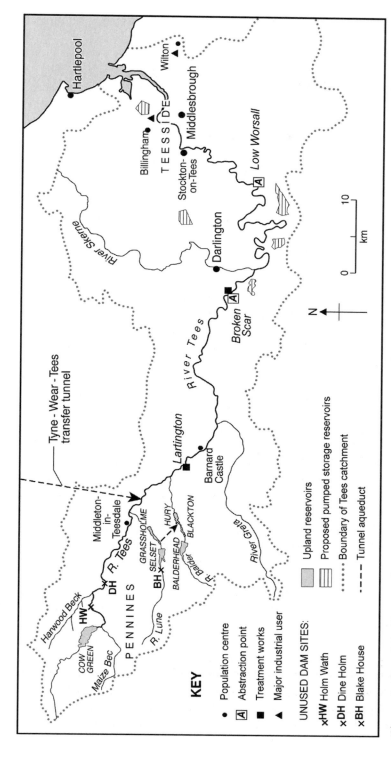

Figure 1. Teesdale Reservoirs

Corporations and their ability to obtain long-term loans, bankruptcy would have loomed.

Twelve years after taking over the water company, the Mayor of Middlesbrough and Chairman of the Water Board was called to account. Pig iron production demanded more water and the Stockton and Middlesbrough Corporations had to apply to Parliament for further powers to abstract even more water from the Tees at Broken Scar to tide them over until one or more of the upland reservoirs had been constructed. At the House of Lords committee hearing, the Mayor was subjected to hostile questioning by Counsel: "Is there a single work; that you undertook, that you have done, or a single promise you have made at this time that you have not violated?" The Mayor was reminded that, as a prelude to the takeover of the private company, he had decried its intention of taking more water from the Tees at Broken Scar; yet now the Mayor was asking to do the same (MRO 1884). Nevertheless, powers were extended following a promise of an early start on the proposed Hury reservoir.

Financial difficulties continued and reports of enteric fever were still being attributed to the drinking of water from Broken Scar (Anon.1890-91). The costs for Hury alone had doubled from the original estimate of £108,637(£7.1M) to £224,933(£14.7M) (MRO1898a). New borrowing powers were required to ensure completion of Hury and Blackton.

James Mansergh must have been a patient man. His plans for the Elan & Claerwen reservoirs had taken 20 years before adoption and his Teesdale scheme, approved in 1876, was only partially built before his death in 1905. His scheme involved relatively pure water being piped from two connected reservoirs: Hury (1894) and Blackton (1896) to a filtration plant at Lartington and then to Teesside. James Mansergh took the provision of compensation water very seriously, although he appeared less sensitive to biological issues. He had planned a third reservoir, Grassholme, in the neighbouring Lune valley mainly to remedy "serious injury" being caused by excessive abstraction (Note 2). Grassholme was connected to Hury by an aqueduct driven through the watershed so that water, above stipulated levels to ensure continuity of compensation releases, could flow into Blackton and Hury and thence into supply. Grassholme was not finished until 1915, built under the supervision of James Mansergh's son.

Financial compensation was substituted for building fish passes over the dams. A donation of £1,250 (£0.08M) "to be expended at such times and in such manner as the Board of Trade in their discretion may think fit for the permanent improvement of the salmon fisheries of the Tees Fishery District" sufficed to grant the Water Board relief from the requirements of the Salmon and Freshwater Fisheries Acts 1861 to 1892. In whatever way the Board of Trade spent the donation, it did not stop the subsequent rapid decline of salmon fishing on the Tees, caused by pollution of the estuary.

An alternative in the search for pure water might have been exploitation of local aquifers but one of the reasons James Mansergh avoided using groundwater was the poverty of its legal protection: "there was no right in underground water unless it could be proved that such water was flowing in a defined and locatable underground channel". This meant that there was no assured compensation if another party drilled a well nearby, reducing supply from the earlier well (Mansergh 1901).

Features of this early phase of dam building included:
- Importance of a champion for the reservoirs who could speak with authority to investigating committees.
- Need for an effective management organisation. The new Water Board took years to build up the necessary finance to implement plans approved in 1876.
- Ease of negotiations with Pennine owners with large estates. Agreements allowed retention of gaming rights, so that sale of the land did not affect landowners' life styles, whilst providing much needed cash to set against increasing taxation and agricultural depression.
- Tenant farmers had little protection except that there might be resistance from the Local Government Board, if it were to be faced with an excessive number of displaced people (more than 10 families).
- Before introduction of chlorination, there was a premium on the relative purity of the upland water.

The second phase of dam-building in Teesdale led by Julius Kennard
Industrial contraction during the interwar years affected the finances of the Water Board so badly that it had to raise the water rates whilst there was much hardship from unemployment. Spens (1948) attributed the lapse in investment in the interwar years also to poor management before an "energetic and enthusiastic" Engineer and Manager was appointed in 1941. Construction of the fourth Mansergh dam at Selset above Grassholme had been planned during the War but no funds were available. Expansion on Teesside of heavy manufacturing, despite its high water demands, was given encouragement by the post-war Government. Imperial Chemical Industries (ICI) opened its Wilton petrochemical works in 1949 and began a programme of rapid expansion (Owen, 1999).

Water demand on Teesside became pressing and the Water Board needed an engineer with both experience in building dams and political skills to win Parliamentary approvals. Heightened political awareness and strengthened defence of alternate land uses faced Julius Kennard on his appointment as consulting engineer to the Tees Valley Water Board in 1952. Promotion of dams in the upland dales now invited opposition from organisations such as the Wear and Tees River Board, the Nature Conservancy, the National Parks Commission, the National Farmers Union, the Ramblers' Association and many other bodies with different priorities for the use of the uplands.

Like James Mansergh, Julius Kennard acted politically both in planning reservoirs, in sounding out opponents and in promoting the schemes. The practice at the time was for the consulting engineer undertaking the promotion to be appointed to design the works approved. This dual role led to preference for supply rather than proposals for demand reduction; and for the form of supply of the type preferred by the appointed engineer, in this case upland reservoirs. Julius Kennard added another reason for his choice: "an underground water scheme will involve the promotion of numerous Orders, which, if objected to, will necessitate local enquiries, and it is more than likely that protective clauses will be included on behalf of existing users"(Kennard 1965).

With a prestigious remit to provide structural solutions to increase supply, Julius Kennard at first followed James Mansergh's plans, developed by his son, for a second reservoir in the Lune valley at Selset, above the Grassholme reservoir, and he reported to the Water Board (1952) that "our survey confirms the information which Mssrs. Mansergh set out in their report dated 1/10/20 suggesting an earth embankment dam of the usual design." He also quoted with approval Ernest Mansergh's views:

> 'Some years ago, and not very many, "compensation water" was looked upon as something bordering on the sacred, and rightly so, because after all it represents a form of property in which others have a right and interest, sentimental perhaps to a very large degree, but nowadays compensation water must be looked at from a more materialistic point of view, not who has a right to the water, imaginary or real, but to whom is the use of the water going to be of the most benefit.'

This more materialistic point of view was endorsed by the Water Board, and drastic reduction in compensation water was sought as a stop gap, until further upland reservoirs could be built.

The Wear and Tees River Board, set up in 1952 with wider terms of reference but excluding water supply, did not view their reasons for wanting more water in the Tees as "sentimental". Water in the rivers was needed to dilute pollution and improve water temperatures for fish, while people visiting the river for recreation wanted to see more water flowing. An unusual battle began between the Water Board and the River Board, all the more extraordinary because of the unremarked conflict of interest of Alderman Charles Allison, who was simultaneously not only Chairman of the Water Board but also Deputy Chairman of the River Board.

In 1953, and again in 1958, the Water Board promoted private bills (Tees Valley Water Bills) seeking temporary reductions in the compensation water. On both occasions, they were petitioned against by the Wear and Tees River Board, without success. Despite "several meetings between representatives of the River Board and the Water Board...unfortunately, no agreement was reached on the several points of difference" (Wear and Tees 1959). Alternatives to the Water Board's plans for further reservoirs in the dales were put forward by the River Board in 1955 but were rejected. (See below: "The Metropolitan solution").

After ensuring the necessary Parliamentary procedures, Julius Kennard oversaw the construction of Selset reservoir, acclaimed by the Water Board as an ample water supply for at least 25 years (Anon. 1955). Yet demand threatened to outstrip supply very soon after building started so that the Water Board wanted further dams.

The enticement of greater discharge encouraged Julius Kennard to stray from the Mansergh scheme, although two potential dam sites remained: at Balderhead above Blackton and at Blake House above Selset, and to investigate the possibilities of dams in the main valley of the Tees. The physical attributes of a large river flowing in a gorge were attractive but not only to an engineer: Upper Teesdale was contested territory. Beautiful scenery was valued by walkers, the dales' improved pastures were important for agriculture. Also most of the land had been designated by the Nature Conservancy as a Site of Special Scientific Interest (SSSI) and the bleakest upper reaches at Moor House had been bought in 1952 as a National Nature Reserve.

Julius Kennard sounded out the Nature Conservancy (NC) over potential dam sites. At first, the NC officers were not alarmed. They took their lead from much revered Professor W.H. Pearsall, F.R.S., who was interested in biological productivity more than preservation. As a member of the Conservancy since 1949, Chairman of the Conservancy's Science Policy Committee 1955-63, architect of the Upper Teesdale SSSI and the Conservancy's land use policy, he wrote to the Regional Officer:
    'I think that it is pretty clear that from the point of view of the naturalists that the project of putting a dam just above Cauldron

Snout is much the better one and I personally would offer no objection to it. I would not offer great objection to the alternative but I am pretty sure that there would be an outcry from the naturalists about this one. It is, between ourselves, logically and geologically the better site and I should not be at all surprised if ultimately adopted. (PRO FT 17/68, 08/05/56).'

But, in November 1956, Julius Kennard met with the Deputy Director, Dr Worthington, and was told that the NC might take strong exception to the reservoir. Worthington noted for the record that Julius Kennard was not interested in Natural History. (PRO FT 17/68 27/11/56).

The upper site above Cauldron Snout, Cow Green, was investigated first to test its geological suitability. The geologist, Edgar Morton, advised the Water Board that the site would not be watertight and should be abandoned. Attention turned to sites below Cauldron Snout, first at Holm Wath just below the cataract and then at Dine Holm further downstream, but above the waterfall at High Force, a major tourist attraction. Morton advised that the narrow valley with dramatic limestone cliffs at Dine Holm could, with some grouting, be suitable for an impounding reservoir. Water augmented by the reservoir could flow by gravity in a pipeline from an intake just below the waterfall to Teesside.

Alarm grew amongst scientists and amenity groups who feared loss of the rare flora, which had made Upper Teesdale internationally remarkable. At the same time as the Water Board was laying plans for a reservoir, an influential paper appeared in the Journal of Ecology (Pigott 1956) analysing why such a concentration of rare species found congenial conditions in Upper Teesdale, far from their usual habitats in high mountains or in the Arctic. The governing committee of the NC on 30/01/57(PRO), agreed "to make the strongest opposition to the proposed reservoir". A letter deploring the proposal was orchestrated for publication in the Times in February 1957, signed by 15 prominent botanists. The stakes had been raised from a local planning issue to a national debate both about nature protection and national policy for industrial water supply.

Communication between the Water Board and the NC appeared indirect at this stage. In July, it was a representative of Durham County Council who told the NC that the Water Board had now confirmed that it would be promoting a Parliamentary Bill in the next session for the construction of an impounding reservoir at Dine Holm (PRO FT17/68 18/07/57).

At last, on 8 October 1957, a meeting was held between the Water Board (Julius Kennard and E.A. Morris), the NC (R. J. Elliott), R. Atkinson

(Durham County Council) and J. Vincent (North Riding County Council). There was little meeting of minds. Elliott reported, "Pressed on the methods that the Board would adopt to meet a recurring water deficit - Kennard's only solution was 'additional reservoirs'...Asked what alternative sources of supplying industries' needs had been investigated - the officers (of the Water Board) present became decidedly hostile" (PRO FT17/68 08/10/57).

On 25 October, the Director-General of the NC, Max Nicholson, wrote "now that the Conservancy have instructed me to fight this Tees Valley case I will do so to the utmost of my ability, and am reasonably confident of success". He had been working behind the scenes, with the National Parks Commission, to tackle the Ministry of Housing and Local Government (MHLG). He recorded:

> 'The most interesting point of all which emerged was that the Ministry and the promoters have given no real thought or study to the alternatives and that they have at least at present no answer which could stand up to examination as to why the reservoir is necessary at Dine Holm or anywhere else (PRO 17/68 25/10/57).'

Then, Max Nicholson had an inspiration: rather than continuing to argue with the Water Board, or to hope that the MHLG would take action, he would approach the Chairman of ICI (1953-60), Sir Alexander Fleck KBE, FRS, DSc, directly. The letter amounted to refined blackmail,

> 'You are likely to be next year's President (of the British Association for the Advancement of Science) at Glasgow when, amongst other things, I understand that the question of water conservation is likely to be discussed...'

He went on to alert Fleck to the threat of the Tees Valley Water Board "irretrievably to destroy this area by inundating it under a reservoir at Dine Holm" and concluded by saying, "we would be very sorry to find ourselves compelled to do battle with ICI without having made every effort previously to reach an acceptable solution" (PRO FT 17/68 01/11/57).

Faced with a potential humiliation on an occasion that should have marked the pinnacle of his scientific career, Fleck readily agreed to meet with Nicholson on 14 November 1957. Nicholson jubilantly reported back "the ICI were ready to put a brake on the Dine Holm project until there had been more opportunity to examine alternative sources of water."(PRO FT 17/68 14/11/57). ICI staff reported dryly on the Water Board's proposed bill:

> 'In view of the expected opposition from outside bodies to the scheme and incompleteness of the investigation of reasonable alternatives, ICI did not feel that they were in a position to support such a bill and this scheme was therefore shelved. (ICI X/11489).'

Julius Kennard reverted to the Mansergh plan for a third reservoir in Balderdale, above Hury and Blackton at Balderhead, despite opposition from farmers (Sheail, 1986), and an extension to the pumping station at Broken Scar. The Daily Express (8/4/61) reported the inauguration of construction at Balderhead and the passionate response of the Chairman of the Water Board who "was very cross about it all". Alderman Allison is reported to have said, "All this fuss is a lot of tommy rot. It is sickening to think that a little flower is more important than the future of Teesside. Who cares if the gentian disappears - it is no good to anyone?"

Meanwhile, the NC was lulled into complacency: the Dine Holm scheme had been averted and the potential reservoir site at Cow Green deemed unsuitable because of permeable rocks. A major flaw in the legal protection of Upper Teesdale remained: Moor House Nature Reserve had been purchased, a further National Nature Reserve had been agreed with the Earl of Strathmore west of the Tees but the land on the east, owned by the Raby estate, included Widdybank Fell with its valued Arctic-alpine vegetation still vulnerable as a "proposed" Nature Reserve with no legal status. The owner, the Hon H.J.N. Vane, later to inherit the title of Lord Barnard in 1964, did not want to comply with the NC's proposal for a nature reserve, perhaps because the barytes mines at Cow Green, closed in 1954, might be reopened should the market for this mineral recover.

To Julius Kennard, this unprotected site, barring the gloomy predictions of leakage by Edgar Morton, seemed more attractive than the last site identified by Mansergh higher up the Lune valley above Selset at Blake House. He sought a second opinion. His son, Michael, with Dr John Knill carried out a detailed site investigation from which they concluded that the high water table on the east side of Cow Green would prohibit leakage through the limestone strata to the adjacent Harwood Beck (Kennard & Knill 1969). With this good news, Julius Kennard recommended that steps be now taken for obtaining statutory powers to construct the reservoir.

The difficulties for the NC were just beginning. Julius Kennard approached them again in August 1964 (PRO FT 17/61 24/08/64) and was at first assured that the Cow Green site was unlikely to be problematic but in fact the proposal to build a reservoir at Cow Green unleashed angry reaction from naturalists in the Northumberland and Durham Naturalists Trust, the Botanical Society of the British Isles and many other environmental organisations. A public subscription was raised to fight the case and, following submission of a private bill in December 1965, the debate continued in the Select Committees of the House of Commons and the House of Lords throughout 1966. The story is told by Gregory (1975). This time, Julius Kennard and the Water Board were victorious: the Board was granted permission by Parliament to build the Cow Green reservoir.

THE RISE OF THE MEGADAM

A decade later, in response to projections of increased industrial demand and in an attempt to avoid adding to the plethora of dales reservoirs, a tunnel was constructed to bring water 45 miles to the Tees from the river Tyne, supported by what was claimed to be the largest man-made lake in Europe, Kielder Water. This scheme made the Teesdale dams no longer essential. In theory, the Teesdale dams could now be decommissioned in favour of water imported from the Tyne. In practice, it is the giant Kielder reservoir with a capacity of 200 Mm^3, double that of all six Teesdale reservoirs, which has remained underused for 20 years, failing in its aim to improve the economic development of the North East by attracting new, water-needy industries. Supply from the Teesdale reservoirs continues as the cost of pumping water from the Tyne to the highest point of the Tyne-Tees tunnel is greater than the cost of supply by gravity flow from the Teesdale dams; also soft water from Lartington is economical for boiler feed. Only twice in its history has Kielder been used to transfer water to the Tees, first in 1983 and then in 1989, (FOE 2003) although water has been transferred as far as the Wear to supplement the underperformance of the Derwent reservoir (Soulsby *et al* 1999).

Planning water resources on such a large scale required political reorganisation. The Water Resources Act 1963 set the scene with the creation of large River Authorities and a national body, the Water Resources Board (WRB), to encourage long-term integration of water supply over wide areas. Rather than continued iteration with the industrial consumers to judge its effectiveness in promotion of economic development, dedicated focus on water supply made it an end in itself and safeguards against overinvestment were weak. Uncritical extrapolation of water demands at the outset was not corrected at later stages when British Steel failed to expand on Teesside. "Over investment for any particular area is indicated when facilities stand idle or else are put to makeshift uses, either to avoid the appearance of idleness or to minimize the losses due to past mistakes."(Hirshliefer *et al*(1960)). Tourism gains from Kielder may be viewed in this light. Short summers, high rainfall, biting insects, restrictions on motor boats and remoteness from centres of population suggest that such a recreational facility would not have been sited in the Upper Tyne valley, if this had been the main aim for such a huge financial investment.

Unlike the financial arrangements in Teesdale, those industries which demanded more water at the Kielder inquiry made little or no contribution to the capital costs of the Scheme, which was funded by loans from the

National Loans Fund £46M(£121M); from the European Investment Bank £63M(£166M) (at interest rates of up to 17 $^7/_8$ % over 25 years) and grants from the UK Government £24M(£63M) and the European Regional Development Fund £36M(£95M) (HoC Public Accounts 1984-85). Brady (1983) claimed that "the financial burden has shifted substantially away from Teesside industries towards the region's other consumers". In 1989, at privatisation, much of the outstanding debt was transferred to Government to make the sale of the Northumbrian Water undertaking attractive. Today, Northumbrian Water Group plc has debts of £1.7bn and receives £11.5M annually from the Environment Agency to operate Kielder (NSL Group 2003).

Environmentally, the assessment is mixed. Omission of a fish pass was justified at the time by substitution of a fish hatchery at Kielder and the hatchery has been successful in reintroducing salmon to the Tyne (Marshall, 1992). Yet there are serious doubts whether the genetic pool from which these stocks are bred is sufficiently diverse for the process to be sustainable (Anon 2002). Transmission of water from the Tees to the Yorkshire Ouse catchment is now physically possible via a pipeline constructed during the 1995 Yorkshire drought but such transfers are opposed by the FOE as dangerous biologically. Instead of importing water from another company, Yorkshire Water has improved conjunctive use of its own resources.

The high costs of the Kielder Water Scheme have weakened support for similar megaschemes. The words of Rocke (1980) ring true "schemes such as Kielder may be the last of their kind for some time".

A CENTURY OF DAM BUILDING FOR SUPPLY TO TEESSIDE: WINNERS AND LOSERS

Determined pursuit of water supply led by water engineers resulted in:
- Successful supply to Teesside industries and domestic users.
- Construction of six reservoirs in Teesdale, without oversupply because of control of funding by the industries benefiting.
- The second phase of 3 reservoirs in quick succession fuelled demands for longer-term planning and a national strategy.
- Expensive and protracted disputes, increasing distrust between water engineers and environmentalists.
- A greatly-modified river environment.
- The Cow Green reservoir, still regarded "as an unforgivable intrusion". (Ratcliffe, 2000). Valued vegetation was drowned and the surroundings affected (Huntley *et al*, 1998).
- The expensive and under-used Kielder Water Scheme, still a drain on the public purse.

Table 1. Impounding reservoirs in Teesdale, also Kielder Water

Reservoir & consulting engineer	Date built. River	Dam dimensions	Full Capacity	Type
Hury *J. Mansergh*	1894 Balder	33m H 374m L	3.9Mm³	Direct soft water supply
Blackton *J.Mansergh*	1896 Balder	21m H 338m L	2.1Mm³	To Hury + flood bypass.
Grassholme *E Mansergh*	1914 Lune	34m H 274m L	6.1Mm³	Compensation + to Hury.
Selset *J.Kennard*	1959 Lune	41m H 928m L	15.3Mm³	To Grassholme
Dine Holm	*Abandoned Tees*		*17.2Mm³*	*Direct*
Balderhead *J.Kennard*	1964 Balder	52m H 914m L	19.7Mm³	To Hury + regulating
Cow Green *M.Kennard*	1970 Tees	26m H 572m L	40.9Mm³	Regulating
Kielder *D.J. Coats*	1982 N. Tyne	52m H 1140m L	200.0Mm³	Regulating

(See Note 3)

AN ALTERNATIVE WATER ENVIRONMENT.

More use of groundwater, demand reduction by improvement of industrial efficiency in energy use, water recycling and elimination of polluting discharges are some of the alternatives raised by critics of this century of impounding dam construction (Kinnersley 1988; Pearce 1982). The quantities of water required might not have been met wholly by such means but a concept raised during the struggles, perhaps too easily dismissed by the water engineers intent on upland dams, is worth revisiting in the light of modern ambitions, such as those raised in the European Water Framework Directive. This was called the Metropolitan solution, basically reducing the spatial extent of the "footprint" of industrial Teesside, following the example of London.

THE METROPOLITAN SOLUTION

Cecil Clay, Chief Engineer of the Wear and Tees River Board, put forward plans more protective of the integrity of the Tees. He suggested conjunctive use of abstraction at Broken Scar with storage in the three existing upland reservoirs and seasonal variation in release of compensation water (HoC 1958). His ideas were supported by Thomas Hawkesley, great grandson of the first engineer of the private Middlesbrough and Stockton Water

Company, who added that water abstracted at Broken Scar would need more treatment and pumping than the upland water but the extra cost would be "a bagatelle on the total annual cost of the undertaking"(HoC 1958) (17). Later, the River Board put forward a plan to the Water Board that added pumped storage reservoir(s) in the Tees lowland to store river water abstracted at Broken Scar or nearby points during high flows. Six possible sites were proposed as shown on the map (Figure 1).

This "Metropolitan" solution, similar to London's supply, with water abstracted from the Thames and stored in large off-river reservoirs at Windsor and Staines, was turned down by the Water Board before they promoted the Cow Green scheme in Upper Teesdale in 1965. Julius Kennard (1965) advised the Board "we are in no doubt that such a scheme should not even be contemplated in the circumstances". He argued that the capacity of the abstraction plant at Broken Scar would have to be extended if high flows were to be abstracted and taken into storage and suggested that a pumped storage scheme might take longer to construct than the Cow Green reservoir. However, it is debatable whether construction in the lowlands would take longer than construction of Cow Green in the Pennines, where the construction season was short because of heavy snowfalls.

A pumped storage reservoir built at about the same time for London's water supply, Wraysbury ($35Mm^3$), provides a comparator with Cow Green ($40Mm^3$). Wraysbury took 5 years to build, (1965-70), and cost £3.7M (£35.2M). Cow Green took 3 years to build (1967-70) and cost £2.5M (£28.6M) (Griffiths 1984). Yet Kennard claimed: "the cost of the reservoir (pumped storage at Teesside) itself could be as much as twice the cost of Cow Green reservoir". WRB (1965) thought two of the six Teesside sites were comparable with Cow Green: at Staindale, and at Cowpen on the estuarine marshes where the building estimate was equivalent to that of Wraysbury, even though costs of construction and land purchase in Teesside were likely to be much less than those in the desirable London suburbs. Other potential problems were listed, none of which deterred the engineers constructing similar off-river reservoirs at London, Farmoor (Oxford) and Exeter. The case was concluded by anticipation of great opposition from the public; in fact, it was the underestimated opposition to the upland Cow Green reservoir that caused two years' delay.

The multiplicity of arguments made against the Metropolitan solution gives an impression of special pleading. Some could be countered; for example one of the sites, on Cowpen Marsh, was not good agricultural land. Even the loss of good farmland did not prevent the building in the area of many service reservoirs for the Teesside distribution network. One of these, the

Long Newton reservoir at 200 ha, a third of the possible size of a pumped storage reservoir, was constructed without opposition, only two miles south of one of the proposed sites at Newbigin. Even if the costs were somewhat higher, reservoir construction near a city offered much needed water-recreation facilities within easy reach of many and, probably, less upset for any families displaced by compulsory purchase because of the greater availability of job opportunities in a suburban area and greater acceptance of industrial development by the public.

The botanist, Professor Donald Pigott (1957), summarised the situation:
> 'The continual expansion of British industry results inevitably in an increasingly urgent competition for space in this crowded island. This would be less serious if industrialisation could be confined to certain agreed areas. But enormous quantities of water are demanded for modern industrial processes and this leads to constant requests for permission to construct reservoirs at points well outside the actual industrial regions.'

If the alternative of off-river pumped-storage schemes had been opened up to public debate, the outcome of the struggles for water in Teesdale might have been very different with habitats of rare plants left unmolested.

CONCLUSION.

A century of industrial expansion in Teesside began with laws requiring compensation for water withdrawn from rivers or for injury to game fish populations and it was a criminal offence to pollute water. Each of these ideals was eroded under pressure, as illustrated in this story but now, with the decline of heavy manufacturing industry in Europe (often re-located overseas to even more water-stressed environments), hopes of an undamaged water environment have returned.

The challenges presented by the European Water Framework Directive will require cooperation rather than the antagonism between engineers and biologists that marred the era of industrial expansion. If the new legislation is to be more successful than the old, many water resource solutions, structural and non-structural, need to be explored before attitudes harden around preferred options. Historical studies of the connections between politics and the environment may illuminate scenario building for a future requiring holistic responses.

ACKNOWLEDGEMENTS
The author thanks Dr Erik Swyngedouw, Professor John Sheail and Mr. Michael Kennard for their help and the Cartographic Office of the School of Geography and Environment for drawing the map. The opinions expressed in the article are those of the author.

NOTES
1. Money has been translated into 2002 purchasing power by Economic History Services www.eh.net/hmit/ppowerbp/
2. In Mansergh's words, reflecting on common law, "no public body may abstract water from a surface stream (other than a large river at a low level) without compensating the owners below, either in money or in water...Further, no riparian may pollute a stream as it passes through his estate, or take water so as to reduce its volume except for fair and legitimate uses upon that estate" (1901).
3. Hury, Blackton and Grassholme engineered by J. Mansergh & Son; Selset and Balderhead by Sandeman, Kennard & Pts; Cow Green by Rofe, Kennard & Lapworth; Kielder by Babtie, Shaw & Morton.

REFERENCES
Anon. (1890-91) 20th Ann. Report of the Local Government Board. *Report of the Medical Officer for 1890*. HMSO, Eyre & Spottiswoode, London.
Anon. (1905). Obituary of James Mansergh, FRS. *Min. Proc. Inst. Civ. Eng.* **CLXI** (3) 350-56.
Anon. (1955). *River Tees Handbook*. Tees Conservancy Commissioners.
Anon. (2002). The Kielder hatchery. *Trout and Salmon* eMap plc. London.
Brady, J.A. (1983). *Water resources in the North East. The Kielder Water Scheme. A decade of change*. Northumbrian Water Authority.
Friends of the Earth.2003. www.foe.co.uk/briefings/kielder_transfer_scheme.html
Gregory, R. (1975). The Cow Green Reservoir. In Smith, P.J. (1965). *The politics of physical resources*. 144-201. Harmondsworth, UK, Penguin.
Griffiths, F.N.(Ed.) (1984) *Recent dams in the U. K.* BNCOLD, London.
Hirshleifer, J. *et al* (1960). *Water supply, economics, technology and policy*. University of Chicago Press, Chicago, USA.
HoC Committee of Public Accounts (1984-85). *Monitoring and control of Water Authorities*. HMSO. London.
HoC (1958). Minutes of Evidence. Committee of Group A of Private Bills. *Tees Valley Water Bill 25 March*. H of Lords Record Office.
Huntley, B. *et al.* (1998). Vegetation responses to local climatic changes induced by a water-storage reservoir. *Global. Ecol. Biogeogr. Lett.*, **7**, 241-257.
ICI X/11489 23 March 1960. *Development of water supplies on Teesside: proposed basis for developmental extensions under the Tees Valley and Cleveland Water Act 1959*. Billingham Archive.
Kennard, J. (1952). *Report to Tees Valley Water Board on the proposed Selset reservoir 14 August*. Unpublished. CEH. Wallingford Archive.
Kennard, J. (1965). *Report to the Tees Valley and Cleveland Water Board*. Unpublished. CEH. Wallingford Archive.

Kennard, M.F. & J.L. Knill (1969). Reservoirs on limestone with particular reference to the Cow Green scheme. *J. Inst. Wat. Eng.*, **23**, 87-136.

Kinnersley, D. (1988). *Troubled water: Rivers, politics and pollution.* Hilary Shipman, London.

Mansergh, J. (1882). *Lecture II* (para 49). School of Military Engineering, Chatham.

Mansergh, J. (1901). Pres. Address. *Min.Proc.Inst.Civ.Eng.*, **143**(1), 2-83.

Marshall, M W (1992). *Tyne waters. A river and its salmon.* H. F. & G. Witherby, London.

MRO (Middlesbrough Record Office. Teesside Archives), (1884). *Stockton and Middlesbrough Corporations Water Bill. 14 May.*

MRO, Stockton & Middlesbrough Corporations Water Board (1898 a) *Estimates of cost of works authorised by the Water Acts.*

MRO, Stockton & Middlesbrough Corporations Water Board. (1898 b) *K Statement showing the profit and loss if interest on works in progress had been charged to capital.*

NSL Group. Annual Report and accounts 2003. Northumbrian Services Ltd.

Owen, G. (1999). *From Empire to Europe. The decline and revival of British industry since the Second World War.* Harper Collins, London.

Pearce, F. (1982). *Watershed. The water crisis in Britain.* Junction Books, London.

Pigott, C.D. (1956). The vegetation of Upper Teesdale in the North Pennines. *J. Ecol.* **44** (2), 545-584.

Pigott, C.D. (1957). The botanical treasures of Upper Teesdale. *New Scientist,* February 21. 12-13

Public Record Office (PRO) FT 17/68.W.H. Pearsall to R.J. Elliott 08/05/56

PRO FT 17/68. Interview Dr Worthington and Mr. J. Kennard. 27/11/1956.

PRO FT 17/68. Letter from Max Nicholson to Prof. Roy Clapham 25/10/57

PRO FT 17/68. Internal memo: Meeting with ICI 14/11/57

PRO FT 17/68. Durham County Council to R. J. Elliott 18/07/57

PRO FT 17/68. Letter from Max Nicholson to Sir Alexander Fleck01/11/57

PRO FT 17/68. R. J. Elliott's report of meeting. 8/10/57

PRO FT 17/68. Internal memo: Meeting with ICI. 14/11/57

PRO FT 17/68. Nature Conservancy Minutes 30/01/57

PRO FT 17/61. Record of phone call. 24/08/64

PRO FT 17/61. Record of meeting. 01/09/64

Ratcliffe, D.A.(2000). *In search of nature.* Peregrine Books. Leeds.

Rocke, G.(1980). The design and construction of Bakethin Dam, Kielder Water Scheme. *J Inst. Water Eng. Sci.* **34** (6), 493-516.

Sheail, J. (1986). Government and the perception of reservoir development in Britain: an historical perspective *Planning Perspectives.* **1,** 45-60.

Soulsby, C.*et al.* (1999) Inter-basin water transfers and drought management in the Kielder/Derwent system. J.CIWEM, **13** 213-223

Spens, C.H. (1947). *Water Supply survey, N. E Development Area. Ministry of Health.* PRO HLG 113/49 & 50.

Tees Valley Water Bill 1953.HMSO. London. CEH Archive

The Times 13 February 1957.

Walters, R.C.S. (1936). *The Nation's water supply.* Ivor Nicholson and Watson Ltd, London.

Water Resources Board (1965) *Water supplies in the area of supply of the Tees Valley and Cleveland Water Board.* Unpub. CEH Archive.

Wear and Tees River Board. *9th Annual Report for the year ending 31 March 1959.*65.

2. The use of new materials

Raising of the Ajaure embankment dam by extending the moraine core with a geomembrane

Å. NILSSON, SwedPower AB, Stockholm, Sweden
I. EKSTRÖM, SwedPower AB, Stockholm, Sweden

SYNOPSIS. The Swedish Ajaure embankment dam is a high consequence dam which is 46 m high and was constructed between 1964 and 1966 with commissioning in 1967.

During the 1980's it was noted that the horizontal displacements in the main embankment dam didn't show any sign of diminishing over a time period. From the time of construction up to the year 2001 the total displacement at the crest was in the order of 500 mm and the creep has continued at a rate of approximately 8 mm per year. In order to stabilize the dam, and allow for future raising of the dam, supporting berms were placed on the downstream side of the left embankment dam in 1989 and 1993.

In addition to the deformation problem the Ajaure Dam required to be upgraded to allow for the new design flood. After comprehensive investigations and studies it was decided to raise the crest of the dam to be able to release the design flood at a water level 5 m above the retention level. The owner Vattenfall used a risk analysis as one input in the decision process to raise the dam. The risk analysis is in the subject of a separate paper for this conference (Bartsch, 2004).

Different construction options were considered, and a geomembrane was finally chosen for the extension of the moraine core. A Flexible Polypropylene (FPP) with a thickness of 1.5 mm was selected. Bentonite Enriched Sand (BES) was used to connect the existing core with the geomembrane. The design of the crest raising was started in 2001 by Golder Associates, UK and continued by SwedPower with detailed design and tender documents. The design and the construction, which was completed in 2002, are described.

Long-term benefits and performance of dams, Thomas Telford, London, 2004, 69–80

BACKGROUND

The Ajaure Dam is situated in the upper part of the River Ume Älv in northern Sweden. The Ajaure embankment dams are classified as high consequence dams, as a dam failure could have disastrous consequences for the downstream hydropower plants all the way to the Baltic Sea.

The dam is similar to that of other Swedish dams built at the same time. However, the downstream slope incline was originally considerably steeper (1V:1.8H and 1V:1.5H near the crest) than that of other Swedish dams. At a late stage of the construction period it was decided to raise the crest of the dam by approximately 1 m. The reason for this was to increase the freeboard along the main part of the dam and to allow for post construction settlements. The raised crest level resulted in steep slopes (1V:1.35 to 1.40H, approx. 35°) in the upper approx. 12 m of the dam. Below this level, the downstream slope has an incline of 1V:1.8H.

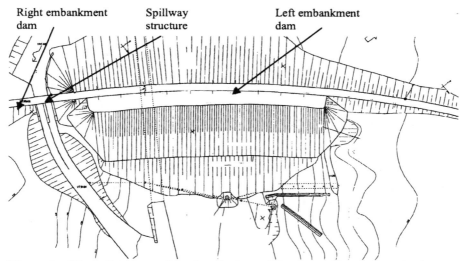

Figure 1. Plan of Ajaure after the placing of the second berm (marked B).

The greatest dam height over bedrock is 46 m. The total length of the dam construction is 522 m. In cross-section the embankment dam has a central impervious core of moraine surrounded by filter zones on the upstream and downstream side. The supporting fill is rock, taken from excavation of the power station, diversion- and tailrace tunnels.

PROBLEM DESCRIPTION

There were two different main problems concerning the safety of the dam. On the one side there was an ongoing exceptional displacement in the left embankment dam. On the other side the new design flood required a rising of the core of dam core by 5 m.

During the 1980s it was noted that the horizontal downstream displacements in the main embankment dam didn't show any sign of diminishing over a time period. The horizontal deformations are significantly influenced by the reservoir as shown in Figure 2.

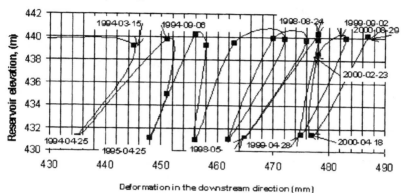

Figure 2. Record of horizontal displacements at the crest

Compared to measurements of deformation in other Swedish dams of similar design, the horizontal deformations were notably large. The present horizontal displacement of approximately 8 mm/year was almost linear and showed no tendency to diminish.

In 1989 test pits were excavated along the downstream toe of the left embankment dam. The rock fill consists of schists and gneiss with a high content of mica. The rock fill had a high content of fines, see Figure 3. It is judged the fines at the lower parts of the fill are partly washed down from higher elevations by the precipitation.

Laboratory shear tests showed that crushing of the material occurred to a large extent. The result of these tests established the weakness of the material at high shear stress, as well as exceptionally low shear strength of the material in the supporting fill. The low safety factor indicated that the shear stresses in situ were close to possible mobilized shear stresses at failure. It was concluded that the high shear stress, in combination with the cyclic loading from the reservoir, results in the progressive crushing of the rock fill and that crushing could be causing the continued horizontal deformations.

Stability analyses also indicated low stability and a stabilising berm of blasted rock was placed against the lower half of the downstream side of the left embankment dam to increase the stability margin in 1990. The berm was 18 m wide and 20 m high, with a total volume of approximately 50,000 m³. The fill material was placed with the slope inclination 1 vertical to 2 horizontal, The berm was also intended to increase the erosion resistance at the downstream toe in the dam. No decrease in deformation was however noted after the lower berm had been placed.

Figure 3. Test pits in the downstream toe of the original supporting fill

A second upper berm with a total volume of approximately 100,000 m³ was placed in 1993. At this time the berm that was placed in 1990 was raised using in the same inclination 1 vertical to 2 horizontal up to a level approximately 1.5 m under the original crest of the supporting fill. The main purpose of the second berm was to further increase the stability margin. Furthermore the second berm was designed to make it possible to raise the crest in the future.

The horizontal downstream movement of the central part of the left dam increased dramatically after the placement of the second berm, and on examination in August 1993, longitudinal fissures on the crest of the dam outside the guardrail were observed.

It has been possible to calibrate reasonably well the deformations that have been recorded while the two supporting berms were placed in 1990 and 1993 with the deformation calculation program PLAXIS. Thereafter the displacement for the raising of the crest from +444 m to +446 together with a 4.8 m higher water level during a design flood was calculated as shown in Figure 4. The sealing element in the crest is not modelled and thus not shown in the figure.

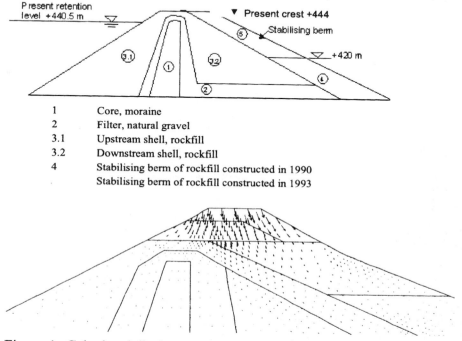

1   Core, moraine
2   Filter, natural gravel
3.1   Upstream shell, rockfill
3.2   Downstream shell, rockfill
4   Stabilising berm of rockfill constructed in 1990
       Stabilising berm of rockfill constructed in 1993

Figure 4. Calculated displacements using the PLAXIS program [ref.1].

The horizontal deformation in the top of the existing left embankment dam core is expected to be minor, in the order of 5 mm, while the vertical displacement in the dam crest is expected to be in the order of 100 – 120 mm. The largest vertical displacements are however expected to occur in the downstream supporting fill. In case of the design flood occurring an additional horizontal displacement of some 130 mm is expected, taking a new suggested crest elevation of +447 into account. The notable displacement will according to the calculations take place outside the moraine core, see Figure 4, which is important since the ongoing horizontal displacement could otherwise cause transversal cracks in the moraine core.

DESIGN

Because of the left dam stability problems and the shortage of suitable moraine material in the area it was decided to raise the core using a geomembrane in a bentonite enriched sand layer set into the existing core crest as the impermeable element. This alternative presented advantages in terms of being less sensitive to displacements, quick construction as well as in cost over other options considered for upgrading the dam. In addition it required a smaller amount of material to be placed on the crest, than if a moraine core had been constructed. The required fill volume was further reduced by introducing an L-shaped concrete wall along the upstream side of the crest to protect against wave run-up. A cross section of the top of the dam is shown in Figure 5.

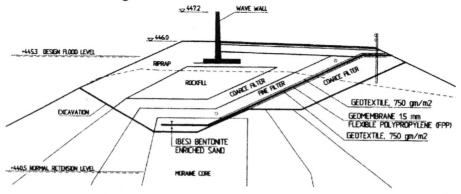

Figure 5. Design of new dam crest on the main embankment dam.

The membrane was embedded in the top of the original moraine core by two layers of bentonite enriched sand. The practicability of constructing the connection to the existing core has determined the geometry of the toe detail. The connection is shown in Figure 5 and is designed to: maintain the integrity of the water retaining structure; seal the seepage path along the surface of the geomembrane; lengthen potential seepage paths to at least half the potential water head; minimize the length of time the excavation exposes the core and to minimize the required depth of excavation.

The slope angle for the membrane was selected for the compaction of the bedding and to avoid a plane of weakness along the membrane and its protective layers. A geotextile was selected as protective layer to achieve a smooth surface for the membrane. The textile maintains the integrity of the membrane against puncture by protruding stone edges. A coarse filter was used as bedding for the textile and membrane.

Assuming that no undetected large-scale damage occurs to the membrane during construction seepages are expected to be small if the sealing in the future will be loaded during a design flood. For seepages that do occur, the coarse filter of gravel provides adequate drainage. The thickness of the coarse filter was determined by practical considerations of traversing compaction equipment. Fine and coarse filters were placed at the upstream side of the membrane.

During construction, settlement of the fine and coarse filters and the rock fill may induce tensile forces in the geomembrane system. In order to minimise the build up of stresses in the membrane the anchoring of the top edge of the geomembrane and protective geotextiles is designed such that the membrane will pull out of the anchorage before tensile forces in the membrane exceed the nominal yield stress. The anchor length is placed on a horizontal surface.

MATERIAL PROPERTIES AND TESTING
Bentonite Enriched Sand, BES
This material is formed by the mixing of sand with bentonite and then adding sufficient water to make the mix suitable for compaction to a high density yet retain some flexibility without cracking. The permeability of the existing core is believed to be in the range 2×10^{-7} m/sec to 2×10^{-9} m/sec. In order to match properties the target design permeability of the BES is 1×10^{-8} m/sec. The sand was single graded with $D_{15} = 0,06 - 0,20$ mm, $D_{85} = 0,20 - 0,60$ mm and $D_{max} = 10$ mm. After laboratory testing it was decided to mix the sand with of 8% (by weight) sodium bentonite with a montmorillonite content > 80 %; 75 % of the bentonite particles should pass the 0,075 mm screen at dry screening; moisture content $8 - 15$ % (tested according to BS 1377); liquid limit > 300 %; and swell > 24 ml/2g after 24 hours.

The BES was tested to determine the optimum density, moisture content and bentonite content in order to achieve the required permeability. Pre-testing of the BES was performed using the following methods: wet screening; sedimentation analysis of material $< 0,075$ mm; and determination of the hydraulic conductivity at 95 % of Standard Proctor density. The mix was then tested in field trials seen in Figure 6.

Figure 6. Field trials with BES. At left the trial surface is compacted in layers of 0.1 m and to the right the material is compacted in two layers of each 0.3 m at water content of 15 %.

Geomembrane

A 1.5 mm thick Flexible Polypropylene (FPP) membrane was selected as this material is more flexible and easy to handle than e.g. HDPE. The FPP is judged to be able to deform around any residual projections in the bedding. A texturing type was available which was necessary in order to develop sufficient friction between the membrane and the geotextiles to maintain stability during construction and operation.

The membrane had the following requirements and was tested according to the following standards: thickness and density (1,5 mm, 900 kg/m^3, ASTM D5994 and D1505A); tensile properties (stress 27 kN/m, elongation 800 %, ASTM D638); tear resistance (90 N/mm, ASTM 1004-90); puncture resistance (300 N, FTMS 101C method 2065); brittle temperature (- 50°C, ASTM D1693); friction angle 29°; carbon black content (ASTM D5994) and carbon black dispersion (ASTM D1603).

Geotextile

The geomembrane was protected from damage by projections and irregularities in the bedding and the coarse filter material, by careful preparation of the bedding surface and a non-woven geotextile with nominal weight of 750 gm/m^2. The same type of geotextile was chosen at the upstream side of the membrane as a protection towards the upstream fine filter.

WORK PROCEDURE

The top of the embankment dam crests was excavated down to 0.6 m below the moraine core and the BES was spread and compacted on top of the moraine, see Figure 7. The excavated material was placed on the upstream side to serve as wave run-up protection during the process of raising the core. The coarse filter was placed, compacted and trimmed prior to the excavation of the existing core. Some blinding of additional coarse filter was performed to fill in voids in the slope face after compaction of the coarse filter. The slope was rolled again after blinding to smooth-face the surface before placing the geotextile on the slope.

The stripped existing core was compacted and then lightly scarified and sprayed with water immediately before placing the BES. A layer of BES was then placed as bedding for the membrane on the excavated core surface, Figure 7. The BES surface was protected against rain and drying using plastic covers while the geotextile was placed on the coarse filter, see Figure 8.

Figure 7. Compaction of the BES bedding layer for the membrane on the existing moraine core. To the right is the compacted coarse filter bedding for the geotextile.

The next step in the construction sequence was to place and weld the membrane on the slope directly on the geotextile, see Figure 9. As the membrane was placed at an inclination of 1V:2H a temporary anchoring was required at the top of the membrane. This was carried out by nailing the membrane in its upper end with 1 m long, ø 20 mm, reinforcement bars. The load of the fill material on the top part of the membrane as described above achieved the permanent anchoring.

Figure 8.   Placing of geotextile on the compacted coarse filter bedding. The textile is temporarily held in place by rocks.

Figure 9.   Placing of membrane on the geotextile. The BES (to the right in the picture) is temporarily covered by plastic to protect the core against the heavy rainfalls that occurred during construction.

Welding was done with double seams in order to be able to test each seam for water tightness, see Figure 10. The seams are required to have at least 75 % of the geomembrane strength at stress at break yield point. In addition to this destructive testing was carried out on a selected part of each gore of the membrane. These parts were tested for peeling and shear resistance.

Figure 10. Membrane welding machine for double seams (left) and air pressure pump for air pressure tests of the membrane seams (right)

A non-destructive air pressure test at a minimum pressure of 200 kPa (2 kg/cm^2) was carried out along the entire lengths of all field seams including patches and repairs, see Figure 10. The requirement was that following initial pressure stabilization the pressure should not drop by more than 10 % in 5 minutes.

Figure 11. Protected connections through the membrane for instrumentation in the downstream filter.

The second BES-layer was placed and compacted above the toe of the membrane surface to complete the connection of the membrane to the existing core. The upper surface of the BES was laid with a fall to prevent ponding, from infiltration, that might soften the BES. The covering geotextile placed on the membrane was extended to cover the BES, to prevent stones being driven into the upper surface of the BES during construction.

No direct compaction of the fine filter on the upstream side of the membrane was done, as it would significantly have increased the contact stresses. The geotextile does however provide sufficient protection to the membrane during compaction of the coarse filter. After the completion of the filters the section was raised using rock fill to the new crest level. Finally the upstream end of the new crest was provided with a L-shaped wall to protect the crest against wave run-up, see Figure 5. The construction of the upstream wall allowed a lower crest elevation and thus allowing a shallower and suitable slope angle for the membrane.

CONCLUSIONS AND LESSONS LEARNED
The construction work with the geomembrane was a very quick operation. In spite of sometimes difficult geometry and many welds for pipes, testing etc. the 3,000 m^2 membrane was completed within a week.

ACKNOWLEDGEMENTS
Special thanks go to Bill Kearsey from Golder Associates, UK for the pre-design carried out for the raising of the crest. Bill also kindly guided the Swedish design team around in a study tour to different construction sites in UK where geomembranes where used in different applications.

The authors also want to thank Vattenfall the owner of the Ajaure Dam for the permission to publish this paper.

REFERENCES
Bartsch M. (2004). FMECA of the Ajaure Dam. *Long term benefits and performance of dams,* Thomas Telford.

Chang Y. & Nilsson Å. (2000). Deformation analyses of Ajaure Dam in Sweden, Beyond 2000 in computational geotechnics – Ten years of PLAXIS International.

Nilsson Å. & Norstedt U. (1991). Evaluation of ageing processes in two Swedish dams, ICOLD, Vienna, Q65, R2.

Design and performance of Elvington balancing and settling lagoons

A ROBERTSHAW, Yorkshire Water Services Limited, Bradford, UK
A MACDONALD, Babtie Group Limited, Glasgow, UK

SYNOPSIS.   The three Elvington Balancing and Settling Reservoirs are each capable of storing 205,000 Ml. of water and were constructed in the period 1992 to 1995 to provide the owner, Yorkshire Water Services Limited, with security of supply to the major treatment works at the site. The reservoirs are of earth embankment construction with the material being won partly from excavation on the site and partly from adjacent borrow areas. A bentonite cement slurry wall was constructed as a cut-off through underlying sand and gravel layers, and the internal face of the lagoons were lined with an HDPE membrane. The total area of liner was around 95,000 m^2.

The reservoirs have been operational for around eight years and the paper will concentrate on the design aspects, in particular the bentonite cement cut-off and geomembrane. A brief description of the overall performance of the reservoirs to date will be given.

INTRODUCTION AND BACKGROUND

Elvington Water Treatment Works is owned by Yorkshire Water Services Limited and is located beside the River Derwent approximately 12km to the south east of York.   The works was originally built for Sheffield Corporation in 1964 but are now one of the main source works for the Yorkshire Grid strategic transmission network which is capable of supplying customers throughout most of Yorkshire.

The primary source of water for the treatment works is the River Derwent although since 1996 water can also be brought to the site from the River Ouse at Moor Monkton approximately 20km away. The treatment works has a maximum hydraulic capacity of over 250Ml/day but the River Derwent abstraction license limits the normal average capacity to

205Ml/day which is approximately one-sixth of the company's daily demand for water.

Due to the strategic importance of the works it was decided in the mid-1980's that the raw water supplies should be protected from short-term pollution of the river and that a storage facility should be constructed which would also have the added water quality advantage of allowing the settlement of solids to take place. The Babtie Group was appointed to design the works in 1990 which were then constructed by Edmund Nuttall Ltd between 1992 and 1995.

DESIGN
General
The site chosen for the reservoirs was on a relatively flat area of agricultural ground immediately adjacent to the existing Works. Three reservoirs, each of 205Ml capacity were required, such that they could be operated in series, with one being filled, one being maintained full for at least 24 hours to allow for quality testing and settlement to take place and one being drawn down into supply.

A number of alternative design options were studied including conventional reinforced concrete tanks, combinations of earthworks and structural solutions, and earth embankment structures using cut and fill techniques to make the best use of the material available on site.

Ground investigations indicated that the sequence of geological strata was generally consistent across the site and comprised:-

Topsoil ; Upper laminated clay; Upper sand and gravel; Clay till; Lower laminated clay; Lower sand gravel; Sandstone (bedrock)

The thickness of the upper sand and gravel layer varied, but in general was no more than 600mm. At some locations it was absent.

Piezometers installed at locations across the site indicated that there was artesian pressure in the upper sand and gravel which responded within a very short space of time to water level changes in the River Derwent, which ran along the eastern boundary of the site. Any design which involved excavation into or through this layer would, therefore, have to accommodate any flows from it or uplift pressures generated within it.

In view of concerns about the suitability of the upper laminated clay for earthwork operations, a trial embankment was constructed, about 30 metres long, 5 metres high, and with side slopes of 1 in 3. Instruments were installed to allow monitoring of the formation during and after construction.

The trial embankment confirmed that the material was capable of being transported and compacted without significant difficulties. Consolidation settlement of the foundation was also found to be fairly rapid, no doubt due to the near horizontal sand lenses within the laminated clay which allowed dissipation of pore pressure into the adjacent excavated area.

Based on the results of the investigation, the decision was taken to proceed on the basis of an earthworks solution with the reservoir basins being formed by excavating down into the clay till, with perimeter and division embankments being formed, founded on the upper laminated clay. A plan showing the general layout of the reservoirs is shown in Figure 1.

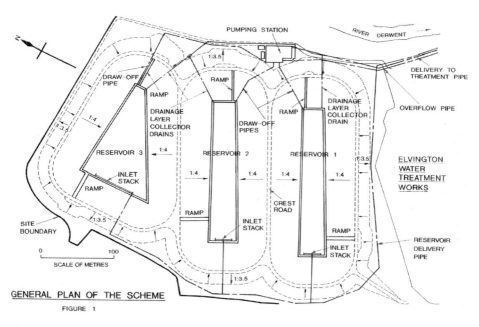

GENERAL PLAN OF THE SCHEME

FIGURE 1

Figure 1: General plan

The decision to adopt an earthworks solution, together with Yorkshire Water's desire to have lagoons that could be cleaned internally led to the generalised lagoon cross-section shown in Figure 2.

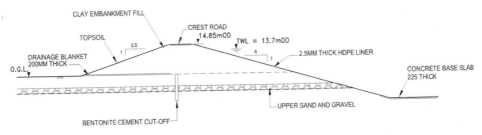

Figure 2: Typical cross-section

Embankments

To simplify construction and to make best use of site won material, it was decided that the embankment fill design parameters would be based on those of the poorest material to be excavated. This was the laminated clay, which had been found to be suitable for earthworking, provided weather conditions were reasonable. The embankments would be of homogeneous construction.

Table 1 summarises the design parameters used for the fill and for the foundation materials.

Table 1: Summary of Design Parameters

Material Type	γ_b kN/m^2	Cu kN/m^2	C' kN/m^2	Ø' Deg.	u'	M/C %
1.Embankment Fill	19.3	50	0	25	0.3	26.5
2.Upper laminated clay	19.3	80	0	25	0.3	27
3.Upper sand gravel	18	0	0	32	0.25	-
4.Clay till	21.9	100	0	29	0.25	14
5. Lower laminated clay	19.3	100	0	25	0.3	27
6. Lower sand and gravel (and weathered rock)	19	0	0	36	0.25	-
7. Bedrock sandstone (unweathered)	25	N/A	N/A	N/A	0.20	-

Table 2 summarises the design conditions considered and the corresponding factors of safety for the embankments.

Table 2: Summary of Design Conditions

Design Condition	Factor of Safety
1.Reservoir water level at 13.70m OD and liner undamaged	
Upstream slope	>4.63
2.Reservoir empty and liner acting as an impermeable barrier	
Upstream slope	1.46
3.Rapid drawdown, liner badly damaged	
Upstream slope	1.12
4.Reservoir full, liner badly damaged	
Downstream slope	1.39

A 200mm thick drainage blanket was installed at foundation level on the downstream shoulder of the embankments, just downstream of the cut-off. The effect of this has been ignored in design conditions, case 4, for the downstream slope, and the analysis can, therefore, be considered conservative.

While seismic effects were not considered in the original design, the situation was reviewed following the publication of "An Engineering Guide to Seismic Risk to Dams in the United Kingdom" (Ref. 1) and its associated application note (Ref. 2). The embankments fall into Category "II" of this guide and as there are no factors particularly vulnerable to damage by earthquake, a seismic analysis was not considered necessary.

Bentonite Cement Cut-off

A bentonite cement cut-off trench, 0.6m wide, was constructed below formation level of each embankment along the approximate line of the embankment crests of all three reservoirs. The purpose of the cut-off was to reduce seepage beneath the embankments and to isolate the foundation and underdrainage system from groundwater in the surrounding land as well as from neighbouring reservoirs. The cut-off, therefore, prevents the river charging the reservoir and closes a potential leakage path from the reservoirs. It also prevents the liner system being subjected to uplift pressure higher than the design values.

The base of the trench was generally the deeper of 1.5 metre below the top of the clay till and 4 metres below top soil strip level. The minimum depth of 4 metres was a requirement to ensure that local variations in the clay till level were catered for.

The design requirements of the bentonite cement hardened slurry were specified as:

- Permeability to be less than 10^{-8} metres/sec under water head of 12 metres at 28 days;
- A minimum strain of 5% without failure by cracking at 90 days;
- A minimum strength of $80kN/m^2$.

Reservoir Liner

The lining system to the reservoirs had to be capable of being cleaned on an intermittent basis to remove silt, etc. In addition, it had to be capable of dealing with uplift forces in the event that the bentonite cement cut-off failed to operate efficiently.

A number of options were considered for the liner including asphaltic concrete, geosynthetic liners, and reinforced concrete slabbing. As Yorkshire Water were keen to be able to run vehicles on the base of the lagoons for cleaning and maintenance purposes, the design adopted was a combination of reinforced concrete slabs in the base and high density polyethylene (HDPE) membrane on the side slopes.

Each reservoir has a reinforced 225mm thick concrete base slab which continues up the internal slope for 2 metres. The slabs were formed in-situ with C40/20 concrete. Pressure relief valves were cast into the bases to prevent unacceptable uplift pressures developing. The slabs were founded on a 250mm thick drainage layer on top of the clay till, connected to a pumped herringbone underdrainage system.

Around $95,000m^2$ of HDPE liner were required and its use in an exposed location such as at Elvington was most unusual. However, in this instance, it was chosen after careful consideration of a number of factors including its ability to accommodate differential settlement along the embankment fill and cut slopes, durability and cost. The cost benefit analysis undertaken for comparing alternative liners, assumed complete membrane replacement after 15 years, although manufacturers were prepared to guarantee the material for up to 25 years.

A 2.5mm thick HDPE membrane was specified. The liner was laid on top of a 200m thick drainage layer of granular material, connected to the base slab underdrainage system. Some of the key material parameters for the liner are given in Table 3.

Table 3: HDPE Liner Technical Data

Properties	Units
Density	0.94 g/cc min
Carbon Black %	2-3%
Tensile Properties:	
- Strength at Yield	16 Mpa/inch width
- Strength at Break	27 Mpa/inch width
- Elongation at Yield	13%
- Elongation at Break	700%
Tear Resistance	289 N
Puncture Resistance	578 N

To counteract uplift forces due to the design wind speed of 45 m/s, the liner was bolted onto concrete anchor beams which run down the internal slopes. At the base, an HDPE connection piece was cast into the concrete base slabs (Ref 3).

Overflow Provisions

The main overflow system for the reservoirs is within the wet well of the reservoir pumping station. Each reservoir has its own double sided weir, separated from adjacent weirs by concrete dividing walls. There are no valves on the pipelines between the reservoirs and the control structure. The water in the wet well rises with the reservoir level up to the weir level of 13.7m AOD. Any discharge over the weir goes into an overflow channel and then into a 1650mm diameter overflow pipe. The overflow pipe discharges into the River Derwent downstream of the supply abstraction point.

The inflow pumps have variable speed drives and will normally be operated at 205 Ml/d i.e. equivalent to the capacity of each lagoon. However, they can be stepped up to a maximum inflow of 324 Ml/d.

Shortly after reservoir construction was complete, it was decided by Yorkshire Water that a second supply source should be added to the system, from their Moor Monkton intake on the River Ouse. The result of this was to increase the normal service maximum inflow from the Derwent and Moor Monkton sources to 355Ml/d, with an absolute maximum of 474Ml/d.

Under the various inflow conditions, the freeboard to embankment crest over stillwater level is shown in Table 4.

Table 4: Reservoir Freeboard

Inflow (Ml/d)	Freeboard (min)
0	1.15m
205	0.62m
355	0.35m
474	0.24m

To provide additional security, high level overflow weirs were constructed between adjoining reservoirs as part of the Moor Monkton contract. Each weir is 7m long and discharges freely down the HDPE slope of the adjoining reservoir. These weirs are set at a level of 14.3m AOD, 600mm above normal top water level.

CONSTRUCTION
General
The contract for the lagoons was awarded to Edmund Nuttall Limited and work started on site in November 1992. The completion certificate for the works was issued in August 1995. The Final Certificate for the lagoons under the Reservoirs Act 1975 (Ref. 4) was issued on 13th June 2000.

Embankment Construction
The earth embankments were constructed over two seasons in 1993 and 1994. A method specification was included for compaction.

The material was generally excavated and placed using tractors and scraper boxes. The embankments were built up in layers 225mm thick. At the start of construction the placed material was compacted by 6 passes of a towed tamping roller but was subsequently reduced to 4 passes. The Contractor also elected to change to using a self-propelled CAT 815 wedge foot roller (dead weight 20 tonnes); the layer thickness remained the same.

Laboratory tests were carried out regularly on samples from the placed embankment fill. The results are given in Table 5. Results showed a surprisingly high, 23%, number of undrained shear strength results which were below the value assumed in the design. The majority of these low results were from areas where laminated clay had been placed and it was thought that the presence of laminations within the samples was causing premature failure under test and did not truly represent the behaviour of the mass fill. The material from each sample was mixed to eliminate these laminations, compacted and retested and results similar or better than the design assumptions for undrained shear strength were obtained.

Table 5: Summary of Earthwork Test Results

	Cu (kN/m^2)	Υb (Kg/m^3)	M/C (%)	Υd (Kg/m^3)
Mean	78	2015	21.1	1671
Maximum	221	2086	25.1	2014
Minimum	31	1953	6.0	1388

Bentonite Cement Cut-Off

The construction of the cut-off was sub-contracted to AMEC Civil Engineering. Although the design mix had to be approved by the Engineer it was the responsibility of the sub-contractor to design a slurry satisfying the specified requirements.

The design mix changed several times in the early stages of construction because the permeability design criteria were not being achieved. A second and sometimes third wall was constructed parallel to the first sections. The accepted cut-off was approximately 1950m long. Mix-specific characteristics are given in Table 6.

Table 6: Mix-specific Characteristics

Reference	Mix Characteristics
Blue	4.5% bentonite Oil Companies Materials Association grade (OCMA) 90 second mix Single Hany mixer
Orange	5.4% bentonite (OCMA) 180 second mix Single Hany mixer
Green	5.4% bentonite (OCMA) 300 second mix Double Hany mix
Pink	4.5% bentonite (OCMA) 300 second mix Double Hany mixer
Yellow	4.5% bentonite Civil Engineering (CE) Grade 300 second mix Double Hany mixer

Constituents common to all mixes were:

- 112kg of ordinary Portland cement to BS12
- 336kg of ground granulated blast surface slag to BS6699: 1986
- 2799 litres of potable mains water

The consistency of the slurry was checked daily for compliance with the specification using a Marsh cone and a mud density balance.

The following tests on the hardened slurry were carried out on a regular basis:

- Unconfined compressive strength to BS1377 Part 7 Test 7 with 2% strain rate at a minimum of 28 days
- Permeability in a triaxial cell to BS1377 Part 6 1990 Test 6 constant head test at approximately 28 days
- Consolidated drained triaxial compression test to BS1377 Part 8 Test 8 for 5% strain at 90 days

Tests carried out on the pink and yellow mixes indicated compliance with the specification. The other three mixes generally failed to meet specified requirements for permeability. The Contractor had used short hydration periods for the bentonite in the blue, orange and green mixes, whereas the final two mixes adopted a minimum 24 hour hydration period prior to mixing. Microscopic examination of the mixes indicated a "balling" effect in the first three mixes which it was felt was caused by a lack of hydration of the bentonite. In addition, the OCMA grade bentonite has a more angular grain shape than the CE grade, which made adequate hydration and mixing times even more important.

HDPE Liner
During the manufacture of the membrane, samples were taken and tested in accordance with ASTMD638. A quality control certificate was issued with the material. The liner was manufactured by the Gundle Lining Construction Corporation. Installation started in September 1994 and was completed in May 1995 with work being suspended between November 1994 and mid-March 1995.

Two types of welds were used on site (Ref. 3). Where panels overalapped, a hot shoe double fusion weld was formed, and where the liner attached to anchors, and elsewhere where the sheets did not overlap, extrusion welds were used. Destructive and non-destructive tests were carried out on both types of weld. The non-destructive tests consisted of air pressure testing the double fusion welds and spark testing the extrusion welds. The destructive tests were "peel" and "shear" tests. In addition to checking the integrity of the site welds, these tests were also used to check the welding equipment on a daily basis prior to it being used on site. A non-destructive test was carried out on every weld.

Instrumentation

Sensors to measure underdrainage flows and geotechnical instruments were installed to provide information throughout the service life of the reservoirs. The inflow into the chambers is measured by ultrasonic sensors upstream of v-notch weirs and the sensor values are relayed to the main control room. There are also level probes in the chambers which set off an alarm if the water level rises above a certain level.

Geotechnical instrumentation consisted of hydraulic, pneumatic and standpipe piezometers to measure pore and uplift pressures and vertical extensometers to monitor formation settlement

Survey pins were installed along the crest to allow the settlement of the embankment to be monitored. The spacing of the pins is generally 20m.

PERFORMANCE

General

One of the recommendations contained in the Final Certificate stated that annual performance assessment reports should be prepared to provide guidance on the significance of the behaviour of the reservoir, its foundations and the surrounding ground as revealed by instrumentation monitoring and visual inspections. Such reports have been prepared for Yorkshire Water Services by TEAM (an amalgamation of MWH and Arup) under the supervison of a Panel AR Engineer. In addition the reservoir was inspected under Section 10 of the Reservoirs Act 1975 for the first time in June 2002. These reports have all confirmed that the reservoir is behaving in a satisfactory manner.

Settlement

In general the settlement of the embankments has been minimal. However, soon after construction was complete, a depression appeared around the top of the magnetic extensometer on the main embankment of lagoon 1. This extensometer had also become blocked soon after installation. The depression covered an area of approximately 2 metres by 1 metre and had a maximum depth of approximately 0.2m causing cracking of the crest road surface and extending under the top of the liner. After a period of close observation it appeared that the settlement had ceased so the area was reinstated in 2000 and no movement has been observed since. The precise cause of the settlement is unknown but it was concluded that the depression must have been caused by local irregularities during installation of the extensometer.

Interior of the lagoons
The design brief for the lagoons stated that the sides and bases of the reservoirs should have smooth surfaces to facilitate the removal of accumulated sludge and other debris. However the geomembrane liner as constructed has a much folded and wrinkled appearance which has proved difficult to clean and detracts from the appearance of the structure. In addition a number of small splits in the HDPE liner have been found which have been easily repaired and have not caused any concern.

All three reservoirs have now been emptied for cleaning. Each time that a reservoir has been refilled after cleaning a rapid increase in underdrain flow has occurred. For example, Reservoir 1 has a normal base flow of less than 0.3 litres / min but after refilling in 1999 the flow suddenly increased to 3.5 litres / min before reducing to its previous value over a period of approximately 3 months. This action has been attributed to self-seating of the pressure-relief valves in the base slabs and the subsequent re-deposition of silt.

CONCLUSIONS
The lagoons at Elvington were innovative in their use of exposed HDPE liner on such a large scale. However, it was an economical material to use, easy to install, and has performed well in service.

The problems that were encountered with the bentonite cement cut-off highlighted the need for adequate hydration times for bentonite and also the differences between OCMA and CE grade materials. Despite this the cut-off appears to be performing well in minimising uplift pressures on the liner system.

The reservoirs are now forming an important element in Yorkshire Water's supply network and are giving increased security of supply to over 4.5 million customers. Long-term liner performance will remain an interest but with three lagoons, remedial or replacement work should be able to be carried out sequentially without a material effect on supply.

ACKNOWLEDGEMENTS

The authors gratefully acknowledge the permission of Yorkshire Water Services to publish this paper.

REFERENCES:

1. Building Research Establishment, *An Engineering Guide to Seismic Risk to Dams in the UK*, 1991
2. The Institution of Civil Engineers, *An Application Note to An Engineering Guide to Seismic Risk to Dams in the UK*, 1998
3. Murphy, (1996*). The Design and Installation of an Exposed High Density Polyethylene Reservoir Liner*
4. Reservoirs Act 1975, HMSO, London

Twenty five years experience using bituminous geomembranes as upstream waterproofing for structures

M. TURLEY, Colas Limited UK
J-L. GAUTIER, Colas SA, France,

SYNOPSIS.
Since 1978, with minor variations in the structure details, more than twenty rockfill or earthfill dams have been constructed over the last 25 years using a thin bituminous geomembrane for the upstream waterproof facing. This paper illustrates, using the examples of the last two dams built in France, the 38 metre high "L'Ortolo" dam in 1996 in Corsica and the 43 metre high "La Galaube" dam in 2000 near Carcassonne, the details of this construction technique which proves to be an efficient and economical alternative to dense asphalt or clay waterproofing structures.

INTRODUCTION

The modern use of bitumen as a geomembrane waterproofing layer for dam facings commenced in the early 1970's with the *in-situ* impregnation of a synthetic geotextile placed onto a prepared substrate, and impregnated with bitumen sprayed onto the material at a high temperature – typically in excess of 180°C. The first application of this form of sprayed geomembrane was in lining ponds in 1973-74 and around the same time, on small dams such as the dam of "Les Bimes" (9 metres high) or the dam of "Pierrefeu" (8 metres high), both of them located in the south east of France.

This, although an effective method of producing a waterproof and seamless geomembrane layer, had major inherent drawbacks both in terms of operator safety and quality control. The process was sensitive to moisture, and the bitumen usage was difficult to control leading to variations in thickness. The introduction of prefabricated geomembranes manufactured using bitumen impregnated geotextile under quality-controlled factory conditions removed these failings, leading to greater confidence in their use. There was a transition to the use of the prefabricated geomembranes in the mid 1970's.

Long-term benefits and performance of dams, Thomas Telford, London, 2004, 94–101

THE PREFABRICATED GEOMEMBRANE

The factory production of a prefabricated geomembrane of bitumen impregnated geotextile, reinforced with glass fleece, addressed the safety and quality problems. Initially, rolls of material 4.0m wide were produced on a small factory-production plant and laid on site in strips. These strips were overlapped at the edges and joined by forming seams in the material using propane gas torches to heat-weld the overlap. A few small dams were waterproofed according to this technique such as "Locmine" in Brittany, (7 metres high) in 1977 and "Gardel" in Guadeloupe, (14 metres high) in 1978.

The geomembrane today

The bituminous geomembrane liner in use by Colas today, Coletanche, is manufactured in a state-of-the-art factory in Galway, Eire, commissioned by the company in 2000. The facility enables a more versatile approach to production of wider rolls (5.15m) available in several thickness grades impregnated with either Oxidised (NTP grade) or Elastomeric (ES Grade) modified bitumen; typically the "NTP" 3 grade is used for dam lining.

Table 1: Physical characteristics of Coletanche NTP3

Characteristic:	Property:
Roll length	65m
Roll Width	5.15m
Material thickness	4.8
Mass per unit area	$5.5kg/m^2$

The structure of the geomembrane

The base structure of the geomembrane is illustrated in Figure 1 and comprises:

- A non-woven polyester geotextile, whose weight per m^2 determines the ultimate thickness of the geomembrane – between 3.5 to 5.6mm. Coletanche NTP3 is 4.8mm thick with a mass of $5.5kg/m^2$.
- A glass fleece reinforcement (which contributes to the strength of the geomembrane and stability during fabrication).
- Total impregnation with a compound including a blown bitumen of 100/40 pen plus filler (NTP grade), or an elastomeric modified bitumen (ES grade)
- The underside is coated with a Terphane film bonded when the membrane is hot, and designed to give resistance to penetration from tree roots.
- Finally the upper surface is coated with a fine sand to a) provide greater traction on a slope, giving greater operator safety and security from slipping, and b) to give protection from the degrading effects of UV radiation.

Among the properties of this geomembrane the significant mechanical characteristics are shown below in table 2.

The material is now used for a wide variety of environmental and hydraulic applications from lining canals and watercourses, groundwater protection, landfill lining and capping as well as for the waterproofing of dams.

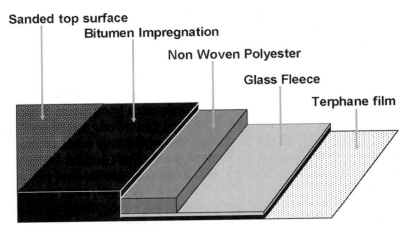

Figure 1: Typical structure of the geomembrane

Table 2: Mechanical characteristics of Coletanche NTP3

Characteristic:	Property:	Test Method:
Tensile strength at break	25,000kN/m	ISO 10319
Strain at break	70%	ISO 10319
Tear resistance	90N	NF G 07-112
Puncture resistance (Static)	500N	NF P 84-507
Puncture resistance (Dynamic)	22J	NF-p 84 502
Permeability (at 0.1 MPa)	7×10^{-14}	Darcy's Coefficient
Conventional watertightness level	$<10^{-4}$ m^3/m^2/j	NF-P 84515
Flexibility when cold	0°C	NF P 84-350
Maximum friction angle	35°	

Application of the geomembrane

The rolls of geomembrane are lifted and unrolled using a purpose-designed hydraulically controlled beam, carried by a tracked excavator. Lining the face of a dam, the material is unrolled down the face of the structure with the excavator remaining at the top.

The geomembrane is laid with a 20 cm overlap which is used to form a welded seam, utilising a propane gas torch to liquefy the bitumen prior to pressing and fusing the two liner sections together to form a watertight seam. This method is also used when fixing to structures, and forming round pipe-work, etc after first coating the surface with a primer.

Testing the seams
Quality control of the seams is undertaken non-destructively using ultrasound to check the integrity of the welded area, either using a single hand-held transducer for small areas or a machine known as "CAC 94", which automatically tests the full seam width using 24 ultrasound sensors arranged in a staggered row to completely cover the seam width, downloading the data onto a computer. Software designed for the machine allows a printout to be produced showing a global view of the weld and any defects: a 20cm weld is required to have a minimum 75% width continuously welded for acceptance.

PROJECTS USING REINFORCED BITUMINOUS GEOMEMBRANE:

Ospedale dam
The first reference project using a prefabricated bituminous geomembrane took place in 1978, with the "Ospedale" dam, located in a remote area of southern Corsica at an altitude of 1,000 metres. This was constructed as a rock fill dam with a length of 135 metres and a height of 25 metres, a 70,000 m^3 embankment volume, with slopes 1.7 to 1 upstream, 1.5 to 1 downstream, and designed to have a capacity of 3 million m^3.

An alternative design proposal
The initial design for the dam was based upon an upstream waterproof facing of a 3 layer hot bituminous-mix, making a total thickness of 24 cm. This would have proved difficult to build, due to the remoteness of the construction project and the subsequent long transportation distances for the hot mix. As an alternative to this design, a proposal was put forward based upon the following structure:

- Two regulating layers of a gravel material impregnated with bitumen emulsion to provide stability, in order to improve the profile of the rock fill embankment,
- A 5 cm thick 3/6 mm open graded cold asphalt mix with a permeability of 10^{-4} m/s, to ensure an efficient under-drainage layer and prevent back-pressure under the waterproof geomembrane in the unlikely event of damage to the membrane layer leading to leakage. This cold-mix material was manufactured on site in a purpose-designed mobile plant, laid using small hoppers lowered on hawsers.
- A 4.8mm thick bituminous geomembrane, Coletanche NTP 3, laid between two geotextiles.
- Finally a mechanical protection made of unbound interlocking pre-cast concrete blocks to reduce risk of under-pressure

There were, additionally, other considerations; the membrane must not slip under its own weight down the face of the dam, and the concrete blocks were to be able to move independent of the membrane.

In order to meet these requirements, the membrane was laid with its smooth Terphane-faced side uppermost, as the friction angle on this side varies from 20 to 28°, compared with 30 to 42° on the sanded side. This proved effective in 1983 when during a huge storm, a large section of the concrete paving slipped away, without the slightest damage to the membrane.

The geomembrane was bonded to a concrete plinth at the foot of the dam using a bituminous primer painted onto the concrete to enable the membrane to be heat-welded. A thin stainless steel strip 5 to 6cm wide was fixed into the concrete through the membrane and covered with one additional thickness of membrane for additional protection.

This alternate design was accepted by the "French Committee on Large Dams", and a paper presented during the 13th International Congress on Large Dams in New-Delhi [Bianchi et al, 1979].

Projects carried out during the '80s and '90s
Following the success of the Ospedale dam project, over twenty further lining projects were undertaken with bituminous geomembranes, based upon either oxidised or polymer modified bitumen, with only minor variations in the design, among which can be noted: "La Riberole" dam, for a reservoir to feed a hydroelectric power plant, with a small height of only 8 metres, but at an altitude of 1,625 metres in the Pyrenees; at "Verney" in the French Alps where the membrane is used for a different purpose, to improve the water tightness of the upstream drainage bed; the dam of "Gachet", on the island of Guadeloupe, 14metres high, where the membrane is protected by unbound pre-cast concrete block protection, and still showing very good behaviour despite the tropical weather conditions; the dam of "Mauriac", in the centre of France, which is a constant level reservoir where therefore only the upper part of the upstream facing is protected with precast concrete slabs, anchored by stainless steel cables in the upper part of the dam [Clérin et al, 1991]. Finally the last two large dams built in France, which show the use of a bituminous geomembrane on a quite different scale: "L'Ortolo" dam, in Corsica in 1996, and La Galaube" dam, near Carcassonne in Southern France in 2000, which are described in more detail below.

"L'ORTOLO" DAM
A rock fill dam of 155,000 m^3, 157 metres long, a width at the foot of 120 metres, and 38 metres in height from the base. With slopes of 1.7 to 1 upstream and 1.5 to 1 downstream it has created a reservoir of 3 million m^3. [Tisserand et al, 1997]

The design:

The upstream waterproofing
- 10 cm 25/50 ballast impregnated with bitumen emulsion at 3 kg/m^2

- 10 cm cold asphalt mix 4/10, with a 7 to 8 % void content and a permeability between 10^{-4} and 10^{-5} m/s, manufactured on site
- Coletanche NTP 3, covered by a 400 g/m^2 geotextile fixed at the base with a 6 mm thick and 60 mm wide stainless steel plate, bolted every 15 cm into the plinth, covered with a double thickness of membrane. The geomembrane was fixed at the crest of the dam.

The protection layer
- Fibrous in-situ cast concrete slabs 14 cm thick, 3 metres wide

Comments

Two major events occurred during the construction of this project, which delayed the completion for several months, and which led to changes in the design of the next dam. Storms and strong winds on January 10th 1996 lifted 1,500 m^2 of membrane during its installation phase. The effect of the wind was amplified by the venturi effect above the crest, and under pressure generated through the dam which was permeable to downstream winds.

This storm was followed, on February 2nd, by a 100 year flood of 270m^3/s that led to a 0.5 m high flow of water through the spillway. The dam filled up in one night, and began to leak, with a flow of 5 to 6 m^3/s through the rock fill (i.e. 2litres/s/m^2), and then emptied within 48 hours through the dam and outlet. No damage was noticed to the embankment, with only slight movement of a few rocks, and a minor settlement of 6 cm on top. However, several weeks were then needed to clean the whole area, and remove the mud and all the debris carried by the flood.

The main conclusion from this project was that the permeability of the asphalt layer below the geomembrane needed to be reduced, both to increase the suction effect in case of strong winds, and to decrease flow rates in case of flood during construction before the membrane is completely installed [Tisserand et al, 1998].

"LA GALAUBE" DAM

This rock fill dam is constructed of 800,000 m^3 of mica schists excavated from the site, on a foundation of granite and a reinforced concrete plinth upstream. With a length of 380 metres, a height of 43 metres above the foundations and slopes of 2 to 1, (26°) it is the highest dam in the world with a bituminous geomembrane as an upstream impervious face and was built in 2000 on the Alzeau River in the south of France. The reservoir created is at an altitude of 700 metres with a 68 ha surface, and a capacity of 8 million m^3. [Gautier et al, 2002]

The design

The upstream waterproofing

- 10 cm layer of non-bound material, with a 0/20 mm grading, impregnated with bitumen emulsion, to regulate the slope. This required 5,000 tonnes of limestone from a nearby quarry, deposited from the crest and leveled using 2 laser-guided bulldozers.
- 10 cm layer of cold asphalt mix, with a 0/10 mm grading, to ensure the final regulating of the upstream face before laying the geomembrane. This was to be a semi-impervious layer designed to a) reduce leakage flow through the waterproofing structure in case of accidental damage to the geomembrane, and b) to overcome the problems experienced in Corsica, of premature flooding prior to completion. Laboratory studies were carried out to determine the recipe of the cold mix asphalt required in order to achieve permeability around 10^{-6} m/s, as well as ensuring that a workable mix could be achieved.

Figure 2: Work in progress at La Galaube

- as with the previous dam at L'Ortolo, 4.8mm thick Coletanche NTP3 geomembrane was used, protected by a geotextile above to reduce the effect of underpressure in case of rapid emptying of the reservoir. As it was important to have no transverse seams in the geomembrane on the slope of the structure each individual roll was manufactured to the required length to match its final position on the dam face, with a unique reference number allocated to its position. A roll of this grade of membrane is normally 65m; some rolls, required to be in excess of 100 metres, had an ensuing weight of over 3 tonnes.

The protection layer
• Fibrous concrete manufactured on site was cast into aluminium formwork to produce 10cm thick slabs 5 m x 10 m with open joints.

The completed structure was delivered in November 2000, after less than a five month period for the waterproofing phase, and a cost of €1,500,000 (£1,050,000) for an impervious surface of more than 22,000 m². This allowed the owner to start the filling of the dam before winter, leading to the first outflow through the spillway 18months later.

CONCLUSION

The continuous monitoring of dams with the upstream face waterproofed using a prefabricated bituminous geomembrane, some of them for more than 25 years, has demonstrated a good performance over this period. The behaviour of these structures over the period since lining has shown no reduction in watertightness with for example constant flow rates of 2.4 l/hour/m² at l'Ospedale dam, or 0.9 l/hour/m² at Mauriac dam through the liner into to the drainage layer beneath, results that are very acceptable to he clients. [Tisserand et al, 2002].

Through this knowledge and the experience gained along all these projects, the use of a bituminous geomembrane as an upstream waterproofing facing on earth fill or rock fill dams has now proved an efficient alternative solution to inner clay barriers or dense bituminous mix facings, with installation costs that can be comparable and even more interesting in remote locations.

REFERENCES

Bianchi, C. Rocca-Serra, C. Girollet, J. (1979) *Utilisation d'un revêtement mince pour l'étanchéité d'un barrage de plus 20 mètres de hauteur.* 13ème Congrès des Grands Barrages – New Delhi

Clérin, J. Gilbert, C. Bienaimé, C. Herment, R. (1991) *Massif Central : a rockfill dam with a bituminous geomembrane upstream facing,* Travaux no. 665

Tisserand, C. Breul, B. Herment, R. (1997) *Feedback from Ortolo Dam and i's forerunners.* Geotextiles-geomembranes, Rencontres 97

Tisserand, C. Breul, B. Gimenes, Y. Antomarchi, E. Herment, R. (1998) *Le Barrage de l'Ortolo,* Travaux no. 746

Gautier, J-L. Lino, M. Carlier, D. (2002) *A record height in dam waterproofing with bituminous geomembrane: La Galaube dam on Alzeau river,* 7th International Conference on Geosynthetics, Nice

Tisserand, C. Poulain, D. Royer, P. (2002) *Durability of geomembranes in embankment dams.* Colloque CFG – CFGB – Saint Etienne

Watertightness and safety of dams using geomembranes

A M SCUERO, CARPI TECH S.A., Switzerland
G L VASCHETTI, CARPI TECH S.A., Switzerland

SYNOPSIS. Impervious prefabricated geomembranes are a well-established technology to provide or restore watertightness in dams. Since 1959, they have been installed on all types of dams worldwide. The paper gives an overview of how the geomembrane systems have been applied and perform on different types of dams. Some significant case histories on concrete gravity dams, masonry dams, fill dams, RCC dams are presented.

INTRODUCTION

Synthetic impervious geomembranes were first used on a dam in 1959. In pioneer installations, all types of geomembranes available in the market were adopted. Selection was based on local availability, on aggressiveness of marketing, on personal knowledge and information. As years went by and field results became available, selection of the geomembrane could rely also on ascertained performance. Materials that performed poorly were abandoned in favour of more flexible, robust, and durable ones. According to the ICOLD 2003 database listing 232 large dams incorporating a geomembrane for watertightness, PVC, mostly coupled to an anti-puncture geotextile, is most frequently adopted. Some other geomembranes are practically no longer used in modern applications: the last reported installation of in-situ fabricated geomembranes on a dam occurred in 1988, and of HDPE geomembranes in 1994.

Pioneer installations were performed on embankment dams. In a few years, improvement in technology allowed installation on more demanding sub-vertical facings (Scuero & Vaschetti, 1998).

Today, all types of dams have a geomembrane as an impervious element, either as a repair method or incorporated in the dam since original construction: these are summarised in Table 1.

Long-term benefits and performance of dams, Thomas Telford, London, 2004, 102–116

Table 1: Total dams by type of works

Type of dam	New construction	Rehabilitation	Unknown
Fill	75	51	39
Concrete	1	34	5
RCC	21	1	5
Total	97	86	49
Percentage	47	53	-

Table 2: Total dams by type of position of geomembrane

Geomembrane is	Total dams	Percentage
Exposed	87	39.2
Covered	135	60.8
Unknown	10	-
Total	232	100

Europe, with 111 dams out of 232, is still the leader, as already reported in 1998.

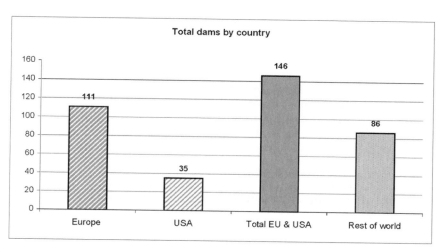

Figure 1: Total dams by country

RECENT TRENDS
The state-of-the-art systems are now widely covered in available literature and will not be discussed in this paper. One recent trend is a further increase

in the adoption of drained geomembrane systems as compared to undrained systems. Among the advantages that the drained system can provide, there is the possibility of progressively dehydrating the dam, as further reported below, and of monitoring efficiency.

Another marked trend is the increase in RCC dams that incorporate a geomembrane: 27 (Scuero & Vaschetti, 2003) out of approximately 260 RCC dams, accounting for more than 10% of the total.

A recent development is also the adoption of leak detection systems based on fibre-optic cables. One of the earliest applications of this system to monitor the efficiency of geomembrane liners was made in UK, at Winscar dam (Carter et al., 2002).

The following case histories aim to give an overview of the most remarkable recent features of the geomembrane system on the various types of dams.

REHABILITATION OF CONCRETE AND MASONRY DAMS
The deterioration of concrete dams is caused by the environment (temperature changes, wetting-dehydrating and freeze-thaw cycles, impact by ice, debris, transported materials, chemical action of water) or by abnormal behaviour of the structure (expansive phenomena of concrete, problems with foundations and differential settlements). Concrete cracks and loses imperviousness, water infiltrates the dam body, and subsequent washing of fines may cause carbonation and clogging of the drains. As the drains cannot efficiently perform their function, seepage extends to the whole body of the dam and saturation of concrete occurs. Increase in pore pressure causes deviation from the initial design conditions, and stability of the structure may be at stake. In dams subject to alkali-aggregate reaction, increase in the water content aggravates the reaction. Rehabilitation generally aims to stop water infiltration and further deterioration of the structure.

On concrete dams, the liner adopted is generally a PVC geomembrane coupled during fabrication to an anti-puncture and drainage geotextile. The liner, supplied in flexible sheets, is generally installed directly over the concrete. Sometimes an additional drainage geonet is installed behind the geocomposite to enhance drainage transmission.

On masonry faced dams, the geocomposite system must meet the demanding requirements of the exceptional roughness of the substrate and its different consistency (stone and mortar in the joints). On these dams, the system is implemented by a transition anti-puncture layer, usually a thick geotextile to achieve a smoother surface without extensive civil works.

Rehabilitation of a concrete dam affected by AAR: behaviour of Pracana after 11 years

Pracana is a 65 m high buttress dam in a seismic region of Portugal, built between 1948 and 1951, and owned by EDP, Electricidade de Portugal. In 1980, the dam was taken out of operation to undertake a thorough investigation of the deterioration related to the continuous concrete expansion phenomena. Investigations ascertained the presence of a secondary alkali-aggregate reaction (AAR), which would be further activated by infiltration of water from the reservoir, creating a critical scenario in respect to sliding conditions along horizontal cracks. Installation of a drained waterproofing liner on the upstream face was deemed necessary to stop water infiltration feeding the AAR, and to avoid the possibility of water exerting uplift in the horizontal cracks, especially in the case of a seismic event.

The exposed drained PVC geomembrane system was installed in the dry season of 1992, in 5 months, concurrent with major rehabilitation works including construction of a new foundation beam and grout curtain, two sets of concrete struts on the downstream face, local grouting of larger cracks and mass grouting of smaller cracks, construction of a new spillway and a new water intake. Since 1992, the behaviour of the geocomposite waterproofing system has been monitored in respect of leakage and its capability of dehydrating the dam, reducing the water content feeding the AAR. At ICOLD's 21st Congress, the owner reported that the dam waterproofing might be assumed to contribute for the reduction of the swelling process (Liberal et al., 2003).

Figures 2 & 3: Pracana 65 m high buttress dam in Portugal, affected by AAR, was waterproofed with an exposed PVC liner in 1992. In 2003 (at right), the owner reports "the concrete dam waterproofing may be assumed to contribute for the reduction of the swelling process".

Rehabilitation of a masonry dam: Beli Iskar, 2002

Beli Iskar is a 50.7 m high rough rock masonry dam located in the highest part of Rila Mountain, in Bulgaria, at 1878.70 m a.s.l. In 1950, the upstream face was coated with "Inertol", between 1976 and 1978 the leaking joints were repeatedly repaired with "Soral", and the face with resin coating. These repairs were washed away. The exposed drained PVC geomembrane system was installed in 2002.

Preparation works included removal of debris and sediments to expose the rock masonry. A 2000 g/m^2 geotextile was installed on the masonry as anti-puncture protection to the PVC geocomposite. The waterproofing liner is SIBELON CNT 3750, a 2.5 mm thick PVC geomembrane coupled during fabrication to a 500 g/m^2 geotextile, and supplied in rolls as long as the section they cover.

The geocomposite is anchored along vertical lines to the masonry by the same patented tensioning assemblies used for concrete dams. The masonry surface was regularised with a layer of mortar under the anchorage assemblies, to provide an even surface for welding of the PVC sheets. Adjoining geocomposite rolls were watertight heat-welded with manual one-track method.

Figures 4 & 5: Beli Iskar 49.7 m high masonry dam, Bulgaria 2002. The exposed PVC geocomposite is installed on an anti-puncture geotextile. The dam is at 1878 m of altitude, ice thickness up to 0.60 m.

The PVC geocomposite is anchored at the perimeter with the same watertight seal, made by compression, adopted on concrete dams. In order to provide an even surface which is crucial to achieve watertightness, the masonry was regularized in the areas of the seals with a gunite layer.

A new plinth was constructed where additional grouting was required. The new grouting plinth was waterproofed with the same PVC geocomposite waterproofing the upstream face. The bottom watertight perimeter seal connects the geocomposite lining to the upstream face with the geocomposite lining the plinth.

EMBANKMENT DAMS

In embankment dams, watertightness must be provided by a material different from the materials constituting the dam body. Traditionally, the barrier to water infiltration had been made with either natural or man-made materials, by constructing either an impervious core using materials such as asphaltic concrete or clay, or an impervious upstream face as in Concrete Faced Rockfill Dams (CFRD) and embankment dams with an asphalt concrete upstream layer.

The use of traditional materials such as clay or asphaltic concrete requires adequate materials selection, placing and compaction. Sometimes appropriate materials are not available at reasonable costs, or available materials do not have the required quality, or construction is difficult, time consuming and expensive. Sometimes inadequate construction, or inadequate weathering resistance, for example with some asphaltic concrete facings, leads to increasing leakage over time. The final result may end up with general poor performance of the dam.

CFRDs in turn involve complicated design, and construction of the face slab and of its complex waterstop system can significantly affect the overall construction time. Time schedules have often been extended considerably beyond what was initially foreseen, especially when contractors have had no previous experience with this type of construction. From a performance perspective, although installation of the concrete layer occurs when placing of the fill has been completed and the anticipated major settlements have taken place, in many CFRD's already built the settlement of the fill continued after the filling of the reservoir. There are dams where the deformation due to the settlement, combined with the deformation of the fill due to the hydraulic head, provoked cracks in the concrete face and/or failure of the waterstop systems. This problem, related to the placement of a rigid element (the waterproofing concrete face) over the deformable fill dam body, is aggravated in dams constructed in seismic areas.

In new construction, the use of synthetic impervious geomembranes avoids the problems connected with design and installation of multiple defence lines of waterstops, with deterioration of waterstops, with connections of the core or of the asphaltic concrete to concrete structures. In case suitable traditional materials are not available at reasonable cost, the geomembrane option can make the project feasible. In new construction as well as rehabilitation, the geomembrane can maintain watertightness in the presence of relative movements of the dam and of differential settlements of the fill.

With very few exceptions, mostly in Chinese dams, the PVC geomembrane is installed in the upstream position. This is technically preferred as it minimises uncontrolled water presence in the dam body, improving safety. Exposed upstream membrane systems also allow easy inspection, and have lower construction times and costs.

Specific aspects to be addressed for rehabilitation of embankment dams are the face and perimeter anchorage systems, which are designed depending on the type and strength of the existing facing (asphaltic concrete or concrete). An outstanding example of rehabilitation of asphaltic concrete facing has been Winscar dam in UK. The following case history addresses a recent repair of CFRD.

Rehabilitation of embankment dams: Strawberry CFRD, 2002
Strawberry, the second oldest concrete faced rockfill dam in the world, is located in the USA. Owned by Pacific Gas and Electric Co. (PG&E), the dam is 43 m high and about 220 m long, and has 9 vertical joints that did not have any waterstops installed. Deterioration of joints caused increasing leakage that over the years became unacceptable to California dam safety officials.

To permanently reduce leakage, PG&E selected for Strawberry an external waterstop system that was a development of the concept adopted for exposed geomembrane systems on entire face dams since 1976. The exposed waterstop system, which is patented, consists of a PVC geocomposite installed at the joints over a support layer. The waterproofing liner intrudes in the active joint at maximum opening of joint under the maximum water head. The geocomposite is watertight anchored along the perimeter and is left exposed to the water of the reservoir.

Differently from conventional embedded waterstops, which allow deformation only of few millimetres in the central portion of the bulb, the external waterstop can deform along the entire width of the PVC geocomposite (typically 40 to 60 cm), hence it is capable of accommodating significant movements that may occur in the joint. The system is the same installed to waterproof the contraction joints of RCC dams (see below).

At Strawberry the external waterstop, patented, consisted of four layers. The first two layers, Layers 1 and 2, anchored along edges on both sides by impact anchors into the existing sound concrete or in new shotcrete which replaced excessively damaged concrete, were a geocomposite made by a polyvinyl chloride (PVC) membrane 2.5 mm thick with a 500 g/m^2 geotextile laminated to it. These first two layers acted as anti-intrusion support on the joints, and were anchored so that they could move independently of each other. Layer No. 3 was a non-woven geotextile anchored along its vertical edges. This layer had an anti-friction purpose, to avoid layer No. 4 being affected by the movements of the bottom layers. Layer 4, the waterproofing liner, was the same PVC geocomposite as used in the first two layers, and was anchored at about 15 cm centres with a stainless steel batten strip along the perimeter edges. Watertightness of this perimeter seal was obtained by a compression system of the same type successfully used and tested up to 2.5 MPa of head. The completed joint liner system was about 1 m wide.

The works were staged in three phases: phase 1, improving the access road, phase 2, installation of the geomembrane system on 6 joints, and phase 3, installation of the geomembrane system on the remaining 3 joints that were not replaced in Phase 2.

Although the Phase 3 work is yet to be performed and the reservoir was filled in May 2002, and the recorded leakage has been about 85 percent below the 1998 leakage rates, more than adequately meeting the acceptance criteria (75 % requested for the all joints) and well below the historic levels.

Figures 6 & 7: On the left, Strawberry 43 m high CFRD, USA 2002. The exposed PVC geocomposite was installed as external waterstop on the failing joints. On the right, Salt Springs 101 m high CFRD in USA. The exposed PVC geocomposite will be installed in 2004 to stop seepage across the upper 2/3 of the dam face.

RCC DAMS
The geomembrane system has been installed in new constructions to provide watertightness to the entire upstream face, or to waterproof contraction joints as an external waterstop, and in rehabilitation, to waterproof failing joints/ cracks.

In new construction, the geomembrane system has been typically adopted in RCC dams of the low cementitious content type, where it allows separation of the static function, provided by the RCC, from the waterproofing function. One main advantage of the geomembrane is that it avoids water seepage at lift joints, reducing design uplift, and the risk that water can hydro-jack the lifts. Also some design constraints can be significantly reduced, such as the need of a conventional concrete layer on the upstream face, and the need for bedding mixes or special paste treatment of the joints. This leads to the possibility of placing one RCC mix over the entire cross section of the dam without the interference caused by bedding mix and conventional concrete. Inferior quantities of cement, pozzolan, fly ash, less stringent properties for aggregates, deletion of provisional sum for cooling, wider weather placement period, can constitute additional benefits.

Two conceptual systems are available: the covered system and the exposed system. The covered system was developed and patented in USA, where the first installation was made in 1984 (Winchester dam, now Carroll E. Ecton dam, Kentucky). The exposed system, an evolution of the exposed geomembrane system developed and used since the 1950s for repair of concrete and embankment dams, was developed and patented in Italy and was first adopted in 1990 (Riou RCC dam, France).

Table 3: Synthetic geomembrane systems on RCC dams

Country	Total dams	Geomembrane on	Position
Angola	1	EF*	Covered
Argentina	1	EF	Covered
Australia	1	EF	Covered
Brazil	1	J**, C***	Exposed
China	3	EF	Exposed
Colombia	2	EF, J	Exposed
France	1	EF	Exposed
Greece	1	J, C	Exposed
Honduras	2	EF	Exposed
Indonesia	1	EF	Exposed
Jordan	1	BF****	Exposed
Mexico	1	EF	Covered
Turkey	1	EF	Covered
USA	10	EF	Covered 9
			Exposed 1
Total	27		Covered 14
			Exposed 13

* EF: entire upstream face
** J: induced joints
*** C: repair of cracks
**** BF: bottom section of upstream face.

New construction: Miel I, Colombia 2002, and Olivenhain, 2003
Miel I is a straight gravity dam constructed in a narrow gorge in Colombia. At 188 m, it is the world's highest RCC dam. The dam crest, at an elevation of 454 m, is 354 m long and the entire upstream face is 31,500 m^2.

To meet the contractual schedule, the original design of an upstream face made of slip formed reinforced concrete was changed to a drained exposed PVC geomembrane system, placed on a 0.4 m wide zone of grout enriched vibrated RCC. Due to the height of the exposed dam face, this double protection was considered necessary (Marulanda et al. 2002). The use of grout enriched RCC allowed applying good compaction of RCC mix at the dam face, assuring a good finishing of the upstream concrete surface. The RCC mix has a cement content of 85 to 160 kg/m^3; total RCC volume is 1,745,000 m^3. Contraction joints are placed every 18.5 m.

The waterproofing liner is a composite geomembrane, consisting of PVC geomembrane laminated to a 500 g/m^2 polypropylene nonwoven geotextile. In the deepest 62 m the PVC geomembrane is 3 mm thick, in the top 120 m it is 2.5 mm thick. The PVC geocomposite was installed over the upstream surface, after removal of the formwork. At the contraction joints, two layers of sacrificial geocomposite, of the same type used for the waterproofing liner, provide support-avoiding intrusion of the liner in the active joint.

The attachment system for the PVC geocomposite on the dam face is made by parallel lines of tensioning profiles assemblies, placed at 3.70 m spacing. Where the water head is higher, from elevation 268 m to 358 m, the stainless steel profiles have a central reinforcement. A seal watertight against water in pressure up to 240 m fastens the PVC geocomposite along all peripheries.

The drainage system, consisting of the geotextile attached to the PVC geomembrane, of the vertical conduits made by the tensioning assemblies, of longitudinal collectors and of transverse discharge pipes, is divided into 4 horizontal sections (compartments). Each horizontal compartment is in turn divided into vertical sections with a separate discharge. Each compartment discharges in the gallery located at its bottom. In total there are 45 separate compartments, achieving very accurate monitoring of the behaviour of the waterproofing system.

To allow installation of the PVC waterproofing system to take place independently of the RCC activities, a railing system was attached to the dam face at a first level of 360 m, approximately 90 m above foundation, and then moved to a second level of 407 m, some 140 m above foundation.

The change in design allowed the schedule to be met. Construction of the dam started in April 2000 and ended in June 2002, for a total of 26 months.

Olivenhain is an RCC gravity dam 788 m long and 97 m high completed in summer 2003 in California. It is the highest and largest RCC dam in USA. The dam is a key element of the Emergency Storage Project (ESP) of the San Diego County Water Authority, owner of the dam. About 90% of water is brought to San Diego from hundreds of miles away, and the aqueducts cross several large active faults, including the San Andreas fault. The ESP is a multi year program that will provide water to the San Diego region in case of an interruption in water delivery deriving from an earthquake or drought, thus allowing time to repair the aqueduct.

The geomembrane system was selected because it was deemed that, following a seismic event, it would have been able to prevent seepage losses through any resulting cracks (Kline et al., 2002). Furthermore, the geomembrane liner and its associated face drainage system were considered to be two features that would tend to reduce the uplift pressure.

The waterproofing system is conceptually the same adopted at Miel I. The PVC geomembrane is 2.5 mm thick, and the drainage system differs in that the number of compartments is lower, and there is an additional drainage geonet over the entire upstream face.

Figures 8 & 9: On the left, Miel I in Colombia, 2002, at 188 m the highest RCC dam in the world. Total leakage from the exposed PVC geocomposite at fully impounded reservoir is 2 l/s, mostly coming from abutments. At Olivenhain 97 m high dam on the right, USA 2003, highest and largest RCC dam in USA, the exposed PVC geocomposite was adopted to assure watertightness in case of a seismic event. At 50% impounded reservoir total leakage from 38,880 m^2 upstream face is 0.18 l/s.

Exposed waterstop: Porce II, Colombia 2000

Porce II is a 118 m high RCC dam constructed in 2000 in Colombia. Its upstream face is formed by curb extruder. The contraction joints, at average spacing of 35 m, are waterproofed with a patented external waterstop system.

In detail, at Porce II the system consists of:
- Support structure: proceeding from the dam body towards the reservoir, support is made by two stainless steel plates having a strip of Teflon in-between to decrease friction and allow sliding of the plates, of one 2,000 g/m² polyester geotextile providing anti-puncture protection against the sliding edges of the steel plates, and of one sacrifice/supporting PVC geocomposite. The steel support impedes intrusion of the waterproofing liner into the active joint, and the flexible components provide extra support and a low friction element, so that the movements of the joint occur without affecting the waterproofing geocomposite
- Waterproofing liner: a PVC geocomposite, SIBELON CNT 5050, consisting of a 3.5 mm thick PVC geomembrane heat-coupled during extrusion to a 500 g/m² polyester geotextile. The geocomposite is exposed and centred on the joint and watertight anchored at the periphery by flat stainless steel batten strips compressing it against the concrete, regularised by trimming the offsets and by a layer of epoxy resin. Synthetic gaskets distribute stress to achieve even compression. At plinth, the PVC geocomposite is connected directly against the rock.

Rehabilitation: Platanovryssi, 2002

The external waterstop system has been used for the underwater repair of Platanovryssi, a 95 m high RCC dam in Greece. Platanovryssi was completed in 1998 and is at present the highest RCC dam in Europe. The dam, of the high cementitious content type, was designed to be impermeable in its whole RCC mass. The vertical contraction joints were waterproofed with 12 Carpi external waterstop system during construction.

On first filling, after mid-December 1999, seepage started increasing, and reached more than 21 l/s at the end of May, dropped again and then increased to a maximum of 30.56 l/s on 10 October 2000. The cause of the leak was what appeared at first as a hairline crack in the gallery, then on the upstream and downstream face, with a maximum opening of 25 mm. The crack was approximately 20 m long. Repair works were scheduled for Spring 2002. Due to an unusually dry season, the owner could not afford to lose the volume of the water to empty the reservoir in order to work in dry conditions. In addition to that restriction, when Platanovryssi is emptied the pumped storage scheme of Thissavros cannot operate, with serious implications to the production system. It was considerably more cost effective to do the work underwater (Papadopoulos 2002).

The system selected for the underwater installation on the crack in 2002 was the same conceptual system that had been selected and installed on the vertical contraction joints during construction of the dam. The materials that constitute the support layers and the waterproofing liner for the crack have a different thickness to those used for the vertical contraction joints, due to the different hydraulic load they sustain at the joints, and that they will have to sustain at the crack.

The repair system installed on the crack consists of:
— Support structure: 2 layers of sacrificial geocomposite, of the same type used for the waterproofing liner, each anchored independently
— Waterproofing liner: PVC geocomposite SIBELON CNT 3750, installed over the joints and left exposed except for the upper 10 m above pool level where it is covered by an independent steel plate which provides mechanical protection. The PVC geocomposite is anchored along the perimeter by a watertight mechanical seal made by stainless steel batten strips compressing it against the RCC.

The waterproofing system was installed in the dry from crest level at 227.50 m to elevation 225 m, and underwater from elevation 225 m down to elevation 208 m. To facilitate underwater works, a special steel frame was constructed and lowered into the reservoir to serve as a template for placement of the perimeter seal.

Works started on April 22, 2002 and were completed on May 23, 2002. The leak through the dam has been fully stopped and the downstream face in the vicinity of the previous leaking joint is now dry.

Figures 10 & 11: On the left, the external waterstops on the contraction joints of Porce II 118 m high RCC dam (Colombia 2000). The same external waterstops were installed in 1998 on Platanovryssi 95 m high RCC dam in Greece. In 2002, the external waterstop was used again at Platanovryssi, to repair a new crack that formed during first impoundment. Installation was performed underwater (on right).

PERFORMANCE

Durability of geomembrane systems is estimated in several decades. Among others ENEL, the Italian National Power Board, is monitoring behaviour in harsh climates of PVC drained geomembrane systems installed since 1976 on 6 of its dams at high elevation (max. 2,378 m). ENEL reports that behaviour is satisfactory, and that the impermeability coefficient has remained quite constant versus time (Cazzuffi 1998).

The leakage rates from the system described are typically very low: at Miel I, in 2003 the recorded leakage at fully impounded reservoir level from the whole geomembrane system is 2 l/s, at Olivenhain, at 50% impounded reservoir total leakage from 38,880 m^2 upstream face is 0.18 l/s.

CONCLUSIONS

The techniques for waterproofing and protecting the upstream face and the joints of all types of dams with drained PVC geomembranes, has reached a high degree of sophistication and reliability. The state-of-the-art system, by constructing a continuous flexible waterproofing liner on the whole upstream face, capable of bridging construction joints and cracks, allows resisting opening of even large cracks in case of differential settlements, of seismic events, of concrete swelling. It can dehydrate the dam from saturation water, and relieve uplift pressures. Installation is quick and can be executed also underwater. With more than 25 years of maintenance free history, it has proven to be durable and reliable, and to have a standard quality not dependent on weather or on a large amount of skilled labour.

REFERENCES

Carter, I. C., Claydon, J. R., Hill, M. J. (2002). *Improving the watertightness of Winscar Reservoir.* Proc. British Dam Society Conference, Dublin, pp 415-430.

Cazzuffi D (1998). *Long Term Performance of Exposed Geomembranes on Dams in Italian Alps.* Proc. Sixth International Conference on Geosynthetics, Singapore, pp 1107-1109.

Klein, R. A. et al. (2002). *Design of Roller-Compacted concrete features for Olivenhain dam;* Proc. 22ndUSSD confer., S. Diego, pp 23-34.

Liberal, O., Silva Matos, A., Camelo, D., Soares de Pinho, A., Tavares de Castro, A., Machado Vale, J. (2003). *Observed behaviour and deterioration assessment of Pracana dam.* Proc. ICOLD 21st International Congress, Montréal, pp 185-205.

Marulanda, A. Castro, A., Rubiano, N. R. (2002). Miel I: *a 188 m high RCC dam in Colombia.* The International Journal on Hydropower & Dams vol. 9, Issue 3, pp 76-81.

Papadopoulos, D. (2002). *Seepage evolution and underwater repairs at Platanovryssi.* The International Journal on Hydropower & Dams vol. 9, Issue 6, pp 88-90.

Scuero A M, Vaschetti G L (1998). *A drained synthetic geomembrane system for rehabilitation and construction of dams.* Proc. British Dam Society Conference, Bangor, pp 359-372.

Scuero A M, Vaschetti G L (2003). *Synthetic geomembranes in RCC dams: since 1984, a reliable cost effective way to stop leakage.* Proc. 4th International Symposium on Roller Compacted Concrete (RCC) Dams, Madrid, pp 519-529.

Downstream Slope Protection with Open Stone Asphalt

A.BIEBERSTEIN, University of Karlsruhe, Karlsruhe, Germany.
N.LEGUIT, Bitumarin bv, Opijnen, The Netherlands.
J.QUEISSER, University of Karlsruhe, Karlsruhe, Germany.
R.SMITH, Hesselberg Hydro Ltd, Sheffield, UK.

SYNOPSIS
Recent investigations and results from tests on the performance of erosion protection systems in Open Stone Asphalt (OSA) on man-made embankments during overtopping conditions are presented. The paper also includes state of the art mix-designs / compositions, installation routines and experience generated from projects in Northern Europe. An analytical statical concept for revetments is presented, which was developed in a research project at University of Karlsruhe, Germany.

INTRODUCTION
A major problem during the construction of a dam that can be overtopped is the ability of the slopes covered with revetments to withstand hydraulic loads (e.g. Rathgeb 2001 and Dornack 2001). Therefore an engineering or geotechnical concept is necessary to ensure the stability (LfU 1997).

It is necessary to protect the downstream slope of such embankments against erosion. Four possible cases of overtopping of an embankment are considered:

Overtopping arrangements
1) Overflow of a man-made reservoir during / after rainstorms.
2) Overflow of flood storage reservoirs, whenever the collected rainfall of a catchment area exceeds the total drainage capacity of rivers plus the flood storage arrangements.
3) Overflow of river embankments, when the drainage capacity of the tidal basin (estuary) of a river has been reduced by a storm surge.
4) Overflow sections of river embankments, when the flood exceeds the design level.

CONSTRUCTION REQUIREMENTS

These overtopping cases are all emergency situations. Since the managed flood control system is only used during an overtopping event, which occurs with a low frequency, the construction can often be hidden in the landscape. The requirement to choose a long lasting and permeable construction material with plastic properties to compensate underground settlements is often combined with the request to restore the existing appearance of the embankment and to support the establishment of vegetation. Excellent hydraulic performance under extreme conditions, together with low or no maintenance cost to the materials are desirable for the erosion protection in stand-by mode.

Open Stone Asphalt

To fulfil the function of a durable, maintenance-free erosion resistant material together with a green appearance, Open Stone Asphalt is often selected. Based on experience gained from seafronts, river and canal revetments and reservoir erosion protection on the upstream faces of dams the material is now being used on the downstream slope of embankments. The bituminous mix consists of single sized crushed stone coated with sand mastic. The design of the mastic coating is based on a sand/filler mixture, overfilled with penetration grade bitumen and with or without added fibres (Smith 1998).

After installation and a light compaction sufficient voids remain for it to be permeable to water and air and to accommodate roots of the surface vegetation.

The porosity of Open Stone Asphalt is almost equal to the original used stone (without mastic coating), so a filter layer is required to avoid loss of subsoil particles when seepage occurs. A variety of filters can be applied: woven or non-woven geotextiles, lean sand asphalt or loose granular material.

Pioneers

After an exceptional storm surge along the Belgian coast, in early 1976, several controlled flood areas were constructed as overflows to the river Scheldt.

The hydraulic models were studied at Flanders Hydraulics in Belgium. Bitumar, the Belgian partner of Bitumarin, was contracted to install the Open Stone Asphalt and specific lab-investigations and studies were carried out at the facilities of Bitumarin in Opijnen, Holland (Leguit 1984).

Erosion resistance of Open Stone Asphalt

The properties of resistance to flowing water with high currents were available from test regimes used to design revetments under wave attack or propeller scour from navigation (TAW 1985).

The main conclusions were:

- (Lith, The Netherlands) - Stationary and quasi-stationary flow: limited, only surface, damage was observed after 34 hours with current velocities of 6m/s. (the maximum possible generated flow of the test facilities)
- (Main-Danube-Canal, Germany) - Turbulent flow: no damage occurred after direct attack by an 800HP cargo ship at full strength for 5 minutes (Kuhn 1971).
- (Opijnen, The Netherlands) - Duration flow by pump surcharge to test construction joints.

Performance of geotextile filters to support Open Stone Asphalt

Various woven and non-woven types of geotextile were tested on filter stability in combination with soil samples extracted from the planned location (Leguit 1984).

The main conclusions were:

- After duration testing, the permeability of most filters was reduced by trapped soil particles.
- A thin blanket of sand to catch silt particles would extend the life performance of the filter construction and would reduce the possibility of erosion under the revetment.

Developments in OSA mix design

Open Stone Asphalt is a gap-graded, underfilled mixture of mastic asphalt and aggregate. The mastic comprises sand, filler and bitumen, and coats the aggregate particles with a layer 1 to 1.2mm thick. The mastic film is resistant to weathering and fixes the open aggregate skeleton together to withstand hydraulic loading. Ongoing research and development has led to several improvements in the mix design procedure (TAW 2002).

Aggregate/bitumen adhesion

The adhesion between the aggregate and the bitumen is important for the durability of OSA as it is the integrity of this bond which gives the material its strength. In the past it was generally accepted that carboniferous limestone had good adhesion to bitumen and this stone was usually used. After problems with the durability of a coastal revetment, the adhesion of different aggregate was investigated, and it was concluded that the chemical composition and the surface texture of the aggregate should also be considered, and that other types of stone could be acceptable.

The current standard is to always check the adhesive properties of the stone with the Queensland stripping tests, using a known control stone.

Mastic Viscosity

The asphaltic mastic coats the aggregate to give the material strength. To coat the aggregate correctly the viscosity of the mastic is very important. At the mixing temperature (typically 140-160°C) the mastic must be sufficiently fluid to coat the stones, but also sufficiently viscous not to drip off the stones causing segregation. The target viscosity for the mastic is 30-80 Pascal seconds (Pa.S) at 140°C, which is tested in the Kerkoven apparatus.

Cellulose fibres

The viscosity of mastic is temperature dependent, so the effects of variations in the mixing temperatures are investigated for the designed mastic. Mastics with a relatively high bitumen content will be more temperature-sensitive than those with a lower amount of bitumen, and so the risk of segregation during mixing due to increased temperature will be greater.

A recent development in the UK is the addition of cellulose fibres to the more temperature-sensitive mastics. The fibres combine with the bitumen to decrease its viscosity and to make it more stable. Typically 0.4-0.5% fibres are added to the mastic, and the bitumen content is increased by 1-2%.

Coating thickness

The proportion of mastic in OSA is generally between 19% and 21% depending on the aggregate grading (40mm or 28mm).

The durability of the OSA depends on a sufficient layer of mastic coating each stone so it is therefore an advantage to calculate the coating thickness rather than rely on a 'rule of thumb'.

For current mix designs the aggregate grading and the flakiness index are used to calculate the surface area per unit weight for the stone. This is then used to calculate the amount of mastic required for each aggregate.

For standard OSA mixes a coating thickness of 1.0mm is used, but this can be increased to 1.2mm for mixes containing cellulose fibres, as the increased stability of the mastic means that a greater coating thickness is achievable without increasing the risk of segregation.

Figure 1: Drilling core extracted from a revetment of Open Stone Asphalt at the project Tielrode (B) with a layer of humus for planting vegetation (Source: ELSKENS 1995)

Performance of Open Stone Asphalt revetment, after overtopping.
Minor cracks were registered in the embankments after overtopping had occurred. A study was carried out to explain the mechanism of failure (Mulders 1983).
The main conclusions are:
Most of the embankments were built like a typical "Dutch Embankment": A sand core, capped with good quality clay. The sand is more permeable to air than the clay, so during a rapid rise in water level the overpressure of the trapped air in the sand core reduces the soil stability and even tries to lift the clay capping during overtopping.

INVESTIGATIONS AT UNIVERSITY OF KARLSRUHE
A collaborative research project "dams and embankments (levees) designed for overtopping" was carried out by the Institute of Soil Mechanics and Rock Mechanics and the Institute of Water Resources Management, Hydraulics and Rural Engineering of the University of Karlsruhe. The main question was how to build dams and embankments of a few metres in height in order to withstand intentional overtopping during a flood. In order to do this, the downstream slope must be adequately protected against erosion. Up to now there have been no design rules, which made it possible to carry out the necessary stability checks in order to obtain technical solutions which are both economical and easy to build. During this project a statical approach was developed to determine the dimensions of a coherent, self-supporting revetment made of Open Stone Asphalt.

Suitability of a coherent revetment

Various steps were examined to prove the suitability of the selected revetment concept:

- Geotechnical aspects: Dimensions of the revetment, determination of the shear parameters as well as proof of the load capacity of the revetment in a tilting flume.
- Hydraulics: Numerical investigations to determine the dimensions and optimisation of the discharge conditions, in particular to guarantee a reliable dissipation of energy at the toe of the embankment.
- Verification of the results by means of investigations on a half dam model on a technical scale.

The proof of the stability of an overtopped revetment can take place on an embankment element for the given conditions (cf. Larsen et al. 1986). Here the stability can be analysed by comparing all relevant forces and resistances (see Fig. 2).

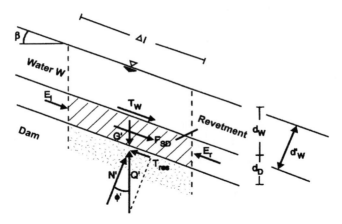

N'		
T_{res}	resulting forces in the shear	
Q'	plane	

Φ'	Angle of friction between the subsoil and the revetment	T_W	Shear stress resulting when overtopped
G'	Weight of revetment under uplift	d_W	Average thickness of the water layer
$E_l \cong$ E_r	Earth pressure	d_D	Thickness of the revetment
F_{SD}	Strength of flow	β	Slope angle

Figure 2: Individual element of an embankment which is overtopped and percolated (Larsen et al. 1986)

An evaluation of the analytical correlation is shown in Fig. 3. In this case the slope is covered with a 12 cm thick revetment layer, and the maximum permitted load for an Open Stone Asphalt revetment is given; the results are shown for different angles of friction in the shear plane.

For practical application, the following becomes clear from the correlations shown, as was to be expected:

- The angle of friction in the shear plane has a significant influence on the permitted load capacity of the embankment.
- On flat embankments (small slope angle) higher hydraulic loads are always permissible.
- On the other hand, steep slopes (e.g. $\beta > 15°$) permit only a very low hydraulic load – almost regardless of the size of the angle of friction.
- The correlation shown applies for the limiting state ($F_S = \eta = 1.0$).

The friction conditions in the shear planes are fundamental for the design and the static proof of the self-supporting revetment system. Thus the shear parameters for the system were quantified without any hydraulic load – an angle of friction of approx. 31° was determined.

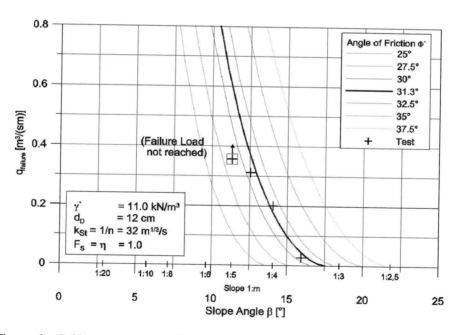

Figure 3: Self-supporting, coherent permeable Open Stone Asphalt revetment – theoretical maximum loads $q_{failure}$ depending on the slope angle β and the angle of friction Φ' (lines) as well as results of experiments from the tilting flume at $\Phi' = 31°$ (crosses)

In order to ascertain the maximum hydraulic load of the selected revetment, investigations were carried out on a slope element in a tilting flume, which was infinitely adjustable between an angle of 0° to 35° (length: 4 m, width: 1.31 m). The results confirmed the analytical calculation approach in the investigated load range; the maximum loads for varying slope angles are entered as crosses in Fig. 3. In the future, the dimensions of revetments can be calculated on this basis for the relevant discharge.

From a hydraulic point of view, it was important to look more closely at the three runoff regimes which arise when a dam is overtopped. These differ from one another characteristically with regard to the development of the flow speeds and the water levels as well as the Froude numbers. Here the dam crest with the transition from a subcritical to a supercritical flow was examined and the slope was evaluated with a supercritical steady flow. Above all, the toe of the slope, where at the transition from a supercritical to a subcritical flow a significant portion of the energy dissipation takes place, was the centre of attention. In preparation for the tests on the physical model one-dimensional numerical calculations were carried out. With the help of these calculations, estimates could be made of the position of the hydraulic jump to be expected and on the quality of the energy dissipation.

Finally all the results were used to design and build a half dam model on a technical scale (see Fig. 5), in which the indeed load situation could be realistically reconstructed. This was done up to a hydraulic load of $q = 300$ l/(sm), in addition the flow conditions in the dambody (sand) were observed and analysed.

An important aspect of the investigations was to optimise the intended energy dissipation at the transition from the slope to the horizontal area, in order to protect the downstream area from erosion. This was achieved by creating a hollow secured with Open Stone Asphalt acting as a scour protection in the transition area between the slope and downstream the toe. The forces and loads arising during energy dissipation are absorbed by the safety element and discharged to the subsoil. Due to the form of the hollow the hydraulic jump cannot move into the unsecured tailwater (see Fig. 6).

Fig 4: Tilting flume – Self supporting Open Stone Asphalt revetment under a load of q = 360 l/(sm) at a slope of 1: 5

Fig 5: Half dam model for overtopping tests at a slope of 1 : 6 during operation

Figure 6: Energy dissipation at the toe of the slope – View from upstream (cf. Bieberstein et al. 2002)

Suggestion for practical application

Based on the statical concept, it is possible to calculate the dimensions of overtopping stretches up to a specific discharge of approx $q = 500$ l/(sm). As a result of the investigations, an alternative design suggestion for the flood retention basin at Mönchzell, as shown in fig. 7, could be constructively derived: The downstream slope of the dam with an angle of 1 : 8 is covered with a layer of geotextile, on which the revetment of Open Stone Asphalt with a thickness of 20 cm is placed hot.

The Open Stone Asphalt revetment is finally completely covered with topsoil and seeded with grass. However, the grass is not included in the statical considerations. In the case of flooding, the fact cannot be ignored that it could become damaged – and in fact could be lost completely, since the topsoil is not necessarily secure. The revetment underneath takes over the securing of the dam or levee as planned during the period of overtopping.

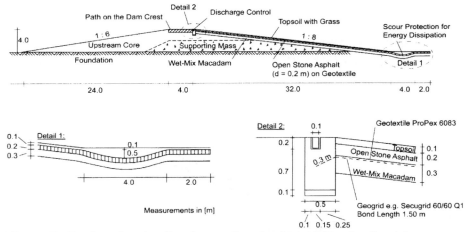

Figure 7: Design for the flood retention basin at Mönchzell with a revetment of Open Stone Asphalt (spec. hydr. load $q = 405$ l/(sm))

DOWNSTREAM SLOPE PROTECTION IN THE UK

Scheme

The Bodmin town leat flood alleviation scheme was designed by Halcrow for the Environment Agency and the main contractor was TJ Brent. The scheme involved the construction of a storage pond above the Cornish town.

The storage pond was created by the construction of a 4 metre high earth embankment with a culvert beneath it to take normal flows. A control structure at the intake limited the flow through the culvert and into the leat through the town, so during flood conditions water would back up and be stored behind the embankment.

In the event of a flood exceeding the design criteria, the embankment was designed to be overtopped over a lowered 60 metre section of the crest, with the water flowing over the downstream slope onto sports pitches below.

The erosion protection required to protect the downstream slope was specified as Dycel 100 blocks placed as pre-fabricated mattresses and secured with in-situ reinforce concrete edge beams.

Open Stone Asphalt (OSA) erosion protection
Hesselberg Hydro proposed an alternative to the concrete blocks comprising a 125mm thick layer of OSA, 250 kg/m^2, placed on a geotextile filter layer.

The OSA layer was thickened by 100mm around the outer edge to increase stability during high flows and at the crest the geotextile was placed in a soil-filled trench to give additional support to the revetment (see Fig. 8).

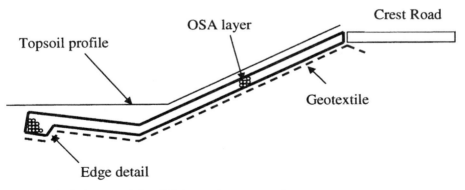

Figure 8: Scheme of the OSA erosion protection

Once the OSA was installed, a layer of topsoil was placed over it and seeded to give the revetment an acceptable appearance.

Overall, 1,000 m^2 was installed complete with crest and edge detail in less than 3 days.

Advantages of OSA

- The in-situ material follows all contours of the stilling basin and is easily placed around manholes and other structures. No cutting of blocks is required.
- A flexible revetment will follow the settlements associated with new earthworks.
- The material is durable and resistant to vandalism and unexpected impacts. In the event of damage to the revetment, the bound nature of the material ensures that damage is not progressive as with a revetment consisting of individual elements.
- The irregular surface voids can support a wider variety of plants than the regular voids in blockwork. The uniformly sized 'pots' in the blocks may suit particular plants which then become dominant.
- A layer of OSA will retain moisture more efficiently than a layer of open concrete blocks where moisture can escape through the holes through the whole block.
- OSA is quick to install and competitive on cost.

Fig 9: OSA erosion protection as   Fig 10: Downstream slope after
        installed   topsoil/seeding

CONCLUSIONS

Overflow sections of dams and also of levees are used for flood prevention purposes all over Europe. Open Stone Asphalt has been used in hydraulic engineering for more than 35 years in many countries. In this paper examples of overflowable dam embankments protected by Open Stone Asphalt have been described. The first such attempts were performed in 1976 in Belgium. The experiences have been presented in short form as well as optimized mix design procedures existing at the present time.

In a research project the University of Karlsruhe, Germany, an analytical statical concept was developed for such coherent, self-supporting and permeable revetments. The results were verified in models on a technical scale and are being transferred into practice right now.

REFERENCES

Bieberstein, A., Brauns, J., Queisser, J., Bernhart, H. (2002): *Überströmbare Dämme – Landschaftsverträgliche Ausführungsvarianten für den dezentralen Hochwasserschutz.* Zwischenbericht: BW Plus

CUR, VBW Asfalt, Benelux Bitume (1995): *Toepassing van asfalt bij. Binnenwateren,* (No.1791995) (Dutch).

Dornack, S. (2001): *Überströmbare Dämme – Beitrag zur Bemessung von Deckwerken aus Bruchsteinen.* Dissertation, Institut für Wasserbau und Technische Hydromechanik, Universität Dresden.

Elskens, F. (1995): *Protecting Overflow Dikes for Controlled Flood Areas in Belgium.* PIANC Conference on Inland Waterways and Flood Control, Brussels, Belgium.

Kuhn, R. (1971): *Erprobung von Deckwerken durch Schiffahrtsversuche.* Wasserwirtschaft 3.

Larsen, P., Blinde, A., Brauns, J. (1986): *Überströmbare Dämme - Hochwasserentlastung über Dammscharten.* Versuchsbericht der Versuchsanstalt für Wasserbau und Kulturtechnik und der Abteilung Erddammbau und Deponiebau am Institut für Bodenmechanik und Felsmechanik, Universität Karlsruhe, unveröffentlicht.

Leguit, N. (1984): *Filter Investigations for Paardewijde, Berlare-Wichelen (B).* Report Bitumarin division R&D.

LfU - *Landesanstalt für Umweltschutz Baden-Württemberg (1997): Dammscharten in Lockerbauweise bei Hochwasserrückhaltebecken.* Handbuch Wasser 2.

Mulders, G. (1983): *Technical Report Failmachanism of Overflow Embankments.* Tielrode - Bovenzande, Bitumarin division R&D.

Rathgeb, A. (2001): *Hydrodynamische Bemessungsgrundlagen für Lockerdeckwerke an überströmbaren Erddämmen.* Dissertation, Institut für Wasserbau, Universität Stuttgart.

Schönian, E. (1999): *The Shell Bitumen Hydraulic Engineering Handbook.* ISBN 0953588505.

Smith, R. (1998): *The use of asphalt in dam maintenance work.* Dams & Reservoirs Vol. 8 No. 3, Dec. 1998.

TAW - Technical Advisory committee on Waterdefences (1985): *The use of asphalt in hydraulic engineering,* (No.37/1985).

TAW - Technical Advisory committee on Waterdefences (2002): *"Technisch Rapport Asfalt voor Waterkeringen"* (Dutch).

3. Risk assessment and reservoir management

Sri Lanka Dam Safety and Reservoir Conservation Programme

LAURENCE ATTEWILL, Jacobs GIBB, Reading
LJILJANA SPASIC-GRIL, Jacobs GIBB, Reading
JAMES PENMAN, Jacobs GIBB, Reading

SYNOPSIS. The history of dam engineering in Sri Lanka dates back some 4,000 years to when ancient Ceylon developed control of the water streams to satisfy the needs of an advanced civilisation. These great works of irrigation are even more impressive, and attract even more interest, than many remains of ancient monuments, palaces and temples. Dam engineering practice in Sri Lanka has been continued to date to include large reservoirs such as Victoria, Kotmale, Randenigala, Samanalawewa.

Under the Dam Safety and Reservoir Conservation Programme (DS&RCP) 32 major dams were inspected and studied by Jacobs GIBB. The scope of the investigations included inspection and technical studies covering seismicity, instrumentation, stability, spillway adequacy and reliability. In addition water quality, sedimentation and catchment land use were assessed. Institutional issues included a review of dam safety legislation, establishment of a data management centre, identification of local research resources and training and skill enhancement for the local engineers.

PROJECT BACKGROUND

DS&RCP of Mahaweli Reconstruction & Rehabilitation Project (MRRP), funded by the IDA and managed by the Joint Committee (JC) has an objective to implement a qualitative management system for all major dams in Sri Lanka in order to improve their safety. The JC comprises the staff from the three dam owners, namely the Mahaweli Authority of Sri Lanka (MASL), the Irrigation Department (ID) and the Ceylon Electricity Board (CEB).

In year 2000, under the MRRP a Risk Assessment study of the 32 major dams in the Mahaweli river basin and adjoining basins was conducted. The study showed that while the modern dams have generally been built to current standards of the world's best-used practices, the same cannot be said for the other dams. Many dams are showing signs of ageing while others have significant deficiencies in monitoring, maintenance, reservoir conservation and other issues. A vast majority of dams including numerous dams managed by ID have not had an overall safety review and risk assessment.

The main objective of the DS&RCP is to assess safety of the selected 32 major dams and to recommend remedial works as well as to assist Sri Lanka in establishing a long term dam safety programme.

PROFILE OF DAMS
The DS&RCP covered 32 dams out of a total dam population in the island of over 300. The 32 dams, whose location is shown in Figure 1, can be categorized as follows:

Mahaweli multipurpose dams
4 of the 32 dams are large modern dams on the Mahaweli river serving both hydropower and irrigation purposes: Kotmale and Randeningala (rockfill), Victoria (arch) and Rantembe (concrete gravity). In addition, Polgolla diversion barrage supplies the Sudu Ganga and associated power stations and irrigation schemes. The five dams are owned and operated by the MASL.

Hydropower dams
6 of the 32 dams are single purpose hydropower dams owned and operated by the CEB. 5 of these dams are concrete gravity dams on the Laxapana river system constructed in the 1950's. The sixth, Samanalawewa is a rockfill dam constructed in the 1980's.

Irrigation dams
The majority of the dams are single purpose irrigation dams and are owned and operated either by the ID of the Ministry of Agriculture or the MASL. 13 of the irrigation dams owned and operated by the ID were originally constructed over 1500 years ago and are still in use after successive rehabilitation and reconstruction campaigns.

Inspections
All 32 dams were inspected early in the programme following a procedure typical for a periodic inspection under the UK Reservoirs Act 1975. Of the 32 dams, all the 14 dams owned and operated by MASL had previously been inspected, by staff of the Sri Lankan consultancy CECB, and reports were available. Irrigation dams are generally inspected monthly or quarterly by ID staff who complete a proforma report. There is no evidence of CEB dams having been previously inspected.

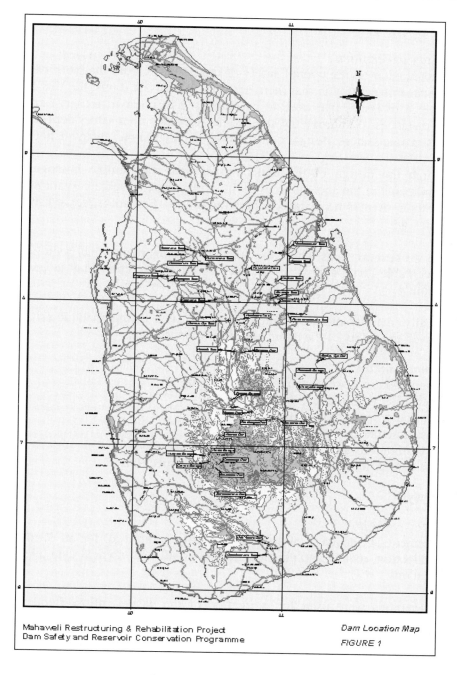

Mahaweli Restructuring & Rehabilitation Project
Dam Safety and Reservoir Conservation Programme

Dam Location Map

FIGURE 1

Figure 1 Location map of dams

CONDITION OF DAMS

Summary of condition
Our conclusion on the overall safety of the 32 dams from the work carried out under this activity is that there are very few unsafe dams, but that there is a range of issues that need to be addressed in order to preserve and in some instances to improve the status quo. Adequate dam safety depends on three separate factors: design, construction, and operation / maintenance.

Although the design of the dams ranges from the simple homogenous embankments of the ancient dams to the sophisticated double curvature arch of Victoria, there is no instance where the safety of a dam is jeopardised by poor design.

There are several dams where the standard of construction has been below an acceptable level, and at several dams poor construction may jeopardise dam safety.

Generally maintenance is barely adequate, and if this situation is not improved the safety of the dams will slowly deteriorate.

Recommendations
Recommendations were made in the report of:
- Remedial works, categorised by priority
- particular maintenance items
- instrumentation and monitoring
- investigations and studies
- the nature, frequency and scope of future inspections

Spillways

Spillway capacity
Assessment of adequacy of spillway capacity comprised, for all 32 dams, the collection, review and detailed analysis of all hydrological data relevant to the dams.

Two methods were used for estimation of the design inflow floods: the statistical approach which is based on historic records of the annual maximum flows recorded at all gauging stations in Sri Lanka and the unit hydrograph method.

The statistical approach is based on the maximum annual flows for each year of record for the 80 gauging stations in Sri Lanka, providing some 2,000 station years of record. The results of the study are presented as a graph of the standardised flood peak versus the probability or return period of the flood (Figure 2). Three curves are presented, as follows:

- Curve no. 1 grouping all Sri Lankan gauging stations together
- Curve no. 2 for areas where the mean annual rainfall is < 2000 mm
- Curve no. 3 for areas of average rainfall
- Curve no. 4 for areas where the mean rainfall is > 3,400 mm

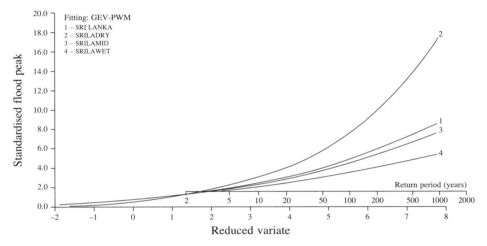

Figure 2: Regional Flood Frequency Curves

Because of the high density of population downstream of the dams, spillway capacity was also checked for the PMF. The PMF inflow hydrographs were obtained by a simplified version of unit hydrograph method and the estimation of the probable maximum precipitation (PMP) over the catchment. The PMP was estimated from the maximum recorded rainfall at each meteorological station over the period of record, which for many stations exceeds 100 years.

The check of the adequacy of the spillways and other outlets of the 32 dams showed that all but three of the spillways had adequate capacity: for these dams extra capacity can be economically and safely provided by heightening the dams concerned.

Spillway reliability
Of the 32 dams, 22 are either wholly or partly dependent on gated spillways for their safety. Of these spillways, 19 are electrically actuated, although most are capable – in theory – of manual operation. Nine of the 22 gated spillway were rated high reliability with no significant remedial works required.

Stability

Embankment dams
Among the 22 embankment dams, only the 4 modern rockfill dams and one zoned embankment had geotechnical information available from the original design stage which proved that the dams were stable. The geotechnical information for the remaining 17 earth embankments (13 of which are ancient) was either very poor or non - existent. Therefore stability of these 17 dams was carried out using an assumed range of lower bound strength parameters.

Based on the stability results 17 dams were grouped into the following three groups:
- Group 1 - FOS<1.3 - Investigation required (4 dams)
- Group 2 - 1.3<FOS<1.5 - Investigation required if high ground water levels or specific defects were identified in the inspections (7 dams)
- Group 3 - FOS>1.5- No investigations required (6 dams)

It was recommended that for four dams from the first group site investigations be carried out and the stability reassessed using the parameters from the investigation. In addition, three other dams from the second group also required investigations because of defects identified during the dam inspections.

Concrete dams
Out of 10 concrete dams, 9 are gravity dams with heights varying from 18.3m to 42m, and Victoria dam, a 120m high concrete arch dam on the Mahaweli Ganga.

Safety of the dams to sliding and overturning as well as the stress at the key points was checked for the normal, unusual and the extreme loading conditions.

Seven dams were found to be stable with an adequate safety margin under all loading conditions. However, three dams, Castlereigh, Nalanda and Norton dam were found not to have sufficient safety margin and appropriate remedial works – improved foundation drainage - were recommended.

Instrumentation
It was found during our inspection that the dams constructed recently were equipped with electronic instrumentation to measure seepage, pore pressure, deformations, deflections, movements, temperature and various other parameters. This equipment, whilst operating well for a number of years, has rarely been serviced or calibrated. Where equipment has failed there

has been little funding available for its maintenance or repair which has resulted in the equipment being abandoned. In some cases, a lack of understanding of a system has led to equipment being abandoned or deemed inappropriate.

The dams that were constructed in mid 20th century have fewer instruments, and the ancient dams usually have no instrumentation at all.

Currently, dam monitoring is undertaken by dam owners and on many of the sites the monitoring is carried out on a regular basis. However, data recording and handling procedures often vary from site to site. The instrument monitoring staff has a basic understanding of the instrument operation but the data handling procedures are not standardised.

Following the inspection, we have recommended and specified additional instruments: these comprise for most dams the collection and measurement of seepage and the provision of survey monuments to enable settlement surveys to be carried out. Standardisation of data recording and presentation was proposed. It was also proposed that the records will be in a centralised data record library within the Data Management Centre in Colombo and will be available via the GIS system.

OPERATION AND MAINTENANCE (O&M)

The perceived shortcomings in present O&M procedures are as much the product of inadequate budgets and the failure of management to recruit, train and financially reward staff of the calibre necessary to operate and maintain large dams, as they are deficiencies in management procedures and practices. This in turn may be seen as being a failure by Government to recognise the importance of the security of the nation's stock of large dams to the national economy, and the threat that unsafe dams pose to the public at large. For this reason, it has been necessary to take full recognition of the initiatives that have been discussed to restructure the main water management agencies, to introduce a new Water Act and to set up a regulatory framework for dam safety. The form that the regulatory framework will take will impose obligations on dam owners that will significantly affect the procedures to be adopted for O&M and safety surveillance.

Prior to the preparation of Guidelines for Improvement of O&M and Emergency Procedures we examined the current practices which are applied within each of the agencies. They are summarised below.

O&M

Procedures for O&M of the large MASL dams are now well established. All of the new dams have O&M manuals prepared by the designers which set out routine procedures for O&M as well as emergency procedures, particularly in the event of a major flood.

Procedures for operation of CEB dams are determined in Colombo to meet energy requirements within the distribution system. The procedure adopted is that gate operating staff are assigned to provide 24-hour cover at each of these dams whenever the water level approaches FSL and continues until the water level has again fallen below FSL.

Operation of the ID dams is regulated by a departmental circular which covers the whole irrigation scheme as well as the headworks.

Emergency Preparedness

Some effort has been made at the big dams to prepare for emergencies, in that key staff have been listed with their home contact details, contact details have been compiled for the emergency services and other key authorities, and lists of emergency service providers have been made. But generally, there has been no attempt to identify risks, to set levels of alarm in response to different emergency situations, or to determine the actions and persons responsible in any set of circumstances. Also, there is no programme of formal training for operating staff in dealing with emergency situations.

Prepared Guidelines for Improved O&M and Emergency Action Plans (EAP)

We proposed that improved management practice for Sri Lanka's stock of large dams requires that the three principal agencies adopt a structured, simple and standardised approach to O&M and Emergency Preparedness. The guidelines were drawn up for preparation of Standard Operating Procedures (SOPs) and EAPs for all dams in Sri Lanka. Prototype documents were also produced that are intended for universal application by the three agencies.

RESERVOIR CONSERVATION

Extent of sedimentation and pollution

In world terms, Sri Lankan reservoirs are not severely affected by either sedimentation or pollution. However the pressures exerted by a rapidly expanding population have resulted in environmental degradation of one third of the total land area. Soil erosion is most severe in the high catchments on steep slopes at mid levels, which are used for market gardening and tobacco production: it is estimated that erosion rates for these land uses are 150t/ha/year, compared with 0-10 t/ha/year for paddy, forest and well managed tea. The actual sediment yield of the catchments varies between 0.5 t/ha/year to 4 t/ha/yr for lowland and upland reservoirs respectively. Of the 32 reservoirs studied only two, Polgolla and Rantembe are seriously affected by sedimentation.

Similarly water quality is becoming a more serious problem because of increasing levels of nutrients, pesticides and effluents entering the watercourses.

Conservation policy

A national conservation policy is required to reverse the adverse trends in sedimentation and water quality in order to protect the countries water resources. Sediment yields will be reduced and the water quality improved by:

- propagation of appropriate land use, including grassing or reforesting steep and high level areas currently used for agriculture, the prevention of overgrazing and the adoption of soil conservation measures
- the adoption of appropriate land use and fiscal policies to improve land tenure systems and discourage the fragmentation of land
- improvement of urban waste water treatment and the disposal of solid waste
- better management of fertilizers and pesticides
- enforcement of the 100m buffer zone of grassland and trees around the reservoir perimeter.

Considerable efforts are already being made in the conservation of the Mahweli catchments, including research, public awareness and farmer training. This work needs to be intensified and extended to all catchments.

TRAINING

Background

Inadequate skill levels were identified as a drawback to overall dam safety. Many of the skilled and experienced operators, technicians and site engineers have left the MASL, ID and CEB for better prospects. The younger operators, engineering and other relevant professional staff, are with limited experience and little exposure to appropriate best practices. It was recognised that there is a lack of a well-structured training and competency assessment programme, and that as a result staff training was an important component of the DS&RCP.

Training Framework

A training framework was produced based on assessments of the workforce capacity of 32 dams and their gaps in skills. The assessments were carried out based on the questionnaires, workshop and interviews with the staff and the senior management of MASL, CEB and ID.

The staff required training was grouped into the following groups:

Group A	Engineers in Charge/Chief Engineer: professionally qualified engineers generally with more than 10 years experience who are potential senior managers
Group B	Civil engineers and technicians engaged in dam monitoring who aspire to become Engineers in Charge or Chief Engineers
Group C	Electrical/Mechanical engineers and technicians who are responsible for the operation of spillway and sluices

A training programme was developed that comprised 9 training modules and technical presentations in 5 technical areas which were delivered by the Consultant. Around 150 staff received the training under this programme, namely 43 staff from Group A, 46 staff from Group B and 95 staff from Group C.

Nineteen local trainers were also identified from all three organisations. The trainers received technical training along with the trainees and in addition they also attended a course in communication and presentation skills. The trainers delivered one training course under our supervision when we had a chance to comment on their performance.

DAM SAFETY MANAGEMENT CENTRE

It is the intention that the three dam owning organisations combine to set up a Dam Safety Management Centre (DSMC), which would be a quasi autonomous body to coordinate the following activities for all dams in Sri Lanka:
- Data management and appraisal
- Emergency technical co-ordination
- Dam survey unit
- Implementation of dam safety programme for 32 dams
- Extension of dam safety programme to other dams
- Monitoring compliance with dam safety code of practice
- Steering group for dam safety legislation
- Training of dam owners staff
- Liaison with IESL and other stakeholders

DAM SAFETY LEGISLATION / CODE OF PRACTICE
As was required by the terms of reference, we prepared a paper outlining the main provisions of future dam safety legislation in Sri Lanka, based on a review of legislation in UK, USA, Sweden and India. The main provisions of the proposed legislation were:
- The dam owner is responsible for the safety of the dam
- A register of dams would be compiled and maintained by the enforcement authority
- Dams would be subject to mandatory inspections by independent engineers
- Recommended remedial works would be mandatory

After much internal discussion the Client decided that Sri Lanka is not ready for legislation and that the proposed provisions should be contained in a Code of Practice. The DSMC will be responsible for monitoring compliance with this Code.

PORTFOLIO RISK ASSESSMENT

Objective
Portfolio Risk Assessment (PRA) provides a rational method of improving the safety of a group or portfolio of dams in the care of a single owner or organization. PRA enables owners to determine
- How much dam safety expenditure is justifiable
- The priority of dam safety measures
- The rate of expenditure
- The risk profile of their portfolio

PRA involves the following steps:
- Engineering assessment of dams
- Assessment of risk posed by dams in their existing state and after dam safety measures
- Definition of dam safety programme

Risk assessment
The risk for all 32 dams was assessed both by the semi-quantitative "Failure Modes, Effects and Criticality Analysis" (FMECA) method and a quantitative analysis in which the probability of a dam failure and the cost of the consequences are expressed numerically.

In the semi quantitative estimate both the probability and the consequences of failure are expressed by a scoring system developed which is based on that and described in the CIRIA publication C542 Report, Risk Management

for UK Reservoirs. In this the probability of failure of a dam can be expressed as the product of at least two factors:
- The probability of an event (slope instability, flood overtopping etc)
- The probability of the event resulting in failure of the dam

Both probabilities are expressed in terms of a score in the range of 1 (very unlikely) to 5 (likely).

In the quantitative assessment, event tree analysis is used to estimate the probability of failure and the consequence of failure is based on an estimate of the loss of life and economic loss from inundation mapping. The results of the risk analysis are shown on the F-N plot in Figure 3.

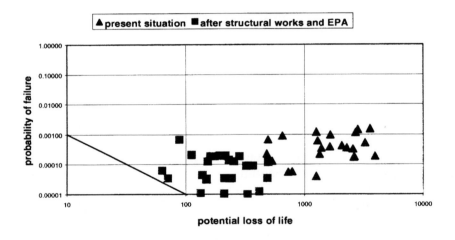

Figure 3 F-N plot

Dam Safety programme
The dam safety programme comprises both structural and nonstructural measures, as follows:

Structural measures

Improvements to spillways and outlets	Rs 434 million
Repairs to upstream slope protection	Rs 265 million
Dam and foundation drainage	Rs 338 million

Non structural measures

Monitoring systems	Rs 43 million
Early warning systems	Rs 67 million

The total capital cost of the entire programme is Rs 1150 million or

US$ 11.5 million.

Evaluation
The evaluation of the economic viability of structural measures uses the
concept of risk cost, which is expressed as product of the probability of
failure and economic loss, to express the benefits.

Because the b/c ratio for the entire programme is low (0.2), consideration
has been given to the early implementation of the most urgent and beneficial
components. A plot (Figure 4) showing the decrease in risk cost with
increasing levels of expenditure on structural measures will assist in
deciding the extent of this initial phase.

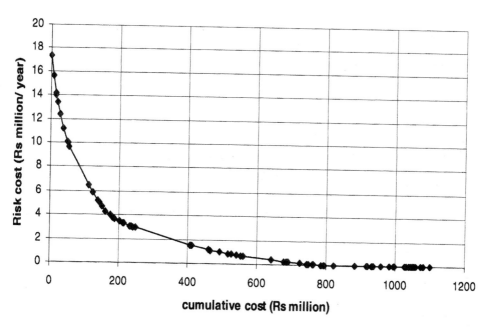

Figure 4: Risk cost vs cumulative cost of structural measures

CONCLUSION
While the full dam safety programme of Rs1,100 million is desirable, 85%
of the dam safety improvements can be achieved with the expenditure of
just half this sum. This reduced programme approaches economic viability
and is recommended.

ACKNOWLEDGEMENTS
The authors wish to thank the Joint Committee for their permission to
publish this paper and their colleagues for their assistance in its preparation.

Condition assessment of Government-owned dams in Finland

P. VUOLA, Finnish Environment Institute, Helsinki, Finland
R. KUUSINIEMI, Finnish Environment Institute, Helsinki, Finland
T. MAIJALA, Finnish Environment Institute, Helsinki, Finland

SYNOPSIS.   Some 480 dams in Finland are covered by dam safety legislation and of these, some 50 dams are government-owned. In spite of shortcomings and a few incidents there has been no complete dam failure in Finland affecting water storage dams that have a significant damage potential in case of failure. To unify the safety level of government-owned dams and to prioritise future maintenance work, the environment administration has decided to carry out condition assessments of dams that have a significant damage potential in case of failure.

INTRODUCTION
Finnish state-owned dams have been built over the last 40 years, and their history is still recent. Sufficient accurate data from different tests is available for many assessment aspects. The data consists of : soil investigations from the planning period, quality control tests from the period of construction, monitoring frost depth, phreatic surface level and seepage flow rate during the period of operation.

Nonetheless, some of the abovementioned data is inadequate or incomplete. Consequently, new testing, monitoring and supervision are necessary in order to obtain proper data for the condition assessment process.

DAM SAFETY IN FINLAND
In Finland, dams have been built mainly for flood control, hydroelectric power production, water supply, aquaculture and for storing waste that is detrimental to the health or to the environment. Most of Finland's dams were built after World War II. Regular monitoring of dam safety by the state-owned power companies began in 1962 and that of state-owned dams (the environment administration) in 1972.

The Act and Decree on Dam Safety were enacted in 1984 to improve the safety of all dams, waste dams included. In 1985, a Dam Safety Code of

Long-term benefits and performance of dams, Thomas Telford, London, 2004, 146–153

Practice was issued to apply the statutory regulations as a practical guideline. This improved the maintenance situation considerably, due to the fact that a basic inspection had to be carried out and a safety monitoring programme created for each dam subject to the Dam Safety Act. The third revised Dam Safety Code of Practice was issued in 1997.

Some 480 of Finland's dams are covered by the legislation. Of these 85% are water storage dams and 15% waste dams. The experts calculate that in the event of an accident, 37 of the dams would endanger human life or health or cause considerable damage to the environment or property (so-called P dams). Most of the dams are embankment dams, and a few are concrete gravity dams. Concrete structures have been used for water regulating structures. Some dams are provided with an overflow structure for high flood situations.

Finland differs markedly from many other countries in topography, soil and climate. Finland is a rather flat country characterised by glacial formations. Typical features of the climate are the long, cold winters, the freezing of the soil and the spring thaw. The ground is seismically tranquil, and there are no earthquakes on a scale to threaten dams.

The emphasis of Finnish dam safety is on the prevention of dam accidents and on the effective reduction of hazards should it not be possible to prevent an accident. Careful design, construction and monitoring of dams and their appropriate maintenance play a key role in preventing dam damage. Long-term changes in conditions and the ageing of structures can be taken into account with regular safety monitoring. Rare exceptional physical conditions, human error or other causes (e.g. internal erosion) may, nonetheless, still lead to dam failure. The objective of the Finnish dam safety system is to restrict any damage that might be caused by dam failure and to prevent loss of human life in the event of an accident. To achieve this we must maintain our dams to a very high standard, have a regular monitoring and emergency action plans designed for P dams to activate the warning function, evacuation and rescuing of the downstream population.

REPAIR WORK ON STATE-OWNED EARTH DAMS

There are some 50 dams owned by the environment administration covered by the dam safety legislation. Eleven of these dams are class P dams (Fig. 1). The basic inspections and further inspections incorporated in the safety monitoring programmes revealed several shortcomings e.g. the following:

- the flood discharge capacity of some dams has been inadequate
- seepage problems

- wet areas and springs behind some dams
- inadequate freeboard against frost in some dam crests
- trees on the dam contrary to the code of practice
- drainage system does not work
- bedrock of some dams needed grouting
- facing of wet slopes needed repair work.

In spite of shortcomings and a few incidents there has been no complete dam failure in Finland affecting water storage dams that have a significant damage potential in case of failure. To unify the safety level of state-owned dams and to prioritise future maintenance work, the environment administration has decided to carry out the condition assessments of its P dams.

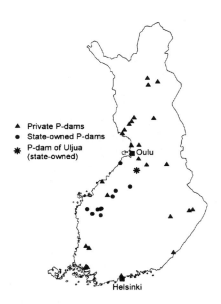

Figure 1: Location of P dams in Finland.

PRINCIPLES OF CONDITION ASSESSMENT
Assessment data
Finnish state-owned dams have been built over the last 40 years, and their history is still recent. Sufficient accurate data from different tests is available for many assessment aspects. This data consists of e.g.:

- soil investigations from the planning period

- quality control tests from the period of construction
- monitoring frost depth, phreatic surface level and seepage flow rate during operation.

Nonetheless, some of the abovementioned data is inadequate or incomplete. Consequently, new testing, monitoring and supervision are necessary in order to receive proper data for the condition assessment process.

An example of proper soil investigation data is presented in Fig. 2. Similar data on dam core permeability and density is available. Nevertheless, these accurate permeability test results do not contain anisotropy data from the core and subsoil. In order to assess the threat of piping, non-homogeneity as well as anisotropy should be taken into consideration.

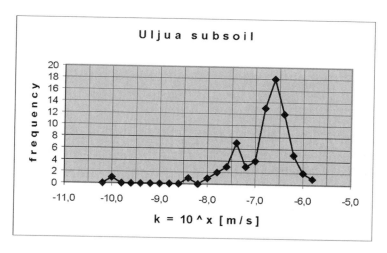

Figure 2: Coefficient of water permeability determined from Uljua subsoil moraine.

As an example, the main cross sections of Uljua dams are presented in Fig. 3. Such designs are typical for many Finnish state-owned dams. The wide core moraine dam construction seems to carry a shortcoming in the shape of poor filtering and drainage on the dry side of the dam. The technical failing of the rock fill dam construction seems to be a core that is too narrow and shallow with a weighted creep ratio that is too low. Both factors lead to an increased risk of piping either through the dam or the subsoil.

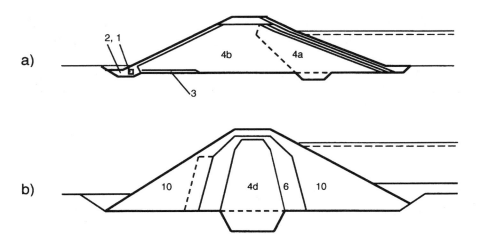

Figure 3: Uljua dams, typical cross sections. a) Arkkusaari 'homogeneous' or wide core moraine dam: 1 = 'stone drain', 2 = sand drainage layer, 3 = bottom drainage layer and filter, 4a and 4b = moraine cores. b) Tulisaari rock fill dam: 4d = moraine core, 6 = filter, 10 = supporting rock fill.

The data we use consists of technical data on one hand, and dam history on the other hand, especially failures. The knowledge of dam history is essential, because the dam itself is a full-scale test. Consequently, a lot of interest is focused on dam behaviour from the period of construction and first reservoir filling until the present. Typical dam incidents include:
- excessive seepage or possibly piping during the first reservoir filling, one in Uljua rock fill dam in the year 1970
- erosion of upstream rock fill blanket during operation
- piping in Uljua rock fill dam in the year 1990 during operation
- inadequate discharge of the drainage system, possibly due to thinness of the bottom filter and clogging of the subsurface drains.

Assessment methods
In order to be able to assess risks, a methodology had to be developed as well as technical criteria. It seems impossible to calculate actual probabilities, and yet the hazard and risk level of different phenomena have to be assessed and compared. Some tools applied or developed for these purposes are:
- application of the Fuller curve to determine the grain size distribution curve of the active portion of soil
- modification of the Foster and Fell filter criteria for practical activity; the principle is presented in Fig. 4

- utilisation of the principle of fuzzy logic; the principle is presented in Fig. 5; in fact Fig. 4 includes the concept of fuzzy logic as well.

Certain tools are considered necessary, because an individual engineering judgment alone may lead to inappropriate deviation in the assessment process. Besides, knowledge of the hazard and risk level is essential when drawing up the repair works schedule.

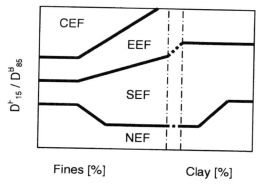

Figure 4: Principle of applied filter criteria chart. NEF = no erosion filter, SEF = some erosion filter, EEF = excessive erosion filter and CEF = continuous erosion filter. All percentages and grain sizes are calculated from corrected grain size distribution curves. D^F_{15} and D^B_{85} represent filter and base material grain size diameters, through which 15 and 85 % respectively, of the material will pass.

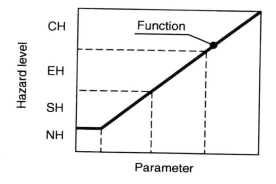

Figure 5: Principle of applied fuzzy logic, based loosely on the terminology used in Fig. 4. NH = negligible hazard, SH = some hazard, EH = excessive hazard, CH = catastrophic hazard.

The aspects to be assessed were classified into four main categories:
a) external erosion, including erosion induced by ice forces
b) internal erosion and subsurface drainage, including particularly:
  • internal stability, self filtering and segregation of each layer
  • filter criteria on the layer interfaces
  • seepage in the dam moraine core, filters and subsoil
  • frost action
  • flow in pipelines, including clogging by ferruginous precipitation
  • adjoining structures
c) slope and subsoil stability
d) additional aspects, including particularly:
  • background ditch drainage
  • vegetation effect
  • supervision and monitoring
  • maintenance
  • emergency action facilities.

Based on preliminary results, internal erosion and subsurface drainage and additional aspects have major roles in the assessment process, while external erosion is merely a matter of engineering. Stability seems to be of minor interest.

Despite the fact that most aspects occur worldwide, there are certain special phenomena featured in Finnish dams. Apparently these phenomena are typical to a larger area, but they are reported rather seldom in literature. These phenomena are:
  • frost action, especially formation of ice lenses, which lead to soil loosening and increasing piping threat
  • ferruginous precipitation, apparently suspended hydrated iron oxide precipitation, which clogs the subsurface drainage system.

The result of the condition assessment process will be a document for each dam including presentation of:
  • history
  • current conditions
  • hazard and risk classification
  • recommendation for repair action.

REFERENCES
Publications of the Ministry of Agriculture and Forestry 7b (1997). *Dam Safety Code of Practice*. Helsinki.
  (http://www.vyh.fi/eng/orginfo/publica/electro/damsafet/damsafe.htm)

Finnish Environment Institute. (2001). *Final report of the EU project RESCDAM (Development of rescue actions based on dam-break flood analysis).* Helsinki.
(http://www.vyh.fi/eng/research/euproj/rescdam/rescdam.htm)

Kuusiniemi, R. and Loukola, E. (1996). *Repairing and Upgrading of Dams.* Symposium in Stockholm, June 5-7, 1996. Stockholm.

Kuusiniemi R. et al. (March 1992). *Internal erosion at the Uljua earth dam.* Water Power and Dam Construction.

Foster, M. and Fell, R. (2001). *Assessing Embankment Dam Filters That Do Not Satisfy Design Criteria.* Journal of Geotechnical and Geoenvironmental Engineering.

Portfolio Risk Assessment in the UK: a perspective

A.K. HUGHES, KBR, UK
K.D. GARDINER, United Utilities, UK

SYNOPSIS.   This paper discusses the merits of Portfolio Risk Assessment (PRA) from the point of view of a practitioner and a dam owner.

INTRODUCTION

The management of reservoir safety in the UK is generally subject to the requirements of the Reservoirs Act 1975 and the assessment methodology applied by Panels of Engineers appointed under that Act. The Health and Safety Executive (HSE) claims jurisdiction over the safety of reservoirs where a business is involved under the powers of the Health and Safety at Work Act 1974, although they defer to Panel Engineers and the inspection system at present.   The HSE also claims jurisdiction over non-statutory reservoirs.

Under the Panel Engineer system, the reservoir inspection is generally based on observational techniques supplemented with historical information such as instrumentation data, previous reports and studies, drawings etc. The Panel Engineer tends to focus on technical matters with the intention of maintaining the safety of the public by preventing a dam failure. The system has a good track record with no failure in the UK causing loss of life since 1925. However, the system only considers the safety of individual dams and does not address the justification and prioritisation of recommended works for owners of multiple dams.   In addition, it does not consider the tolerability of risk and business drivers for identifying and evaluating options for even higher levels of safety.

WHY WOULD YOU CARRY OUT PORTFOLIO RISK ASSESSMENT?

Portfolio risk assessment is a process which can assist owners to manage reservoir safety in the overall context of their business.

The importance of this approach was recognised in OFWAT document MD 161, 'Maintaining Serviceability to Customers' dated 12 April 2000 addressed to 'The Managing Directors of all Water and Sewerage Companies and Water Only Companies' which stated;

'Each company needs to demonstrate how the flow of services to customers can be maintained at least cost in terms of both capital maintenance and operating expenditure, recognising the trade off between cost and risk, whilst ensuring compliance with statutory duties.'

'The Government considers an economic framework related to current and likely future asset performance (serviceability) is likely to provide the best way forward. .... As the (Environmental Audit) Committee recommends , it will be important for this work to investigate the practicability of approaches that are forward looking, taking account of the risk of asset failure (probability and consequences) as well as past historical trends.'

The PRA process specifically addresses the trade off between cost and risk and the compliance with statutory duties through an approach that takes account of the risk of asset failure accounting for both probability of failure and consequences of failure. The PRA approach does not replace or supplant the role of the Panel Engineers, but builds on the Panel Engineers' technical assessments and other information available to an owner. The approach seeks to use estimates of the likelihood of various failure modes, estimates of life and economic losses, and preliminary evaluations against tolerable risk guidelines (HSE 2001) and the owner's business criticality considerations, to identify opportunities for improved dam safety through investigations, and risk reduction brought about by carrying out works at the dam and improved reservoir safety management. Improvements in the effectiveness of detection and response to dam safety incidents by owners and the effectiveness of emergency response by local authorities can also be considered.

THE PORTFOLIO RISK ASSESSMENT PROCESS
PRA is a logical, auditable method of sytematically assessing a stock of dams in its current condition and assessing and prioritising the works required to be done and other measures that would improve reservoir safety, but may not be required under current practice. Some water companies have used this technique, and the prioritised lists and resulting spend profile that comes from it, as the basis of their submission to OFWAT (the regulatory body for the privatized UK water industry). OFWAT had asked that risk assessment be used in the companies funding submissions and therefore the submissions that used these techniques were generally well received.

A risk assessment carried out for a portfolio of dams typically uses data from historic incidents, accidents and failures, together with estimates of the probability of occurrence of extreme floods and earthquakes, to obtain estimates of the probabilities of failures for the failure modes considered. In addition, information from dam break analyses is used to estimate life loss and economic consequences for each failure mode. Remedial measures are defined for each mode of failure to meet UK Reservoir Safety practice and to reduce the probability of failure. Additional measures can be considered to exceed current UK Reservoir Safety practice. An evaluation may then be carried out, including cost benefit analysis, to provide information on the strength of justification for each remedial measure relative to tolerable risk guidelines such as those by the HSE (2001). This also provides data for the prioritisation of these remedial works based on alternative approaches discussed in the next section of this paper. The dam owner must then decide how this information will be used for the reduction of risk.

A number of PRA studies carried out for owners have shown that the process promotes a strengthening of the management of reservoir safety and its integration into all areas of the owner's business such as, the licence to operate, asset management, asset operation and maintenance, risk management, legal and insurance areas.

ISSUES IN USING THE RESULTS FROM A PORTFOLIO RISK ASSESSMENT

Once a Portfolio Risk Assessment has been carried out, many questions arise that can only be answered by the owner. These questions have implications that go far beyond the technical issues that a reservoir safety group typically deals with and therefore representatives from a wide range of departments in the owner's organisation should be involved. The discussion of these questions and some of the suggested answers form the major part of the rest of this paper.

1. **How should the PRA be used to prioritise the remedial works that have been determined should be carried out?**

   - **By probability of failure?** – should the owner take the view that any failure is unacceptable and therefore the dams most likely to fail should be dealt with first?
   - **By consequence of failure?** – some dams, should they fail, might only frighten a few sheep, whereas others might threaten large numbers of people or major elements of infrastructure. It might therefore be prudent to spend money on the dams which have the highest consequence of failure first.

- **By maximising the cost effectiveness of risk reduction?** – the estimated risk (considering by probability and consequences of failure) reduction and cost for all remedial measures can be estimated and the most cost effective remedial works given the highest priority.

- **Using an evaluation against HSE (2001) Tolerability of Risk Guidelines?** - The Health and Safety Executive (HSE 2001) have published guidelines for assessing in the tolerability of risks to individuals and to groups. A risk is sometimes said to be 'broadly acceptable' if it is lower than one in a million per annum. However, the key to evaluating the tolerability of risks is whether the risks have been reduced to be ALARP, or 'as low as reasonably practicable'. The ALARP Principle is an expression of the undertaker's duty of care under common law. The HSE (2001) refers to the satisfaction of the ALARP Principle as requiring a "gross disproportion" test applied to individual risks and societal concerns, including societal risks. The gross disproportion, which should be sought in deciding how far to pursue risk reduction, is between the cost of an additional risk reduction measure and the estimated risk reduction benefit for that measure. HSE (2002) refers to this disproportion as "the bias on the side of safety", "erring on the side of safety", and "compensating to some extent for imprecision in the comparison of costs and the benefits"

- **By some hybrid of the above?** – A suggested hybrid method is shown in Figure 1.

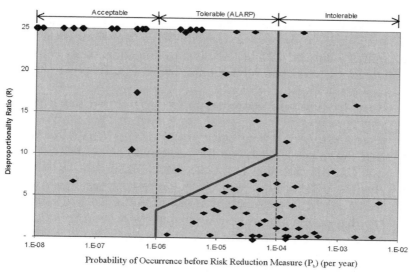

Figure 1. Risk Reduction Measures plotted against. HSE Tolerable Risk Guidelines

This sets limits of tolerability based on HSE guidelines and uses a disproportionality (cost/benefit) ratio and the risk of occurrence before the remedial measure is carried out to prioritise those measures.

- **By doing works recommended "in the interests of safety" at each dam first** – works recommended "in the interests of safety" have to be carried out "as soon as practicable" under the Reservoirs Act 1975 or by the time stipulated by the Inspecting Engineer. The remaining remedial works could then be prioritised by the methods above.

Each approach to prioritisation results in a different rate of risk reduction verses cost relationship. The fastest rate of risk reduction for the investment of funds is achieved by using the cost effectiveness approach, where risk is expressed in average annual terms. However, other factors may be important to consider in establishing a prioritisation. In addition to factors mentioned above, business criticality, or the timing of a capacity expansion construction project, are examples of such factors.

2. **What are the factors that limit the size of the capital programme that can be managed, thus directly influencing the rate of implementation of risk reducing remedial works for a dam owner with a large number of dams?**

- **Limited resources** - Even if the owner had unlimited funds, all works cannot be started at once. Work would be slowed by such things a limited number of site investigation contractors, rigs and engineers, and a limited number of contractors with the relevant experience.
- **Need to maintain water supplies** - Many remedial work projects will require at least a partial drawdown of the reservoir. With the recent history of dry summers many owners would not be prepared to allow work on a number of reservoirs to proceed simultaneously. Equally, if a reservoir is a 'sole source' reservoir, in as much as an area can only be supplied from one reservoir, the owner will wish to wait until the water supply network is reinforced or the risk of losing supply can be minimised in some other way before allowing work to start. It is also often necessary to coordinate works at the reservoir with works at the treatment works.
- **Environmental factors** - Planning approvals, rights of way diversions, the migratory and nesting habits of birds, the presence of toads, newts, badgers, etc., SBA's, SSSI's, Heritage sites, opposition from local inhabitants and landowners and the need and ability to

supply compensation to the river downstream can all affect commencement date and duration of construction works, and thus the priority of the scheme.

3. **Once the prioritised list of works has been agreed upon, how should the Recommendations of the Inspecting Engineer contained in his Report under the Reservoirs Act 1975 be accommodated?**

When the Inspecting Engineer carries out his inspection he is usually unaware of the condition of other reservoirs in the next valley or even in the same valley. Some would say that, traditionally, the Inspecting Engineer has taken the view that his duty is to ensure the safety of the dam that he has been asked to inspect, irrespective of the problems or shortcomings that may exist at other dams in the ownership of the undertaker.

The problem that may arise, following a Portfolio Risk Assessment, is that particular recommendations made "in the interests of safety" by the Inspecting Engineer may not achieve a 'high ranking' and therefore may not be scheduled for a number of years. This may occur because the remedial measure that is responsive to the recommendation "in the interests of safety" may result in only a small reduction in risk relative to its cost, or other words it is not as cost effective as other remedial measures that have been identified for the owner's portfolio of dams. If the undertaker waits too long to act on the Inspecting Engineer's recommendation, this may cause intervention from the Enforcement Authority because the Act states 'as soon as practicable'. The actual 'meaning' of this phrase has not been defined, except that it has been said by some, that money is not a factor to be considered; but it may take a court ruling before it is defined. Certainly, as discussed above, there are other factors that can affect the timing of works from the owner's point of view.

In addition, the recent Water Bill gives powers to the Enforcement Authority to determine what "as soon as practicable" means in certain circumstances. It would seem sensible that owners, and particularly those using a PRA approach, should work with an Inspecting Engineer to determine a time for completion rather than have it imposed on them by the Enforcement Authority.

Thus, the PRA process could produce some conflict or difficulties with the current reporting system unless the Inspecting Engineer 'signs on' to the process. One mechanism to create an understanding is to have a annual briefing of all Inspecting Engineers involved with the owner's portfolio of dams so they understand how the PRA prioritisation lead to the timescales that the owner is proposing for works that are responsive to their recommendations. Inspecting Engineers could then consider this information when setting their timescales or the date of the next inspection. Owing to the way in which the PRA is carried out, it is highly likely that 'recommendations in the interests of safety' will have been identified in advance of the inspection and therefore a risk reduction measure will already have been identified. Any new recommendations by the Inspecting Engineer will themselves have to be assessed and prioritised during an update of the PRA.

Clearly, if there is conflict of any kind, then under the terms of the Reservoirs Act 1975, the owner will be bound to carry out the recommendations "in the interests of safety" irrespective of the fact that the money could be more effectively spent elsewhere to reduce risk to the community based on the PRA.

For example, provision of additional spillway capacity could be recommended "in the interests of safety" for a spillway that will already pass 95% of the design flood without overtopping. The flood that will exceed this capacity could have an annual exceedance probability approaching 1×10^{-6}. At the same dam, the PRA may have identified that seepage failure has a 1×10^{-3} per annum probability of occurring, even though there is no recommendation to improve protection against seepage failure in the Inspecting Engineer's report. This raises several questions, including the following:

- Where should the owner spend his limited available funds?

- Would a referee, as defined by the Reservoirs Act 1975, take account of the PRA if an owner appealed against a timescale set by the Inspecting Engineer?

- If a failure occurred, what might the judgment of the enquiry be if the owner had enlarged the spillway and severe seepage had caused the dam to be washed away?

4. **Is PRA worth the dilemmas that it spawns? Do the advantages outweigh the disadvantages?**

Is the owner's business at greater risk with or without the information provided by the PRA? Should the owner rely solely on the Report and Recommendations of the Inspecting Engineer? Can you hear a barrister in cross examination in court saying would you have carried out work on this dam earlier if you had used a technique called Portfolio Risk Assessment?

Some of the advantages and disadvantages of the PRA methodology are summarised below:

Advantages

1. In the event of a major incident, evidence that the owner had assessed the risks and was carrying out safety measures in a logical sequence may reduce any penalties imposed by the courts when funds had been spent on other dams rather than the dam concerned in the incident.

2. A PRA will allow the risks to the Company and the Community associated with dams to be reduced as quickly as possible.

3. A PRA can provide a persuasive argument to the shareholders and the regulator that increased spending on reservoir safety is justified.

Disadvantages

1. In the event of a major incident, evidence that the owner had assessed the risks at considerable cost, and was carrying out safety measures in a logical sequence, may not be taken into account by the courts, when funds had been spent on other dams rather than the dam concerned in the incident – especially when one considers how the 'expert witness' system works at times.

2. A prioritisation based on a PRA can conflict with recommendations "in the interests of safety" by Inspecting Engineers. Impecunious owners might be put in a position where they have the funds to carry out works that they are obliged to do under the Reservoirs Act 1975 or the high priority items from the PRA but not both.

3. The PRA may reveal unacceptable risks to the owner that they do not have the funds to reduce. Perhaps "ignorance is bliss", but then "ignorance is no excuse" when it comes to the law!

CONCLUSIONS

Despite the issues highlighted in this paper, and the vagaries of the English legal system notwithstanding, the authors conclude that the use of Portfolio Risk Assessment can be strongly recommended as a tool to assist owners to manage reservoir safety in the overall context of their business. The approach follows a logical well thought out process involving evaluation against engineering guidelines and accepted practice, risk analysis, evaluation against tolerable risk guidelines, prioritisation of risk reduction measures, and sometimes prioritisation of investigations to reduce the uncertainties associated with engineering and risk assessments. The process will cause the organisation to think about the relationship of reservoir safety to the business as a whole. Effectively using the information derived from a PRA can result in a corporate reservoir safety management system that is much more effective and efficient, is auditable and more defensible, and is better integrated with other parts of the business, including finance, capital projects, legal and insurance sections.

REFERENCES

Ash, R.A., D.S. Bowles, S. Abbey, and R. Herweynen. 2001. Risk Assessment: A Complex Exercise but a Worthwhile Tool. *ANCOLD Bulletin* 117:97-105, April. Australian National Committee on Large Dams, Australia.

Bowles, D.S. 2001. Advances in the Practice and Use of Portfolio Risk Assessment. *ANCOLD Bulletin* 117:21-32, April. Australian National Committee on Large Dams, Australia.

Bowles, D.S. 2003. ALARP Evaluation: Using Cost Effectiveness and Disproportionality to Justify Risk Reduction. *Proceedings of the Australian Committee on Large Dams Risk Workshop*, Launceston, Tasmania, Australia.

HSE (Health and Safety Executive). 2001. *Reducing Risks, Protecting People: HSE's decision-making process.* Risk Assessment Policy Unit. HSE Books, Her Majesty's Stationery Office, London, England.

HSE (Health and Safety Executive). 2002. *Principles and Guidelines to Assist HSE in its Judgments that Duty-Holders Have Risk as Low as Reasonable Practicable.* http://www. hse.gov.uk/dst/alarp1.htm.

Lewin, J., G. Ballard and D.S. Bowles. 2003. *Spillway Gate Reliability in the Context of Overall Dam Failure Risk.* Presented at the 2003 USSD Annual Lecture, Charleston, SC. April. (Available for download through a link from www.engineering.usu.edu/uwrl/www/ faculty/bowles.html)

Parsons, A.M., D.S. Bowles, L.R. Anderson, and T.F. Glover. 1999. Strengthening a Dam Safety Program through Portfolio Risk Assessment. Invited article *in Hydro-Review Worldwide* 7(4). September.

Hydraulic and operational safety evaluation of some existing Portuguese large dams

EDUARDO R. SILVA, Institute of Water, Lisbon, Portugal.
JOSÉ ROCHA AFONSO, Institute of Water, Lisbon, Portugal.
JOVELINO MATOS ALMEIDA, Institute of Water, Lisbon, Portugal.

SYNOPSIS.   The Portuguese Regulations for Safety of Dams came into force in 1990, after being published as a Decree-law. Since then some rules and guidelines concerning the different stages of the life of dams have also been published, to complete and to help the application of the law.

Following the occurrence of problems during flood events in 1995-96, with the overtopping of a few small dams, and with incidents at some large dams, that could be controlled but, nonetheless, were of great concern, the Portuguese authorities decided to launch a specific program for the safety reassessment of the existing large dams in the country.

For that purpose 11 calls for tenders were made for the study of 38 large dams, concerning all aspects of structural, hydraulic and operational safety, and also including studies of the downstream valleys for dam failure scenarios.

The safety studies were based on all the hydrological data available today, on the original projects and other elements related to dam features, on behaviour records and on site inspections. Some relevant conclusions were reached in these studies. It was also shown that in some cases compliance with current safety regulations had not been met.

The results from the evaluations carried on the hydraulic and operational safety and the actions proposed to lower the risk are presented, where cases of significant hazard at the downstream valley or risk to the structure of the dam are considered to exist. Measures that are currently being undertaken , or will be pursued in near future, are also discussed.

Long-term benefits and performance of dams, Thomas Telford, London, 2004, 163–174

INTRODUCTION

Although dams have an important role in the development of communities they also imply risks, however small, for people and for economical and social activities in the downstream valley that could be affected by the failure of the dam.

The existence of these risks was highlighted in Portugal by the occurrence of problems during flood events in 1995-96, with the overtopping of a few small dams, and with incidents at some large dams, that could be controlled but, nonetheless, were of great concern.

An evaluation of the safety conditions of Portuguese dams is necessary to prevent major accidents due to dam failures or, at least, to mitigate their consequences.

For that matter, due to the lack of human resources within the public administration, a decision for preparing tenders for specialized outsourcing was taken by the Institute of Water (INAG), in 1999. INAG is the Portuguese Authority in dam safety, and has the technical support of National Laboratory of Civil Engineering (LNEC), as defined by dam safety regulations. A total of 11 tenders, concerning 38 large dams, were launched.

As this first group of dam safety studies is concluded, very soon another group of dams will be included in new call for tenders. This plan will continue until all dams of significant or high risk are studied.

PURPOSE OF UNDERTAKEN STUDIES

The major purpose of these studies is to get an accurate revue of the safety conditions of the Portuguese dams regarding the regulation for safety of dams, and to identify the remedial measures that have to be implemented by the owners to improve safety to the new standards.

These studies should include an assessment of the structural and hydraulic-operational safety, a global risk index computation, a possible change of the rules of exploitation of the reservoir, the mapping of the downstream areas affected in case of dam failure, the assessment of downstream risks and the proposal of an alert and warning system.

Also they should propose some immediate measures to be taken to solve simple problems. Major deficiencies or corrective works, including civil works, electric and mechanic equipments, observation systems and alert and warning systems, are identified in the studies but need further analysis and design from the dam owners, to be approved and implemented.

CHOICE OF THE DAMS

The 38 dams that made part of this first group of tenders were chosen after an assessment that included preliminary field inspections. All of these dams had been designed before the new regulations were mandatory, all of them showed several deteriorations and it also was established that for the great majority the associated potential risk was either significant or high.

This group of dams comprised most of the oldest large dams built in the country for irrigation, and also a representative group of more recent dams with that same purpose.

The majority of these dams had shown during the preliminary inspection that there was no observation plan and that no inspections took place regularly, so that the observation activities were very deficient.

Also some of them showed what appeared to be signs of structural problems that needed to be studied to determine and implement the corrective measures.

In some cases the outlet works didn't work and the dams showed lack of maintenance of the equipments and of the structure itself.

The personnel responsible for the exploitation of the dams in some cases had no specific preparation and didn't fully understand how to correctly operate gates and valves.

The chosen dams are mainly situated in the interior northern part of country and in the southern part near the coastline, as can be seen in Figure 1, and their characteristics can be seen in Table 1.

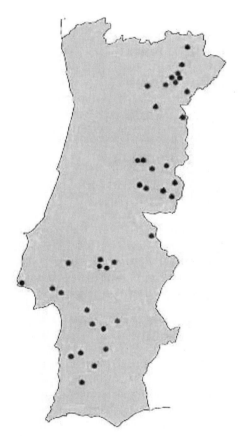

Figure 1. Location of the dams

Table 1: List of studied dams

Dam	End of constru-ction	Age (years)	Type	Use	Height (m)	Reservoir (hm^3)
Alfândega da Fé	1970	33	TE	SI	25	1.60
Alijó	1991	12	TE	S	40	1.74
Alvito	1977	26	TE	SI	49	132.50
Apartadura	1993	10	ER	SI	46	7.50
Azibo	1982	21	TE	SI	56	54.50
Burga	1978	25	TE	I	35	1.80
Camba	1993	10	TE	SI	35	1.10
Campilhas	1954	49	TE	SI	35	21.70
Capinha	1981	22	TE	S	18	0.52
Carviçais	1984	19	TE	S	20	1.20
Cova do Viriato	1982	21	PG	S	24	1.50
Covão do Ferro	1956	47	PG	H	32.5	0.87
Fonte Serne	1977	26	TE	I	18	5.20
Furadouro	1959	44	PG+TE	I	17	0.40
Gameiro	1960	43	PG+TE	IH	20	1.30
Gostei	1993	10	TE	I	35	1.40
Idanha	1949	54	PG	IH	54	77.80
Magos	1938	65	TE	I	17	3.00
Maranhão	1957	46	TE	IH	55	205.00
Marateca	1991	12	TE	SI	23.8	37.20
Meimoa	1985	18	TE	SI	56	40.90
Montargil	1958	45	TE	IH	48	164.00
Monte da Rocha	1972	31	TE	SI	55	104.50
Odivelas	1972	31	MV+TE	I	55	96.00
Pego do Altar	1949	54	ER	IH	63	94.00
Peneireiro	1973	30	TE	S	15	0.80
Penha Garcia	1979	24	PG	SI	25	1.10
Pisco	1968	35	TE	S	24.5	1.40
Ranhados	1986	17	PG	S	41	2.60
Rio da Mula	1969	34	PG+TE	S	17	0.34
Roxo	1967	36	CB+TE	SI	49	96.30
S M Aguiar	1981	22	TE	SI	20	5.40
Salgueiro	1975	28	TE	I	25	1.80
Santa Clara	1968	35	TE	ISH	86	485.00
Toulica	1979	24	TE	IS	16	2.00
Vale das Bicas	1939	64	TE	I	12	2.00
Vale do Gaio	1949	54	TE/ER	IH	51	63.00
Venda Velha	1957	46	TE	I	14	4.64

RESULTS

In Portugal nowadays around 180 large dams according to the ICOLD definition exist, mostly for hydropower and irrigation purposes, many of them more than 30 years old. In the 40's and 50's those that were built for irrigation and water supply were directly promoted by the State, through a specific department that gained a large experience in dam design and construction. More recently, however, we have seen an increasing number of dams constructed as a result of private investment or various public departments and local authorities with a less developed history of dam operation and construction. On the other hand, the operation of irrigation and supply dams has been judged inadequate, with a lack of adequate technical and financial resources identified. As a result a significant number of problems have led to specific direct interventions by dam safety public authorities.

Some cases where safety did not comply with safety regulations

The assessment of the compliance of the hydrologic, hydraulic and operational safety of dam in Portugal has to be made for the return periods imposed by the existing regulations. Those return periods can be seen in Table 2, where it can be seen that "potential risk" has a vital role in defining which one to adopt.

Table 2: Return periods imposed by the RSB

Dam		Potential risk	
Concrete	Embankment	High	Significant
h ≥ 100	H ≥ 50	10,000 to 5,000	5,000 to 1,000
50 ≤ h < 100	15 ≤ h < 50	5,000 to 1,000	1,000
15 ≤ h < 50	h < 15	1,000	1,000
h < 15	-	1,000	500

The potential risk is defined in Portuguese regulations as a measure of the consequences of an accident, not withstanding the probability of its occurrence, and can be sorted by the following levels, according to the loss of human lives and economic damages:
- low, when no lives are in threat and there are few economic damages
- significant, when some human lives can be lost and the economic losses are of some importance
- high, when an important number of lives can be lost an the economic losses can be high

Therefore, to determine which return period to apply to the design flood of spillways one must do the study of the areas affected by the wave resulting of the failure of the dam and determine the number of human lives that could be considered at risk and of the economic losses and infrastructures affected.

In these 38 cases only 2 of them are considered of low potential risk, 7 of significant risk and the rest are considered of high risk. In some cases the results of the studies includes an estimate of the number of lives that could be in danger, the number of homes and a list of other infrastructures that could be affected by the flood wave, such as schools, public services and civil protection structures, roads, railways, bridges and others.

Table 3: Potential risk of the dams

Dam	Scenario	Estimation			Potential risk		
		no. lives	no. homes	Infrastructures	low	significant	high
Alfândega da Fé	overtopping		3	Yes		x	
Alijó	piping	57	22	Yes			x
Azibo	piping	518	192	Yes			x
Burga	overtopping		78	Yes			x
Camba	piping	29	11	Yes			x
Cova do Viriato	sudden breach	175		Yes			x
Covão do Ferro	sudden breach	378	140	Yes			x
Fonte Serne	piping			Yes		x	
Furadouro	sudden breach			Yes	x		
Gameiro	sudden breach			Yes	x		
Gostei	piping	322	119	Yes			x
Maranhão	overtopping			Yes			x
Marateca	piping	216	80	Yes			x
Meimoa	overtopping	1566	580	Yes			x
Peneireiro	overtopping		87	Yes			x
Penha Garcia	sudden breach			Yes			x
Pisco	overtopping			Yes		x	
Ranhados	sudden breach	57	21	Yes			x
Salgueiro	overtopping		58	Yes			x
Toulica	piping	4		Yes		x	
Vale das Bicas	piping	35	13	Yes			x
Vale do Gaio	overtopping			Yes			x
Venda Velha	piping	105	39	Yes			x

Comparing the return periods in Table 4 we can see that generally those determined in the hydrological studies made are greater than the ones adopted in the original studies. The consequence is that the peak flows that result of hydrological studies should be greater than the original ones. But analysing Table 4 we can see that in some cases the peak inflows are lower and in a considerable number of them the peak inflows are almost unchanged. It was seen that it had mainly to do with the new amount of data available today, the new methodologies that are currently used, considered more accurate, and in some cases with mistakes in the original studies that now have been detected and corrected.

In some cases, however, like in Fonte Serne, Magos, Meimoa and Vale do Gaio dams, the peak inflows were over 100% higher.

Table 4: Results of the hydrological studies

Dam	Catchment area (km^2)	Return period		Peak inflow (m^3/s)		Volume (hm^3)	
		initial	new	initial	new	initial	new
Alvito	212,00	100	1000	1300	598		
Capinha	6,30	1000	5000	32.5		0.334	
Cova do Viriato	2,30	100	1000	34	43	0.106	0.299
Fonte Serne	32,00	500	1000	55	125		
Furadouro	3374,00	500	1000	2300	2415		248.0
Gameiro	3255,00	500	1000	2800	2390		240.0
Idanha	362,00		5000	700	1168	43.20	48.40
Magos	105,00		1000	110	279		11.50
Meimoa	61,00	1000	5000	228	505	4.840	14.00
Montargil	1182,00	500	5000	1200	1764	80.00	197.0
Pisco	13,95	?	1000	100.6	105.9	0.362	0.666
Rio da Mula	3,00		1000	22	35	0.060	0.194
Roxo	350,00	1000	1000	740	1232	18.30	35.00
S M Aguiar	128,60		1000		218.6		7.700
Santa Clara	520,00	1000	5000	2000	2482	65.00	100.0
Toulica	26,00	100	1000	80	100	0.614	2.395
Vale do Gaio	509,00		5000	750	1762		
Venda Velha	174,00	100	1000	300	327		18.63

In some other cases the peak inflows remained almost unchanged presenting only variations of about 10%. This happened in 11 dams for which either the design was recent or the studies then showed the cautiousness of the designer.

Looking at the performance of the dams we can see that 30 % of the spillways do not present a discharge evacuation capacity that complies to the new regulation. In these cases construction of a new spillway or modifications of the existing spillway or dam operational constraints are required for re-establishing compliance with current statutory constraints and guidelines.

Toulica dam, for instance, is overtopped in all the studied scenarios, even for 100 years return period flood with 3tc, where tc is the time of concentration of the dam drainage basin. This fact led to a restriction being imposed on the level of the reservoir 2 meters below NPL so it can deal with a 100 year flood with 1tc.

But there are several other cases of insufficient spillway capacity. Montargil, Fonte Serne, Cova do Viriato, Roxo, Meimoa, Magos, Rio da Mula, SM Aguiar, Vale do Gaio, Venda Velha and Pisco dams showed this problem, although the magnitude of it varies significantly.

Table 5: Performance of the spillways

Dam	Spillway type	Gates (y/n)	Spillway capacity (m^3/s)		Discharge evaluation (y/n)	
			Initial	Revised	Satisf-actory	Over-topped
Camba	Surface	no	40	39	yes	no
Carviçais	Surface	no	45	17	yes	no
Cova do Viriato	Surface	no	3.8	18	yes	no
Covão do Ferro	Surface	no	7	20	yes	no
Fonte Serne	Surface	no	36	68	no	no
Magos	Surface	yes	110	195	no	no
Maranhão	Shaft	yes	1600	1987	yes	no
Marateca	Surface	yes	60	77	yes	no
Meimoa	Surface	yes	124	240.5	no	no
Montargil	Shaft	yes	765	1022	no	no
Pisco	Surface	no	43	77.5	no	no
Ranhados	Surface	no	215	140	yes	no
Rio da Mula	Surface	no	7.8	26	no	no
Roxo	Surface	no	64	161	no	no
S M Aguiar	Surface	no	155	200	no	no
Salgueiro	Surface	no	29	20	yes	no
Santa Clara	M glory	no	208	213	no	no
Toulica	Surface	no	17.6		no	yes
Vale das Bicas	Surface	no		107.9	yes	no
Vale do Gaio	Shaft	yes	1000	1200	no	no
Venda Velha	Surface	no	140	236	no	no

In some cases like the Marateca dam and Camba dam the spillways expected performance is near acceptable, with the anticipation of some damages for the revised design floods but without any kind of danger to the structure of the dam.

Results from the evaluation undertaken on hydraulic and operational safety
of dams

One of the main conclusions of these studies relates to some features that
the actual law imposes on hydraulic and operational safety, namely the need
for the gates to be operated locally and from a distance, and to have two
different energy sources available, besides being manoeuvred also manually.

Those requirements apply also to bottom outlets, which are sometimes too
demanding and make it very difficult for existing dams to comply.

The legislation imposes the need for operational manuals at each dam,
which should namely include guides for the reservoir exploitation, as well as
rules related to all the equipment operation and the necessary measures for
maintenance and conservation. The manuals were found to be lacking on the
studied dams. In these studies this lack of information was highlighted,
procedures were drafted and proposed for implementation as soon as
possible.

The operational procedures are of great importance to dam safety because
the operation personnel in most of the studied dams revealed lack of
understanding of the equipment installed and of the right procedures to
operate them, in response to reservoir conditions. This is more dangerous in
cases where spillway gates exist, because it can endanger the dam itself.

The bottom outlets in some cases like Penha Garcia, Cova do Viriato and
Pisco dams were out of order and so, in case of an emergency, it would be
impossible to lower the reservoir. In other cases like Venda Velha, Vale das
Bicas, Toulica or Magos dams the bottom oulets were operating poorly but
allowing some kind of control of the reservoir.

The amount of financial resources needed in some cases makes it difficult
for owners to comply with the legislation.

Actions proposed to lower the risk

As indicated by the studies, measures to lower the risk in some dams led to
restrictions imposed on the operation of the reservoirs such as lowering
normal storage levels, and alternative design of solutions for spillways and
other elements.

This was the case of Meimoa, Fonte Serne, SM Aguiar and Toulica dams,
where reservoir levels were conditioned to prevent damages. Those levels
were determined in each case after discharge evaluations were made for the
chosen design floods and the consequences were assessed.

In other cases, when there were serious problems of reliability and performance of the spillways, it was decided that it was necessary to improve their discharge capacity by modifying the existing one or by designing an auxiliary spillway. This was the case of Capinha, Montargil and Roxo dams.

Some cases of rehabilitations underway

Due to heavy rainfall in December 2000 the spillway of Pisco dam suffered huge damages that threatened the dam itself and motivated emergency intervention. After some remedial work was performed in the spillway, so as to make it endure a small flood, a designed for a new spillway and bottom outlet was made by the consultants involved in the safety studies. Afterwards works were awarded to a contractor and construction now is completed.
Fig. 2 depicts an intermediate phase of the works, with the new spillway completed alongside the old one. Afterwards, a new intake tower and intake and bottom discharge tunnel were constructed at the old spillway section, and the earth fill was remade.

Figure 2. New spillway of the Pisco dam alongside the old one.

Cova do Viriato dam is being subjected to several interventions destined to install a new bottom outlet and a new stilling basin, due to accommodate the increased spillway discharge capacity, and to benefit the water intakes and other supply equipments. Also the gates and valves will be operated locally and from a distance, and they will have two different energy sources for operation, besides manual operation.

Due to the insufficient spillway capacity of Fonte Serne dam a new spillway design for the studied flood was prepared. It will be implemented as soon the owner can call a tender.

Other designs were made to improve safety conditions in hydraulic structures or equipments that will be implemented by the owners as soon as resources are available for that purpose.

<u>Measures initiated and to be continued in the near future</u>
The immediate actions necessary to increase safety resulting from these studies are recommended for implementation as soon as possible to prevent, accidents and avoid endangering lives.

Once the studies have been completed, meetings will take place with owners to discuss all the new available information and to decide on measures to be undertaken.

Most of the concerned owners for the studied dams are irrigation associations, which have some difficulty in obtaining funds to perform necessary interventions, because the amount of money needed to fully and immediately comply with regulations is beyond their current available resource. For this reason interventions have to be sorted in order of risk and programmed in a structured manner.

CONCLUSIONS
The results of the safety studies made for this first group of dams showed that, in spite of the amount of work that needs to be done so that the dams comply with existing safety regulations, the global picture is nonetheless of moderate concern. It is however essential that corrective measures in some structures and hydraulic equipments are undertaken.

It is necessary that dam owners comply with their legal responsibilities, being the Authority's role to guarantee that they do it. For dam owners, and Engineers who are responsible for supervising dam operation and safety, it is fundamental to acquire the knowledge of the problems and implications related to their dams. They require to have the resources in place to implement safety and operation procedures to ensure that the necessary interventions and tasks can be carried out in a phased approach.

To protect lives and to prevent economic losses in the valleys downstream it is also necessary to develop and implement Emergency Plans. These plans are divided, in Portuguese regulations, into the "Internal Emergency Plans", that concern the dams operation and the downstream nearby areas, where the owners may be responsible for the first actions and warnings, and the "External Emergency Plans", directed by the Civil Protection Departments.

New studies will be launched, aiming at improving the Portuguese large dams' safety, especially for those private and public owned dams in which owners don't have the demanded expertise.

To implement studies recommendations, an increase both in the Authority's organization and in dam owners' safety efforts will be needed. This will also imply an increase in new investments by all entities involved.

REFERENCES
CSOPT (1990). *Portuguese Regulations for Safety of Dams*. Decree-Law n° 11/90 of January 6.
CSOPT (1993). Code of Practice on Dam Design. Order n° 846/93, Lisbon, September.
CSOPT (1993). *Code of Practice on Observation and Inspection of Dams*. Order n° 847/93, Lisbon, September.
CSOPT (1993). *Code of Practice on Construction of Dams*. Order n° 246/98, Lisbon, April.
Hidrorumo, Hidrotécnica. *Safety studies of Alvito, Odivelas and Fonte Serne dams*.
COBA, Hidroprojecto, GIBB. *Safety studies of Monte da Rocha, Santa Clara and Roxo dams*.
WS Atkins, GAPRES. *Safety studies of Vale do Gaio, Pego do Altar and Campilhas dams*.
Hidrorumo, Aqualogus. *Safety studies of Idanha, Toulica, Capinha and Cova do Viriato dams*.
FBO. *Safety studies of Gostei, Azibo, Camba and Alijó dams*.
FBO. *Safety studies of Ranhados, Marateca, Meimoa and Apartadura dams*.
FBO. *Safety studies of Covão do Ferro, Venda Velha and Vale das Bicas dams*.
COBA, Hidroprojecto. *Safety studies of Maranhão, Furadouro and Gameiro dams*.
Aqualogus, Tetraplano. *Safety studies of Montargil, Magos and Mula dams*.
COBA, Hidroprojecto, GIBB. *Safety studies of Alfândega da Fé, Burga, Salgueiro and Peneireiro dams*.
Hidroquatro, CENOR. *Safety studies of Carviçais, Santa Maria de Aguiar, Penha Garcia and Pisco dams*.

Reliability principles for spillway gates and bottom outlets

G.M. BALLARD, Consultant, UK
J. LEWIN, Consultant, UK

SYNOPSIS.   Reliability analysis of spillway gate installations, and to a lesser extent bottom outlets of reservoirs, has been increasingly used in risk assessments of dams. As a result there is now considerable collected experience of the design and operation of different types of components and systems, both qualitative and quantitative. The qualitative experience has led to general acceptance of some fundamental principles of design and operation in order to achieve good reliability. The paper discusses some of the more important principles, using examples from spillway gates which have been assessed for reliability by the authors. A common approach to attaining reliability is the provision of redundant equipment, yet the occurrence of common cause failures (CCF) – and the need to provide adequate defences against them – is less frequently considered. Attention is drawn to the types of events leading to CCFs and to some potentially effective design defences.

DESIGN
For a system that is required to have a high reliability, the design features of the system can have as much effect on the achieved reliability as the specific reliability of the individual components that comprise the system. This section briefly discusses some of the more important aspects of system design, using examples from existing spillway gate designs as illustrations.

Well Proven Equipment
Where a system is intended to perform an important safety function it is not generally appropriate to use newly developed types of equipment or technologies. The failure experience of newly developed components is limited and the failure modes of the equipment are likely to be imperfectly understood. If the equipment has not previously been used in similar applications or environments then there may be unpredicted problems which cause the component to fail in an unexpected manner. This may lead to further failures as a result of unpredicted interactions between components. Also, components based on new technology suffer from the absence of improvements which accrue as that technology matures and benefits from manufacturing and operating experience.

These factors can have a major impact where an individual component is used many times in an installation.

Long-term benefits and performance of dams, Thomas Telford, London, 2004, 175–186

When updating or replacing equipment on an existing spillway gate installation, particular areas of concern include bearings and bearing materials, PLC control equipment, and lubricants.

Single Failure Criterion

A safety critical system should be designed so that, if possible, the failure of any single component will not prevent the system performing its function when required. This principle is based on the relatively high probability of a single failure occurring compared to the significantly lower probability of two or more concurrent component failures.

While this may be relatively easy to achieve with electronic, electrical and, to some extent, mechanical systems, it is more difficult or impossible to achieve with structural and civil aspects. This difference is mitigated by the respective failure characteristics of the different system types. Electronic and electrical equipment is prone to sudden failure which cannot easily be prevented by condition monitoring or preventive maintenance. Structural and some mechanical systems may be expected to exhibit failure modes which involve progressive degradation mechanisms that, in principle, should be amenable to prevention by monitoring and preventive maintenance. Therefore the single failure criterion is less critical for structural and some mechanical systems than for electronic and electrical systems.

When a component does comprise a single failure point for a system then special care has to be applied to the design, quality assurance and performance monitoring of that component. The principle of using well proven equipment becomes even more important. Equally, the ability to monitor the component to ensure continuing satisfactory performance is essential. In addition consideration should be given to the existence of any sudden failure modes that may arise for that component and how these failure modes can be mitigated by good design or operating practice.

For an existing spillway gate installation of typical design, the situation assessed against this principle may resemble the following:
- The electrical power system is partly duplicated but there are a few single failure points
- The gate control system has a number of components that are single failure points, e.g. control transformer, rectifier, limit switches, etc.
- The single brake is an example of an electro-mechanical component that mostly has degradation type failure modes but may also have sudden failure modes due to loss of electrical power
- The drive train is almost exclusively a series system, with any single component failure leading to failure of the whole system

- The gate itself is a structural system with no redundancy, as are the spillway piers and other civil structures

Judged against this principle, the design clearly has serious deficiencies.

"Fail Safe" Design

The failure of any component within the system should, if possible, move the system towards a "safe" state. For many protection systems there is a "safe" state which is acceptable and component failures should cause the system to move towards that condition.

For spillway gates the situation is significantly more complicated. The purpose of the gate is both to retain the reservoir water level and to pass the water depending on the situation that arises. Neither state – "gates open" or "gates closed" – can be considered "safe". The gate control system has some features that are used to protect the gates from damage but these may inhibit opening of the gates if they fail to operate as intended. Again there is no unambiguous "safe" state, although in a flooding emergency the requirement to open the gates may be more important than safeguarding them from damage.

A specific example involves the limit switches that control gate travel. Overtravel limits are provided to prevent equipment damage. However if one of the limit switches fail in a specific mode, open or closed depending on the logic of the control circuit, then the gate cannot be moved unless the interlock can be overridden. The other failure mode of the switch is "safe" for gate operation but may lead to equipment damage. Two alternative design strategies might be appropriate in this situation. The first would be to provide a redundant arrangement of limit switches such that no single failure would lead to either potentially "unsafe" state. The second (less preferred) would be to provide duplicated switches to prevent equipment damage, but offer an override facility which could be used if the gates need to be opened in an emergency.

Redundancy and Diversity

The main protection for any system against failure of individual components is the use of redundancy and/or diversity. Frequently this takes the form of providing two or more identical parallel lines, each of which can perform the required function on its own. Thus the electrical system on a typical spillway gate installation may have two parallel electrical feeders from the main 440V switchboard all the way through to the gate breakers. Either circuit will provide power to the hoist motors should the other fail. All that is required is a manual changeover of supply breaker on the gate control panel. An automatic changeover system could be implemented by use of appropriate sensors if required. Further examples are the use of an alternative drive motor (not frequent in practice) should the main drive motor for a gate fail, and the provision of a standby diesel generator to maintain electrical power on failure of the commercial grid supply.

The basic effectiveness of redundancy in improving reliability performance arises because of the failure logic of such systems. If the probability of failure of either one of two duplicate channels is p, then the probability of concurrent failure of both channels is p^2, e.g. if $p=10^{-2}$ per demand for one channel then the system failure probability is $p^2=10^{-2}\times10^{-2}=10^{-4}$ per demand.

In redundant circuits the mode of operation may follow a variety of patterns depending on the exact system type and operation. Where only a single channel can operate at any one time there needs to be provision for an automatic or manual changeover to a standby channel in the event of failure of the first channel. For monitoring or control systems all channels can operate simultaneously and a voting logic can be used to determine how the various channel outputs will be used to define the system output. For example, the parallel gate limit switches are both fully operational at all times and the voting logic is that either limit switch tripping trips the hoist system. For more complex systems involving three or more parallel channels then 2 out of 3 voting arrangements can be used to reduce the occurrence of spurious control/alarm action due to component faults while maintaining a high reliability.

Identical parallel channels can be susceptible to common cause failures, so the use of diverse parallel channels should be considered. In this arrangement, both channels provide a route for the system to function but they use different equipment and/or operating methods to achieve the end result. A simple example would be the use of, say, a vane type limit switch in a parallel channel while a lever arm switch is used on the primary channel. The use of diverse equipment in redundant channels makes it less likely that multiple failures of equipment, affecting both redundant channels, will occur concurrently.

Common Cause Failure (CCF)
The use of redundancy to improve reliability relies on the fact that failure of the individual redundant channels is independent. That is, if one channel fails then the probability of failure of the other channel remains at p, the value it was before the first channel failure. This is not an unreasonable assumption and satisfactorily represents many failure events. However the assumption breaks down when the same cause, a common cause, leads to failure of multiple parallel channels.

To illustrate the effect, suppose that an individual channel has a probability of failure on demand of $p=10^{-2}$ and that p divides into two components, p_I the proportion of random failure modes and p_C the proportion of common cause failure modes. Then the failure probability for a two channel redundant system is not p^2 but $p_I^2+p_C$. If p_C is only of order 5% then the reliability of the parallel system is not 10^{-4}, assuming independent failures, but $0.95 \times 10^{-2} \times 0.95 \times 10^{-2}+0.05 \times 10^{-2}=5.9 \times 10^{-4}$, that is, worse by a factor of ~6. Even if p_C is only 1% then the system failure probability is still worse by a factor of ~2 compared to the fully independent case.

Analysis of many CCF events in the past has suggested that a reasonable working estimate for p_C for a well designed redundant system is approximately 10%, and that specific CCF defence measures will be required if this proportion is to be reduced to any significant extent.

Consideration of the mechanisms that lead to common cause failure (CCF) events indicates that the two most common problems are design errors that have led to unintended interactions between channels or create common weaknesses, and operational errors – particularly in maintenance – that have instigated multiple failures. Other causes, perhaps more widely recognised, are common adverse environmental conditions and external hazards such as fire, lightning or explosion.

A typical spillway gate design is susceptible to a range of common cause failure events including environmental and external hazards, maintenance errors and design interactions. Defences against the causes of CCF events that should be considered when designing and operating systems include:

(1) Design
- Review all stages of the design with the specific target of identifying potential CCF interactions and eliminating or protecting against them
- Equipment or functional diversity such that different equipment or operating principles are used in the redundant channels
- Fail-safe design to ensure that there are no failure modes which can lead to a dangerous CCF
- Well proven equipment so that the failure modes of equipment are well understood
- Protection and segregation of redundant channels to reduce the potential for environmental or external hazards affecting multiple channels
- Derating and simplicity to ensure that equipment is not operating at the limits of its design specification and that the performance of the overall system is capable of comprehensive analysis

(2) Operations
- Comprehensive commissioning trials in order to fully verify equipment performance; comprehensive monitoring, recording and analysis of operating experience
- Ergonomic interfaces to reduce the potential for both simple operational errors and misunderstanding as to the state of the system
- Well thought out and presented procedures for all important activities
- Thorough training and regular practice in realistic exercises

(3) Maintenance and Testing
- Equipment designed to facilitate full testing of all functions without undue interference with the state of the equipment
- Well assessed and presented procedures that can act as a checklist for all relevant important actions
- Staggered maintenance of parallel channels so that redundant equipment is not maintained at the same time

Most of these features are common to the specification for the design of any reliable system, but the potential for CCFs may require special consideration. Examples from typical spillway gate installations illustrate the issues involved:

(1) Environmental CCF
Most of the equipment on a spillway was protected from the weather by sealed enclosures; electrical cables ran to and from these enclosures in steel conduits. If the seals on the enclosure are poorly designed or deteriorate with age then moisture can enter the enclosures and the cabling conduits. There was significant evidence of cable failure due to conduit corrosion and cable degradation as a result of moisture ingress. While concurrent failure of the parallel cabling on the power feeders was not thought likely, at least two factors were of concern. Firstly, the gates were typically all connected to one power feeder and the other feeder was tested infrequently, so one of the feeders could be in a failed state for a significant period of time. Secondly, the gate tests typically involved moving gates under a normal motor load, whereas in an emergency the motor currents could be significantly higher.

The defences in this case could include the following:
- Improved design of water seal; regular preventive maintenance of seals
- Gland seals on all cable entry and exits to reduce the ingress of moisture to the conduits
- Segregation of the control cabinet power feeders so that the failure of one water seal would not affect both power feeders

- Regular and staggered testing of both power feeders both electrically and operationally so that the operating state of both feeders was regularly confirmed
- Occasional testing of the motors with a dummy load that more closely represented the worst conditions of emergency use

(2) External Hazard CCF

Duplicated power feeders run in steel conduit from the 440V switchboard to the spillway gate control cabinets. The conduits run close together over extended distances, crossing structural expansion joints and metal walkways. The conduits did not appear to have any heat protection or slack when crossing structural joints, earthing of the conduits and equipment was often not to modern standards and no lightning protection was installed. If any one of the feeders was damaged due to mechanical interference, fire, seismic shearing, lightning, etc., it is probable that the other feeder would be damaged at the same time.

The defences in this case could include the following:
- Spatial segregation of the cable runs so that the two power feeders would be unlikely to suffer from the same physical event
- Improved protection of the conduits from external events
- Provision of a diverse means of operating the gates, e.g. a portable diesel driven engine that could be connected to the gate drive train

(3) Design CCF

On some spillway installations the motors drive the hoist gear train via worm reduction gearboxes. Some of these boxes, which operate at quite high speed, are small and get very hot during operation. They have breather vents, which are simply holes in the top of the boxes, and water ingress has been a recurrent problem. The water both degrades the lubrication of the gears and has led to significant problems with the shaft oil seals and bearings. Both the main drive motor for any gate and the alternative drive motor operate through identical types of worm reduction gearbox and a systematic problem with this type of box could lead to failure of both alternative drive trains. On one project 4 out of 14 gates had been tagged out for emergency use only because of degraded worm reduction boxes.

The defences in this case could include the following:
- Derating of the worm gearboxes to ensure that they operate well within their design capacity and are thus more tolerant of poor conditions
- Improved attention to environmental protection by fitting breathers with desiccant filters to reduce water ingress
- Prompt action on observed degradation so that the concurrent existence of degraded equipment can be minimised

- An alternative design of redundant motor arrangement that does not share common types of equipment

(4) Design CCF

The design of the spillway gates on some projects incorporated a gate bottom flange which would make the gate prone to severe vibration under certain opening conditions. The operators were not aware of the potential gate vibration problem and were unsure how to react to the occurrence of severe vibration. On one project where vibration had occurred it was attributed to water hammer and not thought to be significant. Continuing severe vibration could lead to failure of the gate hoisting cables or anchorage points, and possibly structural failure of the gates. The condition could affect all the gates if they had to be opened during a major emergency.

The defences include:
- Use of well proven equipment which has a recorded experience in the relevant application and environment
- Design review at project inception to identify potential weaknesses in design or operation of the equipment
- Monitoring, recording and analysis of operating experience to identify potential problems, followed by effective action to remedy them

(5) Operational CCF

In an emergency, spillway gates must be opened to prevent the dam being overtopped. Generally operational staff will receive instructions about the extent and timing of gate opening. However if communication is lost staff will be expected to open the gates themselves using a set of emergency procedures. Interviews with staff at some projects revealed that they had little understanding of these procedures, had in most cases never used them in any training or emergency exercise, and had a number of misconceptions about the correct operation of the gates. If communications were lost in a real emergency, a significant delay in opening the gates could prove critical. The performance of operational staff could affect all gates at the installation and could have breached any redundancy provisions in the design.

The defences that may be relevant to this situation include:
- Provision of clear, well presented emergency procedures and a requirement that these be practised on a regular basis
- Performance of regular emergency exercises simulating a range of emergency scenarios to which project staff must respond appropriately
- Training and certification of operating staff at all projects; regular re-certification requiring demonstration of adequate knowledge and experience

Revealed Faults
The design intent should be for any component failure to become apparent to the operators as soon as possible after it occurs. The objective is to minimise the time for which a system remains in a failed state without any repair action being initiated. For normally operating systems this requirement may be straightforward, but for protective systems operating in a standby mode it requires more consideration. The most common technique is to employ monitoring and alarm systems such that appropriate sensors will detect anomalous conditions and alert the operator.

For spillway gates much of the equipment is deactivated between tests and is therefore not amenable to monitoring. However the electrical supply systems can be monitored and alarmed, particularly where the supply to the gates is separate from the supply to the dam offices and the staff may be unaware of a power trip.

Despite the difficulty of continuous monitoring there is value in considering a monitoring system which is activated when power is applied to the gates for a test. Not all features of the gates may be exercised during testing and a monitoring system could alert the operator to potentially degraded conditions such as low oil levels, high gearbox temperatures, or high earth leakage currents which could be indicators of incipient failures. The electrical continuity of all the circuits could be checked, as could some aspects of the integrity of equipment such as limit switches, protection devices and controls.

Testing
Standby protective systems such as spillway gates may be idle for extended periods. In the absence of fault monitoring systems, component degradation and failure only becomes apparent at the time of an actual demand. Assuming that component failures occur randomly over time, the probability of the system being in a failed state increases approximately linearly with elapsed time since the last demand.

Regular testing ensures that the operability of the system is checked on a much shorter timescale and that system repair can be carried out before an actual demand on the system. An effective test programme must provide for testing all aspects of the system at appropriate intervals. The test interval should reflect the likelihood of potential failure modes, as represented by the failure rate for that part of the system. Care should be taken to ensure that the test programme examines aspects of the system that may have unrevealed failures, where components are not used on a routine basis but comprise a back-up or protection function for use only in specific situations.

With reference to a typical spillway gate installation:

- The test programme should include standby provisions such as an alternative motor arrangement
- If bypass features exist to protect against failure of, e.g., limit switches, then these should be tested regularly; similarly the correct functioning of items such as reset buttons on current overload trip should be verified
- Alternative power supplies such as diesel generators or trailer mounted emergency power supplies should be tested by operating a number of gates; where relevant, it is particularly important to test the interface arrangements for coupling the generator into the power supply circuits

OPERATION

Ergonomic Design
While spillway gate equipment is designed to operate effectively and reliably, it must also be designed to be operated easily. On installations which are manually operated, with no automatic control, the equipment and especially the control systems should reflect good ergonomic practice.

Major design elements for control systems include:

- Controls should be systematically laid out and clearly and unambiguously labelled; controls arranged on a mimic diagram of the system are often effective
- The controls should show clearly the state of the system using lights or other displays as appropriate
- If the system has interlocks, inhibits, protection etc. which can disable the system operation, the state of these should be clearly shown
- If a piece of equipment is in a failed state then this fact should be made clear to the operator by appropriate sensors/alarms/displays
- Any overrides or bypasses intended for irregular use should be protected from accidental use by appropriate means such as key operated switches
- The actual state of the equipment, rather than the state of its control element, should be shown wherever possible (a motor running light should be based on measured rpm, current drawn etc. rather than inferred just from voltage to the motor terminals)
- The operation of the controls must reflect the physical limitations of operating staff; e.g. displays should be visible and easily readable when the relevant controls are being operated, controls should be easily and comfortably accessible and well illuminated where appropriate, manual operations should be within normal manual strength limits

These features are required in order that staff can operate gates reliably, often under stressful conditions when it is easy to make slips and mistakes.

On some existing spillway gate installations the following issues arise:

- The gate controls are generally simple and the layout is therefore straightforward, but on some older plant the labels on control buttons can be illegible, causing a major problem for inexperienced staff
- The control panels may have no indication of the current state of the hoisting system; there may be no indication of electrical power to and from the breakers, no indication of power to the motor or the brake, and no indication of the position of any of the limit switches
- Gate hoist mechanisms incorporate protection systems related to the gate and the electrical equipment, but control panels often provide no information on the status of these interlocks or protective devices (if a gate, when last closed, tripped out on the overtravel limit switch it would have to be backed out using an override button until the overtravel limit switch has cleared, but the operator would have no indication of this)
- There may be no condition monitoring in the form of alarms or sensors, so the operator may have no indication of equipment failure other than lack of response from the system

Operating Procedures
All significant operating tasks, especially those performed infrequently or under stressful conditions, should have clear, well-written procedures to guide the operating staff. The procedure should be simple and straightforward, containing only essential text and diagrams.

The procedure should:
- Explain simply under what circumstances it is to be used and how the operator can determine the relevant circumstances, e.g. what readings to take, how to find them, who to communicate with, etc.
- Explain simply what it aims to achieve, e.g. why the procedure is being performed, how its success or failure can be measured, what data the operator can use to assess the procedure, etc.
- Lay out in flow sheets the sequence of actions required. At each stage the state of the equipment should be specified, with instrument readings if appropriate. References should indicate where ancillary information can be found, addressing issues such as what may go wrong during the action, how it can be identified and how to recover the situation. If there are several separate objectives these should be clearly distinguished.
- Where diagrams or graphs are required the procedure should state simply and clearly how they are to be used, what data is required as input and where it is available, what value should be read from the graph and how it should be used
- Where communications are required the procedure should identify who is the contact, how to reach him/her, what information will need to be given, and what information/instructions need to be received

- Instructions should be in large type, visible in poor light, encapsulated for use outdoors in inclement weather conditions; a copy should be kept in the action location in addition to a clearly identified central location

While much of the above may seem obvious, the authors have visited many projects where operating procedures failed to conform to these guidelines.

Training
All staff who are expected to operate the gates during an emergency should be trained and should regularly practice gate operation. They should be certified as competent to operate the gates after initial training and re-certified on a 3–5 year basis to ensure that they maintain their competence. Re-certification should be conditional on demonstrating a good level of practical experience in routine gate operations and participation in a reasonable number of emergency exercises.

Emergency exercises could vary in scope from simply practising the use of various standby facilities such as the alternative motor drive or the diesel generator, to a larger scale exercise in which a full scenario is simulated and staff have to act in real time. A full scale emergency exercise should be undertaken at least once every three years, and should involve practising both communications with the administrative control centre and the independent action that could be necessary if such communication is lost.

CONCLUSIONS
The benefits of reliability assessment are both qualitative and quantitative. There are clear principles of design and operation which will lead to improved reliability in practice. As a broad generalisation for systems intended to provide some type of standby function, where the appropriate reliability measure is probability of failure on demand, a well designed and operated system should be able to achieve a reliability of $\sim 10^{-3}$, a high integrity system intended for a safety critical function should aim to achieve a standard of $\sim 10^{-4}$, and only an exceptionally carefully engineered, designed and operated system is likely to achieve a reliability of $\sim 10^{-5}$.

Spillway gate installations are safety critical structures. A number of gate systems assessed by the authors have not achieved a reliability standard of 10^{-3}. Sometimes they have been an order of magnitude or more worse. This might be expected from installations that were designed and constructed 30–50 years ago, but the same trend has been found in gates commissioned in the last 15 years. While certain design and operation principles may appear self-evident, many of the installations visited by the authors have fallen far short of the recommendations laid out in this paper.

FMECA of the Ajaure Dam - A Methodology Study

M. BARTSCH, SwedPower AB, Stockholm, Sweden

SYNOPSIS. In 1998 Vattenfall decided to introduce the use of risk analysis in dam safety in Sweden, by issuing two pilot studies on the Seitevare and Ajaure embankment dams. The objective of these studies was to demonstrate methods to be applied for risk analysis on dams. SwedPower performed the Ajaure study in collaboration with BC Hydro International. Incorporated into this assignment were also a number of technical investigations in order to improve the knowledge base of the dam.

In 2000-2001 a second study was performed focusing on development of the application of FMEA/FMECA and other available methods and on staff training, while still relying on the information gathered during the 1998-1999 study. This "Methodology" study is summarised in this document.

The initial step of the FMECA of the Ajaure Facility was to set up a system model and break it down into subsystems and components by the use of block diagrams. The component failure modes their root causes and effects were analysed and documented using fault trees and pathway diagrams. The FMEA was extended to an FMECA for a few components to demonstrate the proposed technique for criticality analysis. The analysis was summarised in FMECA tables complemented by more extensive component data sheets.

The study concludes that the FMECA framework provides a suitable framework for working with dam safety issues at dams. Other methods, such as, functional modelling, pathway diagrams, event and fault tree analysis should be integrated as considered necessary with regard to the characteristics of the sub-system at hand. In fact, coupling of various methods can be looked upon as a promising direction for further development in the area.

It is envisaged that studies of this type will be performed for a limited number of dams in the Vattenfall portfolio.

Long-term benefits and performance of dams, Thomas Telford, London, 2004, 187–196

BACKGROUND

General

Ajaure is a high consequence dam according to the Swedish dam safety guidelines. The 50 m high rock-fill dam is situated in the upper part of Ume River and was constructed 1964 to 1967. The dam has exhibited a number of unanticipated performance characteristics since construction, which include progressive horizontal downstream deformation, and overtopping of the spillway walls. Also, with regard to a revision of the Inflow Design Flood it has been concluded that Ajaure at present has insufficient spillway or surcharge capacity. Therefore the decision has been taken to raise the dam, which also would be beneficial for dams downstream. (The design of the raising of the dam has been performed in parallel to the risk analysis and is presented in the adjacent paper by A Nilsson and I Ekstrom, SwedPower.)

The dam owner Vattenfall (the former State Power Board) decided to consider the issues within a risk management framework and sought the assistance of BC Hydro International (BCHIL). BCHIL assisted Vattenfall and SwedPower (consulting engineers within the Vattenfall Group) by providing guidance on the application of a version of its evolving failure modes and effects analysis (FMEA) process to the Ajaure Dam risk management issue.

A preliminary FMECA was performed in1998-1999. The study included some technical investigations in order to improve the knowledge base of the dam. The present study performed in 2000-2001 relies on the information gathered during the first study. Focus has instead been on methodology and staff training issues. BCHIL was again sub-contracted by SwedPower to provide assistance to accomplish this Methodology Study.

The Ajaure assignment was one part of a two-part initiative by Vattenfall to introduce the use of risk analysis in the dam safety discipline in Sweden.

Problem Characterisation and Method of Problem Analysis

The task has been to characterise and evaluate the risk posed by Ajaure Dam with the view to develop a safety management system, which demonstrates that the risks are being effectively controlled. The risk characterisation process should permit the identification of the relative contribution of different hazards and deficiencies to the overall risk. The process should also permit the assessment of the changes in risk profile associated with modifications to the dam and/or risk reduction alternatives.
Essentially this project involves:
- a methodical approach to hazard and risk identification and their characterisation;

- modelling the ways in which hazards may be realised with resulting harm;
- identification of how the hazard sequence might be arrested or the effects mitigated.

In principle, the risk can be characterised in a relative sense in terms of a criticality index comprised of three indices that reflect the potential for a failure mode to initiate, the sequence progressing to failure; and the consequences of failure. Also, and again in principle, uncertainty can be characterised by assigning ranges to the indices instead of individual values as appropriate.

Objectives

An important objective in Vattenfall authorising a second risk based analysis for Ajaure Facility is further training of its engineering staff (SwedPower) in emerging methodologies. Another focus of the FMECA analysis of Ajaure Dam is characterisation and evaluation of the risks that have been identified in the Ajaure SEED by monitoring and surveillance and through operating experience.

FMEA, general

Failure Modes and Effects Analysis (FMEA) is a method of analysis whereby the effects or consequences of individual component failure modes are systematically identified and analysed. While the actual analysis is inductive, i.e. is based on the question *"What happens if a component or element fails?"*, it is first necessary to 'break the dam system down' into its individual components or elements. Once the system has been de-aggregated the failure modes of each of the fundamental elements can be identified.

Once the failure modes and/or root causes have been identified, the effects of the failure mode on other components of the subsystem and on the system as a whole are systematically identified. The analysis is usually descriptive and information is normally presented in tabular and/or spreadsheet form. FMEA clearly relates component failure modes and their causative factors to the effects on the system and presents them in an easily readable format. A thorough understanding of the system under analysis is essential prior to undertaking an FMEA.

APPLICATION OF FMEA AND FMECA ON AJAURE FACILITY

General

The application on the Ajaure facility comprised the parts where the principles of FMEA were the primary focus of the study.

- The FMEA, in which the facility was broken down and analysed in a structured manner
- The FMECA, in which criticality ratings were assessed for a few components, and
- Derivation of global failure modes related to the Water Retaining Structures, the Discharge Facilities and the Spillway Gate Control.

FMEA process

General
In this application the FMEA process can be said to consist of three basic parts:
- System and subsystem breakdown
- Component details
- Failure modes and effects

The FMEA process was extended to FMECA by adding a fourth part:
- Criticality ratings and criticality index

The analysis has been documented on FMECA worksheets, where each of the four basic parts listed above is found as column headings. Under the heading "System and subsystem details" the functional subsystems were broken down into their physical parts, i.e. from subsystems stepwise down to components.

Under the heading "Component details" the design function(s), the design and performance parameters and the performance details have been listed. For each component the <u>design parameters</u> that characterise its performance have been identified. Input has been collected from designers and design data. The next step has been to collect information on the <u>performance details</u> with the intent to map out the design and construction adequacy. Important input has been gathered from the SEED report, performance records and complementary investigations.

In the third section with the heading "Failure modes and effects" the first step was to list the functional failure modes for each component. Here the failure mode of a design function is identical to the loss of the design function. Fault trees have been used to document the relation between root causes and the failure mode.

For each primary failure mode the potential failure sequences, i.e. the pathways to dam breach, have been explored. Also the possibilities to stop the failure sequence from progressing all the way to dam breach, the ultimate effect, have been documented.

As described above the FMEA findings are documented in FMEA tables in a worksheet format. More exhaustive information on each component is compiled in a "component data sheet". On the component data sheet each of the headings are identical to those of the FMEA table are listed. To illustrate the failure sequences graphical pathways showing the chain of events from component failure mode to dam failure are included. Fault trees have been used to illustrate the interrelationships of root causes to component failure modes.

FMEA application to the Ajaure dam

In a system context the Ajaure facility belongs to the "Super System" of Ume River. Upstream of Ajaure the systems of Överuman Regulation Dam and Klippen Power station are situated. Downstream Ajaure there are 14 hydropower facilities, of which the Gardiken Facility is situated immediately downstream. This is illustrated in Figure 1 below.

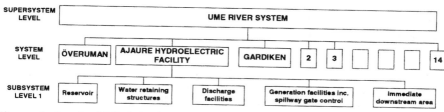

Figure 1. Logical Model of Supersystem, System and Subsystem Level 1

The function of the Ajaure facility (global system) was defined as to "retain water in the reservoir with control of the outflows". It's ability to generate power has been omitted from the study. The motivation is that the focus has been purely on dam safety. As an effect of this the study of the Subsystem Generation facilities has been limited to the Spillway Gate Control.

The "global system" failure mode to be analysed has been defined as "dam breach and release of reservoir water". Component failure modes that cannot initiate a sequence of events that may lead to dam breach have not been covered in this analysis.

In the FMEA the Ajaure facility was broken down into five principal subsystems; Water Retaining Structures, Discharge Facilities, Spillway Gate Control, Reservoir and Immediate Downstream Area. The focus of the analysis was on the first three subsystems. Since they have great differences regarding their structure (continuous versus discrete components, man made versus geological formations, etc) and functioning (continuous loading

versus work on demand, etc) slight differences in the methodology have
been used for the three subsystems.

Spatial and functional models were developed to facilitate the analysis, see
example in Figures 2 and 3.

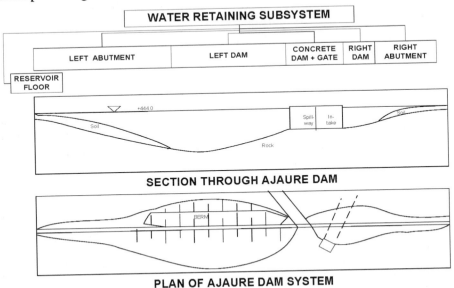

Figure 2. Plan and section of the Ajaure Dam

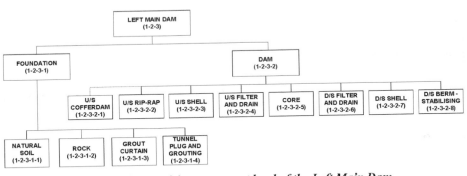

Block diagram of the component level of the Left Main Dam

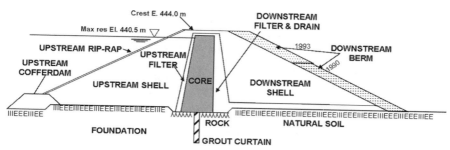

Figure 3. Spatial Model of Left Main Dam

The subsystems were broken down to the component level. Detailed component data sheets including pathway diagrams for identified failure sequences were elaborated for a number of components. Such a sheet for the downstream shell is summarised in Figure 4.

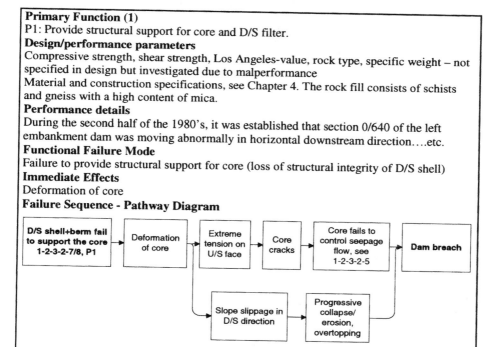

Figure 4. Summary of the Component Data Sheet for the D/S Shell

The design function of the subsystem "<u>Water retaining structures</u>" has been defined as to "retain water in the Ajaure reservoir with a controlled (small) seepage flow and with controlled discharges as required". Every element that has significance in making the system act as a continuous water barrier has been considered to be part of the water retaining structures. However the study was limited to one of the subsystems, the "Left Main Dam". Further, gaps in the present knowledge of the mechanics of the functioning of embankment dams and in the available information on site specific data such as material properties and performance characteristics render this case study is primarily a demonstration of how the principles of the FMEA methodology can be applied.

The design function of the subsystem "Discharge facilities" has been defined as to "convey water in a safe way from the upstream reservoir, through the dam, and to the river downstream of the dam". All elements of significance in making the system perform the spillway function are included in the sub-system.

The design function of subsystem "Spillway Gate Control" has been defined as to "be able to activate spillway gates in a controlled manner given a requirement to pass flows". Every element that has significance in making the system activation of the spillway gates possible has been considered to be part of the spillway gate control. The system components can be grouped in three overall aspects of the spillway gate control, for which functional models were developed

- Information flow and means of activation
- Power supply for spillway gate motors
- Power supply for measuring equipment, remote control, and station control equipment

The availability in many systems is influenced by human intervention (such as design, operation, test, maintenance etc.). There is therefore a logical connection between human reliability and technical reliability. Both human and technical availability is also determined from factors that lie outside the direct work situation. The organisation design is such an overall context and has therefore both direct and indirect influence on the basis for human and technical availability.

FMECA process
The FMEA process has been extended to FMECA by addition to the FMEA tabulation of:
- Criticality ratings and criticality index

The criticality analysis allows us to rank the importance of the failure modes by assigning criticality indices for the probability of occurrence of failure and the severity of the failure consequences. Here a qualitative approach, that does not require detailed frequency data, has been chosen. A relative index scale with five steps 1-5 has been put up. Here a set of three criticality indices has been assigned for each failure mode. They represent:

- Failure mode initiation - the potential for the failure mode to occur
- Failure sequence progression - the potential for the failure sequence to progress to ultimate failure
- Failure consequences - the severity of the consequences caused by ultimate failure.

Based on the three "criticality indices" a risk index has been calculated by multiplying them together. This risk index can be used to rank the potential failure modes according to the combined influence of their index of vulnerability and the severity of their failure consequences.

However, in order to cover differences in component function "demand" a fourth column has been added to provide context to the "criticality indices, e.g.:

- Event likelihood - frequency (1/year) of event that requires the component to function.

The process of assigning criticality indices involves weighing of evidence that supports a postulated failure mode (hypothesis) against evidence that contradicts the postulated hypothesis. Where the available information/evidence is incomplete a range has been assigned to the index. A wide range indicates that there are large uncertainties in the analysis due to lack of information/evidence. Here it is important to point out that a high number does not necessarily mean than that there is a weakness in the dam. It may also mean that there is a great lack of knowledge about the phenomenon in question, suggesting actions such as further investigations and/or a continued analysis. The "weight of evidence" explaining and motivating the assigned criticality indices has been documented.

Global Failure Modes

In the FMEA the system has been broken down into manageable bits and analysed. As an extension of the FMEA, an attempt is made to put the bits back together again, and return to the overall function of the Ajaure facility. This is done by working backwards in the pathways to failure, from dam breach towards the component failure modes. The end-branches (just before dam breach) of the pathways to failure interfaces with a global failure mode. Grouping together of the pathways' end-branches results in a few principal types of global failure modes, with connection to the three primary subsystems:

- Failure by slope instability, crest collapse and leakage/internal erosion, originating from deficiencies in the left main dam in the water retaining structures

- Erosional failure of D/S slope, originating from unsafe passage of discharge flow past the dam, initiated by deficiencies in the Discharge facilities or the Spillway gate control
- Overtopping, originating from failure to control the reservoir water level by discharge, due to deficiencies in the Discharge facilities or the Spillway gate control

For the identified global failure modes the global pathways, or when more appropriate the global fault trees, have been derived from the pathways used to model the effects of the component failure modes.

CONCLUSIONS AND LESSONS LEARNED

The training component of the SwedPower staff is deemed to have been successful, regarding methodologies with regard to risk analysis of dams, as well as training in sound engineering practices in general.

The applied FMEA methodology is regarded to provide a suitable framework for working with safety issues at dams. However, FMEA do not provide a stand-alone method or procedure but other methods such as pathway diagrams and fault tree analysis should be integrated in the application. Further development of the coupling of various methods and the criticality analysis would be beneficial to make the application more straightforward.

The elaboration of global failure modes provides a means of joining the results from the more disciplinary analysis of the various sub-systems. The global failure mode diagrams serve as logical maps displaying the relationship between various component functions and their role along the failure pathways.

The criticality ratings provide insights into what the engineers consider to be the principal issues concerning seriousness of issues and extent of uncertainty. The outputs from the criticality analysis process serve well as a basis for reasoning concerning the management of the risks.

Another conclusion is that complementary technical investigations providing site-specific data are often required to make the FMEA meaningful.

Agent-based dam monitoring

V. BETTZIECHE, Ruhrverband (Ruhr River Association), Essen, GER

SYNOPSIS.   The monitoring of security relevant structures is a task of growing importance in civil engineering. Large structures such as bridges and dams demand the use of precise measuring systems and the collaborative work of engineers, geologists and geodesists. Considering the time and labour consumed by the acquisition, processing and analysis of measured data, concerned authorities, operators and companies are trying to automate these operational procedures. The existing computer-based solutions focus on remote monitoring and neglect a collaborative analysis of measured data. However, an appropriate and effective monitoring system must conduct all of the tasks performed by experts involved in monitoring. The Institute of Computational Engineering of the University of Bochum, in co-operation with the Ruhrverband (Ruhr River Association), is developing a dam monitoring system based on software agents. The nucleus of the system's conceptual design is based upon the autonomous and collaborative analysis of measured data, associated with intelligent agents adopting the part of the experts generally involved in dam monitoring.

INTRODUCTION

Dam monitoring is based on precise measuring systems and the collaborative work of engineers, geologists and geodesists. Considering time and costs of acquisition, processing and analysis of measured data, an automated management of these procedures is desirable. Most of the existing computer-based solutions focus on remote monitoring, presentation and electronic transfer of measured data. To this end, they do not consider the cooperation between the experts involved in monitoring. However, an appropriate and effective monitoring system has to pay attention to the individual tasks performed by the experts. Furthermore, the distributed collaborative data analysis and safety assessment has to be captured through the system established.

The Institute of Computational Engineering of the University of Bochum, in cooperation with the Ruhrverband (Ruhr River Association), is currently

Long-term benefits and performance of dams, Thomas Telford, London, 2004, 197–206

developing a dam monitoring system based on software agents. Software agents represent an innovative, powerful as well as robust software technology allowing not only the implementation of distributed applications but also complex interactions. Consequently, the agent-based dam monitoring system is capable of supporting the collaborative work of the involved experts and incorporates the distributed work flow of data analysis and safety assessment. Thus, the complete work flow of dam monitoring is mapped onto a multi-agent system: regularly performed tasks (i.e. measuring at specified locations at the dam) are carried out by specialist agents. By contrast, the involved human experts are assisted by means of personal agents, which support these experts in performing their specific tasks and allow a direct communication with the multi-agent system.

Software agents - autonomous, mobile and intelligent software programs - provide all the necessary characteristics to innovate and accelerate the development of distributed applications. They represent powerful and robust software technology for implementing distributed-collaborative work flows and complex interaction.

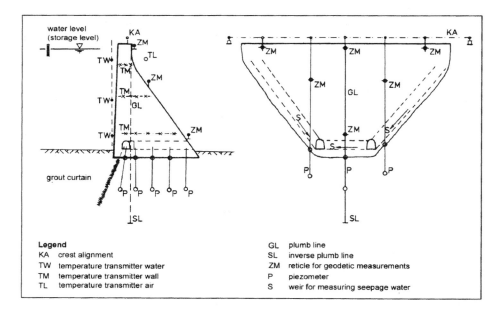

Figure 1: Example for the configuration of measuring devices [1]

DAM MONITORING - TECHNICAL ASPECTS

The aim of dam monitoring is to provide indicators for anomalous structure behaviour in order to be able to take necessary countermeasures in due time and without any reduction in safety. In Germany, the legal basis of dam monitoring is found in the German Code E DIN 19700 (2001). Further, recommendations for measuring devices have been published by the German Association for Water Resources and Land Improvement (DVWK) [1]. An example for the configuration of measuring devices is shown in Figure 1.

The concept of dam monitoring is based on the systematic acquisition of all the relevant parameters, which concern static, hydrologic and operational safety. Therefore, each dam structure must be provided with a measuring and control system, which, then, has to be adapted to the type, size and location of the structure.

The conceptual design of a monitoring system has to consider the following principles:

- Dam and bedrock form a unity, which is embedded in a natural environment.
- An anomalous structure behaviour can occur either gradually or quickly.
- When an anomalous behaviour occurs, the origin should be identifiable by an analysis of the measured parameters.
- Inspection by qualified personnel is indispensable.

In addition, the monitoring system must be adapted to the characteristics of the dam structure and take into consideration the corresponding measuring categories. At arch dams it is important to monitor displacements, and at gravity dams pore pressures are of particular importance in addition to displacements.

An automatic monitoring system rests on extensive electronic measuring equipment. The equipment consists of two essential components: transmitters (sensors) and data recorders (data loggers). Recommended transmitters are i.e.:

- temperature sensors,
- ultrasonic sensors for measuring seepage water,
- laser for measuring displacements,
- vibrating wire piezometers for measuring pore pressure.

The sensors are installed at specified positions (figure 1) inside the structure or the bedrock and they are controlled by electronic equipment (e.g. data loggers) sending electronic impulses. After having received an impulse, the sensors return a signal which can be a measurement of voltage, resistance or frequency. The electronic equipment scales the signal into a value, and either stores it in an internal memory or transfers it to a local database. Data stored in an internal memory can normally be received via a COMMS port (RS232). This interface also allows that the electronic equipment can be programmed from a host computer. An automatic monitoring system is customarily completed with a local computer, usually placed in a control room near the dam. As the redundant data storage is essential in dam monitoring, measured data is stored in a local database and additionally transferred to a central database [4,5].

DAM MONITORING - ASPECTS OF ORGANISATION

Dam monitoring is not only based on instrumental supervision. Several experts have to take care of the data, i.e. they have to analyse the data. The experts view on the monitoring data may be very different, based on their profession and job. For example, a geologist may view at these data differently than a geodesist or a civil engineer.

At the Ruhr River Association the monitoring data have to pass several states of controls as it is shown in figure 2 (left side). Different experts have their view on the data, while on one hand the processing frequency decreases with every processing step, from temporal intervals of one day up to one year, on the other hand the time interval of the viewed data also decreases from an short interval to the whole existing data.

Each step can be briefly described as follows:
1. Data acquisition:
> At the reservoir the crew supervised the dam according to the monitoring plan. This includes daily measurement (manual or automatic) of the relevant parameters, in particular the rate of flow, water level, water pressures, displacements, changes of temperature and others.

2. Check of plausibility:
> Just in time, the manual or automatic measured data are checked with respect to their plausibility. These checks are based on the data of the measurement of the day before or on alarm values and are done by the measuring crew itself.

3. Check of short-time behaviour:
> In the week of the measurement the data are checked by the responsible engineer at the reservoir-administration. He compares the data with the measurement of i.e. the last week to find out

anomalies in the short-time behaviour of the dam. After this the data are transferred to the dam safety department of the company.

4. Check of long-time behaviour:

At the head office of the company several specialist have their own view at the data. At the geological department the data concerning groundwater flow and seepage are checked. Geodesists and Engineers will inspect the movement data. This investigation will be done once a month, in order to find abnormal behaviour in a long-time view of the data.

5. Safety assessment:

Once a year the responsible engineer has a view over all the data collected. His task is to supervise the measurements and to analyse the data by using statistical tools and computer models. At last he has to compose the annual report, documenting the safety of the dam, not least for the surveillance authorities.

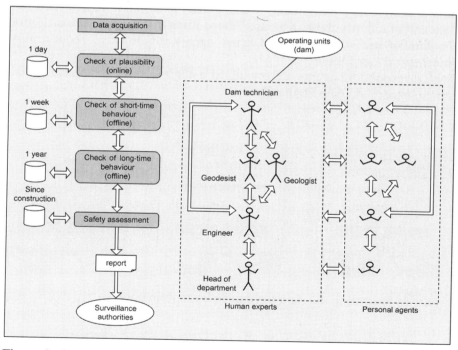

Figure 2: Conceptual design of dam monitoring

CONCEPTUAL DESIGN OF THE DAM MONITORING SYSTEM

The analysis of the described work flow applied by the Ruhrverband indicated that there is a chain of five tasks regularly performed by the responsible experts or in collaboration with other experts. Thus, the basic principle of the conceptual design is to map the regular performed tasks, the individual experts and the interaction between themselves onto a multi-agent system. By that, the software agents can be divided into two categories: specialist agents mapping the regularly performed tasks and personal agents mapping the experts involved in dam monitoring. The conceptual design for the organisation of the agents is shown in Figure 2 (right side).

In order to provide smooth communication between the human experts and the multi-agent system (MAS), each human expert involved in dam monitoring is allocated with a personal agent. This software agent represents the interface "human/MAS" and has to be proactive. A proactive agent is able to realise its environment, to recognise the situation represented by the data and to identify the human user. Depending upon the situation it informs the human user or contacts to other agents, in order to request further information (see Figure 3).

The corresponding agents are organised using the same relationships as the human experts do (see Figure 2).

Figure 3: Informing the dam technician by mobile phone

IMPLEMENTATION OF THE DAM MONITORING SYSTEM

Considering the agents to be applied in the dam monitoring system, there are some basic requirements to be met by the conceptual design of the agent architecture. In the following an appropriate agent architecture is developed by focusing on the basic requirements of agent-based applications.

Interaction, and in particular its basis, **communication**, is an essential element of the networked and collaborative systems of the present time and future. Capable solutions must provide several communication protocols for different requirements. For example, in some environments the HTTP protocol is required in order to avoid firewall problems.

The inter-agent communication within the MAS is to be realised according to FIPA specifications [8], since FIPA is one of the central standards in the agent world. Furthermore, this approach allows inter-platform communication with other FIPA-compliant agents on various platforms.

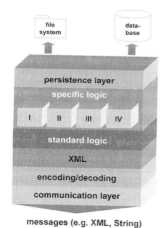

Figure 4: Conceptual design of the agent architecture

When dealing with complex problems, the agents have to be provided with logic. In the current architecture, the logic elements are divided into two categories: **standard logic** and **specific logic** (knowledge). In this particular case the standard logic contains the ontology of the domain "dam monitoring", by which the agents possess the required vocabulary and basic knowledge in order to communicate and to execute simple tasks.

Via modules the individual knowledge of the involved experts can be integrated very easily. This approach enables the user to adapt the agent to new tasks, goals or environments, too. In other words, the agent becomes more "intelligent" [6,7].

The last layer of the given architecture is a **persistence layer** to keep the state of the agent persistent. In case of a system crash this layer helps to identify the actual state of the agent and to continue the work without any loss of time.

Control of automatic measuring devices - Logger API

As an important factor, a capable computer-based monitoring system must cover the applied electronic equipment. In order to control the measuring devices installed within the structure, there are two popular solutions: systems based on process control systems and systems based on data loggers. In the following, only data loggers are discussed. From an objective point of view, they are the better and more transparent solution for dam operators in planning, use and maintenance.

Since the control of the data loggers depends on the specific communication protocol and the instruction set predetermined by the specific manufacturer, a Java-based programming interface, called Logger API, was developed to encapsulate specific loggers. Specific loggers can be added to the developed library without expenditure.

Data processing and visualization – the evaluation module

Data processing and visualization are provided by an evaluation module which has been conceived thus far with a web-based front end. The web-based paradigm has been chosen such that an acceptance test could be performed in practice in a simple manner and so that no further client-sided software would be necessary.

The task of this web-based evaluation module is to read the acquired data of the dam monitoring from the database and to evaluate, edit and prepare the data in a user-oriented way, graphically and/or tabularly [3].

• *Visualization component*

The visualization component (**view**) acts as a graphical user interface which allows database inquiries, administration of users, etc. (*inputs*) on the one hand and visualizes the requested result quantities in different data formats (*outputs*) on the other hand

An additional task of the visualization component is the representation of the requested data in the format indicated by the user. Output objects are instantiated in order to produce the appropriate outputs depending on the desired format (see Figure 5).

• *Database adapter*

In order to be able to attach several (replaceable) databases, the **model** is realised as an exchangeable database adapter. The assigned tasks are to generate a connection between the database(s) and the controller component, to pass on inquiries which concern measured data to the database, and to return the received results to the controller component

Due to the modularity and expandability of the evaluation module developed, this module can be used in a multi-agent system, for example as a wrapper agent, in order to read measured data from a database. A further possibility is the employment of the module as an interface agent, i.e. as an interface between a human and a multi-agent system which converts "clicked" mouse events into messages understandable for agents.

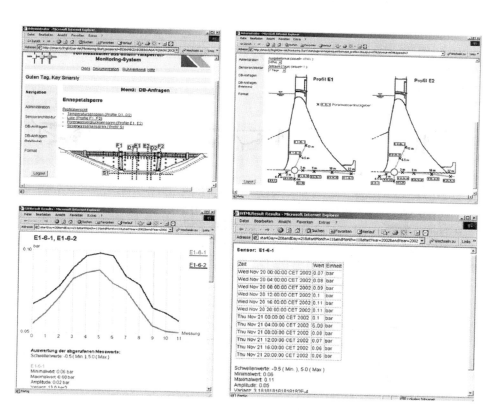

Figure 5: User-specific visualization of the measured data

CONCLUSIONS

Applying software agents, the Institute of Computational Engineering, in co-operation with the Ruhrverband (Ruhr River Association) is taking an innovative approach to developing a modern dam monitoring system, which is capable of supporting the collaborative work of experts involved in monitoring.

The conceptual design of the organisation and the architecture of the agents to be applied in the multi-agent system have been shown. By substantiation, the implementation of two important modules: the logger API and the evaluation module, has been explained. Actually, these two modules represent a conventional, web-based monitoring system. The measuring devices installed within a dam can be controlled online, and measured data can be read from the databases and processed according to user preferences.

The multi-agent system is designed to map the distributive-collaborative work of the concerned experts and to integrate their specific knowledge about dam monitoring and dam behaviour. This conceptual design differs significantly from conventional monitoring systems and represents an innovative approach which is capable of demonstrating the enormous potential of agent-based applications.

ACKNOWLEDGEMENT

The author would like to thank Prof. Dr. D. Hartmann, I. Mittrup and K. Smarsley of the Ruhr-University of Bochum, Germany, who developed the computational solution for the analysis of dam monitoring data.

REFERENCES

1. German Association for Water Resources and Land Improvement (DVWK): Measuring devices for checking the stability of gravity dams and embankment dams, *DVWK-Merkblätter zur Wasserwirtschaft (DVWK-Guidelines)*, Volume 222, 1991.
2. Mittrup, I.: *Design of a web-based facility monitoring application for dam monitoring*. Diploma thesis, Institute of Computational Engineering, Ruhr-Universität Bochum. Bochum, 2002. http://www.inf.bi.rub.de
3. Smarsly, K.: *Development of a web-based framework for evaluation, processing and visualization of measured data of a dam monitoring system*. Diploma thesis, Institute of Computational Engineering, Ruhr-Universität Bochum. Bochum, 2002. http://www.inf.bi.rub.de
4. Bettzieche, V.: Experiences with the monitoring of dams. Scientific Reports. *Journal of The Mittweida University of Technology and Economics*. Volume III, Mittweida, 1/1997.
5. Bettzieche, V., Heitefuss, C.: *Monitoring as a basis of cost-effective rehabilitation of an old masonry dam;* European ICOLD Symposium, Norway, 2001.
6. Ferber, J.: Multi-Agent Systems: *An Introduction to Distributed Artificial Intelligence*. Addison-Wesley, Munich, 2001.
7. Maes, P.: Modeling Adaptive Autonomous Agents. In: *Artificial Life Journal*, An Overview, edited by Christopher G. Langton, MIT Press, Cambridge, 1995.
8. Foundation for Intelligent Physical Agents (FIPA): *FIPA Specifications*. http://www.fipa.org

Armenia Dam Safety Project

JOHN SAWYER, Jacobs Ltd, UK
LAURENCE ATTEWILL, Jacobs Ltd, UK

SYNOPSIS.   The Armenian dam safety project involves the technical investigation of 64 dams during the period June 2002 to July 2003. The scope of work includes:
- Fieldwork: Dam inspections, Site investigations (4000m of drilling), Topographic survey, microseismic survey
- Studies: Hydrology, Flood routing, Dam break, Stability analysis, Seismic hazard assessment, seismic analysis
- Risk assessment
- Rehabilitation preliminary design and costing
- Dam safety plans (Operation & Maintenance, instrumentation, early warning systems and Emergency Preparedness Plan)

The dams include irrigation, water transfer and hydropower schemes and range from 1.5m to 83m high with both embankment and concrete gravity structures.

The paper gives an overview of the project and its challenges. Particular project issues include working across national and engineering cultural boundaries, obtaining information on existing schemes, and using a risk based assessment procedure for prioritising rehabilitation works. Particular technical issues include the refurbishment of neglected mechanical equipment and the rehabilitation of a 65m high dam that collapsed during construction.

INTRODUCTION

The Project Implementation Unit (PIU) of the Committee on Water Economy Management of the Republic of Armenia is implementing a national Dam Safety Project to increase utilisation of the present water reservoirs and to protect the downstream population and infrastructure in the case of a dam break. The safety assessment of 24 large reservoirs was completed during 1999 – 2000, and a preliminary Rapid Investigation of a further 60 dams was carried out in 2000 by Hydroenegetica Ltd of Armenia.

Long-term benefits and performance of dams, Thomas Telford, London, 2004, 207–219

This paper considers the follow up project to the 'Rapid Investigation', which studied a total of 64 reservoirs between June 2002 and July 2003. The project is funded by an IDA loan to the Armenian Government and has been carried out by Jacobs Ltd of the UK with the support of Hydroenergetica and Georisk of Armenia.

The importance of dams in Armenia is very high. Some 24% of National electricity demand is generated by hydropower stations. The remaining balance is generated from thermal stations powered by both nuclear reactors (31%) and fossil fuels (45%), all the fuel must be imported. Hydropower is important therefore not only because it is cheap and clean but also because it provides a secure source of power. The water stored in the reservoirs irrigates 2,870km^2 which reduces Armenia's dependency on food imports with consequential security, social and economical benefits. Dam safety is therefore of national significance, and not just to the population living immediately downstream of the dams.

The majority of the dams have been in operation since the 1960's and 1970's, with some in use since 1940. Based on several factors that include the dam height and the reservoir storage capacity, the reservoirs have been divided into the following groups:

- Large reservoirs (12 dams, 15m to 85m high)
- Small Reservoirs (33 dams, 1.5m to 20m high)
- Artificial lakes (17 dams, 0 to 5m high)
- Partially constructed (2 dams 14m to 21m high)

Six of the large reservoirs are hydropower dams and are under authority of the Ministry of Energy. Two large dams were originally built for mining organizations and are not in operation, the other dams are irrigation or multi purpose dams and are owned by Jrambar CJSC, which is a state organisation responsible for irrigation facilities.

SCOPE OF THE PROJECT
The scope of the Consultant services is as follows:
1. Dam Investigations: reveal the structural and non - structural defects based on dam inspections, topographical and geotechnical site investigation results as well as hydrological, geotechnical and seismic studies.
2. Determine the degree of risk for each dam.
3. Recommend rehabilitation measures.
4. Prepare dam safety plans, which include instrumentation, operation and maintenance (O&M) plans and emergency preparedness plans (EPP).
5. Recommend early warning systems where appropriate.

The project therefore covers not only all the technical issues relating to the reservoirs, but also the interface with the operators, owners, emergency services and general public. The investigations and studies into the 'artificial lakes' were more limited than those for the remainder of the reservoirs due to their low hazard, but covered the same general scope.

DAM INVESTIGATIONS
Field Investigations
Only limited information exists regarding the construction of each dam, and typically the information available is design data rather than construction records. For many of the smaller dams no records were found at all. Thus, although an archive search was carried out, it was necessary to carry out field investigations on most of the reservoirs including field inspections, topographic survey and mapping, and geotechnical site investigations.

The field inspections were generally carried out by expatriate dam specialists accompanied by local technical staff and where possible by the operators. An inspection report was produced for each reservoir, and this was then used to establish the requirements for further investigations, particularly the geotechnical site investigations. Topographic survey was carried out by local contractors.

The site investigation involved almost 4000m depth of boreholes and trial pits. Both disturbed and undisturbed samples were taken for characterisation and strength testing in local soil mechanics laboratories. Two local contractors worked under the supervision of local and expatriate geologists.

The terrain of Armenia is very mountainous and the winter is severe, making access to remote areas impossible for several months. The most remote reservoirs are only accessible in the late summer. So far as possible all reservoirs were inspected and the site investigations completed in the Autumn of 2002. Some follow-up work was carried out in late spring of 2003. Security concerns limited access to some border reservoirs. The Turkish border of Armenia is manned by Russian troops and the border with Azerbaijan is unstable, so access to major dams on these borders was restricted. Inspections were carried out on these dams but site investigations were not possible.

Hydrology
Two methods were been used to analyse the flood inflows into the reservoirs. The first, the SNIP method, is based on standard Russian techniques and is in general use in the country. The second, a statistical method using all annual maxima flow data recorded in Armenia, has been used worldwide to check more particular methods (the Regional Method).

The Soviet Norme (SNIP)

This has two approaches, depending on the information available:

1) Applying analytical distribution functions for annual exceedance probabilities where sufficient hydrogeological data is available for the catchments.

2) In the absence of observed data, the peak flood of a given return period is calculated using a formula in terms of m^3 per km^2 which has terms for basin area, rainfall, geographic characteristics and vegetation.

The Regional Method

The basic hydrological records available for analysis in Armenia comprise the annual maximum flows for 102 gauging stations. The average record of over 40 years ensures that a reasonable sample of floods is available at these sites. By combining the records at different sites it is possible to estimate relations between basin characteristics and the mean annual flood, and also a relation between the mean annual flood and the flood of a rare frequency or long return period.

The relationship between mean annual flood (MAF) and the flood for a given return period (Q_T) was determined to be:

Return Period, yrs.	100	500	1,000	2,000	5,000	10,000
Q_T/MAF	3.23	4.7	5.47	6.33	7.65	8.79

A regression between mean annual flood, (MAF), and basin area (AREA) and annual rainfall (AAR) provides a significant relation between the variables:

$$MAF = 2.53 \times 10^{-6} (AREA)^{0.782} (AAR)^{1.764}$$

These two relationships were then used to assess the MAF and Q_T for each reservoir at the relevant return periods.

Comparison of the SNIP and the Regional method

The Armenian Standard (SNIP) was found to give higher estimates of peak flow for smaller catchments (up to $100km^2$). For very large catchments ($100,000km^2$) the regional Method gave a slightly larger estimate, with reasonable agreement between the two methods in the middle range. See figure 1 below which compares the 1000 year flood estimates.

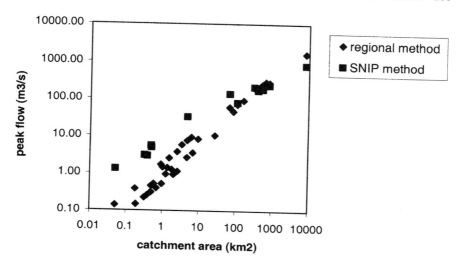

Figure 1: Comparison of 1:1000 year flood estimates

Flood Routing

Flood routing studies were carried out making use of either inflow hydrographs based on SNIP hydrology and SNIP rules for the return periods to be considered, or inflow hydrographs based on Regional Method hydrology and ICOLD recommendations for return periods where this gave larger floods [only the large reservoirs were affected].

Dam Break

Dam-break modeling was used in this project both for input to the Risk Assessment and the Emergency Preparedness Plan (EPP). The dam-break assessment was carried out in three steps:

i.)   Initial screening carried out by identifying the potential flood paths on 1:100,000 scale mapping. In the case of some small reservoirs this indicated that the flood wave presents no hazard, passing through no populated areas and joining river channels which are large relative to the size of flood. In this case no further study is needed. In most cases this initial phase defines the extent of flood route which requires further study.

ii.)   'Quick Dambreak'. This is a spreadsheet based method of analysis which predicts the flood size and characteristics and from which inundation mapping is prepared. The approach was developed from the methodology given in CIRIA Guide C542, Risk Management for UK Reservoirs. For this analysis 1:25,000 and 1:50,000 mapping has been used as this is all that was available to the project.

iii.)   BOSS DAMBRK.   This is commercial software using more sophisticated analysis methods. For this project it was used for the analysis of the most critical reservoirs and to calibrate the results of the 'Quick Dambreak'.

The output from the Quick Dambreak analyses were inundation maps, coloured to show the flood damage parameter velocity x depth, with tables showing flood depth and width, and the time to peak flow at points along the flood path. The analysis has the great advantages of simplicity and ease of use. It has enabled the assessment of all the dams within the project to an adequate level.

The results of DAMBRK were compared with the Quick Dambreak results and demonstrated that within the tolerances of the mapping available the output was satisfactory for risk assessment and emergency planning.

Geophysical Investigations
Seismic refraction survey was carried out at Marmarik dam. The results were used for the assessment of seismic intensity magnification due to the site specific soil conditions.

Electrical resistivity was measured along two profiles at the Landslide N4 at Marmarik dam. The results were used, together with the drilling results, to determine the thickness of the landslide material.

Landslide hazard studies
Desk studies were carried out of potential landslides around Marmarik and Bartsrouni reservoirs. The work involved analyses of satellite images and aerial photos that were taken in 1948, 1976 and 1986.

Four potentially hazardous seismogenic landslides were identified within the Marmarik reservoir area that may influence the dam safety. The impact of the landslides onto the dam safety was assessed and special design provisions were made as a part of the rehabilitation works. They are described in detail in the paper on Marmarik dam.

Bartsrouni dam was constructed on a large, ancient landslide. Recent landslide activities have been demonstrated by numerous scarps. The dam has already been partially destroyed by landslide movements and it is anticipated that future movements will continue to damage the dam.

Seismic Studies

Seismic studies for the dams include the assessment of the seismic stability and assessment of liquefaction potential of fill and the foundation material.

Seismic stability analysis has been carried out using the methodology given in the Seismic Design Standards of Republic of Armenia (SDSRA) –, II.2.02-94 for all dams.

The susceptibility of loose, saturated sands and silty sands in the foundation and dam body to liquefy during an earthquake was carried out according to the methodology given in the Japanese standard.

Seismic design parameters have been selected based on the SDSRA for all dams, and on a Site Specific Seismic Hazard Assessment (SSSHA) for seven critical dams. The selection of dams was based on the level of acceleration assessed in the SDSRA, dam height and the results of the site investigations. The SSSHA was carried out using both a probabilistic and a deterministic approach. The results are given in Table 1 below, and indicate the significant seismic hazard in Armenia.

Table 1. Site Specific Seismic Hazard Assessment (SSSHA) – Design Accelerations

| | Design Peak Horizontal Acceleration, g | | | |
| | OBE | | MDE | |
	Ground	Crest	Ground	Crest
Marmarik	0.32	0.6	0.44	0.82
Shenik	0.12	0.30	0.25	0.61
Tsilkar	0.34	0.49	0.49	0.65
Landjaghbiur –1	0.22	0.25	0.68	0.74
Hors	0.24	0.39	0.69	0.71
Gekhi			0.35	0.675
Akhuryan (concrete)	0.4		0.7	

Stability Analyses

Stability analyses for the embankment dams were carried out using the computer programme SLOPEW (GEO- SLOPE International) based on data from the topographical survey and on the site investigation. The load cases analysed are in accordance with SNIP standards. They include consideration of the upstream and downstream slope under static and seismic loading; and full supply level, maximum flood level and rapid drawdown cases.

For all except four of the dams, the factors of safety obtained in the stability analyses for the static condition were higher than the minimum required.

Stabilistation works were designed for the four sub-standard dams. For some of the dams, factors of safety obtained for the seismic condition were less than unity. However, the displacements that could be generated were assessed to be negligible.

The concrete gravity dams were assessed by using a spreadsheet based analysis. Static and dynamic stability cases were assessed under a range of water levels and uplift assumptions. The dams were generally shown to have satisfactory stability under static conditions, but were liable to some local overstress in seismic events. One dam, which had an unauthorised spillway raising, was shown to have inadequate safety margins unless the spillway was restored.

Summary of Defects
A wide range of defects relating to design, construction, operation and maintenance were identified. In many cases these could be attributed at least in part to the results of the break up of the Soviet Union. Typical defects included:
- Deliberate blockage of the spillway to increase freeboard.
- Inadequate spillway capacity / freeboard.
- Structural repairs required to spillway or outlets.
- Damaged or deficient riprap or wave protection.
- Outlet valve refurbishment required.
- Slope stability inadequate.
- Leakage through embankment.
- Leakage through reservoir floor.
- Unsafe access to equipment.
- Refurbishment required to hydromechanical equipment.

On the basis of the assessment of defects, remedial works were recommended and outline designs prepared. Detailed design is to be carried out by Armenian consultants. In a limited number of cases 'emergency works' were recommended immediately following the inspection. One reservoir was recommended to be drawn down and abandoned (Bartsrouni, built on a landslide), others were recommended to be maintained at a low water level pending remedial works.

RISK ASSESSMENT

Methodology

The approach that has been used in the assessment of risk of all the dams is a semi-quantitative method in which both the probability and the consequences of an event are ranked from low to high and the relative risk levels indicated by the position on a matrix. This method has been adapted from CIRIA Report C542. The following stages are required:

i) identification of failure modes (instability erosion etc.).
ii) comparative assessment of probability of failure (probability of event x probability of this leading to failure).
iii) comparative assessment of consequence or impact of failure (population at risk and economic loss).

All factors are quantified on scales of 1 to 5 or 0 to 4, leading to semi-quantitative assessments. The risk index is the product of the total impact score and the risk score. A comparison of this score for all dams will provide a ranking showing where the priorities for remedial works lies. In addition, if the risk assessment is repeated for the case where it is assumed that the recommended remedial works have been carried out, the reduction in the combined score will enable a quantitative assessment of the benefit of the remedial works to be made.

The risk profile of the dams, as measured by the risk index, is presented in Figure 2. This Figure also shows the reduction in risk that will be achieved by the implementation of the Emergency Preparedness Plans and the remedial works.

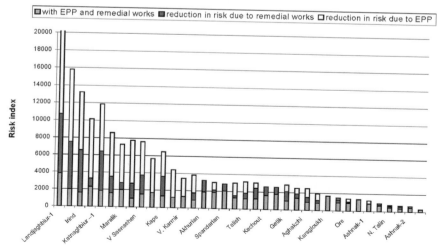

Figure 2: Risk Profile

Cost effectiveness

As a means of assessing cost effectiveness, the reduction in risk index has been divided by the corresponding cost for both structural measures (remedial works) and non structural works (safety materials and emergency preparedness plans) to give benefit/cost ratios for structural and non-structural works.

Figure 3 shows the ranking in terms of the benefit/cost ratio of remedial measures. The average benefit cost ratio is 29 and the range is from 128 (V Sasnashen) to 0.8 (Kechout). One effect of this ranking is to highlight the significant benefit that can be gained from relatively minor works ($17000 at V Sasnashan compared with $1.3 million at Kechout).

Figure 4 shows the ranking of the dams in terms of the benefit/cost ratio of non- structural measures. Not only is the ranking of the dams quite different but the average benefit/cost ratio, 96, is much higher than the remedial measures cost benefit ratio and also the range, from 12 to 427, is more extreme. This indicates that non-structural measures can be regarded as providing better value for money but it is important to bear in mind that this depends on the efficacy of the EPP's which will require commitment and ongoing expenditure to maintain.

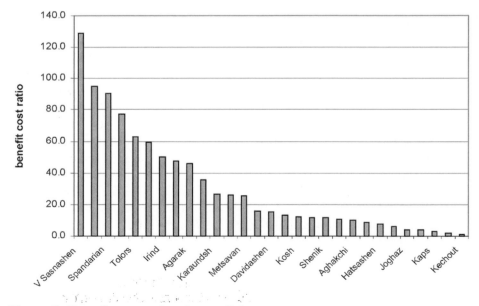

Figure 3: Benefit cost ratio of remedial measures

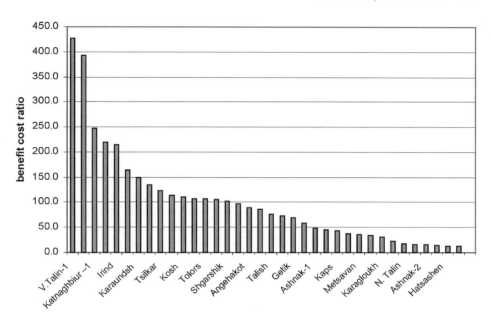

Figure 4: Benefit cost ratio of non-structural measures

DAM SAFETY PLANS
Dam Safety Plans were prepared for each dam. These include recommendations for instrumentation and monitoring, for operation and maintenance and where relevant an emergency preparedness plan (EPP). The plans were tailored to the particular reservoir, and reflected the size and hazard potential for each dam. The recommendations were generally for simple and robust instruments to monitor reservoir level, leakage and movement, typically just a staff gauge for the water level, V-notch weirs for toe drainage measurement, and survey monuments on the crest for the smaller embankments, with foundation piezometers to measure uplift in the concrete dams. Nine dams presenting a hazard to communities immediately downstream have been identified and an automatic water level alarm recommended to give warning in the event of the spillway discharge exceeding the design capacity. Equipment and materials for emergency works have been identified, to be maintained at each regional depot and each major dam. The proposals have been costed, including the requirements for routine supervision and inspection of the reservoirs, to allow the owners to budget for the long term implementation of the Safety Plans.

The EPP for each dam makes use of the technical studies, particularly the dambreak and inundation mapping, and then relates this to the emergency services and civil authorities. Local specialist consultants were used for these aspects as they require particular knowledge of local organizations.

PROJECT CHALLENGES

The project involved considerable challenges, most of which in some way related to communication. Particular issues included:
- Access to remote sites in difficult terrain and an extreme climate.
- Language: the Armenian engineers work in the Armenian and Russian languages, but technical vocabulary is primarily Russian.
- . Engineering culture: the Armenians have historically worked within a tightly regulated system of Soviet Normes (SNIP), rather than to Western approaches. This affects not only design philosophy but also practical details of site investigation and testing and construction techniques.
- Construction records: it proved impossible to obtain reliable 'as-built' information for the majority of the reservoirs, largely due to the effects of the break up of the Soviet Union.
- Communication between the UK and Armenia: time zones, awkward flights, poor telecommunications and internet connections all add difficulties.

In this context it is essential to have Russian speaking technical staff and to adjust Western technical methodologies to suit the SNIP based designs and investigations. If all the geotechnical testing equipment in the country is to Soviet standards, there is little point in insisting on Western ones. It is also essential to have expatriate staff who will respect and adapt to the local culture, while bringing the benefit of their own experience.

CONCLUSIONS

The study has identified substantial remedial works required to the dams of Armenia. The use of a semi-quantitative risk assessment methodology has given a prioritisation of the remedial works. This has been used to substantiate a request for IDA funding. A programme of remedial works is now in progress based on priorities assessed in this study.

The project also delivered Dam Safety Plans for each reservoir which gave recommendations for instrumentation, monitoring regimes, maintenance and emergency planning. Implementation of these plans will require a significant long term organisational commitment, but will go a long way to limiting the need for future major remedial works programmes.

REFERENCES

CIRIA C542 (2000). *Risk Management for UK Reservoirs.* CIRIA, London.

Spasic-Gril, L, Sawyer, J.R. (2004). Marmarik Dam Investigations and Remedial Works. *Long term benefits and performance of dams.* Thomas Telford, London.

Reservoir management, risk and safety considerations

J L HINKS, Halcrow Group Ltd.
J A CHARLES, Consultant

SYNOPSIS. Risk assessment techniques are being increasingly applied to portfolios of reservoirs in the UK and overseas. While hydrological and mechanical/electrical risk can be reliably evaluated using modern techniques, geological and geotechnical risks are more difficult to quantify. The calculation of seismic risk might appear fairly straightforward, but it poses a number of challenges because a severe earthquake may discover weaknesses in the dam or reservoir rim that were not identifed before the event. At larger dams with gated spillways, the probability of mechanical/electrical malfunction can be significant. A simple methodology for the quantification of each major class of risk is described with the aim of calculating a probability of failure for each dam. This can then be multiplied by a figure representing the financial consequences of failure in order to yield an annualised figure of the magnitude of the risk, which can then be used in ranking the portfolio.

INTRODUCTION

Risk analyses have been increasingly used for engineering applications over recent years. In 1982 a House of Lords Select Committee recommended that the techniques should be applied to reservoir safety and this led to the publication, in 2000, of CIRIA Report No C542 entitled "Risk Management for UK Reservoirs".

The paper describes techniques of risk analysis for reservoir safety that have been developed for use in the Balkans, the Caribbean and elsewhere. The methodology has many similarities to that in the CIRIA Report but adopts a definition of risk which is in use in Canada (Hartford, 1997) and Switzerland:

Risk (£/year) = consequences of failure (£) x probability of failure (per year)

The methodology differs from that in the CIRIA Report in that it seeks to quantify likelihood as an annual probability and consequences in terms of £ or $. The advantages of this approach are:

(a)   the risk can be expressed in £/year and represents the premium that would be payable in a perfect market to insure the dam

(b)   a portfolio of dams can be ranked according to the calculated risk that they pose

(c)   account can be taken of all the undesirable consequences of dam failure including interim costs (e.g. provision of temporary water supplies) and the cost of rebuilding the dam.

The disadvantages of the approach include the following:

(i)   the difficulties of putting reliable probabilities to certain types of failure (e.g. internal erosion)

(ii)   the need to allocate a monetary value to the loss of a human life

(iii)   the inability to handle uncertainty other than through sensitivity analyses.

PROBABILITY OF FAILURE

The historical annual probability of failure of large embankment dams up to 1986 is given by Foster et al (2000a) as 4.5×10^{-4} per dam-year and this reduces to 4.1×10^{-4} per dam-year if construction failures are excluded. This figure should be compared with the statement by Hoeg (1996) that the probability of failure of embankment dams had reduced over a period of 30 or 40 years from 10^{-4} towards 10^{-5} per year. Charles et al (1998) have shown that in the period 1831-1930 in Great Britain the occurrence of a failure causing loss of life was 3×10^{-4} per dam-year. However, since the introduction of reservoir safety legislation in 1930 and up to the time of writing, no failures have occurred in Great Britain which have caused loss of life.

Probability of failure may be taken as the sum of the probabilities of failure due to the following causes:

- hydrological failure
- geological/geotechnical failure
- mechanical and electrical failure
- seismic failure

Foster et al (2000b) give the following breakdown for the causes of failure of large embankment dams prior to 1986:

	% of total failures
Overtopping	46
Piping through embankment	31
Piping through foundation	15
Piping from embankment to foundation	2
Slope instability	4
Earthquake	2

Internal erosion thus accounts for 48 % of the failures of embankment dams. Although the term "piping" is used by Foster et al, 2000a and 2000b, piping is just one particular form of internal erosion and the three categories of piping listed above doubtless include other forms of internal erosion failures that strictly speaking were not piping failures. Where failure has occurred it will often be impossible to determine the precise mechanism of internal erosion.

Although mechanical/electrical failure does not feature in the above list from Foster et al (2000b), a more detailed list in Foster et al (2000a) indicates that 13% of failures are associated with a spillway gate. Where large dams with gated spillways are under study this mode of failure cannot be ignored.

Failures due to earthquakes represent only 2 % of the total, but it should be remembered that there are difficulties in defining failure. Dams are frequently badly damaged in earthquakes without an uncontrolled release of water taking place. This may be partly because irrigation dams are sometimes full for only a couple of weeks per year. For the Nihon-kai-Chubu earthquake in 1983 damage equivalent to failure was defined as follows (Gosschalk et al, 1994)

- sliding of slope
- longitudinal crack more than 50 mm wide
- transverse crack
- crest settlement more than 300 mm
- leakage of water

Hydrological failure
Overtopping is believed to have been responsible for about half of worldwide embankment dam failures and most of the deaths (ICOLD, 1997). This statement is supported by the statistic, quoted by Foster et al (2000b), that 46 % of embankment dam failures are attributable to overtopping.

A relationship will often be needed between return period and percentage of probable maximum flood (PMF). The growth curve in Figure 1 is derived from the figures quoted in "Floods and Reservoir Safety". It is only approximate and should probably not be used overseas without careful checking.

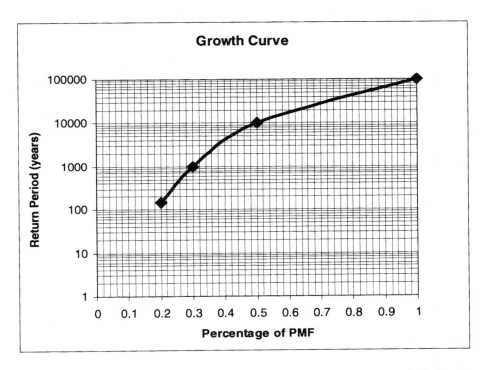

Figure 1. PMF Growth curve for UK (from "Floods and Reservoir Safety")

ICOLD Bulletin 109 argues that where the spillway is designed for, say, the 1,000 year flood the true probability of failure for hydrological reasons will often be an order of magnitude less. This is thought to be for the following reasons:

- the reservoir may not be full at the start of the storm
- wave freeboard may not be taken up by waves
- the dam may be able to withstand some overtopping.

Bearing the above in mind it should be possible to put a probability to overtopping leading to dam failure in a period of risk of, say, 100 years.

Figure 2. Orifice spillway at 51m high dam. The dam is used partly for flood control

Geological/geotechnical risk
Foster et al (2000b) attribute 48 % of embankment dam failures to internal erosion and, when taken across the whole portfolio of dams, the average probability of geological/geotechnical failure will be about the same as the average probability of hydrological failure. About half of all internal erosion failures through the embankment are associated with the presence of conduits. This has been confirmed in a study of internal erosion in European embankment dams where the ICOLD European Working Group on internal erosion in embankment dams found that in almost half the cases where failure occurred, or where failure almost certainly would have occurred very quickly if the reservoir had not been rapidly drawn down, the problem was associated with a structure passing through the embankment (Charles, 2002).

Work by Foster et al (2000b) give the average frequency of failure (during the life of the dam) due to piping through the embankment by dam zoning categories for large dams up to 1986. Some of these figures are reproduced below:

	Average frequency of failure (x 10^{-3})
Homogeneous earthfill	16.0
Puddle core earthfill	9.3
Earthfill with rock toe	8.9
Concrete face earthfill	5.3
Earthfill with filter	1.5
Zoned earthfill	1.2

It should be noted that 49 % of internal erosion failures occurred during first filling of the reservoir, 16 % during the first 5 years of operation and 35 % after 5 years operation.

In areas of steep topography particular account needs to be taken of the risk of landslides into the reservoir such as that which caused the loss of over 2,000 lives at Vaiont in Italy in 1963 (Hinks et al, 2003). This event was particularly disastrous because of the high loss of life (LOL) in the village of Longarone downstream where 94 % of the 1,348 residents perished.

Mechanical and electrical failure

The principal mechanical/electrical risk is the failure of spillway gates to open. However the following also need to be considered under this heading if not elsewhere:

- Non-operation of spillway gates because of human error
- Blocking of spillways with debris
- Non-operation of bottom outlets

During the 1987 floods in south-eastern Norway the percentages of dam owners experiencing problems were reported as follows:

Power failure	50 %
Communication Problems	23 %
Spillways not opened	19 %
Damaged Access Road	17 %
Clogging of spillways	10 %

The above illustrates the high risk of power failures during extreme events; in some environments it may be appropriate to assume that the primary power source will definitely fail. Because of this spillway gates are always provided with a standby power source the reliability of which may itself be questionable. In a recent survey the probability of failure on demand was assessed as between 0.2 % and 1.0 % depending on the details of the particular installations.

For the dam to fail the failure on demand clearly needs to be accompanied by a flood and it may be that the greatest risk to the dam is from the non-operation of all the gates in a flood of relatively modest return period.

Figure 3. 24 metre long by 5 metre high spillway gate. Synchronization between the two ends is not reliable and the gates are at risk of twisting.

Human error in the operation of spillway gates is an important factor since operators will often be reluctant to cause certain flooding downstream. This will particularly be the case if they are subject to high level political pressure not to open the gates. This needs to be factored into the risk calculations.

Blocking of spillways with debris is not strictly a mechanical/electrical problem but there have been a number of serious incidents causing major damage and/or loss of life (Hinks et al, 2003).

The non-operation of a bottom outlet is unlikely to be the main cause of the failure of a dam but it may be an important contributory factor. The problem is often the accumulation of silt or debris in front of the outlet.

Seismic failure

Most of the dams that have failed completely as a result of earthquakes have been small homogeneous dams in Japan, China and India. Another important category of failures are tailings dam, particularly in Chile where there were devastating failures in the earthquakes of 1928 and 1965. For conventional large dams those of greatest concern are those constructed on liquefiable foundations or using liquefiable fill.

CONSEQUENCES OF FAILURE

The methodology provides a mechanism for reducing the consequence of failure to a single number. For the ranking of 33 dams in Albania, Hinks and Dedja (2002) used the number of houses at risk. This worked quite well for relatively small irrigation dams up to 30 m high but is not adequate for large dams where the cost of replacing the dam itself could run into hundreds of millions of pounds. The answer is to calculate the total cost of failure including:

- loss of life.
- loss of housing and commercial property
- agricultural and infrastructure losses
- loss of dam and power station

With the aid of dambreak analyses it should be possible to quantify the above losses, although there may be complications due to uncertainty over the water level in the reservoir at the time of failure.

Loss of life

A particular difficulty arises in determining an appropriate notional cost to allocate to the loss of a human life. It has been suggested that it is inappropriate to put a value on human life and this viewpoint can be readily understood, particularly where the value chosen is much too low. However, it is emphasised that in the context of reservoir risk management, the allocation of a notional cost to the loss of a human life is being done solely to assist in ranking a portfolio of dams by risk and is not meant to reflect on the intrinsic worth of human life.

For overseas work the authors have assigned a notional cost to the loss of a human life by taking the Gross Domestic Product (GDP) per capita of the country concerned and capitalising it at an appropriate rate of interest. In the UK this methodology would give a sum of about £335,000 at 2004 prices assuming capitalisation at 5% rate of interest. This compares with a cost of £1 million to prevent a fatality quoted in the HSE booklet "Reducing Risks, Protecting People" (HSE, 2001). Probably the appropriate notional cost to put on the loss of a life in the UK is somewhere between these two values. However, doubling the assumed cost of human life will often make little difference to the order of ranking by risk.

It is worth noting that priorities for remedial works at a portfolio of dams can be ranked without the need to put a predetermined cost on the loss of a human life. If the cost of remedial works is known at each dam, it is possible to work out what the cost of human life would have to be to justify the expense of those remedial works at each dam. The dams can then be ranked giving the highest priority accorded to the dam where the cost to prevent a fatality is lowest.

In addition to determining the value of each life it is necessary to determine loss of life (LOL) as a proportion of the population at risk (PAR). A number of authors have addressed this issue and various formulae have been proposed which take account of warning time (WT):

- For WT < 15 mins   LOL = 0.5 (PAR)
- For 15 mins < WT < 1.5 hrs LOL = PAR $^{0.56}$
- For WT > 1.5 hrs   LOL = 0.0002 (PAR)

The data from which the above formulae were obtained were all for developed countries and mostly for the United States. LOL may well be greater in developing countries where there is less personal mobility. DeKay and McClelland (1993) have pointed out some of the limitations of these formulae.

Loss of housing and commercial property
The costs of a dambreak associated with damage to housing can be roughly estimated by taking a standard value for each dwelling. If greater accuracy is required higher values can be put on larger houses and lower values on smaller ones.

For some years various levels of damage have been defined as follows in terms of velocity (m/sec) x depth (m) – see Binnie & Partners, 1991:

V x d < 3 m^2/sec   inundation damage
3 m^2/sec < V x d < 7 m^2/sec   partial structural damage
V x d > 7 m^2/sec   total structural damage

The above relationships may understate the damage caused and it is worth noting that in the 2000 floods when 10,000 properties were flooded, the total damage was estimated at £1.3 billion, ie £130,000 per house (Watts, 2003). This compares with a figure of £ 63,000 per house for flooding in Melton Mowbray in 1998 (Kavanagh, 2003)

Agriculture and infrastructure losses
A dambreak is likely to do permanent damage to fields and agricultural infrastructure near to the dam whereas only temporary damage is likely further downstream. Depending on the season there may, however, be extensive damage to crops. Roads and bridges may also be washed away and financial allowance may need to be made for their replacement as well as for the short-term disruption to commerce whilst the bridges are reconstructed.

Loss of dam and power station
For the valuing of dams and power stations, parametric equations have been developed using dam height, dam length, reservoir capacity, installed capacity of power stations etc. This is, clearly, a very simplified approach but it has proved to be more successful than trying to update figures for the original cost of the facilities. The parametric equation used for 24 large dams in the Caribbean was:

$$\text{Cost (\$m)} = 0.65 \times MW + 0.13 \times Mm^3 + 0.52 \times h + 0.065 \times L$$

Where
MW is the installed capacity at the power station in MW
Mm^3 is the capacity of the reservoir in Mm^3
h is the height of the dam in metres
L is the length of the dam crest in metres

Whilst the above equation uses readily available parameters and has proved reasonably successful it cannot be recommended for wider use without careful calibration for the stock of dams to be considered.

Where power stations are underground or a long way downstream of the dam it may be tempting to exclude the cost of their replacement from the estimates on the grounds that they are unlikely to be destroyed. However, if the dam fails, the power station is unlikely to be of much use for several years and expensive alternative generating capacity may have to be installed.

For dams in cascade it will often be necessary to assume that failure of the upstream dam will take those downstream with it.

Other costs
Where dams provide water supply to cities the cost of disruption may be high both in terms of the health of the citizens and in respect of the

development of an alternative source. These, and similar costs, need to be taken into account.

CONCLUSIONS
The methodology described in this paper is suitable for the ranking by risk of a portfolio of dams. The accuracy of the probabilities of failure in absolute terms will depend on the care taken in calculating those probabilities and on the budget available for the exercise. This will, in turn, be dictated by the purpose for which the results are required.

In the words of Cummins et al (2001):

Whilst the precise probabilities and consequences will never be known because each dam is unique and there is a lack of applicable data, these risks can be compared with others faced by the community.

This is just one advantage of seeking to calculate absolute probabilities which form a common language with engineers working in disciplines other than dams.

ACKNOWLEDGEMENTS
The authors would like to acknowledge valuable advice and assistance given by Professor J.Lewin

REFERENCES
Binnie and Partners (1991). Estimation of flood damage following potential dam failure: guidelines". *Foundation for Water Research*, March.
Charles J A (2002). Internal erosion in European embankment dams. Reservoirs in a changing world. *Proceedings of 12th British Dam Society Conference at Trinity College, Dublin,* pp 378-393. Thomas Telford, London.
Charles J A, Tedd P and Skinner H D (1998). The role of risk analysis in the safety management of embankment dams. The prospect for reservoirs in the 21st century. *Proceedings of 10th British Dam Society Conference at Bangor,* pp 1-12. Thomas Telford, London.
Cummins P.J, Darling P.B, Heinrichs P and Sukkar J (2001). The use of portfolio risk assessment in the development of a dam safety programme for Council-owned dams in NSW. *ANCOLD Bulletin No 119*, December.
DeKay M L and McClelland G H (1993). Predicting loss of life in cases of dam failure and flash flood. *Risk Analysis,* Vol 13, No 2 pp 193-205.
Foster M, Fell R and Spannagle M (2000a). The statistics of embankment dam failures and accidents. *Canadian Geotechnical Journal,* Vol 37, No.5, October, pp 1000-1024.

Foster M, Fell R and Spannagle M (2000b). A method of assessing the relative likelihood of failure of embankment dams by piping. *Canadian Geotechnical Journal,* Vol 37, No.5, October, pp 1025-1061.

Gosschalk E.M, Severn R.T, Charles J.A and Hinks J L (1994). An Engineering Guide to Seismic Risk to Dams in the United Kingdom, and its international relevance. *Soil Dynamics and Earthquake Engineering,* Vol 13, pp163-179.

Hartford, D.N.D (1997). *Dam risk management in Canada – A Canadian approach to dam safety.* International Workshop on Risk-based Dam Safety Evaluations, NNCOLD, Oslo.

Health and Safety Executive (2001). *Reducing Risks, Protecting People.*

Hinks J.L and Dedja Y (2002). Rehabilitation of irrigation dams in Albania. Reservoirs in a changing world, *Proceedings of 12th British Dam Society Conference at Trinity College,* Dublin, pp 289-301. Thomas Telford, London.

Hinks J.L, Lewin J. and Warren A.L. (2003). Extreme events and reservoir safety. *Dams and Reservoirs,* Vol 13, No. 3, October, pp 12-19.

Hoeg K (1996). Performance evaluation, safety assessment and risk analysis for dams. *International Journal on Hydropower and Dams,* Vol 3, No.6 , pp 51-58.

Hughes, A.K et al (2000). *Risk management for UK Reservoirs.* CIRIA Report C542.

International Commission on Large Dams (1998). *Dams less than thirty metres high – Cost savings and safety improvements.* Bulletin 109.

Kavanagh, S, 2003, *Personal Communication*

Oosthuizen C, van der Spuy D, Barker M.B and van der Spuy J (1991). Risk-based Dam Safety Analysis. *Dam Engineering,* Vol II, Issue 2.

Watts P J (2003). *Personal Communication,* April.

Development of a probabilistic methodology for slope stability and seismic assessments of UK embankment dams

M. EDDLESTON, MWH, Warrington, UK
L. BEEUWSAERT, System Geotechnique Ltd, St. Helens, UK
J. TAYLOR, MWH, Warrington, UK
K. D. GARDINER, United Utilities, Warrington, UK

SYNOPSIS. The introduction of the "An engineering guide to seismic risk to dams in the UK" in 1991 has led Inspecting Engineers to pay greater attention to the seismic risk of the dams they inspect. For owners of large stocks of dams, such as United Utilities (UU), this has resulted in the need to investigate a large proportion of their dams. In order to proceed in a structured way, UU commissioned a Panel of Experts to advise on a methodology to investigate and analyse their embankment dams and to establish the need for detailed investigation and/or remedial works.

Since the publication of the methodology, which was based on a pilot study of five dams, over 30 further embankment dams have been investigated using the approach. This has not only verified the appropriateness of the initial methodology but has also provided a database of geotechnical information. This information has allowed the methodology to be refined to incorporate probabilistic, in parallel with deterministic analyses. Deterministic analysis suffer from limitations such as the inability to consider variability in the input parameters. Also, there is no direct relationship between factor of safety and probability of failure. Probabilistic slope stability analysis allows for the consideration of variability in the input parameters and it quantifies the probability of failure of a slope. It can be performed using the Monte Carlo method, where a re-running of the analysis is performed using new input parameters estimated from the mean and standard deviation values of the chosen parameters. A distribution of factors of safety is then obtained which can be related to risk of failure. A methodology has been developed to incorporate the results of deterministic and probabilistic analyses, which aligns with current thinking regarding risk assessments.

Long-term benefits and performance of dams, Thomas Telford, London, 2004, 232–246

INTRODUCTION

In order to ensure that a consistent and systematic approach was adopted to investigate the seismic stability of its large stock of embankment dams UU commissioned Bechtel to develop a methodology for seismic investigations in conjunction with a Steering Group of eminent dam engineers (Rigby et al, 2002). The methodology was required to comply with recommendations by Inspecting Engineers under the Reservoirs Act 1975 following the publication of "An engineering guide to the seismic risk to dams in the UK" (Charles et al, 1991) and its associated Application Note published by the ICE and DETR(1998). The methodology utilised conventional effective stress testing and classical soil mechanics theory for the development of slip surfaces. It was recognised that there are alternative approaches but it was considered that this approach would provide information suitable for long term use and for comparison with other studies. The original methodology was introduced in 2000 and has since been used as a basis for the analysis of over 30 of UU's embankment dams.

Since the introduction of the methodology the emphasis placed on risk management has increased (Hughes et al, 2000a and 2000b, Kreuzer 2000). This is leading the dam community to consider the methods used to evaluate embankment slope stability risk. For example Johnston in his Binnie Lecture (2002) commented:

"For the past half century the factor of safety calculated by a limit equilibrium analysis has been the accepted method of assessing stability. Now limit equilibrium's role as the sole or even the best method of analysis is being questioned. The factor of safety faces two challenges. Firstly, from finite element analysis which provides the ability to calculate how a dam will settle (or rise) and move upstream/downstream and how the stresses will change as a response to changing loads. The other challenge comes from advocates of probabilistic risk assessment who suggest that the factor of safety approach disguises the fact that even well built dams are a hazard. The probabilistic approach argues that, since failure cannot be completely ruled out, engineers should define and aim for a target probability of failure."

Bridle (2002a) further suggested that:

"Probability is part of the language of risk, much used and understood by managers and non-engineers. Giving them advice using risk language would therefore help them reach the right decisions about dams and dam safety. Use of this language would help us to consider how safe our dams are, which is important when it comes to the fundamental question of 'are they safe enough? It would also overcome the esotericism of our 'factor of safety' language, which means different things in different contexts."

This paper builds on the experiences of applying the UU methodology and explores the possibility of extending it into probabilistic analyses that align more closely to current thinking on risk management of dams.

DETERMINISTIC APPROACH

Deterministic slope stability analyses compute the factor of safety of a slope based on a fixed set of conditions and material parameters. If the factor of safety is greater than unity, the slope is considered to be stable, if the factor of safety is less than unity, the slope is considered to be unstable or susceptible to failure. Guidance on factors of safety for slope design of new embankment dams is given in "An engineering guide to the safety of embankment dams in the UK" (Johnston et. al, 1999). This approach is adopted in the current methodology with the factors of safety varying with the level of confidence in the data available as detailed in Table 1.

Table 1. Factors of safety used in the deterministic approach

Level of information available/Need for remedial action	Factor of Safety
Based on desk study information and decision charts for deep and shallow slips	at least 1.7
Based on assumed conservative parameters	at least 1.6
Based on the analysis of sufficient field and laboratory testing data	at least 1.5
Remedial works for deep slips	less than 1.3
Urgent attention required for deep slips	less than 1.2

Deterministic analyses suffer from limitations such as the failure to consider variability of the input parameters and inability to answer questions like "how stable is the slope?". Also, there is no direct relationship between the factor of safety and the probability of failure. In other words, a slope with a higher factor of safety may be no more stable than a slope with a lower factor of safety, depending on the nature and variability of the slope materials. For example, a slope with a factor of safety of 1.5, with a standard deviation of $0.5°$ on the angle of shearing resistance used in the analysis, could have a much higher probability of failure than a slope with a factor of safety of 1.2 with a standard deviation of $0.1°$ on angle of shearing resistance. The effect of variations in soil properties is illustrated in Figure 1.

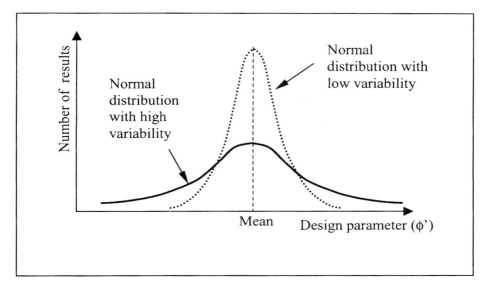

Figure 1. Variability in soil parameters

In the original methodology "worst" credible soil parameters are used in the analyses. The choice of parameters used needs to be considered in relation to the design methodology adopted. CIRIA Reports C580 and 104 (Gaba et al, 2003 and Padfield and Mair, 1984) dealing with retaining wall design define three levels of design parameters for different situations as indicated in Figure 2. As will be discussed later the probabilistic approach generally uses most probable parameters.

Recent investigations, undertaken on UU embankment dams, have allowed an assessment to be made of the effective stress shear strength parameters of a variety of embankment materials. A summary of the results for 10 dams is presented in Table 2. It should, however, be noted that whilst this is useful data, in statistical terms it still only represents a relatively small population. The selection of appropriate parameters is key to the use of both deterministic and probabilistic design methods.

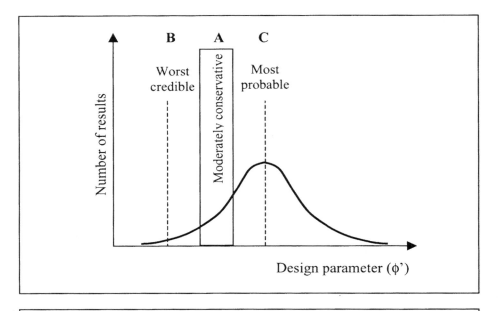

A - The term Moderately Conservative is a conservative best estimate.
Experienced engineers most often use this approach in practice.
B - The Worst Credible value is the worst that a designer could
realistically believe might occur.
C - The Most Probable value is essentially the mean value excluding
obviously anomalous values.

Figure 2. Definition of design parameters as defined CIRIA Reports C580
and 104

Table 2. Soil parameters from selected embankment dams

DAM	Material	Mean	Standard Deviation	No. of samples	Worst Credible Value
	Core	30.8	5.2	10	22
1	Shoulder (clay)	33.5	2.9	17	29
	Foundation	32.4	3.1	11	27
	Core (clay)	30.0	0.8	6	29
2	Shoulder (granular)	32.8	N/A	1	N/A
	Foundation	26.9	1.8	14	24
	Core	32.3	4.5	9	25
Cascade 1	Shoulder (clay)	30.0	3.8	23	24
(3 dams)	Shoulder (gravelly clay)	40.2	2.8	4	36
	Foundation	27.9	2.4	17	24
	Core	28.0	2.4	9	25
6	Shoulder (clay)	28.4	2.4	79	25
	Foundation	27.8	1.8	59	25

DAM	Material	Mean	Standard Deviation	No. of samples	Worst Credible Value
	Core	32.7	3.7	6	27
7	Shoulder (clay)	35.0	3.9	9	29
	Shoulder (gravelly clay)	37.2	3.0	5	32
	Foundation	27.6	3.4	9	22
	Core	31.5	3.3	6	26
8	Shoulder (clay)	31.5	4.1	2	25
	Shoulder (gravelly clay)	42.0	4.3	11	35
	Foundation	34.2	4.4	3	27
	Core	27.6	1.7	4	25
9	Shoulder (gravelly clay)	37.5	2.6	10	33
	Foundation	31.2	4.5	7	24
	Core	31.8	1.8	4	29
10	Shoulder (gravelly clay)	40.8	1.8	3	38
	Foundation	37.2	2.0	2	34

PROBABILISTIC APPROACH

Probabilistic slope stability analysis allows for the consideration of variability in the input parameters and it quantifies the probability of failure of a slope. Probabilistic slope stability analysis can be performed using the Monte Carlo method. Basically, the method consists of re-running the analysis many times by inputting new parameters estimated from the mean and standard deviation values of the chosen parameters. A distribution of factors of safety is then obtained as indicated in Figure 3.

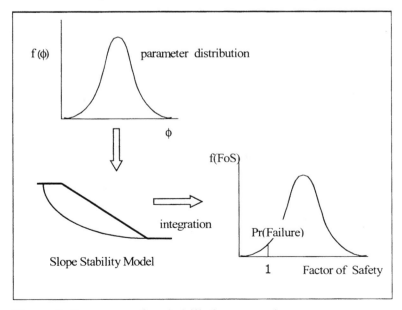

Figure 3. Summary of probabilistic approach

Probabilistic analysis can be performed on proprietary slope stability software such as GEOSLOPE, SLOPE/W. When employing such software the following considerations apply:

i. The use of a probabilistic analysis will not affect the deterministic solution. The software computes the factor of safety of all slip surfaces first and determines the critical slip surface with mean parameters as if no probabilistic analysis is chosen.

ii. A probabilistic analysis is performed on the critical slip surface only.

iii. When the analysis is completed, the factors of safety presented are the minimum, mean and maximum factor of safety of all Monte Carlo trials.

iv. In a probabilistic analysis, the input value of a parameter represents the mean value and the variability of the parameter is assumed to be normally distributed with a known standard deviation.

v. During each Monte Carlo trial, the input parameters are updated based on a normalised random number. The factors of safety are then computed based on these updated parameters. By assuming that the factors of safety are also normally distributed, the software determines the mean and the standard deviations of the factors of safety. A probability distribution function for the factor of safety can then be generated.

vi. The number of Monte Carlo trials required is dependent on the level of confidence and amount of variability in the input parameters. Theoretically, the greater the number of trials, the more accurate the solution. It is important that a sufficient number of trials be carried out. One way to check this is to re-run the analysis with the same number of trials; if the two solutions are different, the number of trials should be increased until the difference becomes insignificant (minimum number of trials is likely to be of the order of 5000).

vii. The probability of failure is the probability of obtaining a factor of safety less than 1.0 and is obtained from the probability distribution function (PDF).

Typical outputs are shown in Figure 4. Figure 4a) shows a situation of a low factor of safety and high probability of failure typical of a pseudostatic analysis of a downstream embankment slope where the analysis is used to estimate deformations using Ambraseys (1972), Ambrayseys and Menu (1988) and Swannell (1994). Figure 4c) shows the situation of a slope with an acceptable factor of safety and a very low probability of failure. Figure 4b) however gives a borderline factor of safety. The question that needs to be addressed is whether a probability of failure of 1 in 2000 is acceptable in relation to the consequence of failure.

ACCEPTABILITY CRITERIA
A number of acceptability criteria based on probability of failure have been found in the literature (based on mean parameters) as detailed in tables 3 to 7.

Table 3. Acceptability Criteria - Smith (1986)

Conditions	Criteria for Probability of Failure	Equivalent Event
Earthworks	10^{-2}	1 in 100
Earth retaining structures	10^{-3}	1 in 1,000
Onshore foundations	10^{-3}	1 in 1,000
Offshore foundations	10^{-4}	1 in 10,000

Table 4. Acceptability Criteria - Santa Marina et al. (1992)

Conditions	Criteria for Probability of Failure	Equivalent Event
Temporary structures with low repair cost	10^{-1}	1 in 10
Existing large cut on interstate highway	10^{-2}	1 in 100
Acceptable in most cases except if lives may be lost	10^{-3}	1 in 1,000
Acceptable for all slopes	10^{-4}	1 in 10,000
Unnecessarily low	10^{-5}	1 in 100,000

Table 5. Acceptability Criteria - Rettemeiere et al.(2000)

Conditions	Criteria for Probability of Failure	Equivalent Event
Likely	10^{-1}	1 in 10
Possible	10^{-2}	1 in 100
Not Impossible	10^{-3}	1 in 1,000
Unlikely	10^{-4}	1 in 10,000
With a degree of probability verging on certainly unlikely	10^{-5}	1 in 100,000
Totally Unlikely	10^{-6}	1 in 1,000,000

Bridle (2000b) related Probability of Failure to the ALARP principle ("as low as reasonably practical") where risks are considered acceptable only if all reasonable practical measures have been taken to reduce risk.

Table 6. ALARP Criteria - Bridle (2000b)

Conditions	Criteria for Probability of Failure	Equivalent Event
Unacceptable	10^{-3}	1 in 1000
ALARP	$10^{-3} - 10^{-6}$	1 in 1000 to 1 in 1,000,000
Negligible	10^{-6}	1 in 1,000,000

Table 7. ALARP Criteria - HSE framework tolerability of risk, (2001)

Conditions	Criteria for Probability of Failure	Equivalent Event
Intolerable	10^{-4}	1 in 10000
Tolerable (ALARP)	$10^{-4} - 10^{-6}$	1 in 10,000 to 1 in 1,000,000
Broadly acceptable	10^{-6}	1 in 1,000,000

The published data indicates a considerable range of values where a balance is needed between both the probability of failure and consequence of failure using for assessment techniques, such as Failure Modes, Effects and Criticality Analysis (FMECA) or Location Cause and Indication methods (LCI) as outlined in the CIRIA report on "Risk management for UK Reservoirs" (Hughes et al, 2000a).

There is some consensus that a probability of failure of 10^{-4} (1 in 10,000) is considered a generally acceptable criterion for slopes where there is a potential for loss of life. Alonso (1976) equates this to the commonly accepted deterministic factor of safety of 1.5 for new build embankment dams. However Christian et al (1994) report probabilities approaching 1 in 1000 for a factor of safety of 1.5.

For a general and conservative approach, which could be considered in parallel with consequence of failure considerations, it is proposed that more stringent criteria be used in preliminary analyses as detailed in Table 8.

Table 8. Suggested acceptable values of Probability of Slope Failure

	Suggested Acceptable Probability of Failure
From desk study information	Less than 2×10^{-6} (1 in 500,000)
Measured dam specific parameters	Less than 1×10^{-5} (1 in 100,000)
Remedial works required	Greater than 1×10^{-4} (1 in 10,000)
Urgent attention required	Greater than 2×10^{-4} (1 in 5000)

These are currently suggested values only and are being evaluated along side the conventional deterministic factors of safety already in use in the existing methodology. It must also be borne in mind that shear strength is not the only parameter that should be considered when using probabilistic methods. Variations in groundwater conditions, inundation of downstream slope due to heavy rainfall, poor drainage or overtopping and the effects of climate change will all need to be taken into account.

PROPOSED METHODOLOGY FOR PROBABILISTIC SLOPE STABILITY ANALYSES

In order to evaluate the possible advantages of the use of probabilistic methods of slope stability analyses of embankment dams, a hybrid deterministic/probabilistic approach is being evaluated for the embankment dams currently under investigation as detailed below.

Choice of parameters

For each parameter (ϕ' and others as required) determine the mean and standard deviation from available testing information.

Probabilistic analysis (mean and standard deviation parameters)

Carry out slope stability analysis including the probabilistic approach to determine the Factor of Safety based on mean parameters.

Deterministic analysis (worst credible parameters)
For each parameter determine the worst credible value. As a guide the worst credible value is sometimes defined as:

mean – (1.64 x standard deviation).

This means that 5% of values are potentially lower than the selected worst credible value. This is similar to the approach used in structural design, in particular for concrete structures e.g. characteristic strength, and also discussed in Eurocode 7 (Driscoll and Simpson, 2001, Cardoso and Fernnandes, 2001, Hicks and Samy, 2002, Samy and Hicks 2002). It should be noted that the choice of 5% is arbitrary and should reflect the risk the designer is prepared to accept on the statistical parameters value, and a degree of engineering judgement is therefore required. Perform deterministic analysis for worst credible parameters and report factor of safety based on worse credible values.

Check slip surface between probabilistic and deterministic analyses
The slip surface geometry obtained from worst credible parameters could potentially be different to that obtained with the mean parameters. If so, re-run the probabilistic analysis with mean parameters on that particular slip.

Report Probabilities of failure and Factors of Safety
Compare and report results obtained.

- Probability of failure from Monte Carlo analysis
- Factor of safety based on worse credible values based on deterministic analysis

A flowchart summarising the proposed methodology is given in Figure 5.

CONCLUSION
The adoption by UU of a rigorous methodology for the seismic investigation of their embankment dams has afforded the opportunity to accumulate and collate a significant common data set for some of its stock of older embankment dams. This has allowed for a detailed comparison of the properties and performance of its assets to enable it to begin to align the findings of conventional deterministic slope stability analyses with probabilistic risk assessment methods. Such an approach allows dam owners to evaluate how safe their dams are in terms of probability of failure. If this is considered in conjunction with the consequence of failure, it will also allow a more rigorous review of the trade off between cost and risk which should improve dam safety management using techniques such as Portfolio Risk Assessments, as described by Hughes and Gardiner (2004).

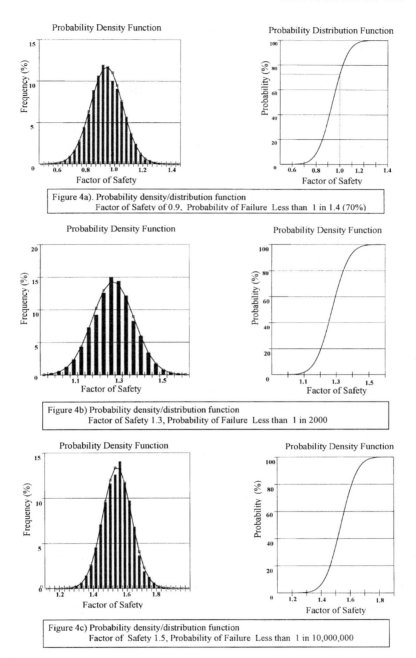

Figure 4a). Probability density/distribution function
Factor of Safety of 0.9. Probability of Failure Less than 1 in 1.4 (70%)

Figure 4b) Probability density/distribution function
Factor of Safety 1.3, Probability of Failure Less than 1 in 2000

Figure 4c) Probability density/distribution function
Factor of Safety 1.5, Probability of Failure Less than 1 in 10,000,000

Figure 4. Typical Probability density/distribution functions

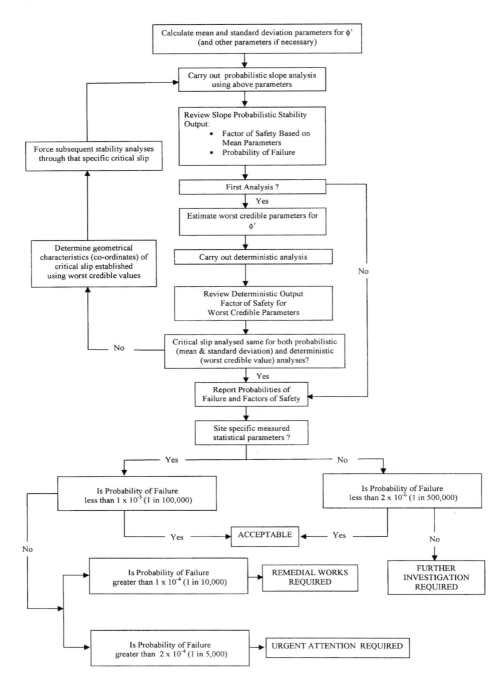

Figure 5. Flow Chart showing the combined Deterministic and Probabilistic Approach

REFERENCES

Alonso, E.E. (1976). Risk analysis of slopes and it's application to slopes in Canadian sensitive clays. Geotechnique Vol. 26, 453-472.

Ambraseys, N.N. (1972). Behaviour of foundation materials during strong earthquakes. Proceedings 4th European Symposium on Earthquake Engineering, London, Vol 7, 11-12. Sofia, Bulgarian Academy of Sciences

Ambrayseys, N.N. (1998). Earthquake induced ground displacements. Earthquake Engineering and Structural Dynamics (Special Issue) Vol. 16, 985-1006.

Bridle, R.C. (2000a) Dams 2000, British Dam Society, 10th Biannual Conference, Bath, Discussion, 104.

Bridle, R.C. (2000a) Dams 2000, British Dam Society, 10th Biannual Conference, Bath, Discussion Conference Closing Remarks, 155.

Cardoso, D.A.M. and Fernandes M.M. (2001) Characteristic values of ground parameters and probability of failure in design according to Eurocode 7. Geotechnique 51(6) 519-531.

Charles, J.A, Abbiss, C.P, Gosschalk, E.M and Hinks, J.L. (1991). An engineering guide to seismic risk to dams in the United Kingdom. Building Research Establishment Report Publ. BRE, Watford.

Christian, J. T, Ladd C. C. and Baecher G. B. (1994) Reliability applied to slope stability analysis. Journal of Geotechnical Engineering, ASCE. 120(12) 2180-2207.

Driscoll, R. and Simpson, B. (2001) Eurocode 7: Geotechnical Design. Proc. ICE, Civil Engineering, Vol. 144 Special Issue 2. 49-54.

Gaba, A.R, Simpson, B, Powrie, W, Beadman, D.R. (2003). Design of retaining walls embedded in stiff clay. Construction Industry Research Information Association (CIRIA) Report No. C580.

Health and Safety Executive, (2001) Reducing Risks, Protecting People.

Hicks, M. A. and Samy, K. (2002) Reliability based characteristic values : A stochastic approach to Eurocode 7. Ground Engineering 39, No 12, 30-34.

Hughes, A.K, Hewlett, H. W. M, Morris, M, Sayers P, Harding A, and Tedd, P. (2000a) Risk Management for UK Reservoirs. CIRIA, Report C542, 213pp.

Hughes, A.K, (2000b) UK Developments in risk management for reservoirs. Transactions of the 20th International Congress on large Dams, Beijing, Vol. 1, 67-70. Invited contribution to Q 76:

Hughes, A.K. and Gardiner K. (2004) Portfolio Risk Assessments in the UK; A perspective. Dams 2004, Thomas Telford, London.

ICE/DETR (1998). An Application Note to An Engineering Guide to Seismic Risk to Dams in the United Kingdom. The Institution of Civil Engineers. Thomas Telford Ltd, London.

Kreuzer, H. (2000) The use of risk analysis to support dam safety decisions and management. General Report. Transactions of the 20th International Congress on large Dams, Beijing, Vol 1. Q76 –R41, 769-896.

Johnston, T. A. (2000) Binnie Lecture "Taken for Granted", Dams 2000 British Dam Society, 10th Biannual Conference, Bath, Discussion, 143.

Johnston, T.A, Millmore J.P, Charles, J.A. and Tedd, P. (1999) An engineering guide to the safety of embankment dams. BRE Report 363, Watford.

Padfield, C.J. and Mair, R.J. (1984). Design of retaining walls embedded in stiff clay. Construction Industry Research Information Association (CIRIA), Report No. 104.

Peterson, J.L. (1999). Probability analysis of slopes. MSc Thesis University of West Virginia.

Rettemeie, K., Falkenhagen, B. and Kongeter, J. (2000) Risk Assessment – New Trends in Germany. Transactions of the 20th International Congress on large Dams, Beijing, 625 - 641

Rigby, P.J, Walthall, S and Gardiner, K. D. (2002) A methodology for seismic investigation and analysis of dams in the UK. Reservoirs in a changing world, Thomas Telford, London, 126-140.

Samy, K. and Hick, M. A. (2000) A numerical solution to characteristic values for geotechnical design. Proc. 10th UK Conference of the Association for Computational Mechanics in Engineering. Swansea, 137-140.

Santa Marina, J. Altschaeffi, A. and Chameau, J.(1992). Reliability of slopes incorporating quantitative information. Transport Research Record 1343, 1-5.

Smith, G.N. (1986). A suggested method of reliability analysis for earth retaining structures. TRRL Report No. 46.

Swannell, N.G. (1994) Simplified Seismic safety evaluation of embankment dams. Dams and reservoirs Oct. 4(3), 17-19.

Ridracoli Dam: surveillance and safety evaluation reported on internet page

P.P. MARINI; P. BALDONI; F. FARINA; F. CORTEZZI - Romagna Acque, Forlì, Italy
A. MASERA - Enel.Hydro, ISMES Division, Bergamo, Italy

SYNOPSIS.   During a period of several seismic events that took place in January 2003 in the valley downstream of the Ridracoli arch gravity dam, inhabitants and local Authorities requested information about the safety conditions of this important structure. To satisfy such an expectation, the Manager of the Romagna Acque, owner of the dam, launched a project aimed at providing such information. Communication through the Internet web was decided and an Internet page was prepared, reporting the safety conditions of the dam, with respect to hydrologic, hydraulic, static and seismic aspects and the resulting surveillance activities.
Methodologies and operative techniques are today mature and available for an effective evaluation and surveillance of dam safety, and for presentation to the resident population living downstream of the dam. Data collected at the Ridracoli dam site by several monitoring systems are in fact automatically processed and interpreted in order to evaluate the different aspects affecting the safety of the dam and the protection of the downstream valley.

The experience gained using automatic monitoring and a knowledge based support system is used to obtain on-line evaluation, explanation and interpretation of dam's behaviour, identifying surveillance activities to manage anomalous trends or to minimize critical situations due to flooding.
All information are summarized and presented on the Internet page. In addition, for the people living in the downstream area, the presentation is available on a video, located at the City Hall.

INTRODUCTION
The selected approach and the methodology takes advantage of the automatic monitoring systems (which encompass hydrologic-hydraulic, static and seismic structural aspects) and of the on-line analysis of structural dam behaviour, compared to theoretical models, in order to identify safety

anomalies, if any. From these analyses the management of surveillance is defined requiring *ad hoc* inspections, collection and analysis of further information in order to define the safety condition of the dam.

The operational procedures for surveillance management have been evaluated by the National Board Authorities for Dams and by the Protezione Civile (Department of Civil Protection), defining the conditions and the thresholds that could induce alert conditions for the dam and for the downstream valley.

RIDRACOLI DAM

The Ridracoli arch-gravity concrete dam (height 103.5 m and crest length 432 m) closes a very wide U-shaped valley in the Tuscan-Romagna Apennines in Italy. The storage reservoir is intended for water supply to 37 communities in the Forlì and Ravenna Provinces, including the main towns and the San Marino Republic.

Figure 1: Ridracoli dam

The reservoir was filled completely for the first time in 1986 and nowadays the dam is commissioned for normal operation.

MONITORING SYSTEM

To control the Ridracoli dam site, that is the catchment area, the structure, the foundation, the reservoir banks and the slopes of the downstream rocky formation, a large monitoring network has been installed during the construction of the dam. An automatic monitoring system, centralized in the warden house via cable, reads most of the measurements (259 sensors are automatically recorded, on a total of 971). Many instruments were installed for a detailed monitoring of the structure's behaviour during construction and the first filling phase. In the current normal operation, the on-line surveillance of dam performance is based on a subset of measurements.

In parallel to the on-line system, the off-line surveillance activities performs analysis of the measurements, automatically or manually recorded, verifying the dam's behaviour, by the comparison to the prediction of the three-dimensional F.E. model.

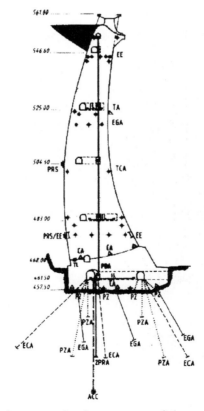

Figure 2: Ridracoli dam: monitoring system of the crown section

HYDROLOGICAL AND HYDRAULIC ASPECTS

Hydrological and hydraulic aspects are fundamental with respect to the safety of a dam. The reservoir capacity is 33.06 Mm^3, the catchment area is about 37 km^2. At the Ridracoli dam site a monitoring system has been installed for reservoir monitoring and management, in particular for the management of the water supply and to foresee flood events. On the basis of the measured data and of the water balance in the reservoir, the inflow and the outflow are computed and both displayed in the Internet page.

If high floods are expected and, in any case, if the outflow is excess of 50 m^3/sec (that corresponds to 10% of the spillway capacity) those responsible for the dam have to start up the extraordinary surveillance and alert the Civil Protection Dept. The surveillance condition and the dam safety conditions is reported on Internet.

STATIC STRUCTURAL BEHAVIOUR ANALYSIS

A decision support system (named MISTRAL) was installed in 1992 on a personal computer connected to the automatic monitoring system in the acquisition Centre, located in the warden house near the dam.

Mistral is a decision system for evaluating, explaining and filtering the information collected by the most important instruments connected to the automatic monitoring system, providing on-line interpretation of the behaviour of the structure in order to support the activity of the personnel responsible of the safety surveillance, requiring his intervention in case of anomalous situation, if any.

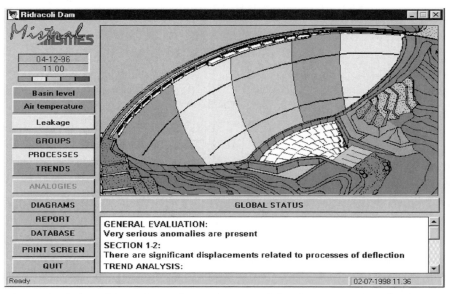

Figure 3: Mistral Interface: General state of the dam (test situation)

The on-line system makes it possible to verify the state of each measurement with respect to threshold levels (physical threshold, measured rate of variation and reference structural model), using knowledge about the significance and, reliability of each instrument, and evaluates the current state of the dam and of any elementary structural part, identifying any anomalous process and verifying the reliability of the measurements by consistency checks. The Mistral system currently operates taking into account the data collected hourly by 40 instruments.

Mistral displays the results of the analysis through a colour-based graphical interface that represents the state of the measurements, of the processes, of each section and of the entire structure under evaluation giving relevant explanations.

If the processed "global status" of the dam corresponds to alert conditions, (level 5 or red colour in the Mistral interface) the extraordinary surveillance enters in force and the Civil Protection Dept is alerted. The surveillance and the dam safety conditions are reported on the Internet page.

SEISMIC STRUCTURAL BEHAVIOUR ANALYSIS
The seismic monitoring system is made of four accelerometric stations and by one seismic station (each station has instruments installed in the three directions). The system allows to measure both the input ground motion and the structural answer of the dam.

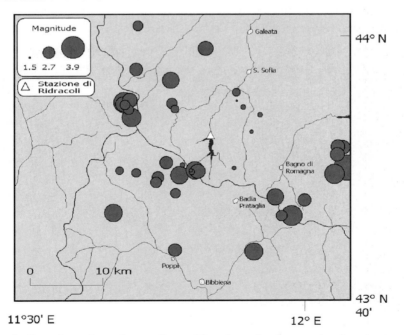

Figure 4: Local earthquakes collected by the seismic monitoring system

In the last period (1995-January 2003) the system collected 128 earthquakes that exceeded the trigger threshold. 63 were far from the dam site and 65 local (epicenter distance nearby 25 km from the dam site, as suggested by Dam's Authorities).

If the peak ground acceleration of the earthquake, measured at the base of the dam, is higher than 0.20 g (that corresponds to the seismic value obtained by the physical model that shows the beginning of cracks in the upper part of the dam, near the spillway sill), those responsible for the dam have to start up extraordinary surveillance (such as *ad hoc* inspection and collection of the whole measurements) and alert the Civil Protection Dept. The surveillance condition and the dam safety condition is reported on the Internet page.

The recorded seismic measurements, are periodically stored into the historical data base and processed to analyze the dam's behaviour, in comparison to the calculated one by a three-dimensional F.E. model and to the dynamic response retrieved from the vibrating test data.

INTERNET PRESENTATION
In the Internet home page of Romagna Acque much information is available relevant to the company, the water supply system and the dam.

Figure 5: On-line images from the dam

In addition to the quality parameters of the water, the production and distribution of drinkable water, many data about the dam are reported, as illustrated in the following figures. The information reported is up-dated every hour.

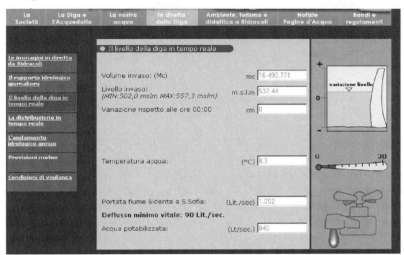

Figure 6: On-line water level in the reservoir

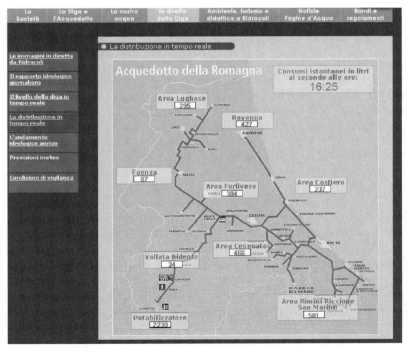

Figure 7: Water supply distribution

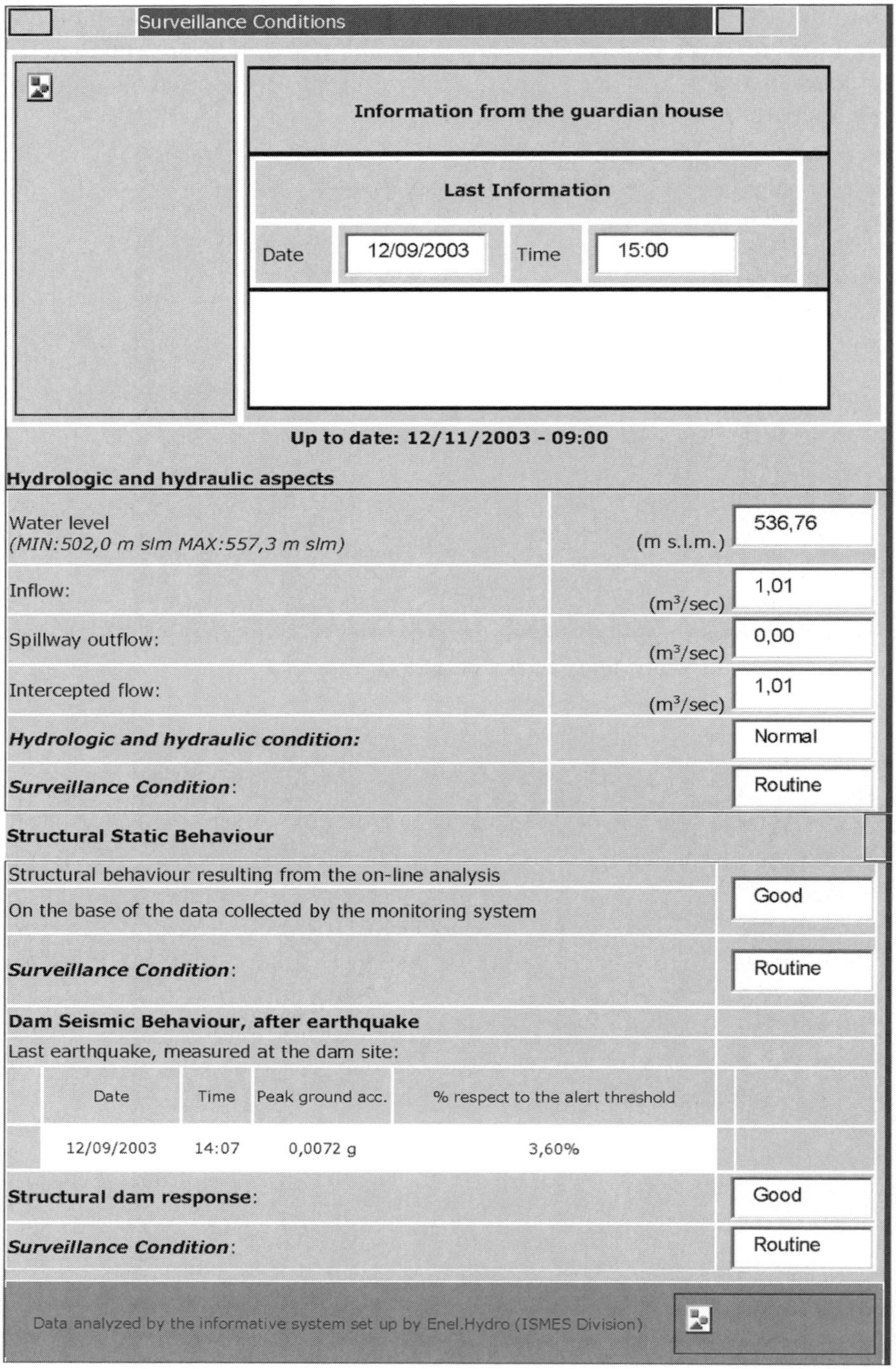

	Surveillance Conditions	

Information from the guardian house

Last Information

Date	12/09/2003	Time	15:00

Up to date: 12/11/2003 - 09:00

Hydrologic and hydraulic aspects

Water level *(MIN:502,0 m slm MAX:557,3 m slm)*	(m s.l.m.)	536,76
Inflow:	(m³/sec)	1,01
Spillway outflow:	(m³/sec)	0,00
Intercepted flow:	(m³/sec)	1,01
Hydrologic and hydraulic condition:		Normal
Surveillance Condition:		Routine

Structural Static Behaviour

Structural behaviour resulting from the on-line analysis On the base of the data collected by the monitoring system	Good
Surveillance Condition:	Routine

Dam Seismic Behaviour, after earthquake

Last earthquake, measured at the dam site:

Date	Time	Peak ground acc.	% respect to the alert threshold	
12/09/2003	14:07	0,0072 g	3,60%	

Structural dam response:	Good
Surveillance Condition:	Routine

Data analyzed by the informative system set up by Enel.Hydro (ISMES Division)

Figure 8: Safety and Surveillance Conditions (translated from Italian)

In the previous figure, the dam surveillance condition, together with the safety evaluation, is reported with respect to hydraulic, static and seismic safety assessment.

On average, the website is visited 40-50 times each day.

The local administration and residents downstream of the dam have given positive indications even though they report that some of the information provided is not always easy to interpret. In particular, reference is made to the difficulty of interpreting the definitions of *Surveillance Conditions*. In view of this feedback, the website Introductory Page is now in the process of revision by the incorporation of additional explanatory notes.

CONCLUSION

Monitoring and data analysis are primary parts in managing the safety of dams by risk assessment methodology. At the Ridracoli dam the on-line data analysis and the surveillance management have became a part of the safety procedures of the dam. The results of such activities are available to the population living in the downstream valley, by Internet network and by video installed in each City Hall.

This is the first time in Italy that the results of risk assessment methodology has been used on-line and available on the Internet.

REFERENCES

R. Riccioni: *Automated monitoring of Ridracoli dam: Organizational aspects*, (Q.52) 14th Icold Congress, Rio de Janeiro, 1982

Icold Bulletin 68: *Monitoring of dams and their foundations – State of the art*, Appendix E – Report by the Italian National Committee

F. Bavestrello, A. Masera, P.P. Marini: *Ridracoli dam: Design, Construction and Behaviour during the first filling*, (Q.60) 16th Icold Congress, San Francisco, 1988

S. Lancini, A. Masera, F.G. Piccinelli, F. Farina: *Ridracoli dam- A Decision Support System for Managing Dam Surveillance*, (Q.60 – R.30) 20th Icold Congress, Beijing, 2000

Reservoirs Act 1975 - Progress on the implementation of the Environment Agency as Enforcement Authority

I.M. HOPE, Environment Agency
A.K. HUGHES, KBR

SYNOPSIS. The Water Act 2003 has established a new role for the Environment Agency (the Agency); that of the Enforcement Authority for the Reservoirs Act 1975 in England and Wales. Currently some 140 Local Authorities fulfill this role.

The Agency is preparing to commence this new role from 1st October 2004 and this paper describes the process being followed and progress to transfer this new duty. It also sets the scene for the subsequent increase in the role of the Enforcement Authority driven by the Water Act 2003, namely the requirement for Undertakers to produce Flood Plans for reservoirs.

INTRODUCTION

The Reservoirs Act 1975 was implemented between 1986 and 1987, (Charles, 2002) and only applies to 'large raised reservoirs' with a capacity greater than 25,000 cubic metres of water that do not fall within the scope of the Mines and Quarries (Tips) Act 1969.

The Act covers some 2600 reservoirs in the United Kingdom, 2000 of which lie in England and Wales. However, the jurisdiction of the Agency is confined to England and Wales. The Water Act 2003 has not affected the 600 reservoirs that will continue to be regulated by the 32 Enforcement Authorities in Scotland.

In England and Wales some 140 Local Authorities (Unitary Authorities, County Councils and Metropolitan Boroughs) are currently responsible as Enforcement Authority for 2000 reservoirs. For local authorities this role has attracted a varying response, often co-ordinated by differing departments. This has led to an inconsistent application of the Act and has ultimately driven the need for consistency led by one body. This requirement was recognised in a review of the Reservoirs Act 1975 and

Long-term benefits and performance of dams, Thomas Telford, London, 2004, 256–267

reported to the British Dam Society (BDS) by Simms and Parr (1998) at the 10th BDS Conference held in Bangor in September 1998.

THE WATER ACT 2003
The Water Act 2003 transfers the responsibility for Enforcement to the Agency. It also establishes the requirement for Undertakers to prepare Flood Plans when directed by the Secretary of State. In addition the provisions of the Reservoirs Act 1975 are to apply to the Crown. Further details can be viewed on the Defra website www.defra.gov.uk.

THE ENVIRONMENT AGENCY
The Environment Agency is a 'Non Departmental Public Body' that reports to the Department for Environment Food and Rural Affairs (Defra). Its vision is for 'A better Environment in England and Wales for present and future generations'. Its role is to be an efficient operator, modern regulator, influential advisor, effective communicator and champion of the environment. These are underpinned by its Values which include an outcome driven approach, being firm and fair and open to change.

The Agency's functions are extensive and its main operating role is Flood Risk Management. Part of its massive Flood Risk Management infrastructure includes 119 flood storage reservoirs that come under the remit of the Reservoirs Act 1975. It also has considerable regulatory powers, responsibility and experience. For example it is responsible for over 1600 authorisations in process industries, more than 100,000 consents to discharge and over 7500 waste management licences. Further details on the Agency can be found on its website:- www.environment-agency.gov.uk.

PRINCIPLES OF MODERN REGULATION
Society demands high environmental and safety standards. The business world rightly expects greater regulatory efficiency, whilst minimising bureaucracy so that compliance costs are kept to a minimum. These potentially conflicting demands can be met by a modern regulatory regime.

Five principles have been set out by the Better Regulation Task Force (2003) to achieve this aim and they are:-

- **transparent** - rules and processes which are clear to businesses and local communities
- **accountable** - by reporting regularly on actions and performance
- **consistent** - by applying the same approach and standards within and between sectors and over time

- **proportionate** - by allocating resources according to risks
  (or risk based) involved and scale of outcomes which can be
  achieved
- **targeted** - the outcome must be central to the planning
  (or outcome focused) and assessment of performance

The Agency has included a sixth principle:-

- **practicable** - to provide clarity to business on how they
  comply

Through the application of the principles of Modern Regulation the Agency aims to be perceived as an effective regulator and to achieve a high degree of public confidence in its activities. The Agency believes in firm but fair regulation and has developed an Enforcement and Prosecution Policy to reflect this. Included in this policy are the factors to be considered in deciding whether or not to prosecute. The Reservoirs Act 1975 empowers the Enforcement Authority to take both Civil and Criminal proceedings. In preparing for this new role the Agency will apply its Enforcement and Prosecution Policy which enshrines the principles of Modern Regulation and apply them to the enforcement of the Reservoirs Act 1975.

ROLES UNDER THE RESERVOIRS ACT 1975
The Reservoirs Act 1975 is principally designed to be self regulating with the onus on the Undertaker to keep records, manage the dam and its infrastructure to a specified operating regime, and procure all necessary services and works. This is in line with the role of a regulated party defined by the Principles of Modern Regulation. The Reservoirs Act 1975 clearly establishes the role of the Undertaker and defines its responsibilities. The term 'Undertaker' has been specifically adopted in preference to 'Owner' as the role of Undertaker, i.e. person or persons that use and control the reservoir, may not always be the owner.

The Reservoirs Act 1975 also recognises distinct roles of engineer (Panel Engineers),each of which are required to be re-appointed to their respective panels every five years. Only a qualified civil engineer who is a member of the appropriate panel can carry out the statutory requirements of the Act relating to engineering aspects of construction, on-going supervision and inspection.

ROLE OF THE ENFORCEMENT AUTHORITY
The Enforcement Authority has a legal duty to ensure that the Undertakers' self regulatory regime is fully compliant (i.e. effectively a compliance audit

role) and can take necessary actions to secure compliance. The main duties of the Enforcement Authority include:-

- Maintaining a register of reservoirs, and making this information available to the public.
- Ensuring that the Undertaker has appointed a Supervising Engineer.
- Ensuring that the Undertaker commissions regular inspections of the dam by an Inspecting Engineer.
- Enforcing the Act by influencing, warning, cautioning and ultimately prosecuting non-compliant Undertakers.
- Commissioning essential works required in the 'Interests of Safety' in the event of non-compliance and recouping full costs incurred from the Undertaker.
- Acting in an emergency if the Undertaker cannot be found.
- Producing a Biennial Report to Defra and to the Welsh Assembly Government (WAG) of enforcement actions taken.

PROJECT MANAGEMENT OF NEW DUTY

The Agency has fully embraced the principles of project management to introduce all new duties to its business. This ensures that business aims are fully delivered to the specified time, cost and quality. For the introduction of this new enforcement role a Project Board has been established chaired by the Head of Flood Risk Management. The Project Board comprises key Agency personnel, a representative from: Defra, the British Dam Society (who is also an A.R. Panel Engineer), the Reservoirs Committee of the Institution of Civil Engineers; a current Enforcement Authority, and the Technical Manager – Reservoir Safety.

A project team, led by a Project Manager, reports regularly to the Project Board. The team, all of whom are part time consultees to the project, represent key elements of the Agency's business which include the following:-

CIS (Corporate Information Systems)	Finance
Legal	Customer Services
Enforcement Processes	Personnel
Records Management	Media Relations
Debt Recovery	Emergency Management
Procurement	Planning Guidance

A comprehensive approach to project implementation has been adopted because of the impact that this change project will have across the Agency business.

PROJECT PLAN

The Project Plan has four elements supported by a comprehensive communications strategy:

1. Process definition
2. Delivery of suitable information technology
3. Acquisition of existing records
4. Recruitment of permanent staff

Process definition includes definition of the new duty, development of a Vision, statements of policy, business processes and work instructions supported by guidance and training.

A business case is currently under review for suitable information technology software termed the Reservoir Enforcement and Surveillance System. The acquisition of existing records is dealt with in more detail below, as is the proposed permanent structure.

VISION FOR RESERVOIR SAFETY

The Project Board has endorsed the following Vision for the overall guiding principles for the execution of the new duty:

"We will assure Reservoir Safety by robustly applying the principles of Modern Regulation in our enforcement of the Reservoirs Act 1975"

The Project Board have received proposals defining the strategic objectives for Reservoir Safety and how these objectives will be translated into performance measures.

The next stage of this process will witness the new team owning these strategic objectives. These objectives will define their business plans and training plans, which will be underpinned by the core values of the Agency.

Crucially, performance monitoring will be instigated against objectives set. The new performance measures derived for the team will also feed into and ultimately be evaluated against the Agency's Water Management Directorate's Balanced Business Scorecard. The Scorecard is designed to provide a high level summary of performance for use principally at Director level and provide early warning of under performance.

ELECTRONIC DOCUMENT MANAGEMENT

The business world is making increased use of electronic document management. This was for example reflected in a paper by Stewart (2002) where he expanded on a system developed for Severn Trent Water to hold all documentation electronically. Increased transfer to Electronic Document Management (EDM) is also reflected in the Agency's EDM strategy, which in part is driven by the targets set by the Modernising Government White Paper in 1999. EDM is also regarded as the most effective way to manage reservoir records in the future. It is planned that the prime method of receipt of notifications, certificates, etc., will be electronically. This contrasts with the current method of the majority of reservoir record retention - paper files. When all the current paper records are amalgamated they will require some 30 filing cabinets to retain them, hence the need for EDM.

TRANSFER OF DATA & INFORMATION
In order to establish the quantum of work and resources required to transfer data and information several Enforcement Authorities were visited at the commencement of the project. The visits were hosted by the lead officer for the authority and register and files reviewed.

Following review, a 3 phase approach to the process of transfer of data and information was adopted:

Phase I   A trial of the process based upon 10% of the total volume (which was completed by December 2003)

Phase II   The capture of the remaining 90% of records (refined by learning from Phase I) and due for completion by July 2004

Phase III   The 'Mop Up' of documents created/filed by existing Enforcement Authorities after Phases I & II.

This approach was designed to ensure that:-

1.   From the 'go live date' of 1 October 2004 all necessary information would be electronically available to the Agency.
2.   Potential issues would be considered and resolved, and important knowledge would be acquired at the earliest possible date.
3.   All enforcement processes would be in place, trialed and operable before the duty commenced.

A questionnaire was despatched to existing Enforcement Authorities to determine:-

- confirmation of contact and reservoir details
- how records were held
- what processes, and computerised systems, if any, were in use
- advice and guidance that should be passed to the Agency

The questionnaire also made it clear that the Agency would not commence its role until 1st October 2004, but wished to ensure it was fully sighted on emerging enforcement issues.

Currently there is no single, comprehensive register of reservoirs for England and Wales. The process of transferring the individual registers from the existing Enforcement Authorities and subsequent checking is enabling this to be compiled.

RESERVOIR ENFORCEMENT AND SURVEILLANCE SYSTEM (RESS)
A consistent, national and effective business tool is required to enable the Agency to effectively undertake the enforcement role.

The system adopted needs to achieve the following four key business needs:-

- A reservoir register to hold structured information about reservoirs (i.e. inspections, actions, etc.)
- Electronic Document Storage and Management
- Standard letter generation
- Workflow support for Regulation business processes

Five options have been appraised that range from a manual process wholly reliant on paper records and increased staff numbers to the adoption of an existing Agency Permit Administration System (P.A.S.) which incorporates automated workflow. A standalone Electronic Document Management (EDM) has also been considered. The options have been assessed against a range of criteria that include:-

- Agency Environmental Vision and Technical Strategies
- Modernising Regulation
- Business Risk
- Wholelife costs (i.e. development, maintenance and operating activities)

The foundation for all the options considered is the processes that are defined by the legislation. It is from these processes that an automated system will be designed or a manually based team trained and managed.

BUSINESS PROCESSES
A Guide to the Reservoirs Act 1975 (ICE, 2000) defines six core procedures for compliance and enforcement. By fundamentally reviewing the Reservoirs Act 1975 some 18 distinct procedures have been established to resolve potential non-compliant acts or offences.

From these procedures (or business processes) the activities and performance objectives of each Agency department can be defined and interaction with outside parties, e.g. Inspecting Engineers, consistently managed.

The Agency Management System (AMS) provides a standard structure for all business processes. Once business processes are developed and approved, they are published and available on the Agency's Intranet. The types of AMS business processes that already exist range from guidance on the resource allocation for waste licence pre-application to the process to be adopted in managing and maintaining the Agency's own reservoirs. The application of AMS to all business processes defined by the Reservoirs Act 1975 will ensure a consistent and transparent approach.

COMMUNICATIONS AND STAKEHOLDER MANAGEMENT

The management of the interface with key reservoir industry organisations is regarded as crucial to the successes of this project and to the future effectiveness of the role. It is essential that the Agency project the profile that meets both the aspirations of the reservoir industry and Government. At an early stage of the project, Stakeholder Mapping was employed as a tool to define and manage the communication strategy for both internal and external stakeholders. To date, constructive working relationships have been established with:-

- Government – including Defra, WAG and the Scottish Executive
- The Health and Safety Executive
- Panel Engineers
- Key Undertakers
- Institution of Civil Engineers (ICE) and the British Dam Society

Comprehensive engagement with key Agency departments is in the process of being developed. An intranet site has been established and an internet site is planned before October 2004.

This paper later reflects on the increased profile to the reservoir industry and wider interest groups that the Agency will have with Reservoirs. The Agency is currently developing a National Customer Contact Centre (NCCC) based in Rotherham which will be fully briefed to handle routine enquiries from the public, for example educational enquiries.

As an active Enforcement Authority, the Agency will be pro-actively engaging with the profession and wider reservoir community in seeking both compliance with the Reservoirs Act 1975 and fostering improvements in approach.

As with any change management process the need for effective communication throughout the process is paramount. In establishing the pro-active relationship with the reservoir industry, the Agency is keen to ensure that changes are well forecast and a 'no surprises' culture engendered. The proposed automated workflow system will establish a continued dialogue with respective parties at relevant stages, for example: checking that a periodical inspection has been arranged, checking that measures in the interests of safety have been completed or checking the appointment of the Supervising Engineer. For some Undertakers this will represent a significantly different approach as they establish an ongoing relationship with the new Enforcement Authority.

NEW TEAM STRUCTURE
In undertaking this new duty, the Agency will build on its existing strengths which include significant enforcement expertise and local awareness through Flood Risk Management. A core team will be formed in Exeter to provide the routine surveillance and the enforcement capability. This will provide a 'One stop shop' approach to the compliance monitoring and enforcement role. The Technical Manager – Reservoir Safety will provide leadership and management to the team that will comprise two key elements:

1.   Surveillance
2.   Enforcement

The surveillance team will manage the reservoirs register, initiate all routine correspondence, and populate the RESS and handle detailed enquiries. The enforcement team will co-ordinate all enforcement across England and Wales and manage the portfolio of enforcement cases. They will take over from the influencing element of enforcement led by the surveillance team and be responsible for the warning, cautioning and prosecuting stages. The services of Panel Engineers will be procured to advise and work with the Agency in order to achieve the appropriate regulatory outcome in accordance with the Reservoirs Act 1975. The Enforcement process will be assisted by a representative from the Agency's Area Flood Risk Management Regulation Team. This local representative will provide the essential 'eyes and ears' on the ground. One of their first roles will be to check that the Register of Reservoirs is complete for their Area.

TRAINING AND DEVELOPMENT
A key element of the project will be the provision of training for all staff with a role to play in the enforcement of the Reservoirs Act 1975. The Agency is progressing towards Investors in People (IiP) accreditation and thorough training and development for its staff is seen as crucial.

To properly enforce such a comprehensive Act all potential scenarios are being worked through as part of design of training. As an illustration, Powers of Entry, and Police and Criminal Evidence Act (PACE) training will be provided to ensure that local staff are fully able to apply Section 17 of the Reservoirs Act 1975 which states a person duly authorised in writing by an Enforcement Authority may at any reasonable time enter upon the land on which a reservoir is situated. This is also particularly important in view of the potential recourse to compensation that could arise from a 3rd party by virtue of Section 18 of the Reservoirs Act 1975. Section 18 deals with compensation to third parties where the Enforcement Authority exercises any powers conferred on it by Section 17. Compensation is payable by the Enforcement Authority where damage is caused. Such compensation is deemed to be a reasonable expense incurred by the Enforcement Authority and is recoverable from the Undertaker.

FLOOD PLANS

The Water Act 2003 establishes the requirement for Undertakers to provide Flood Plans when directed to by the Secretary of State. Reference is made to this requirement in the Defra letter to Water Company Chief Executives reported in Dams & Reservoirs June 2003. It is proposed that these plans will have to be produced from April 2005 and industry provided with a five year rolling programme for their production. In order for Defra to establish how these plans are to be prepared, their constituent parts, the consultees and their respective roles, a research and development (R&D) project had been let to Kellogg Brown & Root (KBR). By the end of 2003 a proposed format had been established based on the Control of Major Accident Hazard Regulations 1999, together with proposed prioritisation criteria for preparation of plans. Defra propose to embark on a major consultation process with all affected parties (i.e. Undertakers, Panel Engineers, Emergency Planners, etc.) in 2004. This programme fits with the development of the Civil Contingencies Bill.

INCIDENT REPORTING SYSTEM

A further R & D project let by Defra to KBR has been to develop an incident reporting (and investigation) system for UK dams. The outcome from this contract was presented to the BDS on 27th October 2003 when an industry wide consultation process was initiated. The aim of the system is to improve the safety of UK dams by promoting awareness of safety issues, learning from experience of others and identifying research needs. A targeted questionnaire to key representatives of the UK dam industry had produced support for the Enforcement Authority to co-ordinate this role. Currently it is planned that the Agency commence this new role from June 2005.

THE FUTURE

The scene is set for an interesting and increasingly important new role for the Agency as it takes on its responsibilities as the Enforcement Authority. This will bring with it considerable opportunities for improved regulatory efficiency and partnership activities establishing closer links with the reservoir community. For example the Agency will be working closely with Defra and WAG, together with the wider reservoir community, to determine new policies on the application of the Reservoirs Act and influence future policy changes. Through the transfer of enforcement for reservoir safety to just one body, the Agency will achieve a consistent, systematic approach to achieving compliance with the Reservoirs Act 1975 and thus accomplish the Vision established for Reservoir Safety.

ACKNOWLEDGEMENTS

The authors wish to thank the Environment Agency for permission to publish this paper.

APPENDIX

Information Sources

www.environment-agency.gov.uk
www.defra.gov.uk
www2.defra.gov.uk/db/panel/default.asp
www.britishdams.org

REFERENCES

Better Regulation Task Force, 2003. *Principles of good regulation.* Great Britain: Better Regulation Task Force

Charles, J.A. 2002. A historical perspective on reservoir safety legislation in the United Kingdom. IN: P. Tedd, ed. *Reservoirs in a changing world: proceedings of the 12th conference of the BDS held at Trinity College,* Dublin, 4-8 September 2002. London: Thomas Telford, 494-509

Environment Agency, 2003. *Delivering for the Environment : a 21st century approach to regulation.* Bristol: Environment Agency

Environment Agency, 2003. *Making it happen : corporate strategy 2002-07.* Bristol: Environment Agency

Environment Agency, 2001. *The vision for our environment : making it happen.* Bristol: Environment Agency

HMSO, *Reservoirs Act,* 1975, HMSO London

Institution of Civil Engineers, 2000. *A Guide to the Reservoirs Act 1975.* London: Thomas Telford

Sims, G.P. and Parr, N.M., 1998. The review of the Reservoirs Act 1975. IN: P. Tedd, ed. *The prospect for reservoirs in the 21st century :*

proceedings of the tenth conference of the BDS held at University of Wales, Bangor, 9-12 September 1998. London: Thomas Telford

Stewart, J. 2002. Where to keep your dam documents. IN: P. Tedd, ed. *Reservoirs in a changing world : proceedings of the 12th conference of the BDS held at Trinity College, Dublin, 4-8 September 2002.* London: Thomas Telford, 575-580

BIBLIOGRAPHY

Bradlow, Daniel D. 2002. *Regulatory frameworks for dam safety : a comparative study.* Washington: World Bank

Developments in management of reservoir safety in UK

A. J. BROWN, KBR, Leatherhead, UK.
J. D. GOSDEN, KBR, Leatherhead, UK .

SYNOPSIS. The UK government funds a continuing programme of research and development on issues related to the safety of large raised reservoirs in the UK. This paper describes three recent projects carried out by KBR which are likely to have a significant effect on the way reservoir safety is managed in the UK.

The first project was to devise and trial a system for quantitative risk assessment of dams, to allow comparison of threats such as inadequate spillway capacity with other threats to the safety of a dam. This system is to be published in early 2004 as an Interim engineering guide for extended trial by dam owners and dam professionals. The second project arose out of the realisation that in UK there are typically about three incidents a year where emergency drawdown of a reservoir is required to avert failure. The project comprised a feasibility study into the content of an incident reporting and investigation system and how this might be established. The third project comprised a feasibility study to identify practicable means of early identification of internal erosion in old dams.

INTRODUCTION
There have been no dam failures involving loss of life in the United Kingdom since enactment of the first Reservoirs Act in 1930. One of the contributions to ensuring that this situation continues is a research and development programme funded by the UK government (Department of environment, food and rural affairs, Defra), to both carry out original research and disseminate current good practice to all those involved in the management of dam safety.

This paper describes three research projects carried out by KBR for Defra, and comments on how UK dam safety management practice may develop in future.

Long-term benefits and performance of dams, Thomas Telford, London, 2004, 268–281

THE INTEGRATED SYSTEM

The prototype Integrated System of Quantitative Risk Assessment for dams (KBR, 2002) is summarized in Figure 1. The system is intended to be a rapid screening level assessment, suitable for use as part of the ten yearly safety review carried out under the Reservoirs Act 1975 or for a portfolio risk assessment.

The definitions used form the cornerstone of the system, and unfortunately there is currently no agreed common framework of definitions used in the dam industry. Some of the key definitions of the processes used in the System are shown in Table 1.

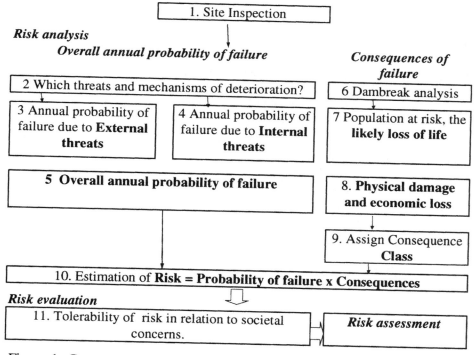

Figure 1 : Process comprising the Integrated System

The selection of threats to quantify is one of the most difficult yet important steps. At the feasibility stage the System contains a methodology for estimating the AP of failure due to the most common threats (namely floods, upstream reservoirs, seismic, wind and internal threats), although it includes a requirement for the user to evaluate the significance of other threats at a particular dam and a facility to add these into the estimate of the overall probability of failure. This is carried out using an event tree similar to the Failure Modes and Effects analysis (FMEA) in BS 5760-5:1991.

Table 1 : Key definitions used in the Integrated System

Term	Definition
Current Condition	Condition of a dam at a particular date as assessed from visual inspection and in some cases physical investigation
Indicators	Measurable outcome from the application of a mechanism of deterioration e.g. deformation, seepage, instrumentation results.
Intrinsic condition	Current physical property or dimension of the dam which can be measured and which affects the outcome of the application of a mechanism of deterioration. Although initially determined by design and construction details; this may change with time due to ageing, neglect, maintenance or upgrading.
Mode of failure	Means by which a failure (uncontrolled sudden large release of water) may occur; four modes are differentiated in the System namely external erosion (including overtopping), internal erosion, sliding and appurtenant works.
Mechanism of deterioration	Process by which the integrity of the dam is undermined. The mechanism can have a quantitative threshold above which deterioration is likely to occur e.g. slope protection designed to withstand waves due to 100 year wind
Threat	Random Event (External threat, such as floods and earthquake) or Potential Internal Instability (Internal threat) that poses a threat to the integrity of the dam.

Annual Probability of Failure

For external threats such as floods the system uses analysis, by adopting the concept of a "Critical" external event, which is an external loading of sufficient magnitude to just cause failure of the dam. The annual probability (return period) of this event is estimated from the relationship between magnitude and return period.

Estimating the probability of failure due to internal threats is difficult, as internal threats do not occur as independent events and it is often difficult to measure the occurrence of the threat. The preferred system for evaluating the probability of failure due to internal threats is to relate the dam condition, in terms of a Current Condition score of 0 to 10, to the annual probability of failure. The annual probability of failure of the worst condition dams due to internal threats is based on performance over the last 25 years (Brown & Tedd, 2003), while it is assumed that the best condition dams have an annual probability of failure due to internal threats of 1×10^{-7}.

A critical element of this methodology is the system for assigning the Current Condition score, which is assessed from indicators of poor performance (e.g. seepage and settlement); the quality of ongoing surveillance; the ability to lower the reservoir rapidly in an emergency and the reservoir operating regime.

Consequences of Failure

It is necessary to quantify the consequences if the dam failed, firstly in terms of areas of inundation and structural damage, and then in terms of the likely loss of life and damage to infrastructure. The system uses published rapid methods of estimating the peak breach flow at the dam (Froehlich, 1995), and how this attenuates down the valley (CIRIA, 2000).

The relationship between the likely loss of life (LLOL) and PAR derived from dam failures and flash floods in the United States was used (Bureau of Reclamation, 1999, which includes allowance for the "forcefulness" of the flood wave and warning time.

The estimation of physical damage is as far as possible based on systems used by the Environment Agency for evaluating potential flood defence schemes; albeit some adjustment is required to take into account the higher velocities and thus greater destruction from a dam breach flood.

Tolerability of Risk

The System plots the probability of failure and LLOL on an FN chart, as both one technique for evaluating the tolerability of risk, and as a means of prioritising dams where several are being considered together (e.g. in a cascade). It also provides a spreadsheet to allow the user to carry out ALARP assessment. This estimates the cost to save a statistical life for a package of works. This value can be compared with the cost of the package of works to assess whether the expenditure is proportionate to the reduction in risk achieved.

Benefits

The following benefits are anticipated on application of the prototype system:

- Explicit consideration of the likely threshold of dam failure can help provide a more considered basis for decision making. It will assist understanding of the margin of safety that is available
- For the first time internal threats can be evaluated in a similar format to external threats
- Permits investment to be targeted where it will do most good i.e. achieve the largest reduction in risk
- ALARP analysis can be a useful tool in identifying the value obtained from proposed investments

An interim engineering guide to an integrated approach to reservoir safety will be issued in 2004 for an extended trial as a screening tool over a period of 5 years. Feedback should be provided to the authors or Defra who intend to carry out a review of the approach at the end of that period.

INCIDENT REPORTING AND INVESTIGATION SYSTEM
One of the contributions to managing dam safety is to learn as much as practicable from near miss incidents, which might have become a failure in different circumstances, and this is the objective of the proposed incident reporting and investigation system (Gosden & Brown, 2004).

Other industries were drawn on in defining the system for dams, where systems for reporting near miss incidents are well established although normally being a statutory requirement.

As part of devising a incident system for UK dams, questionnaires were sent out to a selection of dam owners and panel engineers to obtain their views on the various issues relating to such a system. A questionnaire was also sent out for the third research project described in this paper; devising a method for early detection of internal erosion.

Possible objectives and combinations of output and incident level
There is a wide range of possible combinations of level of detail of analysis and output from the data, and the level of seriousness of an incident which could be included in an incident database, as summarised in Table 2.

The levels of incident that were adopted are as shown in Table 3, being based on those used previously in the BRE database (Tedd et al, 1992) although with some tightening of definitions. The current best estimate of the likely average number of each level of incident per year is also included; being derived from the response to the questionnaire (other than Level 6 incident which is derived as shown).

The practicable options considered are shown in Figure 2, with selection of the preferred option based on the views of UK dam industry obtained from the questionnaire, the likely completeness of reporting, the cost of data collection and processing, and the value of the output in improving dam safety.

Table 2: Possible objectives and outputs from incident system

Objective	Output	Feedback from questionnaire to UK dam industry
Ensure best possible practice is applied to ensure the continuing safety of UK dams	I Lessons learnt	Highest support in principle, 69% of dam owners and 35% of others were prepared to contribute to cost.
	II Trends III Cause and feature of each incident	High support in principle but willingness to pay not tested explicitly
	IV Historic Annual probability	70% of dam owners and 30% of others were prepared to contribute to cost.
Minimise whole life cost of asset	Va Cost and duration of incident	62% of owners were prepared to pay for information on cost, but only 39% for the disruption arising from the incident
	Vb Reliability database	44% of dam owners and 33% of others were prepared to contribute to the cost. Only 7% of dam owners strongly agreed that it was worthwhile
Data collection	VI Number of extreme events/yr	Low priority.

Table 3: Estimated number of incidents a year in UK, with 2600 large dams

Incident Level	Definition	Estimated No/ yr
1	Failure (uncontrolled sudden large release)	0
2	Emergency drawdown or works; serious operational failure in emergency	3
3	Precautionary drawdown, unplanned visit by Inspecting Eng, unplanned works; serious human error	10
4	Works in the interests of safety (Section 10 of Reservoirs Act)	60
5	Physical works not under a higher incident level. Investigation arising out of periodic safety review	30
6	Extreme natural event > 1% annual probability (1 in a 100 year return period)	78 (1% of UK dams/yr. x 3 threats)
7	Other e.g. operational failure	na

Output Table 2	Incident level (Table 3) -Y is combination which is practical, other combinations are not practical						
	1	2	3	4	5	6	7
I	*Option A*			-	-	-	-
II				*Option C*		*Option D*	Y
III	*Option B*						Y
IV							Y
V	-	-	Y	Y	Y	-	Y
VI	-	-	-	-	-	Y	Y

Figure 2: Options considered for combination of Incident level and output

A critical issue is the likely effectiveness of a voluntary reporting system. This was assessed from the responses to the questionnaire sent out to the dam industry. Of the 117 questionnaires, 43% responded to the questionnaire on the incident system and 34% to the questionnaire on early detection of internal erosion, although only 16% of recipients provided case history data for the latter. Of those that responded to the questionnaire on the incident system, 77% considered they would achieve a completeness ≥80% for a Level 3 incident (Precautionary drawdown of the reservoir) and 13% considered they would achieve a completeness ≥80% for Level 6 (Floods> 100yr).

It was concluded that a voluntary system would only attract a proportion of actual incidents, and that based on the response to the questionnaire the likely completeness of reporting of Level 2 and 3 incidents could be between 35% and 85%. Thus depending on the level of reporting, it may be difficult to reliably differentiate trends in safety from changes in reporting completeness. Hence any statistical analysis may be of uncertain value for Outputs 2 and 4, and biased for Output 3. Initially the system will be voluntary. However, depending on the effectiveness it may be appropriate that the system should become mandatory through new legislation.

It was concluded that, based on both the willingness to pay and likely completion of reporting, there is reasonable support in principle from the UK dam industry for Options A and B, but less so for Option C and none for Option D. Option B (which includes Option A) is taken forward as the information to be obtained from the incident reporting system.

Investigation of near miss incidents

For serious near miss incidents it is of value to investigate the incident to maximise what can be learnt, rather than just relying on an incident report. It is proposed that the purpose of the investigation is the same as for the various accident investigation bodies under the Department of Transport; namely to look for the root causes of accidents without apportioning blame or liability.

It was concluded (Gosden & Brown, 2004) that
- The system should investigate all Level 1 and 2 incidents, but the database manager will be given discretion to investigate other incidents that he believes merit investigation
- the investigator should be appointed by, and report to, an independent body. It is proposed that the independent body would not carry out the investigation themselves but appoint a civil engineer, qualified in accordance with the Reservoirs Act, to carry out the investigation

EARLY DETECTION OF INTERNAL EROSION

The objective of this research was to develop techniques for the early detection of progressive internal erosion (Brown & Gosden, 2004). Drivers for this research included recommendations from a recent research project into the feasibility of an Integrated System to assess all threats to dams (KBR, 2000), and a recent serious near miss incident involving an unprotected masonry culvert through an older embankment dam.

The project builds on the work of the European Working Group (Charles, 2001) as well as others (e.g. Vaughan, 2000a, 2000b). The project comprised data collection through both a questionnaire to dam professionals to obtain data on internal erosion incidents, and the use of expert elicitation to quantify parameters which are not readily measurable (Brown & Aspinall, 2004).

Long term strategy

The overall purpose of a strategy for the early detection of internal erosion is to obtain time
- in which mitigation actions can be taken to avert failure (which could include physical upgrading works), and
- if failure cannot be prevented, to warn and evacuate people from the dam break inundation zone

It is implicit that the importance of early warning is greater where the risk of loss of life and/ or damage resulting from a failure is high; namely that the amount of advance warning time should be greatest and the reliability of detection of defects highest where the risk to the public is greatest. This suggests that the strategy for early detection of internal erosion should be risk based. It is considered that in the long term detection should be one of a suite of three risk control measures to reduce risk from progressive internal erosion, namely

a) surveillance (detection);
b) planning of measures to be taken in the event that internal erosion is detected (emergency planning) and
c) the reduction of vulnerability through physical upgrades

Rate of deterioration

Data on the rate of deterioration is available from the questionnaire and expert elicitation (Brown & Aspinall, 2004); with the key variable being T_f, the estimated time from detection of the incident to failure if there had been no intervention. Figure 3 shows T_f by the location of the incident or the type of dam from the questionnaire. This figure shows that the respondents to the questionnaire considered

a) T_f varies by several orders of magnitude, from 1 to over 100 days,
b) incidents associated with culverts and pipes were much more likely to lead to a rapid failure. The median T_f (50% of incidents) for incidents at appurtenant works was 5 days, whilst the median for incidents in the body of puddle core dams was in excess of 365 days (a year).

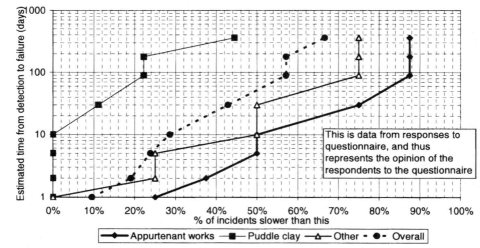

Figure 3 Variation of estimated time from detection of incident to failure (T_f) with incident location or embankment type

Further results from the project are given in Brown & Gosden (2004); with the overarching conclusion being that the understanding of internal erosion processes is still immature. Detailed conclusions include that the rate of deterioration due to internal erosion can be very variable, that there is a threshold leakage for erosion to commence and thus that leakage may occur without internal erosion, depending on issues such as the soil type and magnitude of leakage.

Interim Strategy
Currently there are significant uncertainties in relation to the proposed control measures. For example there are significant uncertainties in estimating the annual probability of failure due to progressive (rapid) internal erosion.   Similarly there are a number of arguments against applying the approach of physical upgrades as a default at the present time (except for very high consequence dams):-
a) Currently it is not possible to reliably predict those dams where internal erosion would be rapidly progressive, rather than steady
b) Pipes and culverts appear to be the largest risk; it is more difficult to upgrade these than the body of the embankment
c) If the mechanisms of deterioration and singularities (e.g. construction features) present at a dam cannot be fully quantified, then upgrades could lead to a false sense of security if they were incomplete in not addressing all potential failure modes. (e.g. if carrying out an upgrade led to a reduction in surveillance this could increase the probability of failure due to progressive erosion)

It is therefore concluded that at present it is more appropriate to concentrate on surveillance, and to link the risk control measures to the consequences of failure, rather than risk, albeit with some provision for adjustment on the basis of an assessment of the vulnerability of a dam to failure. Those dams with higher consequences would justify higher expenditure than those dams where the consequences are limited.

Frequency of monitoring
Four general monitoring regimes are proposed to be applied as shown in Table 4. The proposed "Matrix" to define the monitoring regime, which depends on the consequence class and condition of the dam, is shown in Table 5, whilst the Consequence Class is shown on Figure 4.

The latter is based on the Dam Category for defining the design flood as given in Table 1 of Floods and Reservoir Safety (ICE, 1996); but made more quantitative by changing "could endanger life" to "likely loss of life" and requiring that damage be quantified in £M. It is recognised that the accuracy of the latter should be appropriate to the intended use and generally would only be an order of magnitude estimate.

Table 4: Suggested Guide for in-service dam base monitoring frequency

Parameter	Monitoring regime (Note 1)			
	α	β	γ	δ
Visual surveillance				
Exterior; including Exterior of culverts/ shafts (and Interior where no confined space)	Daily	Daily to Tri-Weekly	Twice Weekly to Weekly	Monthly
Interior of culverts/ shafts, where confined space	Weekly to monthly	Monthly to 3 monthly	3-Monthly to 6-Monthly	Ten yearly
Instrumentation				
Flow of water incl turbidity (Note 2)	As for visual surveillance of exterior			
Telemetry	Recommended	Recommended	Consider	Not applicable
Surface Movement	Yearly	2-Yearly	Consider	Consider
Pre-existing instruments	*For manual reading; where automated readings are available more frequent reading would be appropriate.*			
Piezometers	Monthly to 3 monthly	Monthly to 6-Monthly	3-Monthly to 6-Monthly	Consider
Internal movement/ stresses	Yearly	2-Yearly	Consider	Consider
Parameters required to adjust trigger level				
Rainfall	As for flow of water			
Reservoir level	As for flow of water			

1. These frequencies may need to be varied according to the conditions at, and the type, and size of the dam; these should be determined by the dam owner and his Supervising and Inspecting Engineers.

2. This applies to any flow of water that might be emanating from the reservoir. Where there is concern over the behaviour of the dam then periodic measurements of temperature and/or chemical analysis of the water may be helpful in improving the understanding of the sources of the water.

Table 5 Proposed "Risk Matrix" to define monitoring regime

Condition of dam	Consequence class of dam (From Figure 4)			
	A1	A2	B	C/D
Poor	α	β	β	γ
Average	β	β	γ	δ
Good	γ	γ	γ	δ

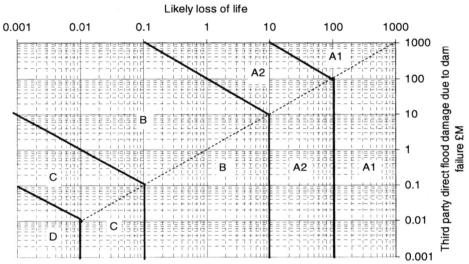

Figure 4 : Proposed Consequence diagram for UK dams

DISCUSSION – THE FUTURE FOR DAM SAFETY MANAGEMENT IN THE UK

There is no reason to be complacent about the good public safety record of dams in the UK, and the projects described will contribute to continuous improvements in the safety regime. Quantitative risk assessment (QRA) is still in the early stages as a management technique, but is likely to have far reaching effects on how risk and uncertainty are perceived and managed, and thus on the nature and extent of physical upgrading works.

In a society which is becoming increasingly litigious it is important that safety management becomes more transparent, and that its application to dams is consistent with the approach in other high hazard industries. QRA should assist in informing the debate on these issues.

New legislation passed in 2003 (The Water Act) will change the enforcement of the Reservoirs Act in England and Wales to a single body, the Environment Agency, and also introduce the requirement for emergency plans for higher risk dams.

Implementation of the incident reporting and investigation system described in this paper should lead to more informed understanding of both the frequency and type of serious near miss incidents and prioritisation of areas for future research.

CONCLUSIONS

The UK government programme of research and development in relation to dam safety continues and provides useful output in terms of how the safety of UK dams is managed. Several recent research contracts have been described and a description of how safety management may change in future given. Further information on the projects described is given on both the Defra and British Dam Society websites. Feedback on the Interim Guide to QRA is welcomed and should be addressed to Defra. Readers are encouraged to use the Incident Reporting System, once in place. Similarly suggestions for future research are always welcomed and may be addressed to Defra.

ACKNOWLEDGEMENTS

The work described in this paper was carried out as a research contract for Defra, who have given permission to publish this paper. However, the opinions expressed are solely those of the authors and do not necessarily reflect those of Defra.

The advice and assistance of two Steering Groups, whose members consist of Alex MacDonald (Chair Group 1), Jim Millmore (Chair Group 2), Howard Wheater, Nick Reilly, Andrew Robertshaw and David Dutton, who reviewed the various research reports on behalf of Defra is gratefully acknowledged.

REFERENCES

Brown & Aspinall, 2004, Use of expert opinion elicitation to quantify the internal erosion process in dams. *In Long Term benefits and Performance of dams*, Thomas Telford.

Brown & Gosden, 2004, Outline Strategy for the management of internal erosion in embankment dams. *Dams & Reservoirs.* Vol 14 No 1

Brown & Tedd, 2003 The annual Probability of a Dam Safety Incident at an Embankment Dam, Based on Historical Data. *Int Journal Hydropower & Dams.*

Bureau of Reclamation, 1999, A Procedure for Estimating Loss of Life Caused by Dam Failure, *DSO-99-06.* Author Wayne Graham. Sept. 1999, p.43.

Charles J A, 2001, Internal erosion in European embankment dams. *ICOLD European Symposium, Geiranger, Norway*, supplementary volume pp19-27

CIRIA, 2000, Risk Management for UK Reservoirs *Report No.C542* p.213.

Froehlich D.C., 1995, Peak Outflow from Breached Embankment Dam. *ASCE Journal of Water Resources Planning and Management.* 121 (1), 1995, pp.90-97.

Gosden & Brown, 2004, An incident reporting and investigation system for UK dams. *Dams & Reservoirs.* Vol 14 No 1

KBR, 2002, *Floods and reservoir safety integration.* Defra research contract. Available at www.defra.gov.uk/environment/water/rs/index.htm

Tedd P, Holton I R. & Charles J A , 1992, The BRE dams database. *Water Resources and Reservoir Engineering,* Proceedings of the 7th British Dam Society Conference, Stirling, pp 403-410. Thomas Telford, London

Vaughan PR, 2000a, Filter design for dam cores of clay, a retrospect. *Conf Filters and Drainage in Geotechnical and Environmental Engineering.* Publ Balkema. Pp 189-196

Vaughan PR, 2000b, Internal erosion of dams - assessment of risks. *Conf Filters and Drainage in Geotechnical and Environmental Engineering.* Publ Balkema. Pp 349-356

Use of expert opinion elicitation to quantify the internal erosion process in dams

A. J. BROWN, KBR, Leatherhead, UK
W. P. ASPINALL, Aspinall & Associates, Beaconsfield, UK.

SYNOPSIS. Expert Opinion Elicitation is a generic term for a number of similar techniques that have been developed to provide quantitative estimates of parameters which cannot readily be quantified through direct measurement or other sampling techniques. The initial motivation for their development was the 1986 Challenger Shuttle disaster in the space industry, and subsequent applications have spread into many other areas: the techniques have been widely used in the nuclear industry, for instance. One particular procedure consists of obtaining responses to a set of quantitative questions from a number of experts, including the range of uncertainty in each response, and then combining these through a weighting procedure to obtain a pooled best estimate of the parameters of interest.

This paper describes an application of that procedure as part of a research contract to improve methods of early detection of progressive internal erosion in UK embankment dams. For some of the parameters, information is also available from a questionnaire circulated to British dam professionals, and the paper compares the outcomes produced by the two approaches. The paper concludes with comments on the future role that expert opinion elicitation could play in providing a better understanding of dam safety issues, in particular in the determination of relevant uncertainties.

INTRODUCTION
KBR are currently undertaking a research contract for the UK government (Department of environment, food and rural affairs, Defra) to "identify a cost effective means of early detection of progressive internal erosion in embankment dams". The terms of reference entail major emphasis on embankment dams which pre-date modern geotechnical engineering (no filters or instrumentation), and that the hazards posed by unprotected pipes and culverts require particular attention. The final output from the project is to be Technical Guidance on the management of internal erosion.

Long-term benefits and performance of dams, Thomas Telford, London, 2004, 282–297

The approach adopted to respond to the terms of reference comprised a questionnaire to dam owners and panel engineers to identify recent case histories of internal erosion, a literature review and expert opinion elicitation. This paper describes the latter from the parameters selected for quantification, through the results it gave, lessons learned and where the technique could be of value in other areas relating to the management of high hazard industries.

EXPERT JUDGMENT AND ELICITATION OF EXPERT OPINIONS

<u>General</u>
In recent years, important changes have occurred in engineering which affect the way in which many safety-related decisions are made. These changes have resulted mainly from the development of risk-based methods for the design and appraisal of engineered systems. One feature of these methods is the objective of quantifying the level of safety in order to estimate the likelihood of engineering failure. The introduction of probabilistic concepts for treating uncertainty requires an engineer to exercise a form of judgment which differs from the conventional professional judgment that he (or she) may have developed during his career through training and practical experience. This alternative form of judgment, which surfaces in all attempts at estimating probabilities, in whatever domain, is generically termed 'expert judgment', and involves enumerating subjective probabilities that reflect an expert's degrees of belief. Hitherto, this subjective element in assigning probabilities has often been treated informally, or ignored altogether, but methodological advances, such as that reported here, are bringing this form of judgment increasingly to the fore.

Various approaches for combining expert opinions are possible (see, e.g., Cooke,1991; Meyer & Booker, 2001), including: *simple averaging, decision conferencing (the committee), the Delphi method, expert 'self-weighting', and the mathematical theory of scoring rules.* It is the latter that has been most refined by Cooke (1991), with his "Classical model" for expert judgment pooling (designated 'classical' because there is a close relationship with hypothesis testing in classical statistics). Cooke's scheme has been extensively tested and used in many areas of science and engineering, including the aerospace industry, nuclear industry, meteorology, hydrology (in the Netherlands), earthquake engineering and volcanology.

Examples of the use of expert elicitations in UK include:

a) O'Hagan (1998), where a consensus approach was used to address future capital investment needs of a major water company, and also in assessing the rock mass permeability at a possible nuclear waste repository at Sellafield

b) Aspinall & Cooke (1998), who describe the use of the structured elicitation methodology and decision-support procedure based on the "classical model" during the Montserrat volcanic eruption crisis, and

c) unpublished work on flight operations safety for British Airways (W.P. Aspinall, pers. comm.).

Classical method

The basis of Cooke's method is that the experts are posed a number of "seed" questions for which the answer is known (or knowable). Their responses are then assessed to obtain scores and individual weights, as defined in Table 1 and illustrated in Table 2; full mathematical details can be found in Cooke (1991). The procedure can be used to greatest benefit when the opinions of several experts (say, five or more) have to be elicited efficiently and promptly - for smaller groups, it may not be justified.

There are some important explanatory remarks in relation to Table 2. Firstly with only two seed questions, the number of degrees of freedom in the Chi-square test for the calibration statistic are too few to obtain results reflecting the accuracy of individual experts – hence Experts 1 & 2 have the same calibration score even though, in this example, one was more 'accurate' in his predictions than the other. Expert 3 falls between Experts 1 and 2 for informativeness, but falls below the threshold level for calibration (with Expert 4) <u>when</u>, as here, the DM's performance is optimised. Expert 4 is highly opinionated, and always fails to make his confidence limits wide enough to score any hits, but there is still a non-zero probability (0.007) that he is actually well-calibrated.

The fully-optimised DM has the highest calibration score, (when it is added to the group, as a virtual expert) but its Informativeness score appears poor because it amalgamates the spreads of all (positively weighted) experts. The DM's overall normalised weight is, therefore, slightly less good than the best real expert in this example, but then the DM's range reflects the collective spread of opinions. When optimised, the DM's 50%ile estimates for both seed questions are very close to the actual realizations, notwithstanding the scatter in the four experts' opinions.

In a real exercise, more seed questions are used for scoring the experts, and different combinations of statistical test power and significance level can be set to constrain relative performance scores across the group and DM.

Table 1. Basis of 'classical model' for combining experts' opinions – terms, scores, weights and factors

Term	Explanation / basis
Item	A 'seed' variable (for calibration purposes) or a question of interest for which an evaluation is sought from a group of experts
Calibration score	Test the hypothesis "This expert is well calibrated" with respect to his peers, on the basis of his estimates for a set of 'seed' variables. The score is the significance level in a chi-square test at which the hypothesis would be just rejected
Informativeness (Inverse is Entropy score)	a) Quantify the individual's 'informativeness' by indexing his cumulative information distribution function for all seed items relative to a uniform 'background' distribution (strictly, an inverse of a chi-square test statistic for closeness of correspondence); b) this 'background' distribution is either uniform linear (suitably truncated) or log normally distributed between quantiles; the latter is typically used when the range of possible values can vary over two orders of magnitude or more
Synthetic decision-maker (DM)	a) constructed from a weighted sum of the experts' responses to the items of interest, item-by-item. b) extracting the DM's distributions for each seed variable, the DM can be treated as a 'virtual expert' and scored against the seed items at different significance levels; the opinion of this virtual expert then can be iteratively re-combined with the real experts.
Expert weights	a) For each expert, the product of his calibration multiplied by informativeness scores across all seed items, normalized so that the sum of all expert weights, including that of the DM, is unity b) The 'classical model' software allows adjustment to the power of the chi-square test and the related significance level setting, which determines the threshold calibration score at which experts are given a non-zero weighting.

Table 2 : Illustration of scores and weights for 4 experts answering (only) two seed questions.

Expert	Experts' opinion ranges	Calibration Score	Inform. score	Normalized wt., incl opt. DM
1	*10, 35, 90* *15, 35, 80*	0.36	0.12	0.05
2	*40, 50, 60* *45, 52, 58*	0.36	1.27	0.52
3	*10, 25, 45* *15, 30, 55*	0.18	0.60	0
4	*80, 90, 95* *75, 80, 85*	0.007	1.60	0
DM		0.94	0.41	0.43
Results	Actual Seed values	5%ile	50%ile	95%ile
DM soln 1	50	*22.8*	*49.7*	*72.3*
DM soln 2	50	*26.4*	*51.8*	*66.8*

The rational mathematical basis for the 'synthetic decision-maker' is one feature of the method which makes it superior to other schemes for pooling judgments, making use of expertise weighted according to the quality of response to the whole set of seed variables. Usually, but not invariably, the DM ends up with a heavier weight than most, if not all, of the 'real' experts. Thus, the concept of the DM can also be described as the creation of a 'rational consensus', for the problem of combining a range of opinions (as opposed to reaching a simple average, democratic compromise or some other variant of egalitarian consensus). That said, in some applications, where suitable seed data are sparse or repeated tests are not possible, the scoring power of the calibration scheme may be weak, and its impact on individual weightings may have to be constrained.

Nonetheless, Cooke's method has at its heart a basis which replicates the formal scientific method, and one of its most valuable attributes is the scope it provides for quantifying realistically the spread of scientific or engineering uncertainty in relation to any parameter of interest. Thus, the procedure is usually framed to elicit suitable lower and upper percentile confidence estimates from the experts (in the present case 5%ile and 95%ile values), as well as a central or 'best' estimate value (which can be the mode, mean or median, depending on the distributional properties being sought). This aspect of the structured elicitation procedure is especially important for those variables for which adequate data do not exist for conventional statistical analysis - where the need for precise differentiation between engineering judgment and expert judgment comes into play.

APPROACH USED ON THIS PROJECT
The approach used on this project was based on that formulated by Cooke (1991), with the best estimate and 5% and 95% uncertainty distribution quantified for each item. To avoid peer pressure biases, the responses of the individual experts are provided independently by each directly to the facilitator, everyone remaining anonymous when the results are reported back to the group of experts. In the present project, the full set of questions had to be completed during the workshop, to avoid compromising the calibration seed questions used to evaluate the 'accuracy' and 'informativeness' of the experts' judgments (given time and opportunity, the experts could have looked up the relevant answers from published papers).

On certain questions of interest for the Defra study, some significant or systematic differences emerged amongst the experts, and the elicitation process was repeated a second time, partly in order that it could be preceded by more extended discussion of the technical issues, but also to further widen the base of experts to include two academics. Eleven experts took part in the second workshop, comprising two owner's representatives (who are both Supervising Engineers), two academics, and seven consultants' staff (six Panel AR and one Supervising Engineer); conduct of the workshop was overseen by the independent facilitator.

PARAMETERS SELECTED FOR QUANTIFICATION
The primary objective was to obtain a separate view from that in the questionnaire on the rate of deterioration of embankment dams due to internal erosion, and thus inform the output from the research project in terms of recommendations of the frequency of surveillance.

One of the key issues was devising a model of internal erosion that could be quantified using both the elicitation and questionnaire. Such a model should ideally include the effect of time, the indicators that internal erosion is occurring (indicators), those factors that determine both the predisposition to internal erosion (intrinsic condition) and how events may progress at a particular dam (event trees). It proved impossible to devise one model that satisfied all these requirements, so three models were constructed, as presented in Brown & Gosden (2004). The questions were devised to quantify elements in each of these models, with the variables of most concern being summarized in Table 3, and issues to be addressed in devising the detailed text of the questions included in Table 4.

Table 3. Groups of variables selected for expert opinion elicitation

		No. of questions
1	Seed questions	11
2	Prevalence of leakage and internal erosion	16
3	Average leakage and erosion rates	4
4	Minimum detectable leakage rate, dam critical flow	5
5	Rate of deterioration i.e. how long from detection to failure	10
6	Contributory factors to rate of progression	14
7	Chance nodes in event tree; i.e. what are the likely proportions of possible types of behaviour?	14
	Total	74

Table 4. Issues in devising questions for expert opinion elicitation

Issue	Adopted
For which dam type(s) the question should be posed	The UK populations of puddle clay core, and homogenous dams. This was on the basis that the data in the BRE database shows that these are the most common types; together comprising 84 % of the UK embankment dam population.
To which dam(s) does the question apply?	Questions were generally posed to apply to the whole UK population of that type of dam.
Clarity of question	The question should be unambiguous. The draft questions were subject to external review by (non-dam) experts familiar with expert elicitation.
How many questions can be included	The first workshop had 11 calibration and 63 elicitation questions, as shown in Table 3. Although this is towards the upper limit of a number for one session, it was achieved, partly, by including a break in the elicitation session.
Content of seed questions	A minimum of 11 questions were required to calibrate the experts. There was some difficulty in finding suitable questions, i.e. those which covered the relevant subject area and for which the majority of experts would not know the answer.

In retrospect it has been realized that the term "vertical puddle clay core" actually describes three separate facets of a dam core, for example a dam which is homogenous in terms of material can have a puddle core (i.e. a core zone where the fill is placed by puddling). Although this issue was raised in discussion during the elicitation, the wording of the questions was not formally updated to reflect this need for precision.

RESULTS OF ELICITATION

Weighting of experts
Although in the results of the first workshop every expert had a non-zero weighting (i.e. contributed to the synthetic DM), it was decided for the second workshop that the weight of the synthetic DM should be allowed to increase towards a maximum, subject to the constraint that a majority of the group (*i.e.* for no less than six of the experts) must retain non-zero weights (see Figure 2 below for an example). This point was reached for a calibration power of 0.5, and a chi-squared significance level of 1%. The net effect of excluding the five lowest scoring experts is to raise the normalized relative weight of the synthetic DM to 0.44, from 0.15 for the first workshop (no non-zero weights). The six surviving (non-zero weighting) experts have weights ranging from 0.19 down to 0.02 (equivalent to a highest-to-lowest weight ratio of $9x$). The synthetic DM would now have more than twice the weight of the best positively weighted individual expert, and $22x$ the weight of the lowest, positively weighted expert.

As a comparison with the weighting from the elicitation, based on performance with the known seed questions, a mutual weighting of colleagues in the group was carried out in the first workshop. There are some significant changes in ranking between the two, for example some experts scoring significantly less well on the performance-based measure than their colleagues might anticipate, while others do much better. This is not an uncommon pattern of ranking in groups of specialists of any discipline: some experts are well-regarded but tend to be strongly opinionated, while other more reflective individuals, who may be considered indecisive or diffident are, in fact, better estimators of uncertainty. In the present case, where the quantification of model parameter uncertainties is one of the main objectives, it is appropriate that the latter experts gain credit for their ability to judge these things well.

Output from process

The 5%, 50% and 95% estimates provided by each of the eleven experts were combined numerically in a computer code version of the classical model to provide a pooled uncertainty assessment for each query variable, using each individual's weight as derived from his calibration and informativeness on the known seed questions.

A typical result is shown in Figure 1 in the form of the experts' range graphs. Figure 2 illustrates both a question with significant variations between experts and also the effect on the synthetic DM results when its weight is allowed to increase by raising the significance level of the calibration test. Figure 3 shows a sequence of how the combined results of the elicitation for one item changed:

- between the first and second workshops,
- after the second workshop, when one outlying expert reconsidered his responses,
- when a change was made to the way in which the synthetic decision maker's effective score was constrained.

Features of note are the significant differences in widths of ranges between experts, and also the commonly wide ranges spanning the pooled 5% and 95% responses, reflecting the significant uncertainty in some of the parameters of interest. For some questions there is a failure of some experts' confidence limits to overlap with others, suggesting significant discrepancies of opinion. This is as illustrated on Figure 2, where the maximum number of experts who overlap at any one value is only four out of the total of eleven experts; additionally there are two groups of opinion about what the appropriate scaling of the value should be. One of the reasons for repeating the elicitation was that the results of the first workshop had produced some items where responses clustered in two disjoint groups in this way, representing 'high' and 'low' schools of thought. This effect had generally disappeared in the results of the second workshop, leaving only marginal instances, as shown on Figure 2.

In Figures 1 and 2, the 5% to 95% confidence spread of the synthetic decision maker spans the whole range of 50% estimates that are provided by the experts when each has a non-zero weight in the analysis. As a result, the DM encompasses the full extent of opinion but then, inevitably, exhibits a much wider confidence range than that of any one expert.

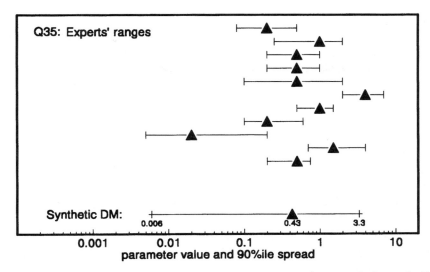

Figure 1. Typical range graph (Q35, median value for population of all UK embankment dams of dam critical flow i.e. uncontrolled erosion flow at which control of the reservoir has been lost)

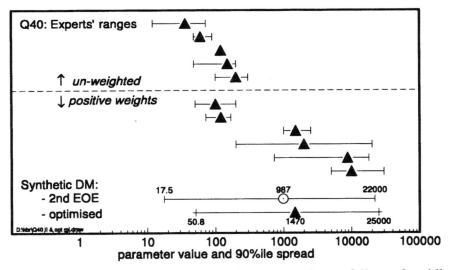

Figure 2. Range graph for Q40 (the time from detection to failure of puddle clay dams due to concentrated leak, in hours; for which only 10% of incidents are slower than this). Note for optimised DM the five lowest scoring experts, above the dashed line, are discounted – note their relatively high opinionation.

Steps can be taken to moderate this effect. If the synthetic decision maker is treated as a virtual expert, and included in the analysis, the calibration test significance level can be chosen so as to optimise the DM's distribution. While reducing the significance level enables all experts to receive positive weight, it does so at the expense of degrading the DM's calibration and entropy scores. Thus, an uncritical combination of expert assessments generally results in very large confidence bounds for the DM, as evinced in Figure 1. In the present case, the significance level was adjusted to the point at which there was still, overall, a majority of real experts with positive scores, as described earlier, thereby reducing the 'noise' of diverging opinions and improving the DM's calibration at the same time. Figure 2 illustrates how the DM's range is reduced slightly, and its 50% value more closely reflects the views the better-weighted experts; however, while some experts are discounted by this decision, similar views survive amongst those with positive weights, so such opinions remain represented in the elicitation.

It can be argued that, even though the DM's 5% - 95% range is typically larger than that of any individual, the spread is more representative of the proper scientific or engineering uncertainty for the variable in question. This is not implausible as some of the experts also present spreads in belief of similar magnitude.

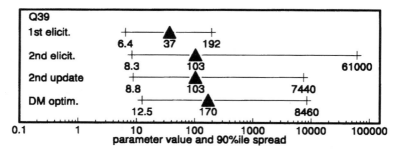

Figure 3. Example of changes between first and second elicitation

The way in which in which the synthetic decision maker's results changed through the various stages of the present elicitation process is illustrated in Figure 3. In this instance, the most marked change arose at the time of the second workshop, when technical issues were re-visited in detail and additional experts added to the panel. A few participants, who gave extreme or discordant values, were then given the opportunity to review their responses, resulting in the revised '2nd update' results. These outcomes were not greatly modified when the DM's weight was allowed to increase at the expense of a minority of the group ('DM optim.'), as just described, above.

Lessons learned
The elicitation process itself was new to all those who took part, and the key aspect that could be improved in future exercises of this kind is to increase ownership of the questions and issues by those taking part. This could be achieved by a longer workshop where the experts themselves assisted in setting the questions to be evaluated. Additionally, discussion could be stimulated by appointing protagonists to argue the case for extremes of possible responses (in some cases, it has been found effective to ask people holding opposing views to play 'devil's advocate', to argue the case for a particular position they themselves don't adhere to - this often reduces strongly-held dichotomies of opinion!).

ASSESSMENT OF RESULTS: ACCURACY AND PREDICTION
This section compares the elicitation responses with data available from elsewhere, and comments on the predictions made by the experts.

Questionnaire to UK dam industry
In parallel with the elicitation, a separate questionnaire was sent to 117 respondents, comprising all owners of more than 15 dams (20 number), a sample of 15 owners of one or two dams, all Panel AR Engineers (56 number), 10% of Supervising Engineers (24 number) and two research bodies. As well as questions relating to personal experience of internal erosion and opinion of the effectiveness of surveillance, requests were made for specific case histories of serious near miss incidents relating to internal erosion. This produced a total of 34 incidents from 19 respondents, and the data obtained are used here for comparison with the results of the elicitation exercise. It should be noted that these data were not available at the time of the first workshop, but a preliminary assessment was available by the time of the second.

Prevalence of leakage
The best estimate, from the elicitation, was that about 10% of puddle clay dams had ongoing steady leakage at each of the body of the dam, along an interface with appurtenant works and through the foundation, with 7% have leakage from the body of the dam into the foundation. Where leakage was occurring it was considered that ongoing internal erosion was occurring at about 10 to 17% of these locations. For homogenous dams steady leakage was judged as less likely (3 to 11% of dams, depending on location), with 7 to 17% of the leakages having ongoing internal erosion.

The questionnaire only provides data on serious progressive (deteriorating) internal erosion, which is likely to be less prevalent than steady ongoing erosion. This reported on average, for the period 1992-2002 three emergency drawdowns and ten precautionary drawdowns a year due to concern about internal erosion. This represents 0.2% and 0.5% of the stock of British embankment dams per year. These confirm that internal erosion is a serious threat.

Erosion and leakage rates

Figure 4 shows the results from three elicitation questions superimposed on a sensitivity study of how concentrated leakage might be expected to vary with crack width for a given crack height and length. The three points for each question represent 5, 50 and 95% uncertainty values. Flow in the crack is laminar up to 0.6mm, then turbulent. The experts' responses appear reasonable when compared with the sensitivity study.

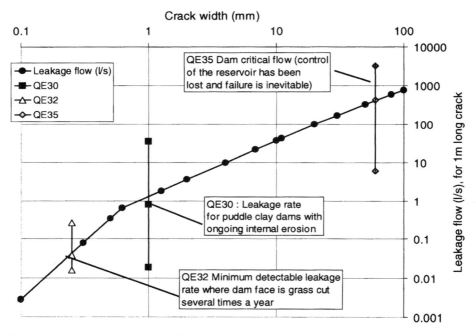

Figure 4. Sensitivity study of concentrated leakage flow to crack width (for flow through a 1m high 3m long crack under 10m head)

Rate of deterioration

Figure 5 shows the experts' opinion of the distribution of the time-to-failure for the whole population of UK puddle clay dams, if progressive internal erosion commenced at every dam, the time-to-failure being defined as that from the moment internal erosion was detected at a level of concern sufficient to call in an Inspecting Engineer to the time at which the dam critical flow rate was reached. Also shown on the figure is the distribution of the questionnaire respondent's opinions on how long before the dam would have failed in that incident, if there had been no intervention.

The significant range for the best estimate is noted, ranging from quicker than a day for 2% of dams to about 4 months for the slowest 2%. However, the response to the questionnaire suggests that the time to failure would have been much slower, with 75% of dams taking longer than 4 months. The significant uncertainty bands for the expert's opinion are also noted.

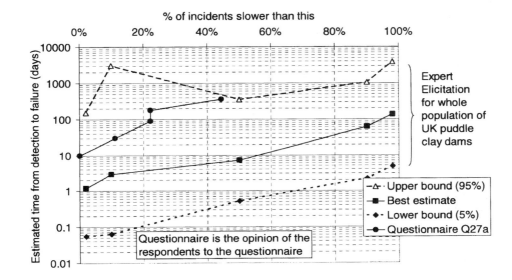

Figure 5. Distribution of time to failure for puddle clay dams

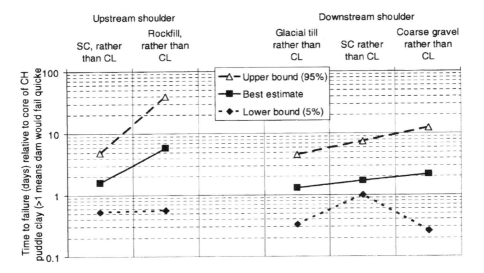

Figure 6. Effect of characteristics of dam shoulders on time to failure

Contributory factors to rate of erosion
The elicitation questions included the effect of factors such as the hydraulic gradient, the plasticity and degree of compaction of core material and properties of the shoulder materials on the time to failure. Typical output is shown in Figure 6. The expert opinion typically gave changes in rate of deterioration of up to 10; this may be low when compared to the ranges in rate of deterioration of several orders of magnitude.

DEBATABLE ISSUES
The understanding of internal erosion processes is still immature, with quantitative methods only available for limited elements. Tools that can help in either quantitatively capturing existing knowledge and experience, or in probing unexplored areas are therefore of value. The elicitation process set out by Cooke is of value in providing rational consensus, in that the opinions of the quiet reflective expert are considered, with appropriate weightings, just as much as those of more dominant personalities.

Elicitation has proved of value in making the wide spread of uncertainty explicit, and in capturing knowledge. The process adopted for this research contract did not fully explore the reasons for the wide discrepancy of results, but this could be pursued in future exercises. Debatable issues raised include:

a) most of the dam experts appear to give uncertainty bounds which are narrower than the true uncertainty, particularly where the uncertainty covers orders of magnitude - however, this trait has been found to be true of technical experts of all kinds;

b) the validity of questions which ask for the spread of a variable over the whole population of a particular dam type. It could be argued that for some of the dams the question is irrelevant, or inappropriate; however, to advance the knowledge of internal erosion further work is required at both a detailed level on specific dams and in understanding of the behaviour of groups of dams;

c) the validity of questions which simplify a complex problem down to focus on only one aspect of the problem, assuming "all other things being equal". For issues governed by two (or more) important interdependent variables this may be an over-simplification.

Possible applications of the technique include research into parameters which cannot readily be quantified, for example floods with an annual probability of less than 10^{-4}/ annum. Additionally in increasingly litigious times the underlying structured basis of the method can provide a valuable record of the way a decision was reached, the impartiality of which could offer both a significant shield against personal liability to individual experts providing critical advice and a transparent decision process for major organisations.

CONCLUSIONS

Expert Opinion Elicitation, a technique first developed for the space industry, was one of the techniques used in an ongoing research contract for Defra to explore current knowledge of internal erosion. It provided a useful set of judgments and insights, including explicit confidence limits, broadly consistent with the findings from the questionnaire to the wider UK dam industry. Significant advantages of the technique are the encouragement which the procedure gives to all participants to express their true engineering beliefs (unbiased by peer pressure). In addition, the combined output from the procedure (the synthetic decision maker) generally provides values for the complete set of questions that are, overall, more coherent and closer to reality than those that would be obtained from any one individual expert, however good.

It is concluded that expert elicitation provides a valuable technique for quantifying those variables that cannot be determined by direct measurement, and for evaluating realistic likely spreads of scientific or engineering uncertainty on engineering parameters.

ACKNOWLEDGEMENTS
The work described in this paper was carried out as a research contract for Defra, who have given permission to publish this paper. However, the opinions expressed are solely those of the authors and do not necessarily reflect those of Defra.

REFERENCES
Aspinall W. & Cooke R.M. 1998, Expert judgment and the Montserrat Volcano eruption. In: *Proc. 4th Intl. Conf. Prob. Safety Assessment and Management PSAM4*, New York, (eds. Ali Mosleh and Robert A. Bari), Vol.3, 2113-2118.

Brown & Gosden. 2004, Outline Strategy for the management of internal erosion in embankment dams. *Dams & Reservoirs*. Vol 14 no 1

Cooke R.M. 1991, *Experts in uncertainty: Opinion and subjective probability in science.* Oxford. Oxford Univ Press 321pp

Meyer M.A. & Booker J.M. 1991, *Eliciting and Analysing Expert Judgement: A Practical Guide.* Rep NUREG/CR-5424. Nucl. Reg. Comm., Washington DC (republished 2001: ASA-SIAM Series in Statistics and Applied Probability, Philadelphia; 459pp)

O'Hagan, A. 1998, Eliciting expert beliefs in substantial practical applications. *The Statistician*, 47 Part 1, pp21-35, Discussion 55-68.

Dam Accident Data Base DADB - The Web Based Data Collection of ICOLD

J-J. FRY, EDF CIH, Le Bourget du Lac, France
A. VOGEL, Risk Assessment International, Vienna, Austria
J-R. COURIVAUD, EDF LNHE, Chatou, France
J-P. BLAIS, EDF CIH, Le Bourget du Lac, France

SYNOPSIS
This paper describes a database (DADB) that includes all information about dam failures which are necessary for the evaluation and the assessment of failure modes and hazards. The DADB currently includes about 900 events, all individually observed and investigated.

INTRODUCTION
Risk estimations associated with dam failures based on statistical studies had been difficult to carry out, because either the information of different data base were contradictory or no data were available. The comparison of different failure rates also faces difficulties, because some failure listings define "failure" as an accident that destroys a dam and renders it useless, while others mean a catastrophic accident, which releases most or all of the impounded water. In 1974 ICOLD published a first failure list, which presented 202 dam failures [2]. 5 years later the results of another investigation showed only 129 dam failures [3].

In 1995 ICOLD updated this compendium [4] by defining a failure as a collapse or movement of a part of a dam or its foundations so that the dam cannot retain the stored water. Accidents during construction were considered to be failures when a large amount of water was released downstream by a river flood which caused the partial or total destruction of the dam, whereby the height of the dam in construction when the overtopping began should have a height of at least 15m or reservoir filling had commenced before dam completion. According to these definitions 179 failure cases were determined, which all concerned large dams, according to ICOLD's definition from 1973 [1].

Therefore no catastrophic failures of dams during construction are considered, as long as the reservoir was empty and also no large slope stability failures during construction, which often led to critical situations for the workers. Not only reservoirs, which impound water, but also tailings dams, impounding tailings or toxic fluids have caused extensive damage in previous failures. The failures of the tailings dams of Buffalo Creek in 1972 caused 125 deaths and in 1985 in the Stava valley, Italy, 268 people died after a similar catastrophe, not to mention the contamination after the failure of the uranium tailings dam Key Lake in Canada in 1984 or the recent release of 100,000m^3 of contaminated cyanide liquid after the failure of a tailings dam in Romania in January 2000 and the subsequent poisoning of drinking water of more than 2 million people in Hungary.

ICOLD recognized the need for a compendium on failure data of such constructions and published for the first time in 2001 a bulletin concerning failure events of tailings dams [5].

Failure causes must be investigated irrespective of the dimensions of a dam or the extent of its hazard. The failure in 1972 of the Canyon Lake dam in the USA, which was only 6 m high, caused the death of 300 persons [6]. Data on failures of small dams include valuable contributions for the assessment of failure modes and causes, as well as for those of large dams [7]. The proposed DADB will be web based and include data on failures of small and large dams as well as failures of tailings dams (Figure 1).

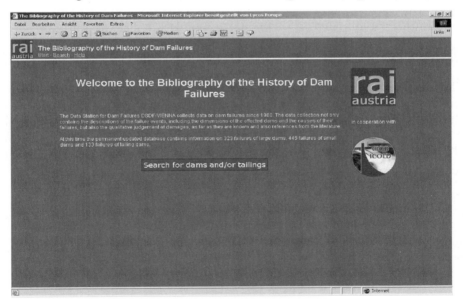

Figure 1: Front page of DADB

CONTENTS OF DADB

Dam information

The opponents of statistical studies based on historical records of dam failures criticise the fact that data of the past would be not homogeneous and therefore the dam failure information not directly comparable. The proposed DADB sweeps away these arguments and offers information about the name of the dam, the country, the date of its construction its purpose, the date of failure, the type of the dam, its height (above ground level and lowest foundation), crest length, crest width, base width, volume, upstream and downstream slope geometry. The type of material (watertightness, upstream shoulder, downstream shoulder, downstream protection), type of spillway (type, width, height, design flood), information about foundation (type, thickness) and reservoir (capacity, normal water level, maximum water level) will also be given. In cases of tailings dams the kind of impoundment is also available.

DADB will relate the failures exactly to all known current dam types. It was therefore necessary to distinguish between 20 different types of dams for water storage and 7 other special kinds of types of tailings dams, according to international regulations and their particular methods of construction.

DADB also provides the user with 7 different uses of the failed dams, which are the storage of tailings, for hydroelectric, flood control, irrigation, water supply, for wood transport or unknown purposes (Figure 2).

Failure information

To avoid probabilistic techniques to estimate dam failure risks and structural reliabilities it was stated that dams can fail through an infinite number of modes, which cannot be fully enumerated [8]. DADB contains the primary failure causes, which were investigated after the dam failures. 13 different failure causes, including the sensitive ones caused by construction or calculation errors or hostile failures are distinguishable (Figure 2).

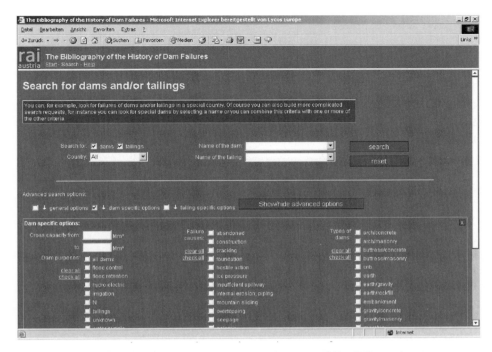

Figure 2: Advanced search page of DADB

The database will include (if known):

- information about breach initiation
- maximum depth above breach
- volume stored above breach invert
- evolution in time of overtopping
- breach height
- breach top width
- breach bottom width
- breach average width
- breach side slope
- breach and empty time
- breach peak outflow
- breach outflow hydrograph
- method of determining peak outflow
- flood peak entering in the reservoir
- flood hydrograph entering in the reservoir
- eroded volume
- outflow volume

Supplementary information

DADB also gives information about the human and the economic damage caused by a dam failure, as far as it was reported. In cases of tailings dam failures the volume of the outflowed tailings or contents and the travel distance is additionally available. Pictures of some events are also available, which show the dimensions of the damage to the dam construction. The pictures of the photo gallery can be enlarged to full size. (Figure 3).

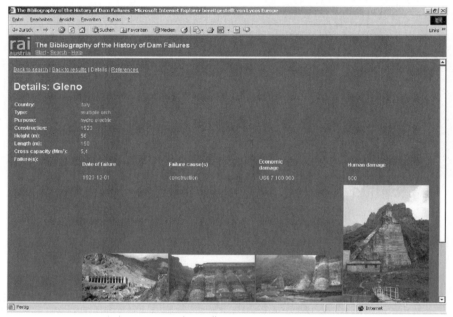

Figure 3: Example of a result page of DADB

Reference information

To verify the origin of the data, a list of references is included for each dam failure, to enable the user to get additional information to that presented on every result page (Figure 4). Most of the references will be also available in form of pdf - files

SEARCH OPPORTUNITIES

On the search page (Figure 2) it is possible to search for every single parameter which is mentioned, but also for all in every combination. The search for height, length and for the storage capacity or for the impoundment is possible for a special rate or in intervals. A click to one of the names of the dams in the summit list of every search operation leads the user back to the failure sheet for this dam, to provide him with the accompanying references, the photos and all the other parameters.

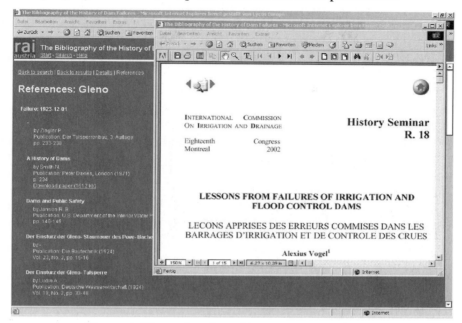

Figure 4: Example of a reference page of DADB

RESULTS

The number of all reservoirs impounding water, which are a standing menace to life and property will be today in the order of 400,000. DADB documents now more than 900 dam failures and 132 of those of tailings dams and will be updated permanent.

CONCLUSION

DADB includes all information about failures of water storage and tailings dams which are necessary for the evaluation and the assessment of failure modes and hazards. Today it includes about 900 events, all individually observed and investigated. The data are also usable for the assessment of failure behaviours and for the investigation of a probable existing failure-cause-specific break-mechanism.

REFERENCES

[1] ICOLD, 1973. *World Register of Dams*. Paris. France.
[2] ICOLD, 1974. *Lessons from Dam Incidents*. Paris. France.

[3] Goubet, A. 1979. Risques associes aux barrages. *La Houille Blanche.* Nr. 8, pp. 475-480.

[4] ICOLD, 1995. Bulletin 99, *Dam Failures, Statistical Analysis,* Commission Internationale des Grands Barrages, Paris, 73 pp.

[5] ICOLD, 2001. Bulletin 121, *Tailing Dams-Risk of Dangerous Occurrences,* Commission Internationale des Grands Barrages, Paris. France.

[6] Vogel, A. 2001. Lessons from Incidents and Failures of Dam Constructions. *Proceedings of the International Conference on Safety, Risk, and Reliability-Trends in Engineering.* pp. 735-740. St. Julian. Malta.

[7] Vogel, A. 2001. Data Collection and Analysis of Dam Failures of Small Dams According to the ICOLD Definition. *Proceedings of the 5th International Conference on Reliability, Maintainability & Safety-ICRMS'2001.*Daliban. China (in print).

[8] Baecher, B., Paté, M.E. and de Neuville, R. 1980. Risk of Dam Failure in Benefit-Cost Analysis. *Water Resources Research.* Nr. 16, pp. 449-456.

Comparison of some European guidelines for the seismic assessment of dams

N.REILLY, Consulting Engineer

SYNOPSIS.   Following the publication of the Application Note to An Engineering Guide to Seismic Risk to Dams in the United Kingdom in 1998, a seismic working group was set up by the Euroclub of ICOLD. The purpose of this was to present and compare the approach to seismic appraisal of dams across Europe. To date guidelines for five countries (Austria, Italy, Switzerland, Romania and the United Kingdom) have been made available. The paper presents the key concepts of these and compares them.

INTRODUCTION

The document "An engineering guide to seismic risk to dams in the United Kingdom" (the British seismic guide) was published by the Building Research Establishment in 1991 as part of a large suite of guidance documents for the design and assessment of dams in that country. There are some sixteen similar semi-official guides applicable to dams in the UK but they are not codes of practice and have no formal legal force. Nevertheless they are widely followed, albeit tempered by engineering judgement in specific cases.

The British seismic guide was received as a very useful advance but there were many who thought its provisions were rather severe in terms of the magnitude of risk that dams were to be tested against. As a result a peer review was set up and this resulted in an additional document, the Application Note to the guide, being published by the Institution of Civil Engineers in 1998. This modified the seismic guide as described below.

In the course of the peer review it was suggested that a working group of the Euroclub of ICOLD be formed to prepare a comparison of practice across Europe in relation to the seismic assessment of dams. This was done and copies of guidance documents (codes in some cases) from five countries have been received and reviewed. This paper presents a brief outline of each

Long-term benefits and performance of dams, Thomas Telford, London, 2004, 305–312

and compares them. The key features are summarised in a comparative table (Table 1).

UNITED KINGDOM

In the UK the key document (Charles et al 1991) was published in 1991 and contains in Part A a brief but comprehensive overview of seismic risk and hazard, drawing parallels with flood risk. It presents a summary of the parameters used to describe earthquakes and reviews the historical seismicity of Britain. The guide goes on to propose the standards to be adopted for the safety evaluation of dams in the UK, both existing and new. The **Safety Evaluation Earthquake (SEE)** is defined as the earthquake which will produce the most severe level of ground motion under which the safety of the dam against catastrophic failure should be ensured. The **Operational Basis Earthquake (OBE)** is also defined but the guide does not concern itself with this.

Dams are allocated a hazard category using the method of ICOLD bulletin 72 (ICOLD 1989) which takes into account reservoir capacity, dam height, number of persons at risk and potential downstream damage. This yields a classification number which puts a dam into one of four categories designated I to IV, IV representing the highest hazard. The guide recommends that category IV dams be tested against a 30,000 year return period event. Alternatively the maximum credible earthquake (MCE) estimated by a site specific study could be used. The MCE is defined as the earthquake that would cause the most severe level of ground motion at the site concerned which appears possible for the geological conditions. The other three categories are to be tested against events of return period 10,000, 3,000 and 1,000 years in descending order. For cases where a site specific study of seismicity was not justified, the guide presented a zone map dividing the country into areas A, B and C and tabulated indicative peak ground accelerations for the range of return periods. For zone A (the most seismically active) the recommended peak ground accelerations (PGA) range from 0.375g for 30,000 years return period, 0.25g for 10,000 years, down to 0.1g for 1000 years.

Part B of the guide contains three chapters dealing with embankment dams. The first chapter (Chap 5) outlines the effects of earthquakes on embankment dams and quotes some examples of UK dams which have been subjected to minor events. (This is supplemented in an appendix by a similar review of world wide incidents). The next chapter outlines the methods of analysis available and the final chapter in this part presents recommendations regarding which methods to apply as a function of height and hazard category.

TABLE 1: KEY FEATURES OF SEISMIC SAFETY ASSESSMENTS

CHARACTERISTIC	UK	AUSTRIA	ITALY	ROMANIA	SWITZERLAND
Status of document	Guide	Guide	Guide	Statutory	Statutory
Hazard designation	ICOLD Bulletin 72	Dam ht, capacity	ICOLD Bulletin 72	Not stated	Dam ht, capacity
Seismic variation	1991: zone map / 1998: contour map	Zone map & contour map	Zone map	Zone map	Contour map
Maximum PGA	1991:0.375g / 1998: 0.32g	MCE: 0.3g / OBE: 0.14g	>0.6g	0.32g	0.03 to 0.16g (for 475 years)
Return periods:					
Cat IV	10,000 yrs/MCE	)Where	>2500 yrs	Top cat: MCE or 800 years	Not applicable / (I) 10,000 yrs
Cat III	10,000 yrs	)applicable	2500 yrs		(II) 5,000 years
Cat II	3000 yrs	)use	1000 yrs		(III) 1000 yrs
Cat I	1000 yrs	)MCE	500 yrs		
OBE	Not stated	200 yrs	Not stated, see text	100 yrs	Not stated
PGA analysis factor*	0.67	Not stated	0.5 to 0.67	Not stated	Not stated
Site specific study	No recommendation	Recommended	Mandatory for cat IV	Recommended	
Seismicity	Very low	Very low	Moderate	High	Very low

*Reduction factor to be applied to PGA for purposes of analysis

Part C deals in the same manner with concrete dams and the quoted appendix reviews worldwide events.

The foreword to the guide stressed that it was provisional in character and would need to be reviewed in the light of experience. As a result of a general view that the risk criteria were unduly severe, a review started almost straightaway, culminating in the Application Note to the guide published in 1998 (ICE 1998). This introduced two main changes. Firstly the zone map was replaced by a contour map giving PGA's for 10,000 year return period events as a result of a nation wide study of seismicity (Musson and Winter 1996). This gives a maximum PGA (in zone A) of 0.32g, which is rather higher than given in the original guide. Secondly the return period for category IV dams was reduced to 10,000 years or MCE.

The Application Note also presents some new information. In the period since the introduction of the seismic guide two large owners of dams had carried out site specific assessments of seismicity for all their damsites. The results of these were summarised and presented. These in general agreed with the Musson and Winter contour map of PGA. The Application Note also presented summary results of a number of seismic assessments of a wide variety of dams, both of concrete and embankment types. It is notable that, to date, despite the great age of many UK dams, no dam has yet had to be strengthened solely for reasons of resistance to earthquake.

AUSTRIA
The Austrian seismic guide is published by the Reservoir Commission of the Federal Ministry of Agriculture and Forestry and is dated 1996. It appears to be part of a broader range of guidelines for dam design. The guide is specifically not a standard but there is provision for its application, procedures and criteria, to be discussed with the authorities. It applies equally to existing and new dams.

The Austrian guide is appreciably shorter than the British guide but it follows similar principles. It follows ICOLD Bulletin 72 in terms of differentiating between OBE and MCE cases but it does not specifically use the bulletin's system of hazard categorisation. Instead it states that for dams >15 m high or capacity >500,000 m^3 then both OBE and MCE should be checked. This would also apply for smaller dams in potentially dangerous circumstances. Otherwise only the OBE case need be considered.

For the OBE a contour map of PGA is presented which has a maximum PGA of 0.14g. The minimum to be considered is 0.06g. For the MCE the guide contains a zone map with PGA varying from 0.11g to 0.3g. However it suggests that in general a site specific study should be carried out.

The guide goes on to give some advice relating to material properties, methods of calculation and factors of safety. It also presents response spectra and time histories for use in analysis and gives guidance on post earthquake inspection.

ITALY
The Italian seismic guide was published by the Dipartimento per I Servizi Tecnici Nazionali of the Presidenza del Consiglio dei Ministri in March 2001 and applies specifically to existing dams. New dams are subject to statutory regulations which since 1959 have included seismic criteria. The seismic guide may be used where it is not possible to apply the current criteria to an existing dam.

In format and philosophy it follows the UK guide quite closely but there are some significant differences which are outlined below.

The system of hazard categorisation follows ICOLD Bulletin 72 but the return period of the events for each category differ markedly. For category IV the return period of the SEE event is specified as not less than 2,500 years or MCE, the definition of the latter being as defined above. For categories III, II and I the return periods are respectively 2500, 1000, and 500 years.

In an appendix, the guide gives some advice on the definitions of high, moderate and low downstream damage. It suggests that high is greater than 1% of gross domestic product (GDP), moderate is 0.1 to 1% and low is 0.01 to 0.1%. Damage less than 0.01% is regarded as none or negligible.

The SEE to be applied is defined by the PGA and there is a legally established map of the country which identifies three seismically active zones and an unclassified zone. For a return period of 2500 years the maximum PGA is given as 0.6g and the minimum (applying in the unclassified zones) is 0.2g. It should be noted that these are the minimum values for category IV dams because of the "not less than 2500 years" criterion mentioned above.

The guide defines the available methods of analysis in a similar way to the UK guide but is more prescriptive in relation to category IV dams which must be subjected to field investigation and dynamic analysis. It also gives more detailed recommendations with regard to material parameters and safety factors and has a section on appurtenant structures.

For the OBE case, the guide recommends using the appropriate zone PGA for category I dams divided by two.

When a dam has been subjected to an earthquake an inspection must be carried out and a report submitted to the authorities. Dams in categories III and IV, as well as those more than 45 m high or retaining more than 10 Mm^3 must be equipped with a seismic monitoring system comprising two strong motion instruments, one at the base and one on the crest.

ROMANIA
The Romanian practice in relation to seismic safety of dams is defined in the "Code for design and seismic safety assessment of dams and hydraulic structures", 3rd edition of March 2002. An English language translation is not available and the following is based on an English precis, hence the level of detail is less than for the other countries' guides. The document comprises a mandatory code plus a detailed advisory guide. It has to be read in conjunction with a code for dams (PE729) first introduced in 1979 by the Ministry of Energy. The latest edition is dated 2001.

Romania differs from the other countries reviewed in that it is seismically very active and a large magnitude event occurred as recently as 1977 (M_L 7.2). The guide contains a useful survey of historical earthquakes in Romania and, despite some very strong events, there has been relatively little damage to hydraulic structures.

The code makes use of two systems of classifying dams which are defined in other documents. The first is "class of importance" (STAS-4273/83) which relates to the economic and social value of the works. There are five classes designated I to V, I being the most important. The other system is "category of importance" (NTHL-021) which relates to the hazard posed by the facility. This grading has four categories, A to D, A being the highest hazard. From the documents available it is not clear how these are derived nor how they are used in combination. However the SEE for categories I/A and II/B appears to be derived by a site specific study with a return period between 475 and 800 years depending on the source of the event. For the lower categories (III, IV, V and C/D) only the OBE case is considered using zone maps giving PGA values for return periods of 100 years. Across the country the PGA varies between 0.08 and 0.32g.

The guide contains detailed recommendations regarding methods of analysis, material parameters and earthquake parameters (response spectra etc). It also addresses appurtenant structures, construction in seismic zones, instrumentation and rehabilitation of dams damaged by earthquakes.

SWITZERLAND

The Swiss seismic guide was published in 2003 as the "Directives relating to the safety assessment of reservoirs subjected to earthquakes" under the authority of the ordinance on the safety of reservoirs (OSOA) dated 1998. It applies equally to new and existing reservoirs.

In format and philosophy it follows the foregoing guides but is appreciably more comprehensive in its treatment of the subject and contains a great deal of theoretical background and bibliography. It also defines in general terms the qualifications and experience required of the engineers who lead the safety evaluation. These are more onerous for the highest hazard category of dam than for the lower hazard ones.

The system of hazard categorisation is based mainly on dam height and, to a lesser extent, reservoir capacity. There are three categories, I (the highest hazard) to III. Categorisation is done by reference to a simple chart of height against capacity. The main determinant is dam height and, broadly, any dam higher than 40 m is in category I and below 10 m is in category III but very large or very small reservoir volumes modify this. For category I the return period of the SEE event is specified as 10,000 years, for category II it is 5,000 years and for category III it is 1,000 years.

The appropriate PGA for the site and return period are given by a series of statutory contour maps for the country and these are supported by response spectra for three types of foundation taken from Eurocode 8. For a return period of 475 years the PGA varies from 0.03 to 0.16g.

The guide defines the available methods of analysis but is generally more prescriptive than the other guides reviewed. Category I dams must be analysed by dynamic methods with material properties obtained by field investigation.

In addition to sections on embankment and concrete/masonry dams the guide has a section on barrages, ie dams containing a preponderance of movable elements. There are also sections on instrumentation and post earthquake inspection. All category I dams are required to have strong motion instruments. Inspections and reports to the authorities are mandatory for all dams following events of specified severity, the threshold event levels being lowest for the highest hazard dams.

CONCLUSIONS

Seismic guidance documents for dams for a range of countries in Europe have been compared. The general approach is similar but there is a divergence on the degree of risk to be accepted for similar categories of dam. This is particularly true of MCE where, despite accepting the ICOLD definition, some countries use a probabilistic approach with a relatively low return period.

ACKNOWLEDGEMENTS
The author wishes to thank the following for supplying national documents:
    Dr Rinaldo Murano, Italy,
    Dr Ion Toma, Romania
    Dr Jost Studer, Switzerland,

REFERENCES
Charles et al. *An engineering guide to seismic risk to dams in the UK.* BRE 1991.
Institution of Civil Engineers. *An application note to the engineering guide to seismic risk to dams in the UK.* ICE 1998.
European Committee for Standardisation. Eurocode 8 *Design provisions for earthquake resistance of structures* ENV 1998-1-1.
Musson & Winter. *Seismic hazard of the UK.* AEA Technology 1996.
Federal Ministry of Agriculture and Forestry: Austrian Reservoir Commission. *Seismic analysis of dams Section 3 Guidelines* 1996.
Presidenza del Consiglio dei Ministri: Dipartimento per I Servizi Tecnici Nazionali. *Guidelines for seismic safety reassessment of existing dams in Italy.* 2001.
Toma, Dr Ion. Technical University of Civil Engineering, Bucharest. *Personal communication summarising the Romanian "Code for design and seismic safety assessment of dams and hydraulic structures",* 3rd edition March 2002.
Federal Office for Water and Geology, Switzerland. *Directives relating to the safety assessment of reservoirs subjected to earthquakes.* 2003.

4. Flood impact and alleviation

European research on dambreak and extreme flood processes

MW MORRIS, HR Wallingford Ltd, UK
MAAM HASSAN, HR Wallingford Ltd, UK

SYNOPSIS. Effective risk management for dams requires an understanding of potential hazards and an assessment of the various associated risks. This requires analysis of potential impacts which, in the case of dambreak, requires an ability to reasonably predict conditions that may result through failure or partial failure of a dam. The IMPACT Project focuses research in five areas related to dambreak, namely breach formation, flood propagation, sediment movement, geophysical investigation and assessment of modelling uncertainty. This paper provides an update on this 3-year programme of work with an overview of some initial findings, particularly in relation to work on breach formation.

THE IMPACT PROJECT

The IMPACT Project (Investigation of Extreme Flood Processes and Uncertainty) is a research project running for 3 years from 2001-2004, funded by the European Commission and supported in the UK by Defra and the Environment Agency. The focus of work is directed at four process areas (breach formation, flood propagation, sediment movement, geophysical investigation) and assessment of uncertainty within modelling tools. These research areas were identified during earlier research (Morris, 2000) as areas where predictive ability was relatively poor, and hence 'weak links' in any risk assessment or emergency planning studies.

Programme of work

Research into the various process areas is undertaken by groups within the overall project team. Some work areas interact, but all areas are drawn together through an assessment of modelling uncertainty and a demonstration of modelling capabilities through an overall case study application. The IMPACT project provides support for the dam industry in a number of ways, including:

Long-term benefits and performance of dams, Thomas Telford, London, 2004, 315–326

- Provision of state of the art summaries for capabilities in breach formation modelling, dambreak prediction (flood routing, sediment movement etc)
- Clarification of the uncertainty within existing and new predictive modelling tools (along with implications for end user applications)
- Demonstration of capabilities for impact assessment (in support of risk management and emergency planning)
- Guidance on future and related research work supporting dambreak assessment, risk analysis and emergency planning

Each work area is briefly outlined below, followed by a focus upon work investigating breach formation. More detailed information on all areas of the project may be found via the project website at www.impact-project.net.

Breach Formation
Existing breach models have significant limitations (Morris & Hassan, 2002). A fundamental problem for improving breach models is a lack of reliable case study data through which failure processes may be understood and model performance assessed. The approach taken under IMPACT was to undertake a programme of field and lab work to collate reliable data. Five field tests were undertaken during 2002 and 2003 using embankments 4-6m high. A series of 22 laboratory tests were undertaken during the same period, the majority at a scale of 1:10 to the field tests. Data collected included detailed photographic records, breach growth rates, flow, water levels etc. In addition, soil parameters such as grading, cohesion, water content, density etc. were taken. Both field and lab data were then used within a programme of numerical modelling to assess existing model performance and to allow development of improved model performance.

Flood Propagation
Work on flood propagation focussed on two different aspects, namely, prediction of flood flow conditions through urban areas and prediction of flood conditions in real topography.

Whilst river modelling has become a routine part of design and analysis of river works, the way in which flooding of urban areas is predicted has not been 'standardised'. A number of different approaches may be taken, such as simulation of streets as flow channels, simulation of key areas as storage reservoirs or simulation of general flow by increased roughness. The objective of this component of work is to compare various approaches and hence identify differences and perhaps the best approach. This work has been undertaken through analysis of both field and lab data. Physical modelling of flow through urban areas provided base data for model comparison.

Sediment Movement

Under dambreak or extreme flood conditions, significant volumes of sediment may move. In the near field, close to a breach or failed dam, sediment will be entrained and carried with the surging flow. In the far field, the nature of flow and sediment conditions may produce significant changes to the river such as lateral widening, braiding or major changes in course. With respect to dambreak assessment and emergency planning, sediment movement and deposition may significantly affect bed, and hence surface water, levels as well as provide an obstruction for access.

Research is underway through a combination of laboratory modelling and numerical simulation. Initial work is focussing upon developing new relationships for sediment entrainment under extreme and varying conditions. It is noticeable that current approaches for predicting breach growth or sediment movement during dambreak all utilise existing sediment transport equations that are typically based upon long term steady state conditions.

Geophysics & Data Collection

This 2-year module of work was added to the IMPACT project through a programme to encourage wider research participation with Eastern European countries. The work comprises two components; firstly review and field testing of different geophysical investigation techniques and secondly collation of historic records of breach formation.

The objective of the geophysical work is to develop an approach for the 'rapid' integrity assessment of linear flood defence embankments. This aims to address the need for techniques that offer more information than visual assessment, but are significantly quicker (and cheaper) than detailed site investigation work. Research is being undertaken through a series of field trial applications in the Czech Republic at sites where embankments have already been repaired and at sites where overtopping and potential breach is known to be a high risk.

The objective of collecting breach data is to create a database of events that includes as much information as possible relating to the failure mechanisms, local conditions, embankment material and local surface materials. Analysis may then be undertaken to identify any correlation between failure mode, location and embankment material, surface geology etc.

Uncertainty Analysis

The objective of work here is to establish the uncertainty that may be present in modelling predictions, and subsequently how this might influence use of the information by the end user. Uncertainty in component model predictions (i.e. breach model, flood propagation model, sediment model etc.) is being established, followed by a combined assessment on an overall case study to demonstrate techniques and conclusions.

Conclusions from the IMPACT Project Research
Conclusions from the IMPACT project research will be presented and discussed in full during a final project workshop, to be held in Zaragoza (Spain) on 27-29th October 2004.

This paper will now focus on work undertaken during Years 1 and 2 of the project within the breach formation theme area.

A FOCUS ON BREACH FORMATION
The objectives of this area of research work were to:
- Collate reliable field and laboratory data demonstrating failure processes for cohesive and non cohesive embankment failure (failure mainly by overtopping, but also through piping)
- Objectively assess existing breach model performance
- Allow further development and validation of breach models to improve performance
- Allow an assessment of the effect of scaling on breach data collection (i.e. field data versus laboratory data)

This was achieved by undertaking 5 field tests (up to 6m high), 22 laboratory tests and an extensive programme of numerical modelling with modellers participating from around the world, as well as within the EC.

Field Work
Five field tests (see Table 1) were undertaken as part of the IMPACT project, although additional tests were also undertaken as part of the Norwegian national research programme.

Table 1.   Programme of field tests

Test	Nature	Height	Failure mode
Test #1	Homogeneous, cohesive	6m	Overtopping
Test #2	Homogeneous, non cohesive	5m	Overtopping
Test #3	Composite (Rock fill shoulders and moraine core)	6m	Overtopping
Test #4	Composite (Rock fill shoulders and moraine core)	6m	Piping
Test #5	Homogeneous (moraine)	4m	Piping

Figure 1 shows material gradings for each of the various test materials and Plate 1 shows Field Tests #1 and #5 at various stages of testing.

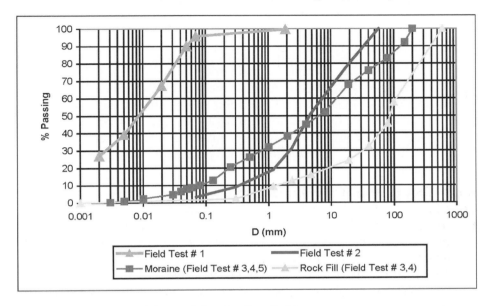

Figure 1   Materials used for the five field tests

Plate 1   Field Test #1 (left) and Field Test #5 (right)

In order to help understand the process of breach formation, a range of data was collected during the tests including water levels, flows and soil properties. Monitoring the rate of breach growth was assisted by the use of movement sensors that were buried within the body of the dam. These sensors recorded the time at which movement occurred, so by recording where the sensors were buried, it was possible to recreate a picture of the breach growth pattern after the failure had occurred.

Initial Findings of Field Work
Whilst data is still being analysed at the time of writing, some initial observations may be made:

Breach Growth & Discharge
Many existing breach models predict discharge by assuming that supercritical flow occurs within the main body of the breach. Flow can then be calculated using a weir equation and the width of the breach. It can be seen from Plate 2 (Field Test #1) that this is not always the case. In this photo it can be seen that the flow through the breach is controlled by a curved weir created by erosion of the upstream embankment face. This 'control section' gives weir flow over a length significantly greater than the breach width.

Plate 2: Weir control section

Lateral Erosion of Embankments
Many existing models assume a uniform and sometimes predefined distribution of erosion of material in order to predict breach growth (e.g. uniform growth of a trapezoidal section). It can be seen from Plate 3 below that lateral growth occurs through erosion of material at the base and sides of the breach with discreet failures of the side slopes leading to growth. Note also in this photo that whilst erosion is occurring at the sides, as indicated by coloured water, the flow through the centre is relatively clear, suggesting minimal sediment transport. Most existing models that calculate sediment transport assume a uniform load.

Plate 3: Lateral erosion of embankments (muddy water adjacent to eroding banks; clear water in centre)

Pipe Formation – Effects of Arching
Plate 1(right) shows pipe formation through a moraine embankment. Even after significant erosion has taken place, the crest of the embankment shows little sign of distress and no subsidence. Throughout growth of the pipe the load of the material above the hole has been distributed across the bank through an arching effect. Reliable prediction of breach growth through pipe formation requires a clear understanding and assessment of this process.

Laboratory Work
A series of tests were undertaken in parallel to field tests in the modelling laboratories at Wallingford. The majority of these tests were designed to reproduce and also extend the range of tests undertaken in Norway. This permitted an analysis of scale effect between field and laboratory experiments (1:10 scale factor), and created a wider range of data sets with which to analyse breach growth and assess model performance.

Two main series of overtopping tests were undertaken using the large 'flood channel facility' at Wallingford. The first series (2002) simulated breach growth through overtopping of non-cohesive material. This related to field test #2. Following an analysis of potential scaling mechanisms, the material used was also scaled at 1:10. In order to create a material with properties matching the material used in field test #2, but at a scale of 1:10, it was necessary to mix 4 different sands. The second series (2003) was undertaken to investigate beach formation through cohesive material. Failure was again by overtopping. The behaviour of cohesive material cannot be scaled exactly without also scaling other loads such as gravity. This option was not available to us (i.e. use of a large centrifuge) hence it was decided that tests would be undertaken using material similar to that used for field test #1 and any scale effects carefully considered. For example, analysis of material condition and hydraulic loading allowed an assessment of the scaling of critical shear stress and material erodibility.

Figure 2 shows the grading curves for both cohesive and non-cohesive tests and Plate 4 examples of each laboratory test.

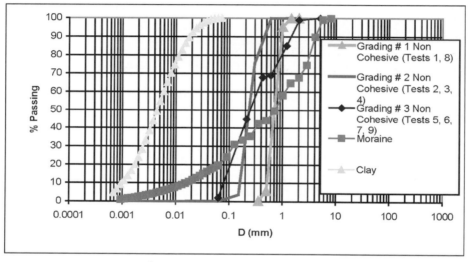

Figure 2   Materials used for cohesive and non-cohesive laboratory tests

Plate 4   Laboratory test of breach formation through overtopping

Table 2 provides a summary of the tests undertaken. Parameters that were varied for the non-cohesive tests included geometry and material grading distribution around a constant D_{50} size. Parameters varied for the cohesive tests included geometry, compaction and moisture content.

Table 2   Objective of laboratory tests

Test	Nature / purpose of test

Non-Cohesive tests; 0.5m high; material gradings 1, 2 or 3:

#1	Facility set-up / trial
#2	Scale of Field Test #2, but uniform material grading based upon D_{50}
#3	Repeatability of Test #2
#4	As Test #2, but breach initiation adjacent to side of flume
#5	Direct replication of Field Test #2
#6	As Test #5, but embankment face at 1:2 instead of 1:1.7
#7	As Test #5, but embankment crest width 0.3m instead of 0.2m
#8	As Test #2 but larger D_{50} for uniform grading of material
#9	As Test #5 but seepage allowed to develop prior to testing

Cohesive tests; 0.6m high; material clay or moraine:

#10	Scale of Field Test#1
#11	Repeatability of Test #10
#12	As Test #10, but constructed with half compaction effort
#13	As Test #10, but constructed to optimum moisture content
#14	Continuation of Test #13
#15	As Test #10 but 1:1 gradient for downstream slope
#16	As Test #10 but 1:3 gradient for downstream slope
#17	As Test #10 but using moraine material

In addition to these 17 tests, a further 5 tests on pipe formation were undertaken. Two of these tests were to aid development of an appropriate failure mechanism to ensure that failure of the piping field tests occurred within a reasonable period of time. The remaining three were testing of pipe formation through 3 samples of real embankment ($\sim 1\text{m}^3$) taken from the Thorngumbald Managed Retreat Site on the River Humber. This work was undertaken by Birmingham University and was also consistent with recommended R&D work under the EA / Defra *Reducing the Risk of Embankment Failure under Extreme Conditions* project.

Initial Findings of Laboratory Work
Whilst data is still being analysed at the time of writing, some initial observations may be made:

Headcut Erosion
Erosion of cohesive and non-cohesive embankments occurs in a different way. Cohesive material tends to erode via a series of steps – called head cutting. This was clearly seen in the field tests, but was also reproduced in the laboratory, suggesting that this process was not affected by scaling at 1:10 (see Plate 4 (left)).

Soil Properties, Condition and Seepage
The effect upon the rate of breach formation of variation in soil properties and / or condition was quite noticeable and in particular, the effects of variation in moisture content for cohesive materials. Changing the moisture content of the cohesive material from 20% to 30% (near optimum) changed the erodibility rate of the material by a factor of 12. This effect was significantly smaller when working with non-cohesive material, where allowing seepage through the bank to establish prior to testing appeared to have a minimal effect upon the eventual rate of breach growth.

Material Grading & Compaction
Many existing models represent embankment material by a single D_{50} value. Tests using different material grades, but each with the same D_{50} value, showed different behaviour, with, as might be expected, a wider grading material offering greater resistance to breaching. Also of significance is the degree of material compaction (or density). In one test, halving the compaction effort resulted in a significant change to the rate of breach formation. Specifically the rate of down cutting increased by x2.5, lateral widening increased by x5 and headcut erosion increased by x1.6.

Numerical Modelling and Analysis
A fundamental objective of the field and laboratory research work was to collect reliable data with which to validate and further develop numerical models for predicting breach formation. At the time of writing, model performance was being assessed through a controlled programme of testing such that field or lab data was only released after initial modelling predictions had been collated. This 'blind' and 'aware' approach to modelling ensured complete objectivity in the assessment of performance.

Whilst some initial results have been assessed, the extent of model performance assessment is not sufficient to allow reporting here. However, full results from this analysis work will be reported later during 2004 and posted via the project website (www.impact-project.net).

BREACH FORMATION: MIDTERM CONCLUSIONS AND OBSERVATIONS

The most striking observation (based upon the field and laboratory test data) is the clear relationship between the breach formation process and the embankment material properties and condition. Whilst this may seem obvious, it is a fact that many existing predictive breach models ignore such information and endeavour to predict the failure process based upon geometry, limited soil property information and hydraulic loading conditions. Whilst tests show that variations in material grading, compaction and moisture content (for example) can affect the rate of material erosion and hence breach growth by factors of more than x10. Where models fail to include even the most basic of soil properties or conditions, then the potential accuracy of their predictions will be significantly constrained.

Failure to account for the way in which breach growth develops will also limit modelling accuracy. For example, it is clear that rockfill embankments behave differently to non or low cohesive earthfill embankments which in turn behave differently to cohesive embankments. This difference applies particularly to the way in which the breach initiates. Most existing models make broad assumptions as to the way in which erosion occurs so as to provide an average rate of formation and hence discharge. Whilst this may be a valid approach for a specific material type and condition (against which the model has to be calibrated), this will lead to inaccuracies when routinely applied as a single solution or model applicable to all materials and conditions.

Future Direction of research

An extensive analysis of breach model performance is currently underway and should be completed by June 2004. This work will also link with an assessment of modeling uncertainty in order to provide the 'end user' with guidance on both the performance / accuracy of breach models, as well as the range of uncertainty that might be reasonable to expect within a model prediction.

In the longer term, it is clear that in order to improve our ability to predict breach growth we will require a much closer integration of soil mechanics and hydraulics analysis. Critical soil parameters that have the most influence upon the initiation and growth of a breach will need to be identified, along with methods for measuring or monitoring these parameters in the field.

REFERENCES

Morris M W (2000). *CADAM – A European Concerted Action Project on Dambreak*. Proceedings of the biennial conference of the British Dam Society, Bath.

Morris MW, Hassan MAAM (2002). *Breach Formation Through Embankment Dams and Flood Defence Embankments: A State of the Art Review*. Stability and breaching of rockfill dams workshop. Trondheim. April 2002.

IMPACT Project workshop proceedings:

Workshop #1	16-17 May 2002	Wallingford, UK
Workshop #2	12-13 September 2002	Mo-I-Rana, Norway
Workshop #3	6-7 November 2003	Louvain-la-Neuve, Belgium

ACKNOWLEDGEMENTS

IMPACT is a research project supported by the European Commission under the Fifth Framework Programme and contributing to the implementation of the Generic Activity on "Natural and Technological Hazards" within the Energy, Environment & Sustainable Development programme. EC Contract: EVG1-CT-2001-00037. The financial support offered by DEFRA and the Environment Agency in the UK is also acknowledged.

The IMPACT project team comprises Universität Der Bundeswehr München (Germany), Université Catholique de Louvain (Belgium), CEMAGREF (France), Università di Trento (Italy), Universidad de Zaragoza (Spain), Enel.Hydro (Italy), Sweco (formerly Statkraft Grøner AS) (Norway), Instituto Superior Technico (Portugal), Geo Group (Czech Republic), H-EURAqua (Hungary) and HR Wallingford Ltd (UK).

Particular recognition is given to Kjetil Vaskinn of Sweco (formerly Statkraft Grøner) for his role in managing the breach formation field tests in Norway and providing data for this paper.

A passive flow-control device for the Banbury flood storage reservoir

J ACKERS, Chief Hydraulic Engineer, Black & Veatch Consulting
P HOLLINRAKE, Senior Hydraulic Engineer, HR Wallingford Ltd
R HARDING, Project Manager, Environment Agency

SYNOPSIS. The Environment Agency is developing a scheme to protect the town of Banbury against flooding, principally by providing an 'on-line' flood storage reservoir on the River Cherwell upstream of the town. A flow control structure will be sited on each of the two branches of the river, incorporating a suitably designed throttle to limit discharges passed through the town in events up to a return period of about 200 years. Construction of the main works of the project is programmed to commence in 2005.

This paper describes the development of the design for the flow-control structures with the aid of a physical model at HR Wallingford. The design is based on a double-baffle orifice capable of maintaining discharges passed downriver within a target range of less than ±10% over a wide range of water levels in the flood storage reservoir.

INTRODUCTION

Banbury lies on the River Cherwell, a left tributary of the River Thames into which it flows in Oxford. The town has a long history of flooding, the most recent major flood being in 1998, with an estimated return period of about 100 years and total flood damage exceeding £12.5M. Flooding in Banbury is the result of the River Cherwell and associated local watercourses having insufficient capacity to convey the runoff from the upstream catchment, and has been exacerbated by development being allowed to take place on the floodplain.

The preferred solution, which was chosen taking account of technical, economic and environmental issues, is to provide an upstream 'on-line' flood storage reservoir, coupled with some local defences in the town. The flood storage reservoir will comprise the following main elements (see Figure 1):

- an embankment of maximum height about 5m and length 2.9km, running parallel to the northeastern side of the M40 and to the eastern side of the Oxford Canal;
- two similar flow control structures, one at the intersection of the embankment with each branch of the River Cherwell;
- service spillways incorporated into the control structures; and
- an emergency spillway incorporated in the embankment between the two control structures.

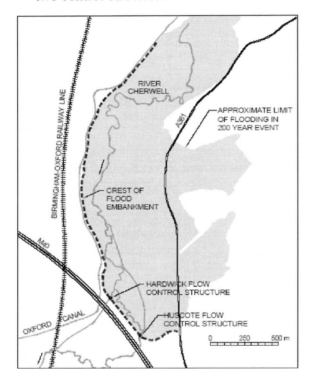

Figure 1   Location plan

The intention is that, in combination, the flow control structures should throttle the river flows to the maximum discharge which can be passed through the town, estimated as 38 m^3/s, impounding the additional flood discharge in the reservoir. The reservoir has been designed to accommodate the volume expected to be impounded in the design 200-year flood event.

When the reservoir is full, the spillways located alongside the control structure will overtop and provide a total discharge capacity approximately equal to that of the unattenuated peak of the 200-year flood. The emergency spillway will allow more extreme floods to be discharged without overtopping the rest of the embankments.

CONCEPT

The Environment Agency was keen that the control of discharges passed downriver from the flood storage reservoir should occur automatically, with no requirement for attendance by their operatives during floods. It was also considered desirable to place no reliance on power supplies or remote operation of the flow control structure. If practicable, a structure with no moving parts would also be preferred.

An ideal flow control device for an on-line flood storage reservoir would allow all discharges less than a target value to pass downstream without starting to impound. As the discharge continues to rise it would then allow the target discharge to pass downstream, impounding all of the excess. Such accurate control is difficult to achieve precisely, even in a fully automated gated system, but would have two advantages if it could be achieved:
- it would minimise the effect on the land within the impoundment area during minor floods (with a return period of up to about five years in the case of Banbury); and
- in the early part of larger floods, it would preserve as much as possible of the available storage volume for utilisation in attenuating the peak of the flood hydrograph, ultimately reducing the total flood storage needed and therefore lowering the peak water level in the flood storage reservoir.

A simple orifice meets the objective of having no moving parts, but results in the discharge rising as the square root of the net head. If a simple orifice is designed to limit the discharge to the target value when the reservoir is nearly full, this results in it starting to impound when the discharge is much less than that target value.

The above objectives led to consideration of the design concepts embodied in the baffle distributor devices which have been used for many years in irrigation systems, in particular the 'Neyrpic module'. Performance information on these devices is given by Neyrpic (1971), Alsthom Fluides (undated), UN/FAO (1975), Bos (1989) and a number of other standard references. The devices are designed to achieve a nearly fixed discharge out of a parent irrigation canal over a range of operating levels in the canal.

Two forms of the device are described in the references, one comprising a single baffle and the other a double baffle, of which the double baffle has the potential to provide a wider range of nearly fixed discharge, so was of particular interest. Although the performance information for double baffle distributors is apparently identical between the references consulted, at least three different variants on the shapes of the baffles are given.

Figure 2 shows the geometry of the device, based on the dimensions and shape quoted by Bos (1989), together with the quoted stage/discharge rating. As the upstream head rises, it impinges on the upstream baffle, which then acts as the control, with the jet clearing the underside of the downstream baffle. As the head rises further it overtops the upstream baffle and a transition to downstream baffle control occurs as the stage rises further. The discharge remains within a band of ±10% for heads between around 0.73 and 1.74 times a nominal design head.

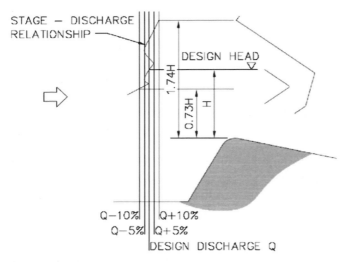

Figure 2   Double-baffle orifice layout and performance (after Bos, 1989)

PRELIMINARY DESIGN
Initial consideration of the outline design for the Banbury control structures indicated that the vertical and longitudinal dimensions should be exactly twice those of the largest standard irrigation distributor module illustrated by Alsthom Fluides (undated). The nominal design head is 2.58m, the nominal design unit discharge is 5.66 m^2/s and the target head range for ±10% is 1.88m to 4.50m. On this basis each bay would be 1500mm wide, giving a nominal discharge of 8.5 m^3/s per bay or 17 m^3/s per structure and therefore a total nominal downriver flow of 34 m^3/s, rising to about 38 m^3/s at the maximum positive deviation of 10%.

In a distinct departure from the designs in the references, it was decided that the invert profile should resemble a Crump weir, with upstream and down-stream slopes of 1:2 and 1:5 respectively meeting at a sharp vertex. Factors in this decision were the simpler construction than the round crest of the original device and the possibility of predicting the lower part of the rating curve (before impingement on the baffles) by the use of the standard formula for a Crump weir.

Prior to model testing, a 'target' stage/discharge relationship was prepared, based on the formula for a Crump weir at low heads and on the relationship for baffle control given by Neyrpic (1971). Although precise compliance with that relationship was not considered an essential outcome of the testing, it was a useful aid for comparing the performance when various details were adjusted during the design development.

MODEL STUDY
The Banbury flow control device will be much larger than the largest of the standard irrigation distributor modules described in the references. Although the reported design and hydraulic behaviour would clearly be amenable to Froude scaling, several factors led to the decision to undertake a programme of project-specific model testing:
- the differences between the references regarding the appropriate configuration for the device;
- a concern that the baffle design shown in Figure 2 (and the other versions) would be vulnerable to debris accumulation;
- a recognition that metal fabrication might not be appropriate for the larger structure and that the use of thicker concrete structural members would have an impact on both the design and the resulting hydraulic behaviour;
- a suggestion in one of the references (Bos, 1989) that the hydraulic performance of the device would exhibit hysteresis, with different ratings for rising and falling stages; and
- the need for a verified rating relationship for use in design.

The model testing commenced with two versions of the double-baffle configuration, as illustrated in Figure 3. One is based on the simplest of the three variants which appear in the references, comprising angled baffles expected to be fabricated in robust steel plate; the other comprises simple vertical baffles, which are thicker than the angled baffles and intended to be suitable for construction in reinforced concrete.

The model design, construction and testing were undertaken at HR Wallingford, with a model geometric scale of 1:12 selected. The model was built mainly from PVC, to provide a suitable boundary roughness (comparable to concrete in the prototype) and the sidewalls in the vicinity of the baffle devices were built in Perspex to allow flow visualisation. Discharges were provided via a centrifugal pump and measured using an electromagnetic flowmeter, giving a basic accuracy of around 1%. Water levels were measured using manual micrometer point gauges reading to an accuracy of about 0.25mm (3mm in prototype terms).

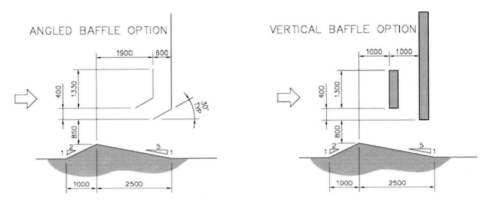

Figure 3   Double-baffle designs for preliminary testing

Each of the two flood control structures was expected to comprise a pair of the orifice devices side by side, separated by a central pier with a semi-circular nose. It was decided to reproduce a complete structure, including the two bays and central pier, in the model, although not the detail of the approach channel and exit channel. In the preliminary tests each bay contained one of the two different versions of the double baffle orifice device, with the approach channel to each bay closed off in turn in order to test one bay at a time. When the design development was complete, the chosen design was built into both bays and confirmation tests undertaken.

The model test programme thus comprised three stages:
- preliminary tests, using the preliminary designs illustrated in Figure 3;
- optimisation tests, in which a series of design adjustments, affecting the baffle positions, elevations and shapes were made and evaluated in a single bay; and
- final tests of the optimised structure in both bays of the model.

The preliminary and optimisation tests were carried out with the downstream water level low enough to avoid any effect on the flow conditions in the structure. The final tests also used tailwater levels derived from flood simulations.

Preliminary tests
The preliminary tests on the configurations shown in Figure 3 showed a close agreement between the test results for the angled baffle and the target relationship (Figure 4), suggesting that the configuration chosen for the angled baffle testing was indeed a valid variant. The flow conditions during

the transition from upstream baffle to downstream baffle control were unstable, with strong air-entraining vortices forming upstream of the upstream baffle, leading to oscillations in the approach channel. The instability was associated with water surface drawdown around the nose of the central pier and the formation of a standing wave immediately upstream of the baffles.

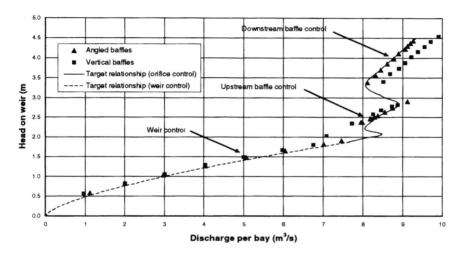

Figure 4   Stage-discharge relationships from preliminary tests

The preliminary vertical baffle arrangement gave a stage/discharge relationship (Figure 4) which diverged from the target relationship somewhat, with an earlier transition from weir to upstream baffle control. The flow conditions during the transition were again unstable, leading to oscillations in the approach channel, but without the strong vortex action found in the angled baffle device. The instability was again associated with the surface drawdown around the nose of the central pier.

Another notable feature was that the initial water surface contact on rising stages was with the downstream baffle, although the resulting effect on the approaching flow profile caused rapid contact with the upstream baffle, which then took over flow control, with the downstream flow surface clearing the underside of the downstream baffle.

Optimisation tests
As a result of the observations in the preliminary tests, it was decided to extend the central pier further upstream and to change its nose shape to a lens, with a 90° internal angle, in order to reduce the severity of the local drawdown and to allow substantial recovery upstream of the weir crest and

baffles. Because the angled version of the device had already closely met the target relationship, efforts were concentrated on optimising the performance of the vertical baffle option, as this was seen to offer a number of potential advantages, if a satisfactory rating relationship could be achieved.

A total of eight different versions of the vertical baffle device were investigated, adjusting the elevations of both baffles and the spacing between them, but in all cases with the upstream face of the upstream baffle 1000mm downstream of the vertex of the Crump weir. The various adjustments made were aimed at achieving two effects:

- a narrow range of discharges under baffle control, with the curve for downstream baffle control lying directly above the curve for upstream baffle control; and
- if possible, a direct transition from weir control to upstream baffle control, without the water surface first impinging on the downstream baffle.

The tests confirmed what was expected, that these two objectives are mutually exclusive – raising the downstream baffle to avoid it impinging first on the water surface inevitably shifts its control curve to the right and therefore increases the spread in the rating relationship.

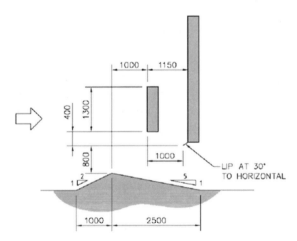

Figure 5   Optimised baffle arrangement

In order to reduce the spread between the curves for upstream and downstream baffle control, it was decided to introduce a small angled plate onto the front of the downstream baffle, resulting in the configuration shown in Figure 5. By making the contraction effect for the downstream baffle more severe than that for the upstream baffle, this had the desired effect, as illustrated in Figure 6. This figure also shows for comparison the relationships for the two designs shown in Figure 3, but now with the central pier extended.

Measurements throughout the preliminary and optimisation tests were made under virtually steady-state conditions, but in the course of rising or falling stages. In no case was any hysteresis effect detected.

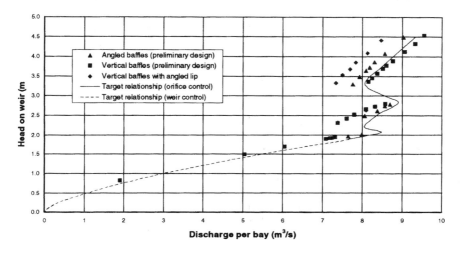

Figure 6   Stage-discharge relationship for optimised design

Although the results for the design shown in Figure 5 lie typically about 10% to the left of the target relationship, this was considered acceptable, because the requisite discharge capacity could be achieved by simply increasing the width of each bay to approximately 1.65m, which would have minimal cost and layout implications.

Final tests
The optimised baffle design shown in Figure 5 was built in both bays and tested under the following tailwater conditions:
- low, as used in the preliminary and optimisation tests;
- rising stage, as on the rising limb of a severe flood hydrograph; and
- falling stage, as on the recession of a severe flood.

The rising and falling stage tailwater levels were taken from mathematical model simulations of the 200-year return period flood, with the scheme in place and using the target rating relationship for the control structures. They are not necessarily representative of all flood conditions under which the flood storage reservoir will operate, but nevertheless give a realistic indication of the general effects of the natural range of tailwater levels on the performance of the control device.

Figure 7 shows the results for all the above cases, including a 'by-eye' best-fit line for the rising tailwater case. The results plotted with solid symbols relate to steady-state measurements, whilst those with open symbols are the results of measurements when steady conditions could not be maintained. (In the latter case, the discharge was measured taking account of the rate of change in the volume stored in the model between the device and the flowmeter.)

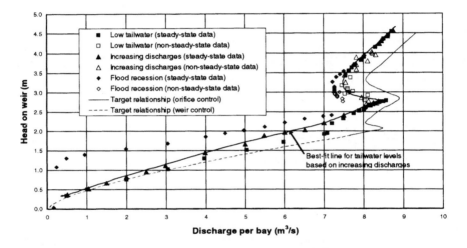

Figure 7   Final stage-discharge relationship for optimised design in two
bays with various tailwater conditions

It may be noted (by comparing the results with those in Figure 6) that, for low tailwater levels, there is only a marginal difference in performance for the twin-bay version compared with the single-bay version. With the tailwater levels based on rising flood stages, the rating is affected by tailwater for heads between approximately 0.7m and 2.2m, but there is no significant difference in the performance of the device for heads between 2.2m and 4.5m. On falling stages, the higher tailwater levels have a modest effect on the behaviour of the device for stages between about 3.9m and 2.5m and a larger effect at lower stages. It should be noted that the hysteresis in this case is wholly driven by the applied tailwater levels and is not a fundamental characteristic of the device.

Plates 1 to 6 show various stages of flow behaviour for the optimised design with low tailwater levels and rising upstream heads.

Plate 1   First contact with
downstream baffle

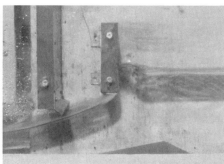

Plate 2   Control quickly transfers to
upstream baffle

Plate 3   Upstream baffle starting to
overtop

Plate 4   Downstream baffle starting
to control flow

Plate 5   Upstream head still affected
by weiring flow over baffle

Plate 6   Upstream baffle virtually
submerged

CONCLUSIONS

A passive flow-control device, based on a Crump weir profile and twin
baffles, has been developed with the aid of a physical model, from the
concepts embodied in the baffle distributor devices used in irrigation
systems.

The device, which includes simple vertical baffles, with an angled lip on the downstream baffle only, is capable of controlling discharges within a band of ±10% for stages between 2.0m and 4.5m, provided that the downstream water level does not influence the flow conditions. No evidence of hysteresis was found.

On rising and falling stages in a simulated 200-year flood, the performance of the device is affected by the anticipated tailwater regime. On rising stages, which affect the utilisation of flood storage, the bottom of the ±10% discharge band is raised from about 2.0m to 2.2m.

ACKNOWLEDGEMENTS
The authors thank the Environment Agency for permission to publish this paper and gratefully acknowledge the assistance of their colleagues who provided assistance and comments on the draft paper. Whilst the model test results and other statements in this paper are believed by the authors to be an accurate reflection of the research work described, they should not be relied upon for any other purpose.

REFERENCES
Alsthom Fluides (undated). *Distributors.* Leaflet ref A6510A (similar to Neyrpic, 1971)
Bos, M.G. (1989). *Discharge measurement structures,* 3rd edition. International Institute for Land Reclamation and Improvement, Wageningen.
Neyrpic (1971). *Distributors.* Leaflet 65 10 71 A. Neyrpic Départment Fluides, Grenoble.
United Nations Food & Agriculture Organization (1975). Small hydraulic structures

Challenges on dam safety in a changed climate in Norway

G H MIDTTØMME, The Norwegian Water Resources and Energy
Directorate, Oslo, NORWAY

SYNOPSIS.   Much research has been done recently on the possible future
scenarios of climate change. The effects on runoff and extreme precipitation
for Norway have been investigated, and the results indicate that there will be
larger and more frequently occurring floods in the future. Some possible
effects on dam safety in Norway are presented as well as some recent
examples of dam incidents caused by unusual climatic conditions.

INTRODUCTION
Over the past few decades there has been an increased interest in climate
change issues. Whether climate change is caused by natural variations or
man-made emissions into the atmosphere (or a combination of these two
mechanisms), we should be prepared to handle possible effects of the
scenarios given by the researchers. A changed climate will evidently result
in changes in the basis for safety evaluation of dams and other hydraulic
structures, and updating of the design flood estimations may be necessary.
Today the need for updating of the design flood estimations in Norway is
evaluated as part of the regular dam safety reassessments, normally every 15
years according to the guideline on inspection and reassessment. That is,
when there are considerable changes in the data series that has been used for
the estimation of design floods, new design flood estimations must be
performed. The guideline on inspection and reassessment is one of several
new guidelines on dam safety that have been published over the last few
years, and more are currently being prepared or are planned in the near
future. The guidelines describe how the requirements in the regulations on
dam safety can be fulfilled. Today the following regulations form the legal
framework for dams, all of them with a legal basis in the Water Resources
Act of 2001;

- Regulations on dam safety
- Regulations on classification of dams
- Regulations on qualification requirements
- Regulations on internal control

The development of the legal framework for dams in Norway is described in a recent article by the author (Midttømme, 2003). The Norwegian Water Resources and Energy Directorate (NVE), which is the regulatory authority on dam safety in Norway is responsible for the development of the new guidelines.

FLOOD ESTIMATION

Around 1980 major changes in the flood estimation methods were introduced in Norway. This resulted in a need for updating of the design flood and probable maximum flood (PMF) estimation for most Norwegian dams. Since then, new flood estimations have been carried out for more than 800 dams (the total number of classified dams is approximately 2300), sometimes followed by upgrading of the dam structure or spillway (Pettersson 1998). A new guideline on flood estimation was prepared as part of the recent revision of the legal framework on dam safety mentioned above. This new guideline requires that flood estimations be classified with respect to uncertainty based on an evaluation of available data. In addition, sensitivity analyses of the flood estimations are recommended. Otherwise, there are no significant changes with respect to methods for estimation of design floods and PMF in the revised regulations and guidelines. The new legislation will therefore not trigger a new general revision of design flood and PMF estimation for dams in Norway. The present method for flood estimation is described briefly in the following paragraphs.

In Norway two floods are defined; for spillway design and dam safety control, respectively:

- The "safety check flood" which must be bypassed safely without causing dam failure. Some damage to the dam may be accepted.
- The "design flood" which is a flood with a specific return period. This flood represents an inflow, which must be discharged under normal conditions with a safety margin provided by the freeboard. The design flood is the basis for the design of spillway and outlet works.

For high hazard dams the PMF is selected as the safety check flood and Q_{1000} (the flood with a return period of 1000 years) as the design flood, see Table 1 below.

The PMF is calculated by use of rainfall/runoff models on the basis of estimates of probable maximum precipitation (PMP). In most cases a snowmelt contribution should be added to the PMF. The design flood, on the other hand, has to be estimated with some kind of frequency analysis.

This is done, either by doing a single site analysis, or by doing a regional analysis. A regional analysis for Norway was prepared in 1978 and updated in 1997 (Sælthun 1997). The updated version introduces a new classification of regions and new estimates for the relationship between mean annual floods and the 1000-year floods. When flow records are insufficient or not available, the design flood can be calculated using rainfall/runoff models and estimates of precipitation events with a 1000-year return period. If possible, the results from this analysis are compared to calculated floods in similar catchments in the same area.

Table 1 Selection of floods in current Norwegian dam safety regulation

DAM HAZARD CLASSIFICATION (CONSEQUENCES)	DESIGN FLOOD	SAFETY CHECK FLOOD
HIGH	Q_{1000}	PMF
SIGNIFICANT/MEDIUM	Q_{1000}	PMF or $1,5 \times Q_{1000}$
LOW	Q_{500}	-

The 1000-year flood is defined in the guidelines as *"the inflow flood, with a return period of 1000 years, that results in the highest water level in the reservoir given particular conditions for operation of spillways and initial reservoir level"*. For routing through the reservoir in order to find the design outflow flood and the corresponding water level, one of the general requirements is that initial water stage in the reservoir be set to the highest regulated water level (HRWL). Transfer tunnels for water into the catchment are normally considered open, while transfer tunnels out of the catchment are considered closed. More details about requirements and methods for flood estimation can be found in the guidelines (NVE 2002) and in an article prepared for the ICOLD European Symposium in Barcelona in 1998 (Pettersson 1998).

SEASONAL AND REGIONAL FLOOD CHARACTERISTICS

Norway is a country with distinct seasonal and regional variations of climate and runoff. The variability in floods over the year and from region to region is exemplified by two catchments shown in the figure below. There are three main causes of natural floods: snowmelt; rain on snow; and rain. Autumn floods are caused by heavy rain and saturated soil, sometimes in combination with melting of newly fallen snow. Spring floods are a result of snowmelt and may be increased due to rain or melt water flowing over frozen ground. Spring floods tend to have longer duration than autumn floods, but there are exceptions. Typical areas dominated by spring floods are Southeast Norway and Finnmark, the northernmost county of Norway (on the mainland). In coastal areas there may be no seasonal distinction between spring floods and autumn floods. Floods may appear at any time of year, and summer is often a low runoff season. Some Norwegian river basins also contain glaciers. Glacial runoff can be dominant in catchments with glaciers covering only a small percentage of the area. Characteristic for these catchments is floods during summer (Sælthun 1997).

Many rivers are regulated and the purpose of most Norwegian reservoirs and dams is hydropower production. Consumption of water from the hydropower reservoirs is usually highest during winter, when there is normally very little inflow to the reservoirs due to precipitation falling as snow. Thus, the large reservoirs are mostly empty during late winter, which is a benefit when there are severe spring floods. On the other hand, the reservoirs are filled during summer and autumn, and offer very little storage capacity for autumn floods.

CLIMATE AND RUNOFF SCENARIOS FOR 2030-2049

As part of the research project RegClim, the global climate scenarios developed by IPCC (The Intergovernmental Panel on Climate Change) has been downscaled in order to prepare for impact studies of climate change in Norway. The most probable scenario, according to the RegClim-project, is that the annual temperatures in Norway will increase by 0.25 C/decade up to 2050 (Iversen, 2003). It may be worth noticing that the increase in temperature is estimated to be highest in the winter months and in the northern parts of the country. There is also an expected increase in annual precipitation for the whole country, and the highest increase is estimated to occur in the western part of Norway in the autumn. A minor increase in wind velocities and the number of storms is also expected, especially in Central Norway.

Based on the results from the RegClim-project, work has been done to estimate annual and seasonal runoff for the period 2030-2049 (Roald et.al. 2002). The runoff scenarios are based on two modelling strategies, i.e. modelling by the use of two different versions of the HBV-model for 42 catchments in Norway. Both modelling approaches are based on the same rainfall/runoff model. A comparison of the change in runoff simulated by the two models, show that the difference is generally quite small. The results differ by 2% or less for 30 of the 42 catchments. With a few exceptions, the results show a general increase in annual runoff for all the catchments for the period 2030-49 compared with the control period 1980-99. The highest increase in annual runoff is estimated to 20% in the western part of Norway. A study of the simulated changes of the seasonal runoff show that the winter runoff will increase significantly in the southern part of East Norway and in catchments along the coast of West Norway. The spring runoff will increase in the inland and the mountainous part of most of Norway, but the highest increase will be in the coastal catchments of Finnmark, the northernmost county of Norway. The summer runoff will decrease in most of the catchments, while the autumn runoff will increase in West Norway and Finnmark. More details are shown in Figure 1 below.

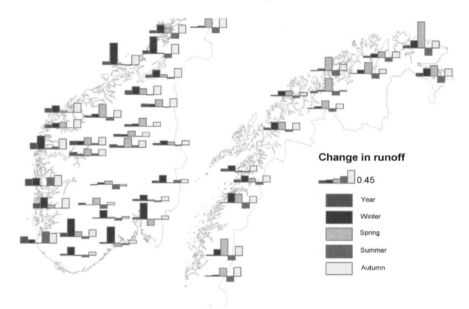

Figure 1: Seasonal changes in runoff for scenario-period 2030-49 compared to control-period 1980-99 (Roald 2003).

Scenarios of extreme precipitation of duration 1 and 5 days due to climate change have also been developed for Norway (Skaugen et.al. 2002). Time series of 1000 years have been generated on basis of downscaled precipitation values from a global climate model. The study indicates an increase in extreme values and seasonal shifts for the scenario period 2030-49, compared to the control period 1908-99. The study is based on simulations at 16 locations in Norway and the results show a significant regional variability.

POSSIBLE EFFECTS ON DAM SAFETY
The majority of dams in Norway were designed and constructed long before the first dam safety regulations were made valid in 1981. Even though many dams and spillways have been upgraded already to meet the present safety standards given in the new legal framework (see above), there is still a need for upgrading and rehabilitation of many dams, also as a consequence of damage due to deterioration and ageing processes. Floods and extreme weather will cause extra strain on dams and spillways, and the latest results from the studies of climate change and effects on runoff and extreme precipitation indicate that we should be prepared for more extreme weather and larger and more frequently occurring floods. For spillways that have

rarely been in use so far, it should be noted that damage in the spillways might occur at discharges much lower than the design flood (Kjellesvig 2002). Recent studies on damages to concrete dams, which is the leading dam type with respect to number of dams in Norway, showed that 5 % of these dams had damages, which were considered to be a threat to the overall safety. It is also worth noticing that 44 % of the Norwegian concrete dams had been repaired already, and that 40% needed repair in the near future. There were also several reports about dams that had been repaired more than one time (Jensen 2001), indicating that the dam owners put too little emphasis in finding the actual cause of damage and/or the best repair method. In total, we may be facing a growing need for repair works in the future for all dam types. However, as there seems to be an increase in winter and/or spring runoff in larger parts of the country, it may be more difficult to find an appropriate time for doing necessary rehabilitation and upgrading works than what is the situation today. The present energy situation in Norway may also worsen the situation. There is an increasing gap between energy production capacity and energy consumption. As Norway is totally dependent on hydropower, it may be a problem to gain acceptance for closing down of hydropower facilities and/or lowering of reservoir levels in order to do necessary upgrading and rehabilitation of dams and spillways.

A significant increase in the design flood values for dams may be the result of climate change, and this will further lead to insufficient spillway capacity and/or freeboard at many dams. Even without climate change, floods larger than the design flood are likely to occur. Reasons for this may be for example short time series or incomplete time series used for design flood estimation. The challenge of floods exceeding the design flood is therefore relevant to discuss in any case, as well as the possibility of experiencing more extreme weather and more frequently occurring floods than what we have experienced so far. Typical problems related to operation of dams during floods and recommendations for handling large floods and more frequently occurring floods are given by Kjellesvig & Midttømme (2001) and Kjellesvig (2002). The main conclusion is simple; upgrade the dams and spillways in order to provide safe bypass of floods exceeding the design flood. Redundancy in spillway systems is also promoted as a safeguard against flood related damages and any following consequences, along with good monitoring systems for early detection of adverse conditions. Another possibility would be to lower the HRWL in order to increase the freeboard, and thereby be able to store more floodwater in the reservoir. A simple solution to the problem of increasing design flood values is always to add extra safety margins when a dam is upgraded in the future as recommended by Bergström (2003) in a recent seminar focussing on climate change within in the hydropower sector. The challenge is probably to persuade the dam owners to select a more expensive solution than required today, even though this may be a sensible and perhaps also economic solution in the long run.

As mentioned above, floods cause an extra strain on dams and spillways, and in some cases the utmost consequence of a flood is a dam failure. A recent example from Norway is the failure of Nervatn Dam in Norway in January 2002. Nervatn Dam was a small concrete dam classified in the low consequence class. The failure was caused by erosion of the right abutment, which resulted in the overturning of a 20m long section of the concrete dam (Pedersen 2002). The stop logs in the spillway had not been opened because the dam owner was too late in reacting to the increasing water level. The flood was caused by extraordinarily warm weather combined with rain (i.e. 250 mm in 4-5 days). The rainfall had an estimated return period of 200 years. The consequences were limited to damage to three downstream bridges including one main road bridge, and the loss of approximately 1 million m^3 of water (50 % of the reservoir capacity). The failure of Nervatn Dam can not be assigned to climate changes directly, but the incident is a reminder of the fact that dams may be more vulnerable to floods, and effects of floods, in the future.

The probability of experiencing more ice-related problems in the future is also of special interest to Norwegian dams. Changes in temperatures may influence on the probability of "ordinary" floods and the more rarely occurring glacier lake outburst floods (GLOFs; also denoted jøkulhlaups) in glacier dominated catchments, as well as on other ice-problems such as icings/aufeis and ice-jams. A recent glacier lake outburst flood at Blåmannsisen glacier in northern Norway is believed to be a result of climate changes (Engeset 2001). The flood was probably caused by a deficiency in ice-mass, that is, the accumulation of winter precipitation (as snow) could not compensate for snow and ice melt during summer. The flood at Blåmannsisen resulted in a 2.5 m increase in the water level in the Sisovatn reservoir, corresponding to a 40 millions m^3 increase in reservoir volume. The water level in the previously glacier-dammed lake, which released water into Sisovatn, decreased by 70-80 m. Due to the low reservoir level prior to the flood, the Sisovatn Dam and downstream areas were not affected (Josefsen 2001). An interesting point in the case of Sisovatn is that a glacier lake outburst flood from Blåmannsisen had not been considered a probable exceptional load on the Sisovatn Dam in the most recent safety reassessment. The consultant performing the safety reassessment of the dam had probably not been made aware of the possibility, whereas the hydrologists performing glaciological investigations had foreseen this possibility several years in advance (Pedersen 2003).

CONCLUSION

It is difficult to conclude on how climate change will affect the safety of our dams in the next few decades, even though some scenarios have been pointed out as more probable than other scenarios by the researchers. The key parameter with respect to dams is the design flood, but the frequency of smaller floods is also of interest. Given the most probable scenarios, we should be prepared for more extreme weather and larger and more frequently occurring floods. The result will probably be an increased need for upgrading and rehabilitation of dams. As a regulator on dam safety, NVE will continue to evaluate the results from the climate change research. Further studies on climate change effects in Norway will naturally be of special interest in this context, as well as other research related to dam safety. The current approach to climate change with respect to dam safety is to evaluate actual changes in the data series used for flood estimation, i.e. to trigger a reaction to any observed changes in the climate (belated wisdom). The guideline on flood estimation points out that flood estimations must be classified with respect to uncertainty based on an evaluation of available data. This classification of the flood estimations is meant to be a support for NVE in the further evaluation of any structural or non-structural measures for the dams. The guideline on flood estimation also recommends sensitivity analyses. The latter can be helpful in order to evaluate possible consequences to dams of specific climate scenarios.

REFERENCES

Bergström, Sten (2003). *Klimat och vattenresurser – senaste nytt från Sverige* (in Swedish). Presentation at EBL-seminar, Stjørdal, Oct.2003.

Engeset, R.V. (2001). *Klimaendringer gir jøkulhlaup ved Blåmannsisen* (in Norwegian), www.nve.no, Nov.2001.

Iversen, Trond (2003). *Sluttrapport RegClim 1997-2002* (in Norwegian), http:\\regclim.met.no, Nov.2003.

Jensen, V. (2001). Survey about Damage, Repair and Safety of Norwegian Concrete Dams. In Midttømme et.al. (eds) *Dams in a European Context*, Balkema/Swets & Zeitlinger, Lisse.

Josefsen, Aage (2001). Personal communication.

Kjellesvig, H.M. and G.H. Midttømme (2001). Managing dams and the safe passage of large floods. *Hydropower and Dams*, issue 2, vol 8.

Kjellesvig, H.M. (2002). *Dam Safety – The Passage of Floods that Exceed the Design Flood*. Dr.ing.thesis 2002:84. Norwegian University of Science and Technology, Trondheim.

Midttømme, Grethe Holm (2003). Changes in the legal framework for dam safety in Norway. *Hydropower and Dams*, issue 5, vol 10.

NVE (2002). *Retningslinje for flomberegninger til § 4-5 i forskrift om sikkerhet og tilsyn med vassdragsanlegg* (in Norwegian). NVE, Oslo.

Pedersen, V. (2002). Personal communication.

Pedersen, V. (2003). Personal communication.

Pettersson, L.-E. (1998). Flood estimations for dam safety in Norway. In Berga (ed) *Dam Safety*, vol 2, Balkema, Rotterdam.

Roald, L.A., T.E. Skaugen, S. Beldring, T. Væringstad, R. Engeset & E.J. Førland (2002). *Scenarios of annual and seasonal runoff for Norway based on climate scenarios for 2030-2049.* Consultancy report no10-2002, The Norwegian Water Resources and Energy Directorate, Oslo.

Roald, L.A. (2003). *Klimaendringer og store flommer* (in Norwegian). Presentation at NNCOLD-seminar, September 2003, Stjørdal, Norway.

Skaugen, T., M. Astrup, L.A. Roald and T.E.Skaugen (2002). *Scenarios of extreme precipitation of duration 1 and 5 days for Norway due to climate change.* Consultancy report no 7-2002, The Norwegian Water Resources and Energy Directorate, Oslo.

Sælthun, N.R. (1997). *Regional flomfrekvensanalyse for norske vassdrag* (in Norwegian). Report 14-1997, NVE, Oslo.

Weedon Flood Storage Scheme - the Biggest Hydro-Brake® in the World

G P BOAKES, Design Project Manager, Halcrow Group, UK.
A STEPHENSON, Proposals Manager, Hydro International, UK.
J B LOWES, Construction Project Manager, Halcrow Group, UK.
A C MORISON, All Reservoir Panel Engineer, Halcrow Group, UK,
A T USBORNE, Project Team Manager, Environment Agency, UK.

SYNOPSIS. The Northamptonshire villages around Weedon in the upper River Nene valley, suffered disastrous flooding in 1947, 1992 and 1998, with Weedon Bec being particularly badly affected. The channel through the village is constricted by historic developments and the opportunity to enlarge the channels was not available. Restricted culverts under the railway embankments downstream compounded the flood situation. To alleviate the problem the Environment Agency and Halcrow Group developed an upstream on-line storage reservoir scheme.

The project includes a 450m long, 6.8m high clay embankment across the valley, with a culvert on the line of the original river channel to carry the controlled outflow. A 150m long concrete-block spillway carries excess flood flows over the embankment. The embankment site has been landscaped to minimise visual impacts and the borrow area has been developed into a large wetland area as a habitat for aquatic flora and fauna.

The key component of the flow control system is a 6.5 tonne, stainless steel Hydro-Brake® Flow Control device located in the dam inlet structure. The Hydro-Brake® was designed by Hydro International to control the maximum outflow rate despite fluctuating head, and incorporates the facility to adjust the controlled outflow between 8 and 12m³/s. The use of the Hydro-Brake® helped reduce the upstream storage requirement and hence the land take and frequency of flooding involved.

This paper provides a description of the options considered during the design stage of the flood defence scheme, details of the actual design and construction of the dam, an explanation of how the Hydro-Brake® operates and the benefits it provides over other forms of flow control.

Long-term benefits and performance of dams, Thomas Telford, London, 2004, 348–359

BACKGROUND TO THE PROJECT

The Problem
The village of Weedon Bec is situated west of Northampton and suffered serious flooding from the River Nene during Easter 1998. The village had no formal flood defences and there was a risk of flooding once every three years. 95 properties were at risk of flooding, 45 were flooded in the Easter 1998 event and many others were affected. The major cause of flooding was the restriction to flow at a road bridge within the village and at the culverts under the railway embankment downstream of the village.

A range of options was considered, but it soon became apparent that all options other than upstream flood storage were unacceptable. Channel improvement to pass flood flows required the existing river channel to be doubled in size, producing unacceptable loss of land and disruption. The cost of enlarging the road bridge and railway culvert would also have been very high. This option also produced an unacceptable increase in downstream flows through several villages and Northampton, which were already at risk of flooding. Containment of floodwater within the river channel would have required construction of flood walls through 30 private gardens. The cost would have been high, there would have been unacceptable disruption to residents and there would be access problems for future inspection and maintenance.

Conveniently, within one kilometre upstream of the village, the river flows through a well-defined valley with little habitation and this forms a suitable location for flood storage.

Scheme Selection
Having determined that flood storage was a viable and acceptable option, studies continued to determine the location of the dam and storage area and the most economic standard of flood protection.

The dam location was determined by consideration of:-
- Minimising the size and cost of the dam while achieving the required storage capacity.
- Avoiding flooding of property within the flood storage area.
- Minimising visual impact.

The location was largely dictated by the position of public roads and the Grade II Listed Dodford Mill, which is adjacent to the river approximately one kilometre upstream of the village. The dam is located approximately 100 metres upstream of the Mill, behind a belt of trees that obscures the view of the dam. Consideration was given to locating the dam 100 metres further upstream at the confluence of two branches of the river, but this would have resulted in a lower, longer, more expensive dam. A smaller dam could have been located further downstream but this would have resulted in the regular inundation of Dodford Mill, making it uninhabitable.

The standard of flood protection provided by the flood storage area was determined by economic evaluation. The project was grant aided by Defra. The economic evaluation, carried out using Defra procedures, determined that the project qualified for grant aid and that the most economic standard of protection would be 1 in 50 years.

DESCRIPTION OF THE PROJECT
The scheme was completed in autumn 2002 and comprises an earth fill dam with a maximum height of 6.8m and a crest length of about 450m. The storage area occupies the valleys of the Newnham and Everdon arms of the River Nene as shown on Figure 3. The capacity of the reservoir to spillway level is 810,000 m^3, providing a 1 in 50 year standard of protection to Weedon Bec. The flooded area at full capacity is 370,000m^2. The in-bank capacity of the river channel through the village of Weedon Bec is 10 m^3/s. The flood storage reservoir reduces the peak flow through Weedon Bec from 26 m^3/s to 10 m^3/s during a 50-year event. Figure 1 shows the dam under construction.

Figure 1: Weedon Dam under construction. The borrow pit is at the lower right. Note retained tree and hedge lines screening the embankment.

Embankment & Cut-off

The underlying geology at the dam site is blue Lias Clay. On the sides of the valley this was found directly beneath the topsoil. In the valley bottom it was covered to depths of up to 4m by mixed alluvial deposits ranging from soft silty clays to shallow sand and gravel beds. The gravel beds were considered to be potentially interconnected to the existing river channel and sufficiently permeable to provide seepage paths beneath the embankment.

The embankment was founded on the surface of the alluvium after topsoil and surface stripping. A cut-off trench was excavated to the Lias Clay beneath the centre of the embankment on either side of the outlet culvert, itself founded on the solid clay in the original channel bed, and backfilled with clay. The gravelly material excavated from the foundation was retained and used to improve the roadway on the dam crest.

The embankment was formed as a homogenous clay bank, generally with 1:3 side slopes, using some 32,000m^3 of firm Lias Clay. Initial plans were to excavate this from a borrow pit on the hillside, but this was rejected in favour of a borrow pit in the valley bottom, reinstated to form a wetland, even though this required the removal and stockpiling of between 1.5 and 4 m depth (12,000m^3) of alluvial overburden. Consideration was given to using the alluvial material in the upstream shoulder of the embankment, but it proved too soft to withstand tracking without drying.

The clay fill was placed and rolled at natural water content to form a hard fill material. Despite this, the clay material has the potential to crack on drying, always a concern on flood embankments normally kept empty. To help counter this, a horizontal geo-mat was incorporated in the non-spillway sections of the bank 0.5m below finished crest level, and the crest was topped with hoggin formed by mixing the clay fill and alluvial sands and gravels from the cut-off trench, stockpiled for the purpose.

The borrow area has now been landscaped to form a lake surrounded by tree and shrub planting. As much as possible of the original, established hedge and tree lines around the site have been preserved, and additional areas around the dam have been planted as woodland to break up the view of the dam from a distance.

Spillway

Located upstream of Weedon Bec, the reservoir is Category A in accordance with "Floods and Reservoir Safety" and was designed to safely pass the Probable Maximum Flood (PMF) which was assessed to be 195m^3/s. The spillway is formed by a 150m long lowered section of the dam crest. The crest, downstream slope and buried stilling basin are reinforced with tied cellular concrete blocks so that the spillway can safely pass the PMF with a depth of some 800mm over the crest and a maximum velocity on the downstream face of less than 8m/s. The downstream face of the spillway section was flattened to 1:4 to achieve this. The concrete blocks have been covered by a sacrificial layer of topsoil planted with grass so that the embankment blends in with the surrounding countryside when viewed from a distance. The non-spillway section of the embankment has a crest level 1.6m above the spillway crest to provide the recommended wave freeboard.

Downstream of the buried stilling basin, the water discharging from the spillway passes through an existing mature hawthorn hedge, on to fields forming a gently sloping flood plain, and from thence back to the river. As the spillway will only operate with the downstream channel already bank full, only minor erosion is expected downstream of the spillway, even in extreme floods, and out-of-bank flooding downstream is expected to be less frequent than at present.

Outlet Structure and Controls
The flow from the reservoir passes through a 2.4m wide by 2.1m high box culvert constructed on the line of the original river channel. Alternative options for controlling flows through the culvert were considered. An essential requirement of any option was that it had to be capable of permitting passage of both fish and small aquatic mammals through the culvert and control structure under normal operating conditions.

Alternative controls considered included:-
- A fixed orifice with an area of $1.2m^2$, limiting the downstream discharge to 10 m^3/s at full head.
- A penstock located at the upstream end of the culvert. This would have initially been set to provide a fixed orifice with an area of $1.2m^2$. Use of a penstock would permit manual adjustment should this prove necessary. However, because of the height of the dam, it would have been difficult to provide a visually acceptable arrangement to allow the penstock to be adjusted during a flood event with the reservoir full.
- A penstock housed in a chamber within the dam so that it could be adjusted manually from the crest during an event.
- An electrically or hydraulically actuated penstock to automatically adjust the penstock as the reservoir filled to maintain a constant downstream peak discharge of 10 m^3/s.
- A float operated radial gate to maintain a constant discharge of 10 m^3/s
- A Hydro-Brake® which provides a reasonably constant discharge up to a maximum of 10 m^3/s.

Controls at the downstream end of the culvert were avoided because this would have pressurised the culvert though the dam, which has the potential to lead to leakage and consequent hydrostatic pressures within the dam fill.

Various of the above options were rejected for the following reasons:-
- Arrangements to allow manual operation of penstocks during a flood event were not considered to be of practical benefit because it would be unrealistic to expect Agency staff to operate them safely during a flood event. The penstock would therefore, in effect, act as a fixed orifice.
- A penstock housed in a chamber within the dam would create a confined space, which was not acceptable to the Agency.
- A fixed orifice would cause unnecessary, frequent and significant flooding upstream of the dam which would limit use of the land for agriculture, which was unacceptable to the affected landowners. Early storage of water did not, however, have a great influence on the height of the dam.
- There is no power supply near to the site for automatic gate operation and to provide this added greatly to the scheme cost. There would also be a risk of power or equipment failure during a flood event.
- There is a risk of failure of operation of equipment only intermittently used or tested, which was unacceptable to the Agency.
- There would be a significant maintenance requirement, which the Agency wished to minimise.
- A float operated radial gate across the culvert exit controlled by downstream water level was rejected because it would have pressurised the culvert and required maintenance.

The Hydro-Brake® was chosen on the basis of its simplicity, low maintenance requirements and relatively low cost for this site. The final arrangement is shown in Figure 2. The Hydro-Brake® restricts the flow more at low head than an automatically controlled penstock, but it does allow a reasonably constant discharge to pass at both high and low heads. A comparison of the stored flood levels and storage areas for the control options is given in the following tables. Figure 3 shows the flooded areas.

The data in these tables show that the use of a fixed orifice rather than a Hydro-Brake® would have only increased the dam height by 300mm. However at low return periods (when there is no need for flood storage to prevent flooding in Weedon Bec) the flood level and area flooded are much lower with a Hydro-Brake® or automatic penstock. With a Hydro-Brake®, in a 1 in 3 year event, the flooded area is limited to the area immediately in front of the dam, which is largely occupied by the borrow pit. With a fixed orifice the flooded area would extend into the surrounding fields.

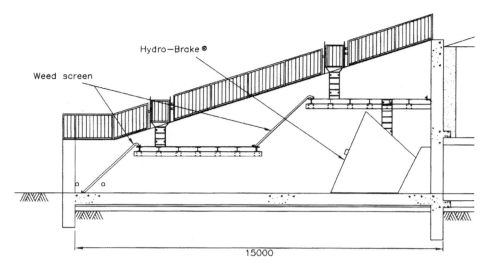

Figure 2: Section through the intake structure and Hydro-Brake®

Table 1: Storage Areas and Levels

Control	1 in 3 Years		1 in 50 Years	
	Level (mAOD)	Flooded Area (m²)	Level (mAOD)	Flooded Area (m²)
Fixed Orifice	89.6	146,145	91.7	417,890
Automatic Penstock	88.2	21,660	91.5	379,610
Hydro-Brake®	88.6	49,635	91.4	369,370

Table 2: Approximate Return Periods at which Storage would Commence

Control	Return Period when Storage Begins
Fixed Orifice	1 in 1 year
Automatic Penstock	1 in 3 years
Hydro-Brake®	1 in 1 year

As can be seen from the above figures an automatic penstock would have reduced the flooded area in low return periods further than a Hydro-Brake® but this was not possible for reasons explained previously.

Adjustment of the Hydro-Brake® is possible so that the peak discharge can be varied from the value determined by computer modelling should this prove to be necessary in practice. The peak discharge can be varied from 8 to 12 cumecs by the removal or addition of stop logs bolted across the Hydro-Brake® inlet.

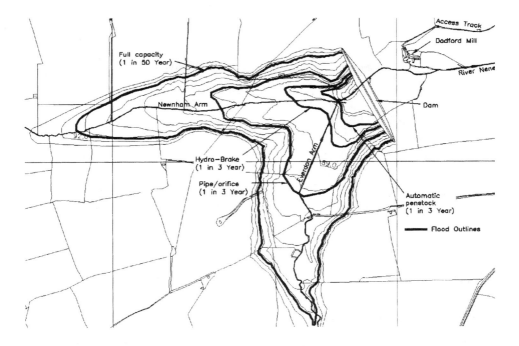

Figure 3: Comparison of flooded areas.

A trash screen has been provided upstream of the Hydro-Brake® and there is also a security screen at the downstream end of the culvert to prevent access by unauthorized people, particularly children.

Water level sensors are provided upstream and downstream of both the trash screen and the security screen, these are linked by telemetry to the Agency's control centres in Peterborough and Kettering. This allows monitoring of water level upstream and downstream of the dam and also shows if there is a difference in water level across the screens indicating that there may be a build up of trash.

USE OF THE HYDRO-BRAKE® FOR FLOW CONTROL

History of development and previous use of the Hydro-Brake®
The Hydro-Brake® is a proprietary gravity operated vortex flow control device designed by Hydro International plc. Outwardly having the appearance of a coil-shaped or conchoidal 'shell', units typically range from less than 1m to over 3m in length. The secret of their proven performance lies in the precise design of their shape, size, inclination and approach characteristics – not in expensive and complicated mechanical engineering.

In the United Kingdom, the first known major use of vortex flow control was to control and dissipate energy in drop shafts. The first commercial application in the UK as an integral part of drainage infrastructure to attenuate flows and alleviate flooding, was in 1980. Worldwide, more than 13,000 Hydro-Brake® Flow Controls are already in use, the majority having been installed on new developments to maintain flow rates equivalent to those of the greenfield site (pre-development run-off rates).

Prior to the Hydro-Brake® at Weedon becoming the 'Biggest Hydro-Brake® in The World', its predecessor had been installed as part of the Ashford Flood Alleviation Scheme over 12 years ago. This unit, which is basically the same shape and type as the Weedon Hydro-Brake® (without the in-built adjustability), has an outlet approximately 1.25m in diameter, whereas the Weedon Hydro-Brake® has an outlet diameter of 1.75m.

Experience to date with the Aldington installation has been very positive with the Hydro-Brake® performing exactly as expected. During the flooding experienced in that area in October and November 2000 the storage area at Aldington was actually overtopped, whilst the Hydro-Brake® discharged at precisely the correct levels. This reservoir was designed to retain floods of up to 1 in 100 year return period with a controlled discharge, illustrating the severity of the rainfall at that time. It has been well documented that had the Ashford Flood Alleviation Scheme not been in place at that time, Ashford would have suffered enormously. Older parts of town, close to the international railway station, would have flooded and about 100 houses would have been under water.

There has been much development of the Hydro-Brake® Flow Control since its original conception over 20 years ago, with constant ongoing testing and research to improve the hydraulic characteristics and develop more efficient units. Several new types have been introduced in recent years providing larger openings thus reducing the risk of blockages, as well as improved head / flow characteristics which reduce the amount of upstream storage required without exceeding the maximum required flow rate.

Hydraulic characteristics

The Hydro-Brake® is a self-activating passive flow control device with no moving parts and requiring no external sources of power to operate it. Instead, it uses the inherent energy in the flow field to control flows in sewerage systems, drainage channels and outlets from storage systems.

As flows build-up, a Hydro-Brake® typically exhibits two distinct modes of operation (see Figure 4 below). In the first mode, termed pre-initiation, the unit behaves like a large orifice, allowing relatively high flow volumes to be discharged at low operating heads. As the operating head increases, the volute of the Hydro-Brake® fills and the upstream water energy is converted into rotary motion within the device. This generates increasingly higher peripheral velocities, which eventually results in the creation of an air core, occupying most of the outlet of the device. In turn, this produces a back pressure that opposes the through flow of water. This second mode, is termed post-initiation, the 'throttling' effect causing the device to behave like a conventional orifice control or throttle pipe having a significantly smaller opening than the outlet size of the equivalent Hydro-Brake®.

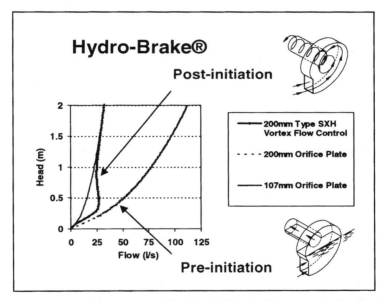

Figure 4: The Hydro-Brake® : Flow and Head Characteristics for the Pre and Post-initiation Phases

The two most obvious advantages in the use of a Hydro-Brake® at Weedon were reduced upstream storage requirements and comparatively larger openings. Other advantages include the lack of power required to operate as well as the absence of any moving parts. These factors coupled with the typical self-cleansing properties of a Hydro-Brake® result in a much reduced maintenance commitment.

Any drawbacks with the use of Hydro-Brake® Flow Controls tend to be either perceived or avoidable. The purchase costs are often quoted as a barrier, but can virtually always be outweighed by the savings in storage requirements and reduced maintenance costs. Another perception is that they are sometimes prone to blockage, especially when used in a foul / combined sewer application. Any flow control in a drainage or sewer system is, by its very nature, a restriction of some sort with an outlet size generally smaller than the system leading up to that point. The passage of objects larger than that opening is potentially a problem with any form of control and it is therefore important that consideration is given to preventing large masses from reaching it. It is true to say that the unique shape of a Hydro-Brake® generally prevents there being a straight path through the control, but with a correctly designed chamber or inlet structure including good benching etc., problems can always be avoided.

SCHEME CONSTRUCTION

An ECC Option C contract for the project was let to Edmund Nuttall Ltd in February 2001 with the flood storage dam comprising one section of a four-section contract for the Environment Agency. The agreed target price of £1.0 million was negotiated in April 2002 following completion of detailed design and having value engineered the project with the contractor.

Construction work commenced in April 2002 with a 34 week construction period. The planned sequence of operations is summarized below, although in practice there was some overlap of these activities.
- Establish site
- Install temporary bridge and divert the river into a temporary channel
- Construct culvert, headwalls and associated structures on the line of the existing channel
- Divert the river back through the culvert and reinstate the temporary river diversion
- Strip surface, excavate cut-off and place fill to the cut-off and dam
- Place erosion protection to the embankment and spillway
- Install Hydro-Brake®, trash rack and security screen
- Place topsoil, reinstate site and landscape

After some delays early in the contract, fine weather allowed the earthmoving to proceed quickly, so that it was essentially completed in October 2002. High river flows then caused delay in installing the Hydro-Brake®, which was completed in November 2002. Completion of topsoiling, seeding and finishing works was delayed until spring 2003.

Installation of the Hydro-Brake® was programmed to take place after placing of the spillway blocks, as this would effectively cause the reservoir to become operational. As it turned out, placing of the Hydro-Brake® under winter rather than summer flow conditions was difficult, and would have been easier if done earlier. This could have been possible by leaving a temporary opening in the upstream headwall to supplement the flow through the Hydro-Brake® and prevent impounding in the reservoir until the spillway was ready.

Environmental Aspects

A mineral extraction licence had to be obtained from Northamptonshire County Council who were the approval body for the borrow pit and its restoration, whereas the building of the dam was subject to planning permission from Daventry District Council who were the approval body for the dam landscaping. The two bodies had different landscape approaches.

The dam landscaping was relatively straightforward. Cellular concrete blocks used on the spillway were topsoiled and seeded. Elsewhere, grass seed was sown on prepared topsoil. An area of this on the upstream face was covered with fibre erosion protection matting over the topsoil. Hedges that had been removed were replaced with new planting, and some additional screening woodland was planted.

The borrow pit area was less straightforward because Northamptonshire C.C. did not want to have a significant body of water in the restoration, although this had been part of the scheme concept preferred by the Environment Agency. As a value engineering exercise, the original restoration plan was modified using input from the main contractor with a specialist landscaping sub-contractor and the Halcrow project environmental scientist. The accepted restoration incorporates areas of native tree planting and wild flower mix seeding which provides a diversity of habitat. This area has now been handed back to the farmer owner who continues to manage it as envisaged although there is no formal agreement relating to it.

CONCLUSION

A detailed study of the options to address the frequent flooding in Weedon Bec identified an upstream on-line flood storage reservoir on the River Nene as the only viable solution. Investigation and design, with due regard to flooding frequency and environmental factors has produced an economic scheme, with minimum adverse impacts on the surroundings, largely using materials available on-site.

The selection of a Hydro-Brake® as a flow control has significantly improved the scheme hydraulic performance, particularly in reducing the frequency of flooding of the storage area, which in this case is actively managed arable farmland. While use of a Hydro-Brake® at Weedon is not unique in flood control schemes, this installation has pushed the boundaries forward in the scale of what can be accomplished using these proprietary devices.

Integrating design with the environment to maximise benefits from a flood storage dam: successful implementation at Harbertonford

W T BRADLEY, Design Team Leader, Halcrow Group Ltd, Leeds, UK
M E JONES, Environmental Assessment & Design Ltd, Exeter, UK
A C MORISON, AR Panel Engineer, Halcrow Group Ltd, Swindon, UK

SYNOPSIS.   Environmental enhancement is not just about changing the seed mix and planting a few trees to hide the structures we build. Designing using the environment rather than seeing it as a constraint can help to produce cost-effective designs that bring great benefit to the environment and raise the public perception of the benefit dam engineering can bring to their lives. So often, the potential of embankment dams to benefit the environment is not taken advantage of. This paper explores what can be done to fully realise the potential of flood storage dams.

The award winning Harbertonford Flood Defence Scheme, described as "the future of flood defence schemes" by Sir John Harman, Chairman of the Environment Agency, was a combination of in-village channel lowering and upstream flood storage (Palmers Dam). This paper focuses on the flood storage element of the scheme and demonstrates the multifunctional benefits that have been delivered through the integrated work of design professionals, driven forward by the flood defence aim.

BACKGROUND

The UK flood defence industry
The flood defence industry is at present the source for the greatest number of Reservoirs Act dams being constructed in the United Kingdom. There are currently over 30 dams either recently constructed or under development. These dams vary in height from under five metres to over 15 metres and are generally earthfill embankment dams providing the temporary flood storage element of flood alleviation schemes. Such structures can provide cost-effective protection and offer flood defence benefits to the rest of the river catchment downstream through slowing floodwater progress and reducing flood peaks.

The Harbertonford Flood Defence Scheme

The Harbertonford Flood Defence Scheme, costing £2.6 million, provides flood alleviation to the picturesque village of Harbertonford, near Totnes in South Devon (refer to figure 1).

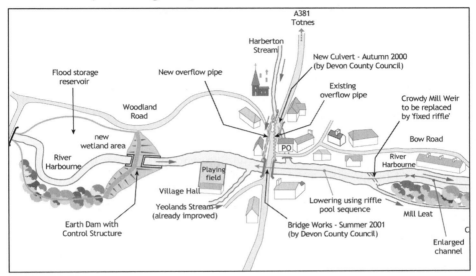

Figure 1: Schematic of overall scheme

Originating on Dartmoor, the River Harbourne flows through the village, where in the centre, by the historic bridge, it is joined by two tributaries. These watercourses have flooded the village 21 times in the past 60 years, including six times between 1998 and 2000. Flow in the River Harbourne varies from less than 1 cumec at low flows, to 28 cumecs at the 10 year flood flow, through to 300 cumecs for the PMF event (Probable Maximum Flood). The flashy nature of the catchment meant there was little warning for the residents of the village to prepare for the flooding and the misery it causes. One elderly resident of the village had resorted to living solely on the upper floor of her house.

This high frequency of flooding and associated damage resulted in considerable disruption and in January 2001, the scheme was accelerated by the Environment Agency (the client) to ensure that remedial works were designed and constructed in time to protect the village against possible flooding in winter 2002/03.

Solutions to the problem had been proposed during the previous thirty years, but all schemes previously put forward were deemed to cause too much damage to the environment and could not be justified. The village of

Harbertonford is designated as a Conservation Area and several listed structures, including the village bridge are also contained within it. Atlantic salmon, bullhead, sea trout and brown trout occur in the river and protected species are also present within the catchment, including otter and common dormouse.

The restrictions on the scheme development and the limited timescale focused the team to look at processes that could be used to achieve the scheme aim of providing a sustainable flood defence solution in place by winter 2002/03. This type of scheme would usually take four years to develop and implement. Only two years were available for this project.

Early studies showed that neither channel improvement nor upstream storage alone were capable of providing an appropriate level of flood mitigation. Storage sites available were too small to store the volumes of water required without unacceptable flooding of the upstream valley and channel works to carry full design flood flows required unacceptably large channels in the village centre or removal of downstream mill structures, which was also unacceptable. However an acceptable scheme was developed from a combination of both approaches.

THE DESIGN PHILOSOPHY

The benefits
Attention focused on five main objectives to ensure the design maximised the long-term benefits delivered. These were:

- maximise justifiable flood defence capability
- ensure reservoir safety
- minimise future operation and maintenance through working with the fluvial geomorphology of the river
- maintain and enhance biodiversity, amenity and landscape value
- minimise adverse effects on cultural heritage value

Through adding site-specific detail to these objectives, a clear framework was established to achieve the scheme aim of sustainable flood defence.

To deliver further long term benefits, the local school visited the dam under construction, and team visits to the school ensured the scheme construction and design were used as an educational resource, informing the children and teachers of the achievements of the scheme and the benefits it would bring. This promoted an understanding of their environment and allowed community ownership.

The design process

In describing the designs developed and techniques used at Harbertonford to deliver multifunctional benefits, it is worth summarising the approach used to generate these, which the authors believe represents a change from traditional practice.

The design philosophy adopted was to work with the environment, seeing it as an opportunity rather than a constraint. This meant using each other's skills most effectively; in particular, bringing the environmental scientist and other specialists to provide input directly into the design and not just comment on issues that needed to be considered, then periodically review designs.

Figure 2 below represents the values and perspectives held within the team. The process centred around the core values of Teamwork, Innovation and Consultation (TIC).

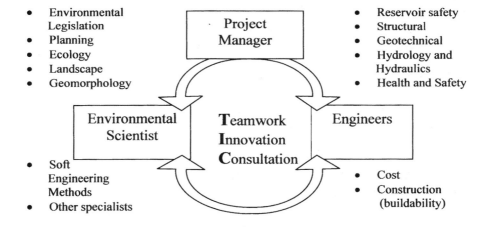

- Environmental Legislation
- Planning
- Ecology
- Landscape
- Geomorphology

Project Manager

- Reservoir safety
- Structural
- Geotechnical
- Hydrology and Hydraulics
- Health and Safety

Environmental Scientist

Teamwork
Innovation
Consultation

Engineers

- Soft Engineering Methods
- Other specialists

- Cost
- Construction (buildability)

Figure 2: The TIC Process

Specialist advice was brought in at key stages throughout the overall scheme design, including notable inputs from the River Restoration Centre. Public consultation was also undertaken throughout the design process, the outputs of which further influenced the nature of the final scheme. This gave local 'ownership' of the scheme, which was considered to be very important for construction and operation phases.

All approvals were gained first time, on a fast-tracked schedule leading to a reduced design development time, earlier design certainty and more efficient designs.

THE DESIGN OF PALMERS DAM

Overview
The dam structure spans a 100 metre width of valley and is up to five metres high, constructed of 10,000m^3 of fill material. A concrete box culvert passes through the dam to carry the normal river flow. The outflow is controlled by two actuated penstocks with an automated control system and the structure can retain 150,000m^3 of floodwater, so is subject to the Reservoirs Act 1975. An overflow spillway occupies the majority of the dam crest.

Dam
The dam is located two kilometres upstream of Harbertonford in a quiet valley area with rural landscape that contains areas remaining from previous quarrying and milling industries. The area is designated as an Area of Great Landscape Value which was an important influence on the dam design. The structure was located away from public areas, but still provided good operational access and minimised tree loss.

The zoned, clay core embankment dam creates an online temporary impoundment of floodwater and by improving the channel capacity through the village from 15 cumecs (3 year flood) to 28 cumecs (ten year flood), the flood storage site available has been used to deliver the greatest flood defence benefit. Any further improvement of the channel capacity could not have been justified on cost and environmental grounds and the dam height was restricted to avoid flooding the public road and upstream properties and this determined the maximum justifiable standard of scheme.

The underlying geology at the dam site is slatey shale, weathered at the surface and overlain in the valley bottom by alluvium, typically to a depth of about 2m, consisting of a mixture of clays derived from weathered shale, and quartz sands. The valley sides consist of shale, weathered *in-situ*. The bedrock is exposed in the riverbed just downstream of the selected dam site.

The single-track, restricted-width public road to the site from the village was seen as a major constraint to construction, and it was decided early in the design process that as much of the material for the dam as possible should be sourced on site. This also minimized traffic impacts in the village.

Materials investigations showed that only a limited quantity of suitable alluvial clay for an impermeable core was available from a field upstream of the dam, typically as a 0.5m thick layer. To minimise the land area disturbed in excavating this, a zoned, clay core embankment dam with alluvial shoulders was adopted.

Spillway design

In view of the flood-prone village downstream, the spillway is designed to pass the Probable Maximum Flood (PMF) flood flow of 300 cumecs. The structural integrity of the dam is protected from the PMF overtopping flow by incorporating rock gabions in a stepped arrangement into the downstream face. These were placed at a gradient of 1 in 2 above the shoulder material surrounding the clay core. Integration of the landscape with the reservoir safety design is achieved by the addition of a zone of non-structural 'sacrificial' material above the rock gabions, creating a varying surface profile of between 1 in 4 and 1 in 8 (figure 3). This changed the visual appearance of the dam considerably.

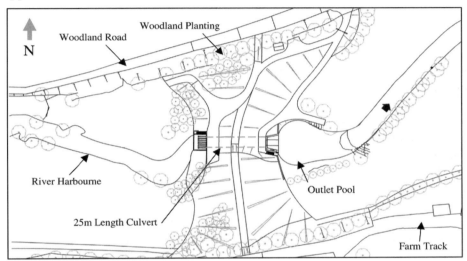

Figure 3: Layout plan of Palmers Dam.

The sacrificial material is covered with a mix of slower growing native grasses and wildflowers. The species chosen increase biodiversity whilst still retaining a good vegetative cover. Erosion resistance has been improved through the use of a 3-D geotextile in the root system. Importantly, the geotextile protection beneath the grass sward allows machine cutting of the grass when needed.

A Grasscrete crest locks the slope protection in place and the spillway occupies the full 100 metre length of dam crest to reduce overflow depth and the corresponding erosive force. Half of the spillway is set 0.5 metres lower and this is located on the southern section of dam, allowing continued operational access to the outlet structure (if needed) once overtopping flow has commenced.

Other design concepts

The upstream slope also incorporates a zone of non-structural material, again creating variable slopes, but also allowing the planting of broad-leaved woodland and scrub. This increases biodiversity and also provides a wildlife corridor linking the woodland on either sides of the valley. Badger protection mesh was placed against the 1 in 2 structural dam slope surface beneath the sacrificial material to protect from burrowing animals.

The sacrificial material represented an additional 5% on the cost of the dam, but since a minimal structural specification was needed, a wider range of materials could be used. This led to the project, including the channel enlargement works downstream, generating less than 10 percent waste to be taken away from site, most of this associated with contaminated material from the clean up of historic petroleum tanks during works through the village itself. The sacrificial material offers protection to the structural material, especially with regard to weather influences and retaining moisture contents.

Initial consideration was given to sourcing the gabion fill from waste in an abandoned slate quarry close to the dam site, so further reducing the amount of material brought to site. However enquiry revealed that the quarry and spoil tip posed difficult access issues and the area was an environmentally important area, so this idea had to be abandoned.

Culvert Design

The concrete box culvert which carries normal water flows through the embankment is designed to a four metre water width, similar to the natural river channel. This prevents throttling of the river flow and maintains the passage of the migratory fish, including salmon and sea trout, but also delays the point at which the culvert causes water to backup, reducing storage before the dam is designed to come into operation and associated sediment deposition.

The culvert is set with a lowered invert to allow a natural gravel bed formation and a minimum 300mm water depth. This follows best practice on migratory fish passage and has led to the creation of a varied bed profile, mimicking the natural river channel. The varying topography of the dam minimised the length of culvert needed to 25 metres, reducing build cost and length of river channel affected. With an internal height of over two metres, good airflow and access to both inlet and outlet structures, the lower risks for both authorised and unauthorised entry into the culvert ensures the culvert has not been classified as a confined space. This gives direct benefit related to future maintenance costs and safety liabilities.

Inlet arrangement

Screening of the culvert is achieved through bars at 600mm spacing. This allows most organic debris to pass through the culvert, fuelling the river system, reducing the risk of blockage and reducing the frequency of clearing operations. Screening to restrict unauthorised access would require a bar spacing of not more than 150mm, posing a higher risk of blockage, which could lead to earlier overtopping and consequential flooding of Harbertonford. This reasoning, together with the reduced safety liabilities of the culvert allowed agreement of the wider spacing. Additionally, the larger bar spacing allowed a smaller structure to be designed which has less impact on the landscape.

Constructing the culvert offline minimised the construction effects on the river and so consideration was therefore given to the best arrangement to guide the river to the new alignment. Previous dams have lined a new channel with hard bank protection such as gabions or walls, but the need for this was questioned. The main requirement is preventing scour of the embankment structure. This is most likely along the previous river path, so keying in a short length of blockstone into the bedrock in this location offers protection and a new channel excavated and lined with riverbed gravels guides the river to its new path. A bed check constructed upstream of the pool made from natural stone prevents long-term nickpoint erosion progressing upstream and contributes, with the natural curve of the river, to maintain a resting pool created at the inlet for fish passage (figure 4).

Figure 4: Photograph of inlet arrangement

Outlet arrangement

To dissipate energy and enable flow measurement, reinforced concrete stilling basins, as detailed by The United States Bureau of Reclamation (USBR) have regularly been used in south-west England. However to maintain accurate flow measurement, regular maintenance is needed to clear sediment deposition and ensure the structures dissipate energy as intended. At Palmers Dam, naturally outcropping rock was used to create a scour pool that improves fish habitat, is self maintaining and provides better landscape and visual amenity value. Water levels in the pool and culvert are maintained using a bed check created using natural slate. This was simple to construct and has performed exactly as intended from hydraulic perspective and visually mimics the existing rock outcrops.

Around the outlet (and throughout the scheme generally), sloping ground has been used rather than vertical drops wherever possible to reduce the need for fencing and allow members of the public who may get into this area a safe means of egress. Bird boxes were installed in the masonry faced walls as part of additional habitat creation.

Flow control system
Flood storage commences during a 1 in 10 year flood event. At a 1 in 40 year event the scheme standard will be reached and overtopping flow will cause the increased in-bank flow capacity through the village to be exceeded and progressive flooding to occur. The maximum storage efficiency making best use of the limited storage volume, the wide culvert, and the minimal screening were all possible due to the decision to adopt a variable gate control structure. This is located on the outlet structure which has minimised the impact on the landscape and allowed access to the gates even when the reservoir is impounding. A watertight culvert capable of withstanding up to a six metre head of water internally was achieved through using a combination of standard waterproofing seals between joints and a casing of mesh reinforced concrete around the box culvert structure.

The vertical penstock gates are normally held fully open, but are closed progressively to limit flows downstream to what can be accommodated within the river channel. The automated penstocks are controlled by a programmable control system which is operated in response to water levels downstream of the dam. Alternatively the gates can be closed remotely by a central flood control room in Exeter, however the gates can only be opened by manual control on site to ensure that the gates are not accidentally opened during a flood event.

Local power supplies were used and additional benefit delivered to the village through arranging the supply to be supported from two separate sources, thereby increasing the dependability of supply to the village as well as the dam. The programmable control system allows changes to be made to the operating regime in the future. Backup power systems have been put in place at the dam to minimise the risks of operational failure.

Sensors detecting the water level at the dam inlet now allow flood warnings well in advance of the level that was previously possible.

Washland creation
The benefits have been extended still further through taking advantage of the potential the borrow area provided. Twenty thousand tonnes of material was needed for the dam construction and this was substantially sourced from the field immediately upstream in such a way to create shallow scrapes. The remainder of the fill came from material excavated as part of the village works. The field was previously semi-improved grassland used for grazing and biodiversity has been increased through creating a wet woodland/grassland nature reserve. This design contributed to achievement of the Department of the Environment, Food and Rural Affairs (DEFRA) high-level biodiversity targets

Figure 5: Photograph of washland area one year after construction

Construction costs were saved through considering the final wetland profile at the outset, using the geotechnical information to target specific areas and depths and minimising earthmoving needed at the end of dam construction. The seeding and tree planting, which greatly surpassed the number of trees lost during construction has already started establishing within a year, creating an area that is regularly visited by the local population and a seating area has been provided to maximise this amenity value.

Wetland creation was only possible through the Environment Agency's decision to purchase the area involved. This aspect of the project has since been presented at the DEFRA annual conference as a case study project in best practice washland creation and showed the benefits of combined flood defence and biodiversity (Morris, 2003). Effective consultation from early stages with the landowners concerned has led to support for the solution and even an offer of assistance in the future management of the area.

CONCLUSION

There are a large number of embankment dams being constructed in the UK for flood defence purposes and consideration of the environment is a key factor associated with these. Palmers Dam was developed using an integrated team of professionals that brought the latest knowledge and experience into the process. This, together with proactive public and stakeholder consultation throughout, brought ideas and encouraged ownership of the scheme, all of which led to maximised benefits to the community, ecology and landscape. The designs also have reduced future operation and maintenance requirements and long-term health and safety liabilities through considered design and integration with the fluvial geomorphology of the river.

The reaction received from the residents at the opening of the scheme has shown the high regard with which they perceive the scheme and the integrated team that delivered it in time to save them from flooding that would have occurred on New Years Day 2003.

ACKNOWLEDGEMENTS

The Environment Agency and DEFRA funded this project. The authors gratefully acknowledge the key individuals within these organisations who shared the vision of the scheme and the benefits that it was to produce.

REFERENCES

Scottish Executive. (2000). *River Crossings and Migratory Fish: Design Guidance, A Consultation Paper.*

CIRIA. (1997). *Culvert Design Guide, Report 168.*

United States Department of the Interior, Bureau of Reclamation. (1987). *Design of Small Dams.* United States Government Printing Office.

Morris, J. (2003). *Managed Washlands for Flood Defence and Biodiversity.* 38th Defra Flood and Coastal Management Conference.

Raciborz Flood Reservoir

LAURENCE ATTEWILL, Jacobs Ltd, UK
EDOARDO FAGANELLO, Jacobs Ltd, UK

SYNOPSIS. The river Odra, which rises in the Czech Republic and disgorges into the Baltic, suffered an extreme flood in July 1997 which was responsible for the loss of 50 lives and over a billion dollars worth of damage in southern Poland. The return period of the flood is variously estimated between 250 and 1000 years.

The paper describes the studies for a flood reservoir to be constructed on the Odra just upstream of the ancient town of Raciborz. These studies include the hydrological studies, the hydrodynamic modelling of a 220km stretch of the river where most of the damage occurred, flood damage studies both with and without the proposed reservoir, environmental impact assessments and resettlement plans, in addition to the engineering studies of the dam itself.

INTRODUCTION
This paper describes the feasibility studies for a flood protection reservoir carried out by Jacobs GIBB in association with Hydroprojekt Warsaw for the Regional Water Management Board in 2002 and 2003. The study area is shown in Figure 1.

Figure 1: The study area

Background
The Odra river rises in the Czech Republic and flows north through Poland to the Baltic Sea. The river has been liable to flooding, 14 floods having been recorded in the last two centuries which have caused considerable damage to the cities, towns and villages of the upper Odra valley – Raciborz, Opole and Wroclaw. These floods have lead to a the construction of a complex system of flood defences including embankments, by pass channels and flood storage areas which are designed to be capable passing floods of up to a 1 in 100 year return period without serious damage.

Figure 2: The river Odra during the 1997 flood event

These flood defences were overwhelmed by the flood of 1997, known in Poland as the Great Flood, which exceeded all previous floods in flow rate and volume.

The possibility of the construction of a flood reservoir at Raciborz was first proposed after the 1880 flood and by 1906 a scheme with a storage capacity of 640Mm³ had been designed. Over the course of the century the scheme characteristics evolved with a gradual reduction in the storage capacity due to the expansion of towns and villages. The present project, the conceptual design of which was prepared by Hydroprojekt, comprises three stages:

1. the Bukow Polder close the Czech border (constructed 2001)
2. the construction of a flood storage reservoir at Raciborz
3. channel improvements, and the construction of a new polder and bypass for the city of Wroclaw

RACIBORZ FLOOD RESERVOIR

Purpose
The primary role of Raciborz Reservoir is to reduce the frequency and severity of flooding in the Upper Odra River. This will be achieved in two ways:

1. Firstly, the reservoir will provide flood storage so that the flow rate downstream of the reservoir will be greatly reduced and the effectiveness of the existing flood defence system in containing the flows will be improved.
2. Secondly the reservoir will delay the timing of the flood peak at the confluence of the important left-bank tributary Nysa Klodzka with the Odra so that the adverse combination of the two floods that was so damaging in 1997 is very much less likely.

Two stages of development are envisaged: the first being the 'dry reservoir' the sole purpose of which is flood mitigation: in the second the reservoir will be partially impounded when important secondary benefits will be navigation, water supply and recreation. The reduction in flood storage resulting from partial impounding will be offset by gravel extraction from the reservoir in the 'dry reservoir' stage.

Layout
The location of the reservoir was selected to be upstream of the town of Raciborz and within Polish territory: the maximum flood level is constrained by a Czech/Polish protocol. The storage volume of the reservoir will be 185 Mm³. In normal operation the reservoir will be dry with the river flowing through a main gated outlet structure into the bypass channel.

Releases will also be made through a subsidiary outlet into the old river that flows through the town. In times of flood the outflow through the outlet structure will be controlled by operating the gates so that excess water is stored within the reservoir. The outflow is varied according to the magnitude of the expected flood and therefore a flood warning system is essential. The strategy of the operation rules is that for any flood, irrespective of return period, the flood storage is used to its maximum extent and the reservoir outflow is selected to achieve this.

The layout of the reservoir is shown in Figure 3.

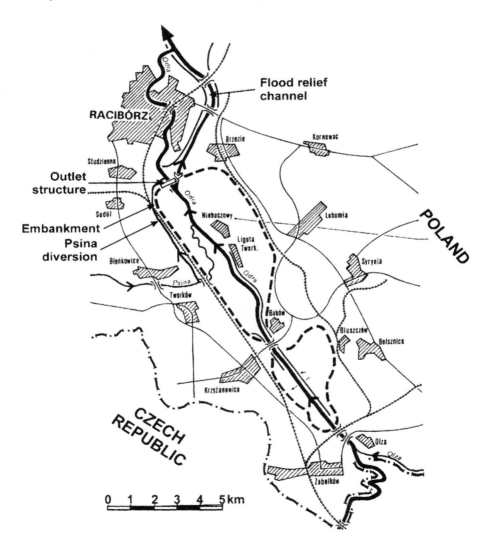

Figure 3: Layout of Raciborz reservoir

Design

Foundations
The reservoir is formed by an earth embankment dam, 22.5km long and with a maximum height of 10m. The dam foundations consist of alluvial deposits comprising alternate layers of silty clay and granular material, with interspersed peat lenses. The top layer will be stripped and the dam founded on the upper cohesive layer. Although foundation seepage is not an issue in the dry reservoir stage - the reservoir will not be full long enough for steady state seepage conditions to become established – it will become a consideration when the reservoir is permanently impounded and therefore a 5 km length of cut off wall through the main granular layer is proposed. In addition the excavation of embankment fill or gravel extraction will be prohibited within a 100 m strip of the upstream toe, the upper cohesive layer forming a natural blanket.

Embankment
The embankment will be constructed over a four year period, from April to October of each year. The embankment which utilizes both the cohesive overburden and the underlying sandy/gravel with the minimum of selection, is essentially a homogenous clay cross section with a substantial drainage bund at its downstream toe. Drainage of the underlying water bearing layer will be provided either by a trench drain or by wells, depending on the depth. A typical cross section of the embankment is shown in Figure 4.

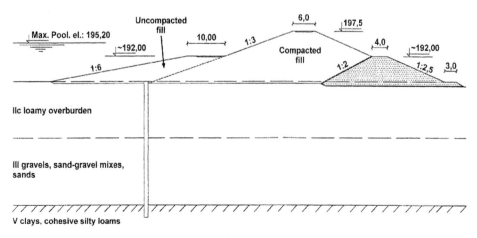

Figure 4: Typical cross section of the embankment

Outlet works and spillway
The outlet works comprise a gated structure consisting of
- a reinforced concrete forebay with a 115m wide apron
- 5 bays each 12m wide with one navigable flat sill at a slightly lower level. The bays are separated by 7.4 m wide piers which house the gate operating mechanism
- vertical lift gates, 1 per bay 12m wide x 8.5m high
- a reinforced concrete bridge over the bays
- a reinforced concrete stilling basin, 109m wide x 55 m long,
- an outlet channel, protected with gabions and riprap, discharging into the river downstream.

Operation
In normal operation the reservoir will be dry with the river flowing through a main gated outlet structure into the bypass channel. Releases will also be made through a subsidiary outlet into the old river that flows through the town. In times of flood the outflow through the outlet structure will be controlled by operating the gates so that excess water is stored within the reservoir. The outflow is varied according to the magnitude of the expected flood and therefore a flood warning system is essential. The strategy of the operation rules is that for any flood, irrespective of return period, the flood storage is used to its maximum extent and the reservoir outflow is selected to achieve this.

SOCIOLOGICAL AND ENVIRONMENTAL ASPECTS
Sociological and environmental issues are dominated by the need for 240 families living in the two villages of Nieboczowy and Ligota Tworkowska to be resettled. The study, which included the formation of an outline resettlement plan, was carried out shortly after the publishing of the World Commission on Dams reports and great effort was made to follow their precepts in this area. An alternative dam alignment excluding the village of Nieboczowy, proposed by the villagers was examined in detail but the reduction in storage volume made this option uneconomic. Despite two public meetings and door to door interviews public opinion remained adamantly hostile, largely due to fears of inadequate levels of compensation.

Compared with this the adverse environmental impacts are relatively minor and in any case are heavily outweighed by the environmental benefits of the scheme to the river valley downstream of the dam.

HYDROLOGY

Data
Flow data from the 20 gauging stations shown in Figure 5 were used in the hydrological analyses. The data record at most stations is at least 50 years.

1997 flood
The 1997 flood was caused by exceptionally intense and prolonged rainfall in the upper catchment: 200mm was recorded in the 5 days from 4th to 8th July 1997 over a wide area with a peak intensity of 585mm over the same period at one station, Lysa Hora in the upper catchment. Peak flow rates are approximate because the river levels so far exceed the calibrated rating curves but were in the region of 3,120 m³/s at Raciborz , and 3,640 m³/s at Wroclaw, the upstream and downstream limits of our study area.

Probabilities
Synthetic input hydrographs for the model for a range of return periods were derived from an analysis of the historic flow data and a prediction of the peak flows for as range of return periods computed according to Polish standards by the Institute of Meteorological and Water Management. These predictions were based on the statistical analysis of historic data at single stations which led to the 1997 flood being assigned a return period of 1000 years.

Regional Analysis
The estimation of the frequency or the return period of rare floods is not easy, as extrapolation from the record at a single station involves uncertainty about the choice of statistical distribution to represent the extreme floods, and also the choice of method of fitting the curve to the records. It was therefore decided to test the sensitivity of the estimation of return period by considering the regional flood frequency approach to flood frequency analysis which has been found to give consistent estimates of the relation between flood magnitude and return period or frequency of occurrence when applied to areas of reasonable hydrological homogeneity.

The method depends on the collection of annual maximum flood series from the gauging stations within a region. The method derives from the approach developed during investigation of floods in the British Isles (Natural Environment Research Council, 1975) and its application to many different regions of the world has been described in a number of papers (Farquharson et al., 1987, 1992, 1993: Meigh et al.,1997; Sutcliffe & Farquharson, 1995).

After inspection of the results, the stations were grouped as follows:
·   the 6 stations on the upper Odra in the Czech Republic

- the 6 stations on the main Odra below Raciborz
- the 4 stations on the Nysa Kłodzka
- the remaining 6 stations on the Odra.

The curves for these four regions are illustrated in Figure 5

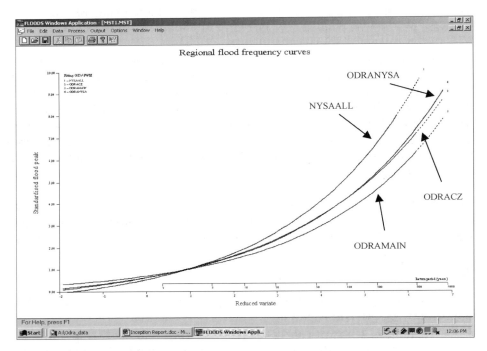

Figure 5: Regional Flood curves

It will be seen that the curves for the upper Odra groupings are very similar up to 50-100 years, but the Nysa curve is higher at longer return periods. The curve for the lower basins is different and is treated separately. The curve for the whole upper Odra is considered as reasonably representative of the whole area.

Comparison of the 1997 flood at individual stations, expressed as Q/MAF, with the regional curve derived from the 16 stations in the upper Odra basin, indicates a lower limit of the range of return periods which could be allocated to this event. This implies that the whole set of records is typical of the region, but it has been shown that the group curves are similar to each other and that the regional curve is similar to that of the upper Vistula. The lower limit for the return period of the 1997 event, on this basis, is about

200 years for the Odra stations from Bohumin down, but rather higher at about 300 years for the Nysa stations.

HYDRODYNAMIC MODELLING

Model description
The Upper Odra hydrodynamic model has been built using MIKE 11, a software package developed by the Danish Hydraulic Institute (DHI).

The model covers approximately 204 km of the main river channel (between the village of Olza, situated near the Czech border and the village of Trestno, immediately upstream of Wroclaw) and includes the floodplain on the right and left bank as well as the two main tributaries: the Nysa Klodzka and the Mala Panew (63.5 km and 18 km respectively). A total of 342 cross sections were included in the model. The cross sections used in the hydrodynamic model are the result of a recent survey carried out after the 1997 flood event.

The MIKE 11 model does not cover the six relevant reservoirs situated on these two tributaries. The effect of these reservoirs operating under the current rules has however been considered in separate flood routing calculations and the resulting outflows included in the model as input hydrographs.

The methodology adopted for modelling the floodplain of the Odra has involved three different techniques:
- extension of the in-bank river cross sections into the floodplain;
- introduction of flood cells so that when the bankful capacity of the main channel is exceeded water spills into them (this technique has been used in particular to model some of the polders);
- simulation of the adjacent floodplain as a 'separate river' using parallel river branches attached to the main channel by lateral spill units (link channels).

A global value of 0.05 of Manning's roughness has been considered in the computations although different values in the range of 0.03-0.100 have been used for many of the modelled reaches in order to reflect the land use and improve model calibration.

The Odra River is maintained as a navigable river. A large number of sluices and lock structures have been constructed along its watercourse as well as bypass and diversion channels in order to improve and increase the capacity of the system during a flood event. There are two main types of hydraulic structure which have been incorporated in the hydraulic model by using the recently surveyed river cross-sections (1997-2000) and the inventory of the existing hydraulic structures between the Polish-Czech boundary and Trestno compiled during the study:

1. weir complexes, which may consist of a variety of broad and sharp crested weirs, barrages, sluice and radial gates associated with culverts and/or bridges
2. polders

Only permanent structures have been taken into account and modelled as fixed (without operation rules). Removable gates have not been included in the model because they are usually removed from the watercourse during flood events. Navigation locks have been simulated with the gates open.

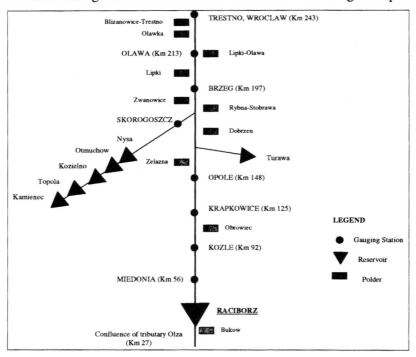

Figure 6: Model schematic

Model calibration

Once the hydrodynamic model was completed, it was calibrated progressively from upstream to downstream using the 1997 flood event by adjusting the roughness, the hydraulic structures and spill (out-of-bank) discharge coefficients so that modeled river levels match those measured as closely as possible. The results of the calibration are presented in Table 1

Table 1 Calibration results

Gauging Stations	Recorded peak level (m OD)	Modelled peak level (m OD)
Miedonia	186.73	186.67
Kozle	171.98	172.02
Krapkowice	165.83	165.88
Opole	154.89	154.70
Skorogoszcz	145.12	145.37
Brzeg	136.50	136.55
Olawa	129.64	129.66
Trestno	121.76	121.71

Model results

The model results, in terms of the reduction of the peak water level attributable to the Raciborz reservoir are shown in Figure 7.

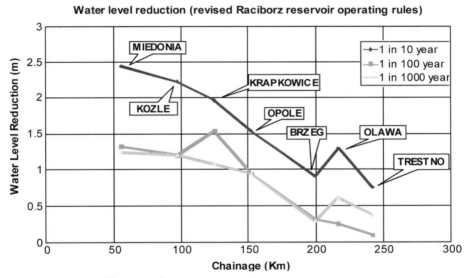

Figure 7: Modelling results.

Sensitivity

The sensitivity of the model results has been examined for the following changes:

- Reservoir volume
- Storm timing
- Operation rules for the reservoirs on the Nysa river

The model was run with a reservoir volume of 154Mm³ (16% reduction on the base case), representing an option which minimised resettlement, and with a reservoir volume of 290Mm³ (50% increase on base case) representing a possible future condition in which gravel deposits within the reservoir are extracted. The modelling showed that on average the effectiveness of the reservoir would decrease and increase by approximately 5% and 15% for the reduction and the increase in volume respectively.

The modelling shows that the river levels are not very sensitive to the relative timing of the main river and tributary flood peaks: a delay of 12 hours in the time to peak of the Nysa result in an increase in water level of 4cm below the confluence.

The sensitivity analysis shows that the flood levels on the Nysa river are sensitive to changes in the operating rules of these reservoirs and to the proposed construction of a new reservoir at Kamieniec Zabkowicki. However these changes will have little impact on flood levels in the main Odra river downstream of the confluence.

BENEFITS

Inundation mapping
The effect of the Raciborz reservoir on the area inundated by floods of the range of return periods considered is illustrated in Figure 8

Flood damages
Flood damages were estimated for the with and without reservoir cases by estimating the areas of each of 20 land use categories flooded in each case and applying damage init rates that were derived from 1997 flood damage data. The results are illustrated, for the range of flood probabilities, in Figure 9.

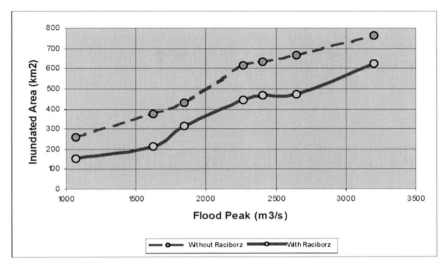

Figure 8: Inundation areas with and without Raciborz reservoir

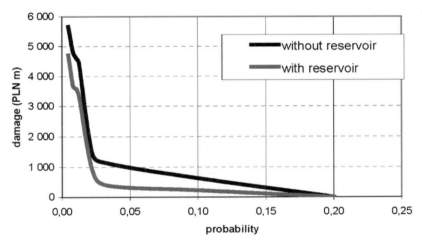

Figure 9: Reduction in flood damage

CONCLUSIONS

The conclusions of the study are that the Raciborz reservoir offers substantial but not complete protection against inundation of the Upper Odra against severe floods. The effectiveness of the reservoir is sensitive to reservoir volume and must be accepted that the proposed reservoir is at the small end of the range of useful volumes: a larger reservoir would be considerably more effective. The level of protection provided by the reservoir is naturally greatest immediately downstream and decreases, especially downstream of the Nysa confluence. However it is likely that Raciborz reservoir together with the implementation of the various channel improvements mitigation measures proposed for Wroclaw will provide adequate protection to that city. The effectiveness of the reservoir will depend on careful operation in which a reliable flood warning and conjunctive use with the Nysa reservoirs are vital.

ACKNOWLEDGEMENTS

The authors are indebted to RZGW Wroclaw for their permission for this paper to be published, and to their Polish and British colleagues for their help in its preparation. Particular thanks are due to Dr J V Sutcliffe (hydrology), Dr John Chatterton (flood damage assessment), Mr Paul Devitt (resettlement aspects) and to the support of Mr. Stanislaw Naprawa, Ms Alina Kledynska and Ms Karina Zachodni.

5. Instrumentation and monitoring

Monitoring of dams in operation - a tool for emergencies and for evaluation of long-term safety

Thomas KONOW, Norwegian Water Resources and Energy Directorate (NVE), Oslo, Norway

SYNOPSIS.   As the number of new dam projects dropped during the late 1980's, the focus shifted from construction to operation of dams in Norway. In this process, the importance of emergency warning and monitoring of long-term behaviour was realized and resulted in a guideline for monitoring and instrumentation in 1996. After 7 years, the guideline has been revised and this paper summaries the content of the revised guideline [NVE, 2003]. Some Norwegian dams have a dam break warning system and a brief history of this system is also described.

INTRODUCTION
Traditionally, monitoring of Norwegian dams has been limited to the initial filling and the first years after commissioning. Surveillance of long-term performance was generally not systematic and limited to random inspection of dams.

The need for monitoring of long-term performance was visualised when the guideline for inspection and reassessment was introduced in 1994 [NVE, 2002]. According to the guideline, a reassessment is required about every 15th year, which includes a detailed evaluation of the dam and appurtenant structures. An element of the reassessment is an evaluation of the long-term performance of the dam, in order to compare the theoretical and the actual behaviour.

In 1994 a guideline on emergency action planning was also published [Svendsen, Molkersrød and Torblaa, 1997]. The guideline was a result of increasing focus on emergency planning and how to reduce the consequences related to an abnormal situation. It is evident that early warning is important to prevent worsening of the situation and to reduce the consequences.

Long-term benefits and performance of dams, Thomas Telford, London, 2004, 387–395

The realization of the importance of instrumentation to monitor long-term behaviour and for the purpose of emergency warning resulted in a guideline for monitoring and instrumentation in 1996. During 2003, the guideline has been revised and this paper summaries the recommendations of the revised guideline. Some Norwegian dams also have a dam break warning systems, and a brief history related to this system is also included.

LEGAL FRAMEWORK

The practice of public supervision with dams in Norway started in 1909. After almost 100 years, the regulatory authority has been transferred to the Norwegian Water Resources and Energy Directorate (NVE) which reports to the Ministry of Petroleum and Energy.

The Dam safety regulations [NVE, 2000a] represent the legal basis for public supervision and safety control of dams, spillways and hydraulic structures (called watercourse structures as a collective term). More detailed specifications and technical safety recommendations are specified in guidelines. The structure of the regulations is given in figure 1, below.

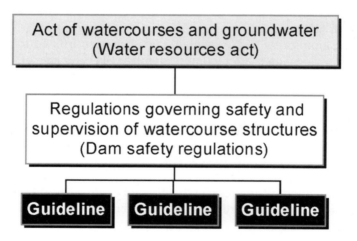

Figure 1: Structure of the legal framework

The guidelines give recommendations on how to fulfil the regulations but do not set any direct requirements. Guidelines are developed and managed by NVE and can easily be revised and amended. The Dam safety regulations and the Water resources act include general requirements and therefore needs a formal approval from the highest level in the administration, a process that is very time-consuming and complicated.

All together, 20 different guidelines are planned [Midttømme, 2003]. Eight of these have been issued, whereof two has been translated to English. The English versions of the regulations and guidelines are available on NVE's Internet site; http://www.nve.no-> English pages-> Safety and supervision-> Legislation

Classification of dams

The regulations define 3 different consequence classes [NVE, 2000b] and requirements for instrumentation are dependent on the classification of the dam. Each class is defined on the basis of the number of houses affected by a dam break, as shown in Table 1. Environmental and economical consequences shall also be assessed as an element of the classification.

Table 1: Classification of dams – definition

Consequence class		Affected housing units
Class 1	Low hazard	0
Class 2	Significant hazard	1-20
Class 3	High hazard	More than 20

PLAN FOR MONITORING

A plan for monitoring is important as a basis for the surveillance. A monitoring plan can be part of the inspection program for the dam, since an evaluation of the collected data often is an element of the inspection procedure.

The plan for monitoring will normally contain the following elements:
1. Overview of the different types of instrumentation on each dam.
2. Description of the different measurements that are being carried out and frequency of the readings.
3. Background on the choice of instrumentation or reasons for lack of instrumentation when this does not coincide with the guideline.
4. Description of the location of each individual monitoring device
5. Specifications on the different instruments.
6. Description of the accuracy of the instruments and expected errors in the recorded monitoring data.
7. Plan for calibration of the instruments where necessary.
8. Plan for testing, inspection and maintenance of the instruments.
9. Limit values to initiate actions in case of an emergency.

Limit values are important to give a warning in an emergency situation. The values are worked out as part of the Analytical Phase, which forms the basis for development of an Emergency Action Plan [Svendsen, Molkersrød and

Torblaa, 1997[3]]. Limit values will specify when to intervene and can for example be defined as;

- water level where access to the gatehouse or gate is interrupted
- water level when overtopping of the core will occur
- maximum expected load that the structure can withstand
- normal or largest acceptable leakage

MONITORING AND INSTRUMENTATION
Generally, the need for monitoring and instrumentation will depend on the dam type and consequence class. The number of instruments and their location must be assessed according to the dam type, height and length; the state of the foundation; normal reservoir levels, reservoir size and other factors in the reservoir area.

It is important that the instruments are reliable, accurate and easy to read off. Care must be taken when installing the instrument and the location must be chosen in order to ensure a correct and adequate reading of the monitoring data. For example, the water level sensor should not be located so that gates or spillways can influence the readings. This should be obvious, but experience show that it is not always taken into consideration, fore example in cases where the gates or spillways are seldom in use.

Recommendations for instrumentation of dams in Norway to monitor long-term behaviour and to provide emergency warning are given in the guideline for monitoring and instrumentation [NVE, 2003]. The guideline also specifies the frequency of the readings. Additional instrumentation and other frequencies than recommended by the guideline will need to be evaluated, dependent on the dam in question.

MONITORING OF LONG TERM BEHAVIOUR
Monitoring of long-term behaviour is generally limited to monitoring of the following elements:

- Leakage
- Pore pressure
- Deformation

Leakage
Variations in the recorded leakage will give an indication on the performance of the dam. Decreasing leakage can indicate increasing pore pressures as a result of poor drainage, while increasing leakage may indicate deterioration of elements within the dam or foundations that will need further investigations. Measurements of leakage should be recorded together

with water level in the reservoir, precipitation and snow melting, as these factors may influence the readings.

Continuous measurements of leakage are of importance to detect sudden changes in the leakage that may not be determined by a measurement now and then. This is of particular interest for embankment dams, but also for concrete dams founded on soils or rock with weak zones.

Pore pressure
Pore pressure measurements of the foundation will generally be required for dams founded on soils or rock with weakness zones. Additional monitoring of pore pressures can prove necessary, particularly on high concrete dams. However, as a result of the glacial history of Norway, the foundation often consist of hard, resistant and durable rock, and potential monitoring will need to be evaluated in each separate case and this is therefore not a general recommendation. Measurement of pore pressures is recorded together with the water level in the reservoir.

Deformations of concrete dams
Measurements of deformations are particularly recommended for arch dams and dams where alkali aggregate reaction (AAR) are detected or suspected. Deformations should be recorded together with measurements of water level and concrete temperature. Measurements of cracks must also be considered, however, this will be based on an individual assessment of the dam in question.

Deformations of embankment dams
Generally, settlements of the crest will be measured at least once every year. Annual levelling of moraine- and asphalt concrete core is also recommended. In this way any sudden changes in settlement can be detected and a more detailed evaluation can be carried out.

In addition, horizontal and vertical deformations of bolts are measured about every 5th year. Suggested distribution of deformation bolts are given in the guidelines. The need for a more detailed survey may be required to identify local deformations that will not be recorded within the grid of deformation-bolts, e.g. deformations caused by beaching or internal erosion. For this purpose, detailed topographic mapping can be made on basis of aerial photographs or sonar. A picture of sonar mapping is shown in figure 2. Sonar may give a better level of detailing than aerial photographs, but will be limited to the upstream face.

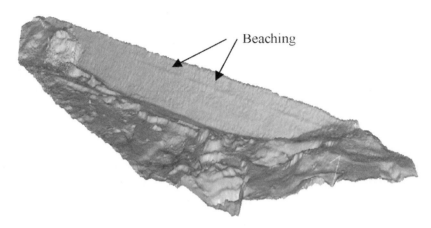

Beaching

Figure 2: Topographic mapping of the upstream face of an embankment dam (Photo: Artec Subsea AS)

Evaluation of the monitoring data
The monitoring data must be analyzed and evaluated continuously and presented graphically so that both long term and short term tendencies are visualized. Some examples of graphical presentations are illustrated in the guideline [NVE, 2003[1]].

The monitoring data need to be assessed and commented on the basis of the following elements:
- Accuracy of the monitoring data.
- Possible changes in trends.
- Factors that may have influenced the measurements.

MONITORING FOR EMERGENCY WARNING
Instrumentation for the purpose of emergency warning is generally limited to monitoring of water level and leakage. Monitoring of the reservoir water level is recommended on all high hazard dams and dams with gated spillways. Monitoring of leakage for emergency purposes is limited to embankment dams with a central core of moraine and asphaltic concrete or dams founded on soils or rock with weak zones.

Where monitoring for emergency warning is required, a continuous reading and transmission of the data is necessary in order to detect any development of a possible abnormal situation.

In some cases, abnormal situations have not been detected as the instrumentation did not work properly. In such cases a surveillance camera on the dam site would have been useful. A camera on site can also reveal additional information that is not necessarily detectable by instrumentation alone and may be of particular use when the control centre is at another location than the dam.

DAM BREAK WARNING SYSTEMS

Some Norwegian dams also have been required to have a dam break warning system installed. [Martinsen, 1995]. This is not included in the guideline for monitoring and instrumentation, as the dam break warning system is only required on some particular dams.

The first formal warning system of dam failure was established during the Second World War. It covered nine river basins, and was based on the use of telephone or radio at the dam itself, and cars along the main roads with alarm sirens. After the war, this system was abandoned.

In some river basins, dam failures could cause disasters of enormous dimensions, in terms of both loss of lives and material damage and in 1966 a working group for dam safety in emergencies was set up. In 1971, the group concluded that a modern warning system should be developed. The decision was based on the following main factors;

- Even though the likelihood of dam failure is relatively small in times of peace, the possibility of it occurring as result of natural reasons, technical damages or damage caused by terrorism or sabotage should not be ignored. The system should therefore also be operated in times of peace.
- Dams constitute targets for attack in times of war. The primary motive for attack would usually be to exploit the destructive effect of the breach wave in the area below.
- In the event of a dam failure, loss of life can be very extensive. A good warning system allows even people living very close to the dam, to be evacuated.

These arguments, combined with a strong local political pressure, led to development of an electronic warning system for dam failure in two river basins.

The dam owners were requested to install and operate the warning systems, as well as the communications and transfer of the warning signals to a first reporting point. From this reporting point, the Civil Defence services assumed responsibility for installing and operating the system of sirens, and for communication between the first reporting point and the sirens themselves. In this way the warning was transferred to a large number of sirens throughout the river basin.

The warning system is based on two different electronic monitoring systems:
- Four single current electric circuits built into the dam. When one circuit is broken it will be detected.
- Downstream the dam, there are four independent floats that measures the water level, and a warning signal is given at a previous defined water level.

To avoid unnecessary warning, the warning system is only activated if a dam break is indicated by at least three of the independent systems.

CONCLUSIONS
The guideline for monitoring and instrumentation [NVE, 2003] is valuable as a basis to determine a minimum of instrumentation for Norwegian dams. However, an assessment of each individual dam should always be made in order to determine the need for any additional monitoring, or in some cases the need for reduced monitoring compared to the guideline.

The guideline gives recommendations on what to do, when and where. However the human factor should not be forgotten, as instrumentation and monitoring is just a tool on the way to achieve better dam safety. Just as important as the actual instruments are how they are operated and how the monitoring data is analyzed. Valuable information may drown in an enormous amount of data. Further, improper analysis of the data may not reveal information that could have been detected. These factors may prove to be the real challenges when monitoring a dam and should not be overlooked.

REFERENCES
NVE (2003), *"Guideline for monitoring and instrumentation of dams"*, NVE, Oslo, Internet: www.nve.no
NVE (2002), *"Guideline for inspection and reassessment"*, NVE, Oslo Internet: www.nve.no
V.N. Svendsen, K. Molkersrød, E. Torblaa (1997), *"Emergency Action Planning for Major Accidents within River Basins in Norway"*, Proceedings ICOLDs Congress; Q75 - page 261, Florence
NVE (2000a), *"Regulations governing the safety and supervision of watercourse structures"*, NVE, Oslo, Internet: www.nve.no

Midttømme, G.H. (2003), *"Changes in the legal framework for dam safety in Norway"*, The International Journal on Hydropower & Dams; Issue 5, page 150

NVE (2000b) *"Regulations governing the classification of watercourse structures"*, NVE, Oslo, Internet: www.nve.no

J.G. Martinsen (1995), *"Dam failure warning systems"*, The International Journal on Hydropower & Dams; Issue 3, page 38

Long-term stress measurements in the clay cores of storage reservoir embankments

K S WATTS, Building Research Establishment Ltd, Watford, UK.
A KILBY, Thames Water plc, London, UK
J A CHARLES, Building Research Establishment Ltd, Watford, UK.

SYNOPSIS.   In 1987 push-in spade-shaped earth pressure cells and BRE miniature push-in earth pressure cells were installed to study stresses within the puddle clay cores of Staines South and King George VI storage reservoirs in west London. The spade cells were installed to measure horizontal stress and the miniature cells were installed to measure both horizontal and vertical stresses. In 1998 spade cells were also installed at various sections in the rolled clay cores of Queen Mother and Wraysbury reservoirs. This paper outlines the monitoring programme and briefly describes the instrumentation and installation techniques. Selected data sets demonstrate the reliability and longevity of the instrumentation. The results show that these instruments can provide valuable long-term information on stress levels within clay cores and, in particular, the effects of reservoir drawdown and refilling on the magnitude of these stress levels in relation to reservoir water pressure.

INTRODUCTION

A survey by Charles and Boden [1] of nearly 100 cases of unsatisfactory performance of embankment dams in the UK suggested that the most serious hazard for old earth dams as they age in service is associated with internal erosion. Hydraulic fracture of a clay core is one possible mechanism which can initiate internal erosion and it has been postulated that hydraulic fracture can occur if the water pressure from the reservoir exceeds the minimum total earth pressure acting on a transverse plane within the body of the core. The state of stress within clay cores is therefore of considerable interest. Charles [2] has reviewed case histories of the deterioration of puddle clay cores and Charles and Watts [3] have described a programme of field measurements to examine the horizontal pressures within puddle clay cores and puddle-filled cut-off trenches of old earth dams.

Long-term benefits and performance of dams, Thomas Telford, London, 2004, 396–407

With uniform ground conditions and a level ground surface it is usually assumed that the vertical total stress can be calculated with sufficient accuracy by multiplying the depth below ground level (z) by the mean bulk unit weight of the soil above that depth (γ). However, there are situations where the vertical stress is significantly different from the calculated overburden pressure. An important example is where there is "arching" involving stress transfer between soils with different stiffness such as between the clay core and shoulders of an embankment dam. In such cases the vertical total stress may be significantly smaller than γz and the horizontal stress will be a complex function not only of depth and unit weight of the soil, but also of the stress-strain relation and stress history of the soil. Reliable determination of vertical and horizontal stress usually requires in-situ measurement.

INSTRUMENTATION
Two types of pressure measuring device have been installed to monitor the stresses within the cores of four embankment dams.

Spade type pressure cells
The use of push-in spade-shaped earth pressure cells ("spade cells") in various types of clay has been described by Penman and Charles [4] and Tedd and Charles [5]. Spade cells have proved to be very simple and reliable for stress measurement although there is a tendency for them to over-read even when the excess pore pressure set up during installation has dissipated. The amount by which spade cells over-read has been investigated and a simple empirical correction of half the undrained shear strength ($0.5c_u$) has been proposed by Tedd and Charles [5]. Ryley and Carder [6] have found that a larger correction is needed where $c_u >150$ kN/m^2 but this is of no significance for the work reported in this paper.

The spade cells used in the investigations were manufactured by Soil Instruments Ltd. and consist of an oil filled steel envelope approximately 200mm long x 100mm wide and 6mm thick. Each spade cell incorporates a piezometer above the pressure cell and the pressures are measured by pneumatic transducers.

Installation of the cells was accomplished by pushing the spade cell about 1m beyond the bottom of a vertical borehole and all the cells were aligned to measure horizontal stress along the axis of the dam (σ_{ha}). Several weeks had to elapse after installation before the decay of the excess pressures, which were generated by pushing the cells into the clay, was complete.

BRE miniature pressure cells

Cells pushed into the soil from the bottom of vertical boreholes can only be aligned to measure horizontal stress. In the situations where the measurement of vertical stress is required, a vertical borehole may provide the only access for in-situ measurement. The BRE push-in miniature earth pressure cell ("miniature cell") is designed to be jacked horizontally into soft clay from a vertical 150mm diameter borehole.

The miniature cell consists of a 2.4mm thick oil filled envelope attached to a wedge shaped slim body. It has a measuring area 44mm in diameter, an overall length of 115mm and a maximum body thickness of 20mm in the direction of stress measurement. The miniature pressure cell operates on similar principles to the larger spade cells.

Miniature cells are installed using a special placing device which is lowered down a vertical borehole. Cells are pushed horizontally about 450mm beyond the borehole wall and multiple installations can be carried out at different elevations within a single borehole. The cell can be pushed into the undisturbed soil with an attitude to measure either vertical stress or horizontal stress. The system has been described by Watts and Charles [7] and Watts [8].

It has been found that, generally, shorter times can be allowed after the installation of miniature cells for the dissipation of excess pressures than for the larger spade cells. Generally the correction for over-read is smaller than that required for a spade cell. No corrections have been applied to the data presented in this paper for spade or miniature type cells.

PROGRAMME OF FIELD MEASUREMENTS

The study of stresses within the puddle clay cores of King George VI and Staines South storage reservoirs in west London commenced in 1987. Further installations were carried out between 1991 and 1997 to investigate potential for hydraulic fracture on several sections at both dams. In 1998 instrumentation was installed in the rolled clay cores of Queen Mother and Wraysbury reservoirs, also situated in west London. All the dams comprise continuous embankments encircling non-impounding reservoirs which store water above the surrounding natural ground level.

Cross-sections of the central parts of all four embankment dams with the elevations of the pressure cells within the clay cores are shown in Figure 1. All cells are located on the centre-line of the cores. The pressures measured at, or close to, reservoir full condition at each of the dams are plotted in relation to the elevation of the cells in metres above Ordnance Datum. The

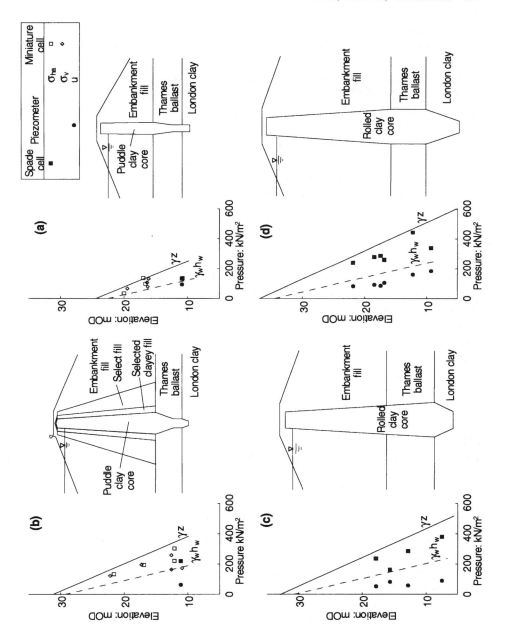

Figure 1: Pressures measured on the centre-line of the clay core at:
(a) Staines South; (b) King George VI; (c) Wraysbury; (d) Queen Mother

readings represent the equilibrium pressures measured after excess pore water pressures generated during installation had fully dissipated and therefore the individual cell readings at a particular dam were not all taken at the same time. The pressure generated by the reservoir water ($\gamma_w h_w$) on the upstream face of the core, the theoretical overburden pressure (γz) within the core and the measured pore water pressures are also shown.

STAINES SOUTH

Staines South reservoir was completed in 1903. It is part of a twin reservoir and shares a common embankment with Staines North reservoir and has a top water level approximately 3m lower than Staines North. The dam has a maximum height of 9m. The central puddle clay core extends 6m to 8m below original ground level to form a cut-off through the Thames ballast and is keyed a short way into the underlying London clay. The plasticity results of the puddle clay plot above the A-line of the plasticity chart and are classified according to BS 5930 [9] as very high plasticity (CV). Undrained shear strengths (c_u) measured from samples from the core were in the range 20-30 kN/m^2.

Miniature cells were installed in the puddle clay core at 8m below crest level to measure vertical stress and horizontal stress in the axial direction at a section where the embankment height was 9m above original ground level. Another miniature cell was installed at 5m below crest level to measure vertical stress. A spade cell was installed to measure horizontal axial stress 13.5m below the crest in the clay filled cut-off trench at the level of the boundary between Thames ballast and London clay. In 1993, a similar installation was carried out at another location comprising a spade cell at 13.6m below the crest and miniature cells to measure vertical and horizontal stresses at 8m and 5m below crest level.

The spade cell and miniature cell orientated to measure horizontal stress along the dam axis (σ_{ha}) in the deepest part of the clay cut-off are in close agreement and show pressures slightly above reservoir pressure. Data from miniature cells installed in the puddle clay at a depth of 8m below the top of the embankment are shown in Figure 2. These typical measurements demonstrate the longevity and stability of the pressure cells over a long period. The vertical stress at this position is less than the horizontal stress.

The maximum level at which the reservoir has been held has varied, but a broadly consistent pattern of earth pressure changes has been observed. The variations in vertical and horizontal pressure due to a number of major drawdown events have been monitored and the drawdowns in 1994, 1997 and 1998/99 when the reservoir was emptied are of particular interest.

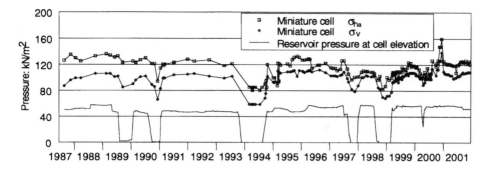

Figure 2: Measurements of vertical and horizontal pressures within the puddle clay core of Staines South at 17mOD.

On reservoir drawdown, the earth pressures have shown substantial reductions and the measured pressures have then risen rapidly in response to reservoir refilling. However, there has been some delay in returning to the pre-drawdown stress levels with final recovery in stress occurring over a period following refilling.

The anomalous peak in the data in late 2000 is more likely to be related to a common readout fault or operator error than to actual ground behaviour. There has been a general convergence of the vertical and the horizontal pressure readings since 1987, principally as a result of a steady increase in vertical stress. This pattern may be associated with the major drawdown events.

KING GEORGE VI

King George VI reservoir was completed in 1947. It has a maximum height of 17m above original ground level. The embankment is constructed of ballast excavated from the centre of the reservoir and contains zones of selected fill either side of a puddle clay core. Although slightly wider, the geometry of the cut-off through the Thames ballast is similar to that at Staines South. The plasticity results of the puddle clay plot above the A-line and vary from high (CH) to very high plasticity (CV). Undrained shear strengths of recovered samples were also in the range 20-30 kN/m^2 and were generally found to increase with depth.

In 1987 a spade cell was installed to measure horizontal stress in the axial direction at 20m below crest level in the clay filled cut-off trench. Miniature earth pressure cells were installed in adjacent boreholes to measure both vertical and axial stresses at approximately 19m and 10m below crest level. In 1991 five additional miniature cells were installed in a single borehole

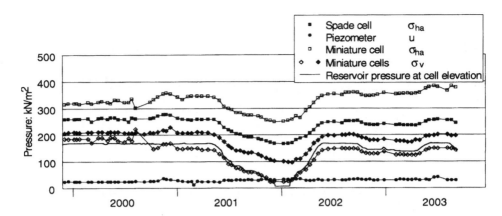

Figure 3: Measurements of vertical and horizontal pressure within the puddle clay core of King George VI at 12mOD.

close to the 1987 installations to measure vertical pressure at 20m and vertical and horizontal pressure at 19m and 14m below crest level. A further four cells were installed to measure vertical and horizontal pressure between 20m and 18m below crest level at another location in 1997.

The two miniature cells which are orientated to measure vertical stress (σ_v) in the clay filled cut-off trench have consistently measured reservoir pressure since, or shortly after, the time of installation. One cell encountered some granular material during installation and did not register excess pressures during or after installation.

In Figure 3 the pressures measured by a spade cell with piezometer and three of the miniature cells installed in adjacent boreholes at an average depth of 19m below top of embankment are plotted along with the reservoir pressure at that elevation. The figure covers the period from 2000 to 2003, during which the reservoir underwent a major drawdown and was empty for about 1 month during the winter 2001/02. The response to such an operational event indicated by the short-term detailed observations illustrates the benefits of instrument stability and regular monitoring.

The profile of pressure reduction measured by the cells closely follows the reservoir drawdown. The measured earth pressures generally recovered to values similar to those before drawdown, but the horizontal pressure (σ_{ha}) measured by a miniature cell located about 1m above the spade cell has continued a rising trend which was evident before the drawdown. All the cells showed some small time delay in reaching a maximum pressure reading some time after the maximum reservoir level had been fully reinstated.

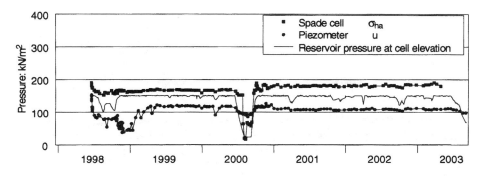

Figure 4: Measurement of horizontal pressure within the rolled clay core of Wraysbury at 15.6mOD.

WRAYSBURY

Wraysbury reservoir was constructed between 1965 and 1970 and comprises a zoned embankment with rolled clay core and gravel shoulders with clay layers within the upper part of the downstream shoulder. The embankment has a maximum height of 17m above foundation level.

The rolled clay core extends 11m below surrounding ground level to form a cut-off through the Thames ballast and keys a minimum of 3m into the underlying London clay formation. The plasticity results of the rolled clay core generally plot above the A-Line and are classed as high plasticity (CH). Water was added to the clay, which was placed to a specified undrained shear strength of about $80kN/m^2$ with air voids not greater than 3% [10].

In 1998 single spade cells were installed at four locations between 15m and 25m below crest level. These installations were carried out to investigate stress levels within the core and changes in horizontal pressure in response to rapid operational changes in reservoir level.

The deepest and shallowest cells illustrated in Figure 1(c) have given remarkably stable readings since installation in 1998. The pressures measured by these cells during reservoir full conditions remained unchanged after a 7m drawdown during August and September 2000. The cell at elevation 12.8mOD has indicated a steady fall in horizontal stress σ_{ha} of about 50kPa or 5.0m water head since 1998. The cell at elevation 15.6mOD, which measured pressures close to reservoir full level after installation has shown a quite different trend as shown in Figure 4. There has been a marked and sustained rise in earth pressure at this cell after the reservoir drawdown while the piezometer at this location has measured a small but sustained fall in pore water pressure.

QUEEN MOTHER

Queen Mother (formerly known as Datchet) reservoir was constructed between 1969 and 1974 and comprises a zoned embankment with rolled clay core and gravel shoulders with clay layers within the downstream shoulder. It has a maximum height of 20m above foundation level.

Core clay properties are similar to Wraysbury. The specified placed undrained shear strength for the clay was also about 80kN/m^2 with air voids not greater than 3%[11].

In 1998 single spade cells were installed at six locations at depths between 15m and 28m below the crest. The installations were also carried out to investigate stress levels within the core and changes in horizontal pressure in response to rapid operational changes in reservoir level.

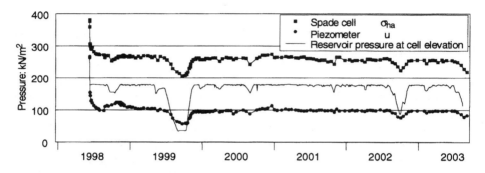

Figure 5: Measurement of horizontal pressure within the rolled clay core of Queen Mother at 17mOD.

The long-term measurements obtained from the pressure cells installed at Queen Mother indicate a rather different pattern of behaviour to that observed at Wraysbury. Figure 5 shows pressures measured by a spade cell and piezometer installed 20m below the crest of the dam. It is one of a cluster of three cells at similar depths shown in Figure 1(d). Initial readings after installation in June 1998 show the dissipation of excess pressure due to installation. The cell was installed during a period when the reservoir was at, or close to normal top water level.

Since cell installation there have been two periods when the reservoir level was reduced by a significant amount and a number of minor reductions in level have occurred over shorter time periods. The plot shows the sensitivity of the pressure cell to changes in reservoir level and hence variations in horizontal pressure within the clay core. During 1999 the reservoir level was reduced by approximately 14.5m and held at the reduced level for just over one month. This resulted in a reduction in σ_{ha} of 65 kN/m^2, equivalent to about half the reduction in reservoir pressure. This reduction was time dependent and the pressure was still falling when the reservoir level was rapidly raised. The rates of change of the earth pressure and the pore water pressure measured by the associated piezometer appear to be closely related.

There has also been a small but steady decline of about 15kN/m^2 or 1.5m head in the earth pressure measured by this cell for reservoir full conditions over the five years since installation. The measurements obtained from this cell are typical for the dam and a very similar pattern of reaction to reservoir drawdown and a steady decline in pressure is repeated for all the cells at different locations along the dam.

DISCUSSION
A considerable volume of data now exists for all the four monitored dams. The oldest instrumentation, which is in Staines South and King George VI, has provided a continuous record of pressures over a period of 16 years. The vast majority of the spade and the miniature cells have given realistic and consistent measurements throughout the monitoring period. The instruments have made possible the measurement of in-situ stress under static reservoir full conditions and with fluctuating reservoir levels.

Under static conditions with reservoir full, the earth pressures within the clay cores generally are significantly above the reservoir pressure at that particular depth. Vertical and horizontal pressures measured at similar elevations in the puddle clay cores of Staines South and King George VI are generally similar in magnitude and no consistent pattern as to a dominant direction has emerged.

The situation in the narrow clay cut-offs of puddle core dams is somewhat different. Stresses at or close to reservoir pressure have been monitored at both King George VI and Staines South.

Pressures at all elevations in the rolled clay cores at Wraysbury and Queen Mother are generally well above the reservoir pressure. One exception is a cell at Wraysbury which was installed at a predetermined elevation within a softened zone. Its elevation is coincident with the boundary between embankment fill and foundation ballast and this may be of significance.

CONCLUSIONS
1. The field measurements have demonstrated that spade cells and miniature cells can provide a reliable means of monitoring stresses in clay cores over long periods.

2. Reservoir refilling following a major drawdown is a critical time for hydraulic fracture and the instrumentation can be used to monitor stresses during this period.

3. The stress conditions in narrow clay cut-offs tend to be more adverse than in the clay cores within the embankments.

ACKNOWLEDGEMENT
The data in this paper is presented with the kind permission of Thames Water plc.

REFERENCES
1 Charles J A and Boden J B (1985) The failure of embankment dams in the United Kingdom. *Failures in earthworks*. Proceedings of symposium organized by Institution of Civil Engineers, March 1985, 181-202. Thomas Telford, London.
2 Charles J A (1990). Deterioration of clay barriers: case histories. *Clay barriers for embankment dams*. Proceedings of conference organized by Institution of Civil Engineers, October 1989, 109-129. Thomas Telford, London.
3 Charles J A and Watts K S (1982). The measurement and significance of horizontal earth pressures in the puddle clay cores of old earth dams. *Proceedings of Institution of Civil Engineers*, Part 1, **82**, Feb., 123-152.
4 Penman A D M and Charles J A (1981). Assessing the risk of hydraulic fracture in dam cores. Proceedings of 10th International Conference on Soil Mechanics and Foundation Engineering, Stockholm, 1981, **1**, 457-462.
5 Tedd P and Charles J A (1983). Evaluation of push-in pressure cell results in stiff clay. Proceedings of International Symposium on In-Situ Testing, Paris, **2**, 579-584.
6 Ryley M D and Carder D (1995). The performance of push-in spade cells installed in stiff clay. *Geotechnique*, **45**, No 3, 533-539.
7 Watts K S and Charles J A (1988). In situ measurement of vertical and horizontal stress from a vertical borehole. *Geotechnique* **38**, No. 4, 619-626.
8 Watts K S (1991). Evaluation of the BRE miniature push-in pressure cell system for in situ measurement of vertical and horizontal stress from a vertical borehole. *Field measurements in Geotechnics*, Sørum (ed.), Balkema, Rotterdam, 273-282.
9 BS 5930: 1999. Code of practice for site investigations. BSI, London.
10 Reed E C (1971). Wraysbury and Datchet reservoirs. *Civil Engineering and Public Works Review*, **66**, June, pp 606-610.

11 Pawsey D B H (1976). The Queen Mother reservoir, Datchet – some aspects of its design and construction. *Ground Engineering*, **9**, October, 27-30.

Glacial risk and reservoir management: the Lago della Rossa reservoir example (Valli di Lanzo, Western Alps, Italy)

A TAMBURINI, Enel.Hydro S.p.A.-Ismes, Seriate (BG), Italy
G MORTARA, CNR-IRPI, Torino, Italy
L MERCALLI, Società Meteorologica Italiana, Torino, Italy
M LUCIGNANI, Enel Green Power S.p.A., Torino, Italy

SYNOPSIS.   The climatic evolution of recent years, characterised by slight winter snowfalls and very high summer temperatures, is causing a progressive loss in ice mass combined with the increase of glacial risk. Besides the widespread retreat of glaciers, one of the most evident consequences of temperature increase is the formation of epiglacial lakes, like the one formed on the left side of the Croce Rossa glacier in 1998. The glacier overhangs the reservoir of Lago della Rossa, the highest in Italy. A complete survey (GPS surveyed strain net, ablation stakes, radar echo sounding, automated air and ice temperature measurement) has been carried out since 1998 in order to establish the main features of the glacier and monitor its evolution. Mathematical and physical models have been applied in order to evaluate glacier stability and future scenarios in case of epiglacial lake outburst.

INTRODUCTION

The formation of ponds and supraglacial lakes even at an elevation higher than 3000 m a.s.l. represents one of the main consequences of the atmospheric warming which presently affects high mountain regions where glacier- and permafrost-related hazards are rapidly increasing (Mercalli et al., 2002a; Mercalli et al., 2002b; Tamburini et al., 2003; Kääb et al., in press).

Once formed, supraglacial lakes tend to expand due to thermokarst process. The appearance of a glacial lake represents a cause for concern about potential glacial lake outburst flooding (GLOF) (Mercalli et al, 2002a).

In order to either prevent or reduce the accumulation of large volumes of water and to prevent dangerous situations, practical measures have been successfully taken in many cases (pumping, syphoning, oblique drilling or tunneling through the ice body, excavation of drainage channels).

In 1818 at the ice-dammed Giétroz lake (Swiss Alps), despite the extreme technical and environmental conditions, a drainage tunnel about 200 m long was excavated by hand through ice (UNST, 1981; Hambrey et al., 1992).

In 1996 at the Gruben Glacier (Swiss Alps) a drainage channel was excavated in order to prevent the occurrence of outburst floods from a periglacial lake (Haeberli et al, 2001).

Drainage channels should be oriented according to the sliding direction of the glacier, in order to increase their effective life and efficiency even in cases where the speed of the glacier itself is high. Once the channel is excavated, water flowing through it rapidly melts ice and increases the width of the channel, so enhancing the excavation process.

As part of glacial lake hazards mitigation at Hualcán (Cordillera Blanca, Peru) in 1993, following installation of siphons, the construction of a 2-m diameter tunnel started. The tunnel was 155 m long beneath a rock bar below the moraine dam (Reynolds et al., 1998).

In the Himalaya remediation strategy to reduce the GLOF hazard at the Tsho Rolpa (Nepal) consisted in syphon pipes to augment the original trial pipe and an artificial spillway ((Reynolds, 1999).

DESCRIPTION OF THE INVESTIGATED AREA

Figure 1: Aerial view of the Lago della Rossa reservoir and Croce Rossa Glacier (CNR-IRPI, 16/09/2001)

The Croce Rossa Glacier (Fig. 1 and 2) is located on the NE slope of Croce Rossa peak (3556 m a.s.l.) in Valle di Lanzo (Western Italian Alps). The steep terminus of the glacier (3380 m a.s.l.) overhangs the Lago della Rossa reservoir, the highest in Europe (2715 m a.s.l.), which is located about 700 m below.

Figure 2: The Croce Rossa Glacier (Tamburini, 05/08/2003)

In 1998 a small supraglacial lake (Fig. 3) formed near the left margin of the glacier at an elevation of about 3450 m a.s.l..

Figure 3: The supraglacial lake appeared in 1998 (Tamburini, 2000)

The presence of the supraglacial lake represents a cause for concern due to the increased risk of triggering ice avalanches into the Lago della Rossa reservoir. If waves should be generated by such events, they could seriously damage the dam and overtop the dam crest, with resulting floods in the downstream valley.

For this reason surveys were immediately carried out on the glacier in order to assess and reduce hazards and a monitoring system was installed.

STUDIES ON THE CROCE ROSSA GLACIER
The following studies were carried out in the Croce Rossa Glacier area:
- Aerial photogrammetric survey, georeferenced with GPS, provided a map at 1:2000 scale of an area including the glacier and the reservoir below
- GPS-assisted georadar surveys from glacier surface provided glacier bed morphology and glacier volume
- A physical model was performed in order to assess the water waves generated by a glacier avalanche in the Rossa reservoir

Moreover a monitoring system for the measurement of ice temperatures and glacier displacement rate has been established.

GPS-assisted GPR surveys: methodology, results and evolutive scenarios
GPS-assisted GPR (Ground Penetrating Radar) surveys from the glacier surface were carried out in December 1998 and completed in April 1999, in order to determine the ice thickness and the depth of the glacier bed. Four transverse and two longitudinal profiles were surveyed in locations determined by the glacier morphology and where access was possible.

The following equipment was used:
- GSSI SIR-2 radar system
- RADARTEAM Subecho 40 antenna, with 35 MHz base frequency
- differential GPS positioning system

Raw GPR data were processed with GSSI WINRAD Software, in order to convert the original time vs. time profiles into distance vs. depth cross sections of the glacier. A map of the glacier bed was obtained in ArcView environment, by creating and contouring a DEM (Digital Elevation Model) obtained from interpolating glacier bed elevation data. By subtracting the glacier surface and glacier bed DEMs in ArcView environment, the overall volume of the glacier and a map of the ice thickness were obtained (Fig. 4). This map was later used for physical model performance.

The most relevant results obtained by GPR investigations are listed below:
- the overall volume of the glacier was calculated as about 1.5 million m^3
- a maximum ice thickness of about 60 m was detected in the central part of the glacier

- a change in glacier bed dip, corresponding to a rather flat area in the central part of the glacier, resulted from interpolating glacier bed elevation data; such a morphological step has been considered as favouring stability of at least the upper part of the glacier (Fig. 5)

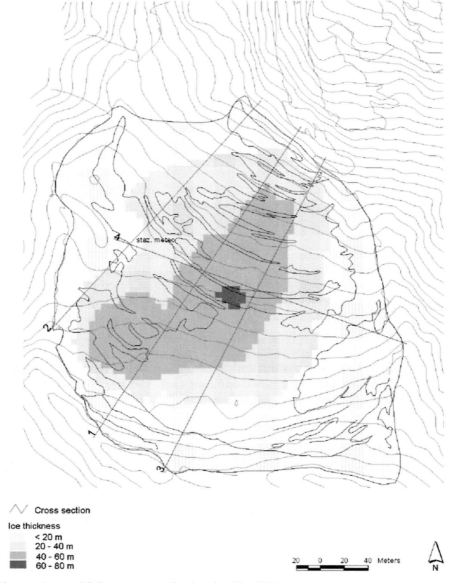

/\/ Cross section

Ice thickness
< 20 m
20 - 40 m
40 - 60 m
60 - 80 m

20   0   20   40 Meters

N

Figure 4: Ice thickness map obtained with GPR survey

Due to the morphology of the glacier bed resulting from GPR surveys, a sudden collapse of the entire glacier was considered less probable than a partial collapse in case of outburst of the epiglacial lake and subsequent sudden increase of water pressure at the base of the glacier. Considering both the major crevasse pattern and the epiglacial lake position, some hypothesis about the unstable portion of the glacier have been carried out. The worst scenario refers to the collapse of a volume of about 500,000 m³ of ice.

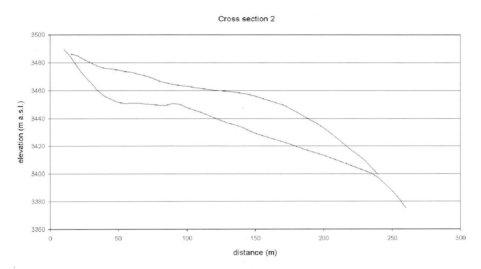

Figure 5: Longitudinal cross section of the Croce Rossa Glacier.

Physical model

A physical model of the glacier and the reservoir below was carried out in the Enel.Hydro labs in Milan, with the aim of assessing the water waves generated by a glacier avalanche in the Lago della Rossa reservoir (Brambilla et al., 2000). Several scenarios have been simulated, with different ice volumes and reservoir levels, in order to define threshold values for the reservoir level. The temporal development of an ice slide generated by manual and "instantaneous" removal of a gate is shown in Fig. 6.

The main results obtained by physical modelling are listed below:
- the time span occurring between the collapse and the dam overflow is lower than 1 minute
- the model provided a table of water level vs. ice avalanche volume, showing that no overflow of the Lago della Rossa dam occurs in case of collapse of an ice volume lower than 200,000 m³
- in the worst case (500,000 m³) only the right part of the dam top is overflown

Figure 6: Temporal development of ice-slide 5, 10 and 16 seconds after the release

According to the results of the model, the maximum water level in the reservoir has been fixed 9 m lower than the maximum water level during the critical season (approximately July to September). Unfortunately, as the maximum water level is generally reached at the end of September, the above restriction represents a significant limitation on reservoir management. For this reason, appropriate measures have been proposed in order to enable a sustainable management of the reservoir itself, based on the improved knowledge of the glacier evolution acquired during about five years of glacier monitoring

MONITORING SYSTEM INSTALLED ON THE CROCE ROSSA GLACIER
The following devices were installed on the Croce Rossa Glacier:
- ablation stakes, set in holes carried out by steam drill device, provide data for mass balance and glacier sliding evaluation
- thermometers, inserted at different depths in holes, provide ice temperature measurement; both ice and air temperature data are automatically acquired and stored locally in a datalogger

Surface displacements are measured with GPS (5 stakes) and EDM (3 stakes) devices. EDM measurements are taken from the Rossa dam.
The five ablation stakes, each one formed by 5 wooden sections for a total length of 10 m, were installed in holes carried out by steam drill device. As the stakes are not visible from the below dam, static GPS measurements are regularly carried out twice a year. A GPS reference station was established on outcropping rock along the left margin of the glacier in order to simplify site operations.

Three stakes equipped with reflecting prism were installed near the glacier terminus, in order to enable glacier sliding measurement from a remote measuring topographic station from directly the below dam. Due to distance (about 1500 m) and difference in elevation (about 1000 m) traditional topographic measurements are affected by a high error, so only distance measurements can be considered reliable for the control of the glacier terminus. Distance measurements can be easily automated by using a motor driven theodolite.

Four ice thermometers were installed in holes carried out by steam drill device near the internal side of the supraglacial lake, at depths of –2, -4, -8 and –14 m respectively from glacier surface. The deepest hole reached the base of the glacier, so providing temperature values at the ice-rock contact.

A cable connection between the ice thermometer and a permanent meteorological station installed near the GPS reference station enables continuous automatic ice and air temperature measurement. The data acquisition rate is 12 per day (every 2 hours).

Mass balance is evaluated on five ablation stakes, the same used for surface movement detection with GPS measurements. Twice per year, when glacier surface displacement measurements are carried out, snow thickness, stratigraphy, density, etc. measurement are taken in order to calculate the mass balance of the glacier.

Finally, photographs of the glacier and epiglacial lake are regularly taken at least twice per year: a comparison with previous photographs allows the integration of instrumental data for an overall monitoring of the glacier evolution.

Monitoring system installation

Two programmes have been carried out in order to investigate the glacier and install the monitoring system operating at present:

- at the end of December 1998:
  - o main GPR survey (completed on April 1999)
  - o installation of ice thermometers at different depths, as described, with a battery powered datalogger
  - o installation of ablation stakes; the stakes close to the terminus were equipped with reflecting prisms at the top, in order to enable EDM measurements from below the Rossa dam
  - o installation of a reference point both for GPS and EDM measurements on outcropping rock near the left margin of the glacier
- on July 2001:
  - o installation of a meteorological station at an elevation of about 3450 m a.s.l., anchored on an outcrop near the left margin of the glacier, powered by solar cells for automatic acquisition and storage of air and ice temperature data

o completion of the ablation stakes network; the stakes close to
the terminus had to be replaced as they got lost

Results of glacier monitoring campaigns

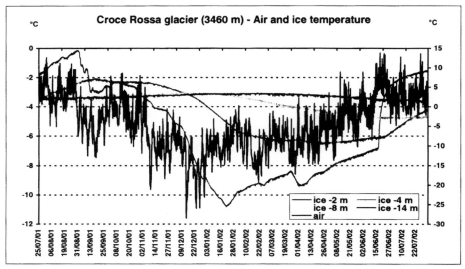

Figure 7: Air and ice temperatures recorded from July 2001 to July 2002

The glacier has been regularly surveyed for five years. The main results
provided by the monitoring activity are listed below:

• the average surface planimetric displacements are about 2 m per year
in the centre of the glacier, lower than 40 cm/yr near the left margin,
where the thermometers are located; such values seem to be constant
within the considered period (Fig. 8)

• up to 8 m of depth temperatures in ice are subject to seasonal
variations; a delay in the yearly maximum value can be observed:
such delay increases with depth, up to 6 months at −8 m from the
surface (Fig. 7). This is in agreement with what was observed by
Paterson, 1981.

• at a depth of −14 m from the surface, temperature is about constant
all the year round, varying from −3.1 to −3.7 °C; this confirms that
at a depth of −14 m ice is not subject to seasonal variation;
moreover, the presence of a cold ice layer at the base plays an
important role in increasing the glacier stability, as low temperature
is responsible for the adhesion of ice to the rock below (Luthi, 1994)

• during 1998-1999 and 1999-2000 glaciological year, an intense
ablation was observed; the effects of the snowy 2000-2001 winter
have been observed during the following two years, characterised by
a slightly positive mass balance. Finally, the exceptionally warm

2003 summer is responsible for a strongly negative mass balance (–2.5 to –3 m w.e.); the above data suggest the opportunity of performing a new photogrammetric survey, in order to calculate the overall mass reduction and upgrade the instability scenarios outlined after the first investigation programme in 1998

- water release from the terminus was never observed; this could confirm the adhesion between ice and rock at the base of the glacier, as previously observed

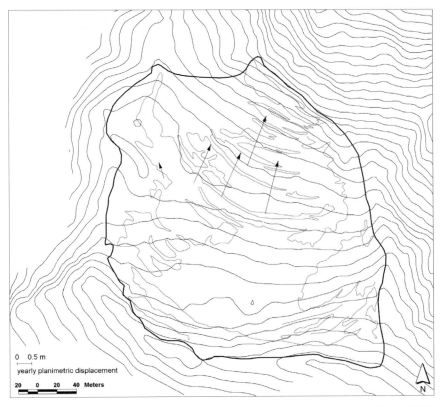

Figure 8: Yearly planimetric displacement vectors (from September 2002 to August 2003)

SCENARIOS FOR THE FUTURE

According to the results provided by monitoring, the behaviour of the glacier seems to be constant for the considered period. The dynamics of the glacier seem not to have been influenced by the high temperatures recorded during summer 2002 and 2003. Anyway, in order to verify the absence of a long term effect of the exceptionally warm 2003 summer, a further surface displacement measurement programme is progressing.

Two main collapse scenarios can be outlined for the Croce Rossa Glacier:

- sudden collapse of the lower portion of the glacier (the original volume of about 500,000 m^3 should be verified) due to outburst of the epiglacial lake and subsequent sudden increase of water pressure at the base of the glacier
- gradual increase of displacement rate at the glacier terminus before the collapse of the lower part of the glacier.

The former can be considered instantaneous, the latter can be forecast by measuring the displacement rate at the glacier terminus. In any case the main hazard is represented by water stored in the epiglacial lake, which should be removed.

FUTURE DEVELOPMENTS

In order to substantially reduce hazard and enable efficient management of the Lago della Rossa reservoir, further investigations and interventions have been planned. The main activities are listed below:

- drainage of the epiglacial lake, which could trigger an unpredictable collapse of the glacier
- performance of a new aerial photogrammetric survey, in order to evaluate the glacier volume decrease and upgrade the collapse scenarios
- set of new ice thermometers closer to the glacier terminus at different depths, including the glacier base, in order to assess the glacier stability; a proper hole must be drilled to a depth of about 50 m
- mathematical model application in order to evaluate the long term effects of increasing temperature on the glacier dynamics
- reservoir level threshold value upgrade, taking into account the reduced volume of the unstable portion of the glacier and the possibility of forecasting the collapse and the discharge of water from the reservoir in case of emergency; water level vs. time curves should be computed for different discharge rates.

To assist glacier monitoring after draining the epiglacial lake, surface displacements and temperature (air and ice) measurements should be automated, in order to enable a real time control of the glacier evolution..

Automatic EDM measurements, by means of a motor driven theodolite, should be carried out in summer and early autumn (June to October), when the reservoir level reaches its maximum and the possibility of glacier collapse is higher. In case of displacement rate increase, the application of on line Voight model approach in order to forecast the collapse time could be helpful for hazard management (Voight, 1988; Voight, 1989).

Ice and air temperature, at present automatically stored in the datalogger of the meteorological station installed of the glacier, should be transmitted by either radio or satellite to the dam, enabling their automatic processing and comparison with threshold values.

Moreover periodic inspections on the glacier, will integrate instrumental measurements and provide data for mass balance evaluation.

CONCLUSIONS

The results of the studies carried out on the Croce Rossa Glacier and the data collected by the monitoring system during the last five years showed that:

- the main cause for concern is represented by the supraglacial lake, which must be drained in order to eliminate the main cause of sudden collapse of the lower portion of the glacier
- mass balance results showed that the overall volume of the glacier has been significantly reduced
- ice temperature values indicate the presence of a cold ice layer at the base of the glacier, which plays an important role in increasing the glacier stability
- the sliding velocity of the glacier is known and seems to be constant during the considered period, with a maximum of 2 m per year in the central part; an increase in sliding velocity can be identified if an automatic monitoring system is operating during the critical season (June to September).

Once the supraglacial lake has been drained, an automatic monitoring system, based on displacement and ice temperature measurement, will enable the possibility to manage the reservoir, according to what happens in other similar situations in the Alps, like the Jungfrau railway below the Eiger Glacier or the Mauvoisin dam below the Giétroz Glacier (UNST, 1981; Luthi, 1994)

REFERENCES

Brambilla S., Pacheco R., Zaninetti A. (2000). *Experimental investigation on laminar highly concentrated flow modeled by a plastic law.* Proc. Of the International Conference on Avalanches-Landslide-Rock Falls-Debris Flows. Vienna January 2000.

Haeberli W, Kääb A., Vonder Mühll D., Teysseire P. (2001). *Prevention of outburst floods from periglacial lakes at Grubengletscher, Valais, Swiss Alps.* J. Glaciology, 47 (156), 111-122.

Hambrey M., Alean J. (1992). *Glaciers.* Cambridge University Press.

Kääb A., Huggel C., Barbero S., Chiarle M., Cordola M., Epifani F., Haeberli W., Mortara G., Semino P., Tamburini, A., Viazzo G. (in press). *Glacier hazards at Belvedere Glacier and the Monte Rosa East face, Italian Alps: processes and mitigation.* Interpraevent 2004.

Luthi M. (1994). *Stabilitaet steiler Gletscher. Eine Studie über den Einfluss möglicherKlimaänderungen. Untersuchungen am Beispiel eines Hägegletschers in der Westflanke des Eigers.* Dipl. ETH Zürich.

Mercalli L., Cat Berro D., Mortara G., Tamburini A. (2002a). *Un lago sul ghiacciaio del Rocciamelone, Alpi Occidentali: caratteristiche e rischio potenziale.* Nimbus 23-24, pp. 3-9.

Mercalli L, Mortara G., Tamburini A. (2002b). *Il ghiacciaio sospeso della Croce Rossa, valli di Lanzo: misure ed evoluzione.* Nimbus, 7, 18-27

Paterson W.S.B. (1981). *The physics of glaciers.* Pergamon Press, Oxford.

Reynolds J.M., Dolecki A., Portocarrero C. (1998). *The construction of a drainage tunnels as a part of glacial lake hazard mitigation at Hualcán, Cordillera Blanca, Peru.* In Maund J.G., Eddleton M M. (eds): Geohazard in Engineering Geology. Geol. Society, London, Engineering Geol., 15-41-48).

Reynolds J.M. (1999). *Glacial hazard assessment at Tsho Rolpha, Rolwalling, Central Nepal.* Quarterly Journal of Engineering Geology, 32, 209-214

Tamburini A., Mortara G., Belotti M., Federici P. (2003). *L'emergenza del lago Effimero sul Ghiacciaio del Belvedere nell'estate 2002 (Macugnaga, Monte Rosa, Italia). Studi eseguiti, tecniche di indagine utilizzate e principali risultati ottenuti.* Terra Glacialis, 6, 37-54.

Ufficio Nazionale Svizzero del Turismo (1981). *La Svizzera e i suoi ghiacciai.* Edizioni Trelingue, Lugano.

Voight B. (1988). *Material science law applied to time forecast of slope failure.* Landslide news, 3.

Voight B. (1989). *A relation to describe rate-dependent material failure.* Science, vol. 243.

The Performance of Thika Dam, Kenya

D. A. BRUGGEMANN, KBR, Leatherhead, UK
J. D. GOSDEN, KBR, Leatherhead, UK

SYNOPSIS.   Thika Dam is situated on the Thika River about 60km north of Nairobi in the foothills of the Aberdare Mountains where the river bed is 1985m above mean sea level (AMSL). The dam forms a major element of the Third Nairobi Water Supply Project which was constructed between 1990 and 1995.

The dam incorporated a range of instrumentation and monitoring systems. The data from the instrumentation was collected regularly for 5 years after construction with periodic reviews of the dam performance.

The purpose of the paper is to show that the dam constructed from halloysitic clay has performed satisfactorily and to compare the behaviour with dams constructed from conventional materials.

INTRODUCTION

Thika Dam is a 70m high earthfill embankment constructed entirely from a residual soil rich in halloysite. The unusual behaviour of residual soil rich in halloysite has been discussed by Terzaghi (1958) in relation to the construction and performance of Sasumua Dam in Kenya. Thika Dam incorporated vibrating wire piezometers, total pressure cells, inclinometers and settlement gauges. Surface monuments were installed on the embankment to monitor surface movements and seepage was collected in a measurement chamber at the downstream toe. The dam is shown in Figure 1

This paper examines the performance of the dam over the first 5 years of operation. The post construction settlement behaviour is examined and compared with settlement data from other dams. Results of seepage measurements from the drainage blanket are discussed and the pore pressure response of the upstream shoulder is also discussed. Comparisons are made with design stage predictions and data from other dams.

Long-term benefits and performance of dams, Thomas Telford, London, 2004, 421–432

Figure 1 – Thika Dam, Kenya

GEOLOGY

The rocks underlying the area are of Pleistocene age and are of volcanic origin being predominantly Pyroclastic tuffs. Two origins of these tuffs can be recognised:
- Pyroclastic flows consisting of fragments of rock dispersed in a medium of fluidised fine material.
- Pyroclastic falls from material thrown into the air by the volcanic explosion.

The remainder of the volcanic sequence comprises flows of phonolite lava. Lavas represent the height of volcanic activity with eruptions occurring from localised vents. Their deposition was sometimes accompanied by air fall activity and thus the phonolite may be found either as massive units or interbedded with the tuffs.

Six periods of volcanic activity can be recognized. The end of each deposition period is marked by a weathered horizon at the top of the sequence. The presence of these residual soil horizons indicates ancient erosion surfaces which were subsequently covered by later volcanic deposits.

The modern drainage pattern has deeply dissected this volcanic sequence and a highly to completely weathered material covers the slopes. Outcrops of rock are restricted to small areas of very steep slope in the valley sides and to water falls formed in the valley floor where streams flow over the more resistant lavas.

DAM DESIGN

Embankment

The embankment is 450m long (curved in plan), 70m high and is constructed of residual volcanic soil. The dam is homogeneous with the exception that the upstream sloping core was placed at a higher moisture content than the shoulders, from which it is separated by a chimney drain. The higher moisture content in the core was intended to make the core sufficiently plastic to maintain high post construction total stresses. Lower moisture content in the shoulders was necessary to minimise construction pore pressures and hence maximise strength. A section through the embankment at maximum height is shown in Figure 2.

During construction higher than expected construction pore pressures were experienced in the downstream shoulder. To ensure construction stage stability 125 vertical sand drains, average depth 15m and 4 drainage blankets, at 5m vertical intervals, were installed in the downstream shoulder. A 1:10 toe weight was constructed from reject material at the upstream toe.

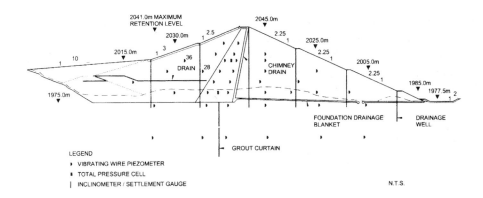

Figure 2 – Cross Section at Maximum Height

Drainage

Seepage through the core and foundation is collected by the chimney drain and a foundation drainage blanket. On the steep abutments the drainage blanket was replaced with a series of finger drains. Further drainage measures were provided by a line of drainage wells along the downstream toe of the valley section of the dam, and by two drainage adits, one in each abutment.

Foundation Treatment

The embankment was founded primarily on residual soil, with typically 3 m of stripping to remove all organic material and provide a suitable profile for filling. On the line of the original river bed, about Chainage (Ch) 290, the embankment was founded on Grade III Lapilli tuff which occurred at a convenient level.

A grout curtain was provided to limit the seepage through the moderately permeable foundations (Lugeon values in the range 5 to 50). The grout curtain was constructed by means of jet grouting in the upper part and by conventional grouting in the lower part. The decision to incorporate a grout curtain through the residual soil was influenced by the lack of precedent for founding a 70m high embankment on up to 35m of residual soil without positive foundation treatment.

This decision was vindicated during construction when an underground cavern, with a volume of at least $8m^3$ was encountered on the right abutment of the upstream shoulder. Similar caverns were exposed in the borrow area upstream of the dam and appeared to occur at depths of up to 10m. The jet grouting of the upper part of the foundation provided security against the possible inter-linking of such caverns.

Further information on the jet grouted cut-off is provided in Attewill et al (1992) and Attewill and Morey (1994)

Instrumentation

The dam was instrumented primarily at three sections; Ch 120, Ch 200 and Ch 290. The section at Ch 290 is at the maximum embankment section, is the most comprehensively instrumented and is shown in Figure 2.

Instrumentation comprised the following:
- 92 vibrating wire piezometers (embankment and foundation)
- 5 inclinometer / settlement gauges

- 26 survey monuments
- 3 total pressure cell arrays (5 in each array)
- 33 observation wells in the abutments
- 18 double installation observation wells along the rim of the reservoir

A number of instruments have failed over the years; 50 piezometers and one total pressure cell array remain functional. Possible causes of these failures were ineffective cable joints and horizontal strain in the embankment. Joins in cables were carried out using proprietary epoxy jointing kits but these may not have been effective in all cases. The cables also passed through materials with different stiffness and a horizontal displacement of up to 330mm was observed in the deepest inclinometer. Despite snaking the cables during installation and the use of a special cable with large strain properties, the strain in some of the cables may have exceeded the failure strain.

Seepage was measured manually by means of V-notch weirs in a seepage measurement chamber which collected flows from the blanket drain.

Material Properties

The material used to construct the dam was predominantly the red to reddish-brown residual soil which formed a mantle up to 6m deep over the borrow area. The results of X-ray diffraction analysis indicated a halloysite content of 60 – 65%. Terzaghi (1958) and Wesley (1973) have noted that halloysitic rich clays exhibit abnormal properties in comparison with sedimentary clays from more temperate regions.

The plasticity index is much lower than that of a sedimentary clay with equal liquid limit. The angle of internal friction and permeability are higher and the compressibility lower than the corresponding properties for a clay with equal liquid limit. Irreversible changes also take place on drying and affect the Atterberg limits, particle size tests and compaction test results. Terzaghi attributed this abnormal behaviour to the clay fraction occurring in clusters or aggregates rather than as individual particles. Moreover, water is located in the voids between the clusters and in their solid structure. The water in the solid structure is inert and has no influence on the mechanical behaviour (Geological Society (1997)).

The Atterberg limits plotted well below the A – Line on Unified Soil Classification System plasticity chart with liquid limit in the range 80% to 100% and Plasticity Index in the range 30% - 40%.

The peak effective stress parameters, as measured in isotropic consolidated triaxial tests with pore water pressure measurement, were
c' (apparent) = 10kPa and φ' = 33°. These parameters were used in the design and were confirmed by laboratory tests carried out during construction.

The field dry density of the fill, depending on whether placed in the shoulder or the core, was in the range $1.1t/m^3$ to $1.2t/m^3$. The field water contents for the shoulder and core were 46% to 53% respectively with corresponding laboratory optimum water contents of 45.5% and 48%.

Further information on the material properties is given in Attewill and Bruggemann (1997).

DAM PERFORMANCE

Embankment construction commenced in October 1991 and was substantially completed in February 1994. The post construction performance of the embankment is discussed in the sections that follow.

Settlement

Settlement was monitored by plate magnets incorporated in the inclinometer installations and by surface monuments. Figure 3 shows the post construction settlement of the crest and the downstream berm at elevation 2025mAMSL. Crest settlements are shown for the surface monuments and the top plate magnet on the crest inclinometer / settlement gauge. The record for the surface monument is shorter than that for the plate magnet because the surface monuments were installed only after the crest road and wave wall construction was completed.

The cumulative settlement of the crest, as measured by the top magnet, since the end of construction was about 400mm or 0.6% of the maximum embankment height. The crest was provided with a 1m camber and thus the settlement was still within design provisions.

The top magnet at the crest shows that there was a slight increase in the rate of settlement about 60 days after the end of construction and since then, at an average rate of about 55mm/year. The value of the crest settlement index, proposed by Charles (1986), was estimated to be 0.003. The index was developed for puddle clay core dams and a range of 0.002 to 0.074 is given in Johnston et al (1999), nevertheless the index for Thika Dam is at the lower end of the range and suggests the dam's performance appears to be in keeping with other types of dam.

Figure 3 – Post Construction Settlement

The value of the drawdown settlement index, proposed by Johnston et al (1999), was estimated for drawdown events during the first 5 years. Five drawdown events provided a cumulative drawdown of 24m. The settlement associated with these drawdown events amounted to 38mm, yielding an index of $0.023mm/m^2$. The index for the individual events varied from $0.013m/m^2$ to $0.058mm/m^2$. These values are generally towards the lower end of the range of values given by Johnston et al (1999) and suggest satisfactory performance.

Seepage

Seepage through the dam is collected by the chimney drain which is connected to the foundation drainage blanket. The foundation drainage blanket was divided at the valley bottom so that flows from each side of the valley were monitored by V notch weirs. The seepage flow is plotted against reservoir water level in Figure 4. The maximum flow from the left hand side was about 1,200litres/minute (20litres/second) and from the right about 600litres/minute (10litres/second) to give a total seepage from the drainage blanket of about 1,800litres/minute (30litres/second). The minimum compensation release required downstream was 15,000litres/minute (250litres/second) and seepage flow made a contribution to the compensation flow.

The seepage estimates made at design stage were based on conventional hand sketched flow nets with the foundation permeability ten times the embankment permeability and an impervious cut-off. The seepage was estimated at 17litres/second.

This seepage is about 60% of the maximum experienced during operation of the dam and reinforces the recommendation of Cedergren (1967) that drain designs should be based on liberal factors of safety. The magnitude of the seepage experienced in the field suggests that care needs to be exercised in the estimation of the foundation bulk permeability from permeability values determined during the ground investigation.

The data shows that seepage from the left hand side was twice that from the right hand side of the valley. This behaviour could be a result of the foundation conditions on the left abutment, where a layer of boulders was encountered in the area of the jet grouted cut-off. The interlocking of the jet grouted columns was unlikely to be as efficient as when installed in residual soil, as on the right abutment. The greater seepage from the left hand side supports Casagrande (1961) that small imperfections in a cut-off can have a major influence on the overall performance of a cut-off.

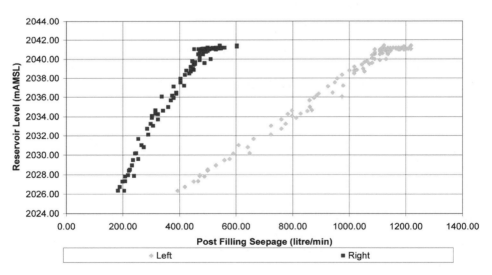

Figure 4 – Blanket Drain Seepage

Piezometer Response

The response of four vibrating wire piezometers situated in the upstream shoulder to an operational drawdown is illustrated in Figures 5 and 6. At the start of the drawdown all four piezometers measured pore pressure closely reflecting the water level in the reservoir. Although the fill material in both the shoulder and the core has relatively low permeability there is very little head drop in the upstream shoulder.

The drawdown took place between July 1998 and April 1999 from near top water level at 2040.8mAMSL to 2028.7mAMSL. The rate of drawdown initially averaged 0.04m/day but after reaching elevation 2034mAMSL it increased to around 0.1m/day.

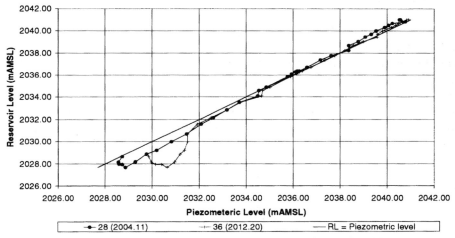

Figure 5 – Upstream Piezometer Response to Drawdown CH 290

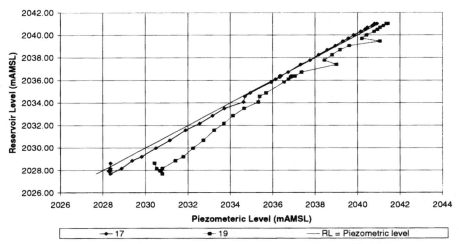

Figure 6 – Upstream Piezometer Response to Drawdown Ch 200

At the slower rate of 0.04m/day the pore pressure response of the four piezometers was similar with Bbar (Δu / $\Delta\sigma_v$) values around 0.95. This value of Bbar reflects efficient drainage with the pore pressure dropping at the same rate as the reservoir. When the drawdown rate accelerated, the response of the piezometers varied with Bbar values dropping to between 0.6 and 0.9. At these values of Bbar pore pressures will lag behind the reservoir level reduction and the factor of safety of the embankment against slope stability failure will reduce. There is no obvious reason for the varying values of Bbar solely as a result of the location of the individual piezometers. This variation is more likely to be a reflection of local variations in fill material and preferential drainage paths.

Two cases for drawdown were used in the original design analysis; emergency drawdown to elevation 2015mAMSL and operational drawdown to 2000mAMSL at drawdown rates of 0.42 and 0.11m/day respectively. The operational drawdown rate is comparable to that observed during drawdown in 1998/99. The r_u values at the end of the operational drawdown are given in the design report and varied at the piezometer locations from 0.2 to 0.4. To make a comparison between the 1998/99 drawdown and the design analysis the r_u values for the 1998/99 drawdown have been estimated assuming:

- Drawdown continues to elevation 2000mAMSL at a rate of 0.1m/day
- Bbar values remain unchanged below elevation 2028mAMSL
- No further dissipation of pore pressures occurs once the drawdown continues below the elevation of the piezometer tip

Using these assumptions the estimated r_u values range from 0.1 to 0.45. These correlate well with the design analysis and demonstrate that the embankment is behaving as predicted and will have an adequate factor of safety.

CONCLUSIONS

The paper has examined the post construction performance of Thika Dam with respect to settlement, seepage and pore water pressures.

The settlement data suggests that adequate allowance for post construction settlement was incorporated at design stage. The settlement indices determined from the settlement data were at the low end of the range published for UK dams.

The seepage flow at maximum retention level was about twice that estimated at design stage. The low seepage at design stage was probably a result of an under-estimate of the foundation bulk permeability from "point" values determined during the ground investigation. The efficiency of the cut-off may also have been reduced by the inclusion of minor imperfections. The need to design drains with liberal factors of safety was confirmed. The observed seepage was about 12% of the minimum required compensation flow and thus contributed to this requirement.

Piezometers in the upstream shoulder indicated that design assumptions of the pore water pressure response during drawdown were consistent with the observed response. This behaviour suggested that there was a satisfactory factor of safety during reservoir drawdown.

The Thika Dam embankment appears to be behaving satisfactorily.

ACKNOWLEDGEMENTS

The authors wish to thank Nairobi City Council for permission to publish this paper. Acknowledgement is also due to the staff of Howard Humphreys (East Africa) Ltd. who diligently and reliably collected, recorded and summarised instrument data over the first 5 years of operation of the dam. The late Dr. Geoff Sims also requires special mention as he carried out the 5 year Inspection of the dam and was always keen to have information published on the performance of Thika Dam.

REFERENCES

Attewill L J S, Gosden J D, Bruggemann D A and Euinton G C (1992), *The construction of a cut-off in a volcanic residual soil using jet grouting*, Water Resources and Reservoir Engineering, Proc. 7th Conference of BDS, Stirling, June, Thomas Telford Ltd.

Attewill L J S and Morey J (1994), *The use of jet grouting for the cut-off of Thika Dam, Kenya*, XIII ICSMFE, New Delhi, India.

Attewill L J S and Bruggemann D A (1997), *Some Experience with the use of Halloysitic soil at Thika Dam, Kenya*, Proc. 19th ICOLD, Q73, R15, Florence.

Casagrande A (1961) *Control of Seepage Through Foundations and Abutments of Dams*, 1st Rankine Lecture, Geotechnique, Vol. 11, No. 3, pp 161 – 182.

Cedergren H R (1967) *Seepage, Drainage, and Flow Nets*, 2nd. Edition, John Wiley and Sons, pp 200 – 201.

Charles J A (1986) *The significance of problems and remedial works at British earth dams,* Proc. of BNCOLD Conference, Edinburgh. Vol. 1 pp 123 – 141.

Geological Society (1997*), Tropical Residual Soils, Engineering Group, Working Party Revised Report,* Ed. Fookes P G.

Johnston T A, Millmore J P, Charles J A & Tedd P (1999) *An engineering guide to the safety of embankment dams in the United Kingdom,* BRE, Garston, UK

Terzaghi K (1958), *Design and Performance of the Sasumua Dam,* Paper No. 6252, Proc. ICE Vol. 9, pp 369 – 394.

Wesley L D (1973), *Cluster hypothesis and the shear strength of a tropical red clay,* Geotechnique, Vol. 23, No. 1, pp 109 – 113.

Masjed-e-Soleiman Dam instrumentation

PAUL J WILLIAMS, Halcrow Group Limited

SYNOPSIS. The 187m high Masjed-e-Soleiman clay core rockfill dam forms part of a 1,000MW hydropower scheme. The dam was constructed between 1996 and 2001 and impounding commenced in late 2000. The instrumentation installed at the dam was designed to meet international guidelines for the primary purposes of monitoring construction, impounding and long term performance. The instrumentation comprises earth pressure cells, extensiometers, piezometers, groundwater observation holes, survey monuments and seismic monitors. In the event a significant proportion of the buried instrumentation failed during construction. A study was undertaken to review the performance of the instrumentation, evaluate available monitoring data and develop the criteria for control of the impounding.

INTRODUCTION

Masjed-e-Soleiman is a clay core rock fill dam situated in a narrow gorge on the lower reaches of the Karun River in southwestern Iran. The Karun River rises in the Zagros Mountains in western Iran and flows southward to the Persian Gulf. A number of dams are constructed along the Karun River and more are planned as shown in Figure 1.

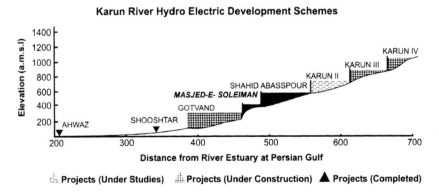

Figure 1: Karun River Cascade Development

The catchment area of Masjed-e-Soleiman reservoir is 27,550km^2. The reservoir has a total storage of 285 million m^3 and a live storage of 90 million m^3 between elevations 363m and 380m. Peak flood inflows for the 1,000 year and 100 year floods are 9,300 m^3/s and 6,800 m^3/s respectively.

The purpose of the dam is to provide river regulation and storage for hydropower generation. Clay core rockfill construction was favored because of the relative seismicity of the area and the availability of suitable clay and rockfill materials nearby.

The dam comprises a rockfill embankment with clay core and upstream and downstream filters. The upstream slope of the dam is at 1^V to 2^H and incorporates a rockfill cofferdam with upstream clay membrane at its toe. The overall downstream slope is at 1^V to 1.8^H and incorporates an access roadway for construction. The clay core of the dam has a minimum width at the crest of 10m increasing in width by 0.4m for every 1m below crest elevation. Each of the filters has a nominal width of 5m. A typical section through the dam is shown in Figure 2.

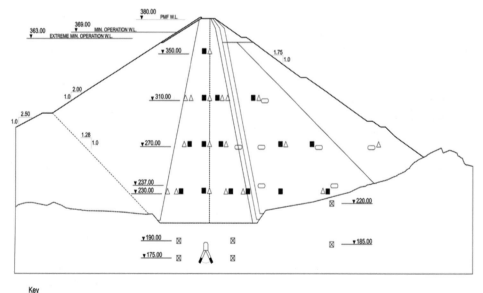

Figure 2: Typical Instrumentation Arrangement at Masjed-e-Soleiman

The clay core was placed at approximately 2% wet of optimum moisture content of 14.2% and compacted densities of 98.5% of maximum were maintained. The core material has a plasticity index of 19.9% and a permeability of between 10^{-8} and 10^{-9} m/s.

The dam foundation comprises alternating layers of permeable sandstone and impermeable claystone with a dip toward the upstream. In general the grout take for the cutoff curtain was low, with the only area of high grout take being on the left abutment between the upper and lower galleries. For this reason the arrangements of the groundwater observation holes in the galleries were amended to give clear indications of seepages in this area.

The site experiences temperature extremes ranging between $+55^{\circ}$C and 0°C. The high temperatures through July and August were particularly disruptive to construction and made it difficult to control the moisture content of the clay fill.

INSTRUMENTATION
The primary objectives of the instrumentation and monitoring systems can be summarized as follows:

- To confirm the design assumptions and predictions of performance at the construction phase
- To monitor performance of the embankment during the impounding of the reservoir
- To confirm safe operation through the life of the dam including the provision of early warning of the development of unsafe trends in behaviour
- To verify the safe aging of the structure

The instrumentation installed, at Masjed-e-Soleiman was evaluated against international guidelines for dam instrumentation including the ICOLD Bulletin No 60 and the US Army Corps of Engineers manual EM 1110-2-1908. In general it would be considered a well instrumented dam if compared to other rockfill dams of a similar size worldwide. Both guidelines promote the monitoring of ground water, pore pressure, movements, deformation and fill pressure whilst recognizing that every instrument system is unique and that a significant amount of engineering judgment must be applied.

The instrumentation installed in the Masjed-e-Soleiman embankment comprises the following equipment:

Foundation Peizometers	Standpipe Piezometers
Embankment Piezometers	Casagrande Piezometers
Earth Pressure Cells	Groundwater Observations Holes
Hydrostatic Settlement Gauges	Seepage measuring weir
Settlement Inclinometers	Earthquake Accelerometers
Surface survey monuments	

The instrumentation for the dam was intended to assist in the evaluation of the performance relating to the following areas:

- Seepage and leakage
- Deformation due to
    - Slope instability
    - Settlement due to internal erosion
    - Consolidation of fill
    - Consolidation of foundation strata
    - Secondary consolidation of fill and foundation
    - Volume change in clay
    - Changes in reservoir levels
- Seismic disturbance

INSTALLATION & INSTRUMENT FAILURES
During the construction of the embankment a significant number of the instruments were damaged or became inoperable. The reasons for the malfunctions included incorrect installation, damage by construction plant, use of incorrect equipment and faulty instruments. Table 1 sets out details of the instrumentation installed and their operational condition at impounding.

The instrumentation for the embankment was largely installed as the filling progressed. The foundation piezometers were installed in boreholes prior to starting the dam filling.

Hydraulic type peizometers were installed in the downstream shoulder together with hydrostatic type settlement gauges. These instruments could not be used to monitor the construction of the dam because the instrument houses could not be constructed until construction of the embankment was complete. This meant that potentially beneficial monitoring data could not be used to analyse the performance of the dam until after impounding.

There were three types of instrument, which showed significant rates of failure. These were the vibrating wire foundation piezometers, the earth pressure gauges and the settlement inclinometers, each with failure rates well outside of the range that would be expected from equipment malfunction, given careful installation.

Table 1 – Operational Instrumentation at the end of construction				
Type	**Location**	**Installed No.**	**Operational No.**	**Defective**
Peizometers – foundation - vibrating wire	Core	14	6	57%
	D/S fill	5	4	20%
Piezometers -dam – vibrating wire	Core	25	18	28%
Pore pressure –dam – hydraulic type	D/S fill	10	Unable to monitor	
Standpipe piezometers	Abutments / galleries	22	22	
Casagrande piezometers	D/S toe	3	3	
Groundwater observation holes	Abutments	15	15	
Earth pressure gauges	Core	39	30	23%
	D/S fill	9	8	11%
Hydrostatic settlement gauge	D/S fill	13	Unable to monitor	
Settlement inclinometers	Core	4 (519m)	4 (340m)	35%
	D/S fill	4 (288m)	4 (263m)	9%
Earthquake accelerometer	Crest/Core/ Gallery/face	1/2 1/1	Installed later	

Foundation piezometers

The vibrating wire foundation piezometers monitoring data is of importance at construction stage to evaluate the stability of the foundation under the loading imposed by the dam. This is particularly important when construction is rapid and pore pressures do not have time to dissipate. With the majority of the piezomters upstream of the grout curtain inoperable, the ability to evaluate pore pressure distribution across the grout curtain at impounding was compromised. It was not considered advisable to proceed to impounding without establishing a method of monitoring foundation pore pressures upstream and downstream of the grout curtain.

A series of vibrating wire piezometers were retrofitted from the foundation grouting gallery to replace those piezometers, which were inoperable. These piezometers were installed by drilling inclined upward holes from the gallery to position a piezometer tip close to the position of the inoperable piezometers. The installation of these piezometers went well and presented no problems. The proposed piezometers were installed before impounding and gave plausible readings.

Earth pressure gauges

The earth pressure gauges provided important information relating to the build-up of earth stresses as the fill progressed. This information was used to monitor embankment stability during the construction period when pore pressures were particularly high and effective stresses low. Fortunately, a sufficient number of the instruments remained operational to allow the stability of the dam to be assessed. The importance of the monitoring data from the earth pressure gauges reduces as the fill continues to settle and construction pore pressures dissipate thereby increasing the effective stress.

Although there are earth pressure gauges available that could have been installed in a borehole from the surface, it was considered that, these were difficult to install and because they were relatively small their effectiveness would be limited. It was decided that it was not cost effective nor technically beneficial to install additional earth pressure gauges.

Settlement inclinometers

The settlement inclinometers would normally provide useful monitoring data on the distribution of settlement within the body of the embankment. The settlement inclinometers consist of inclinometer tubes installed with magnetic ring plates fixed to the outside of the tube at regular intervals. The ring plates are arranged to ensure that they settle along with the fill thus compressing the tube system. A probe is then lowered down the tube to record the relative positions of the magnetic rings. The tubes can also be used as traditional inclinometers to monitor horizontal displacements in any direction. In the event all inclinometers below EL 290m became inoperable and therefore it was not possible to assess the consolidation of the fill. Although there were hydrostatic settlement gauges below this level in the downstream shoulder these instruments could not be read until after impounding when the permanent instrument houses were completed.

It would have been possible to drill boreholes and to install settlement inclinometers in a borehole but this is generally only done for shallow holes. To reinstate the inclinometers in the dam core below EL 290m would have necessitated drilling holes to a depth of 170m through the clay. Such drilling would be very expensive and particularly disruptive to the core itself.

Failures

The reasons for the failure of the foundation and core piezometers and earth pressure cells could not be established from an analysis of the records. Given that in general it was the instruments in the upstream shoulder and in the core which had failed it was considered likely that the reason for loss of readings was due to damage of the connections from the instruments to the monitoring points on the downstream face. The cables and tubing had been laid in sand filled trench across the filter and shoulders and at the transitions between core, filters and shoulder the cable had been 'snaked' in the trench to allow for drawing out due to differential settlement. It was concluded that the friction on the cabling or tubing as the fill continued was too great to allow any movement. This in turn led to high stresses in the cable or tubing and failure at the interfaces where differential settlement occurred.

The reason for the failure of the settlement inclinometers was easier to determine when photographic records were studied and a failed section uncovered. The inclinometer tubes had been joined by use of an outer sleeve as per manufacturers instructions but the installer had failed to leave a gap between the tube sections to allow for telescoping of the tubes under consolidation of the fill. Subsequent settlement caused buckling of the tubes at the joints, which meant that it was not possible to pass the probe down the tube.

EVALUATION OF EMBANKMENT

The instrumentation suite was designed with particular interest in monitoring pore pressures, stress and deformation in the clay core. This was to enable an assessment to be made of the transfer of total stress to effective stress as pore pressures dissipated allowing consolidation to occur in the form of vertical displacement.

Before impounding could proceed it was necessary to demonstrate that the embankment section was stable and that there would be no safety problems associated with the excess pore water pressures that were being experienced. On completion of the embankment construction it was found that pore water pressures in the core were particularly high and that even at the lower elevations there had been very little dissipation of pressure. This meant that with low effective stresses the shear strength of the clay core was also reduced. Figure 3 shows the recorded pore pressure ratio r_u values on the highest section of the dam at the end of construction in 2000, despite the fact that construction at the foundation level had commenced in 1997. It was estimated that it would take a further 1-2 years for the pore pressures to dissipate significantly.

For this reason it was necessary to evaluate whether the reservoir could be impounded immediately after completion of construction and if so at what rate the water level could be raised.

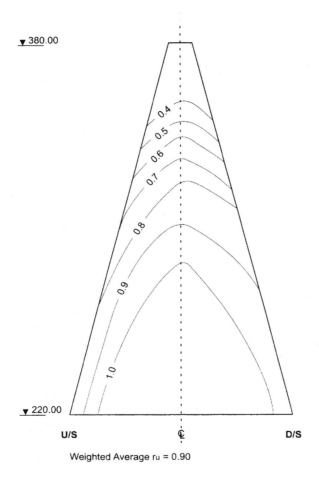

Weighted Average r$_u$ = 0.90

Figure 3: r$_u$ values for the dam core at completion of construction

Of particular concern was the stability of the upstream face of the embankment for the condition of embankment at full height and also for initial impoundment. The slope stability issue was exacerbated by a layer of alluvium under the cofferdam forming the upstream toe, which had a potentially lower angle of shearing resistance. This layer had a thickness of up to 8m and had not been stripped from the foundation when the cofferdam was constructed. The uncompacted toe layer when coupled with the high r_u values in the core meant that it was possible to generate a non-circular slip failure surface through the core and upstream toe foundation, which was barely above unity for the condition of unregulated impounding. A separate study showed that provided the impounding rate was controlled to allow pore pressure dissipation in the core then the short-term factor of safety for impounding could be maintained above 1.3.

The need to make a controlled impounding became more pronounced shortly after embankment construction was completed as the single operable diversion tunnel suffered major damage to the concrete lining when the other diversion tunnel was taken out of service for conversion to a bottom outlet. The situation that developed meant that unless the remaining diversion tunnel could be closed to allow repairs the tunnel would erode further and lead to collapse. It was against this scenario that a balanced impounding procedure was developed. The impounding procedure allowed for initial impounding to EL 303m whilst the bottom outlet was brought into service and then for the water level to be held at this level using the bottom outlet, with its inlet sill at EL 300m, for a period of 3 months to allow the embankment pore pressures to continue to dissipate before filling continued to spillway sill level at EL 350m at a target rate of 0.5m/day.

The early impoundment was complicated by the fact that the upstream shoulder had a clay upstream face to elevation EL 300m. Therefore water ingress into the upstream shoulder was limited to the seepage through the clay membrane until such time as it was overtopped at EL 300m. For this reason the rate of rise immediately above EL 300m was very gradual to allow slow filling of the shoulder and stabilization.

The operational earth pressure cells show a pronounced variation in earth pressure across the core, filters and downstream fill at a number of sections. A typical effective stress distribution has been presented in Figure 4 and represents a series of six working cells. Figure 4 compares the actual minimum effective stresses against the theoretical total stress, calculated as the overburden pressure, and a lower limit of 70% theoretical, which was set as the stable limit for impounding by the designers. This represents a section towards the highest part of the dam at EL 310m some 70m below crest level and is at the time of completion of the embankment filling. Figure 4 clearly shows that the minimum effective stress in the core was considerably less than that in the filters and the shoulders. It was apparent that with effective stresses in the core of only 60% of theoretical overburden stress the core was effectively hanging up on the shoulders which themselves were being crushed with stresses up to 130% of theoretical.

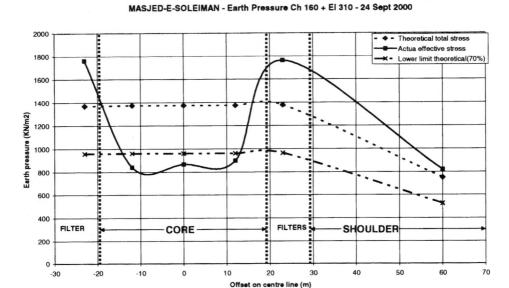

Figure 4: Distribution of effective stress across core and filters

The principal concern with the clay core was that on impounding the hydrostatic pressure of the water in the upstream shoulder would exceed the effective stress in the core material leading to potential fracture. For this reason it was decided to regulate the impounding to ensure that the hydrostatic pressures from the rising water level were not allowed to exceed 80% of the actual recorded minimum effective stress at that level in the core. To achieve this it was necessary to monitor increases in water level and the actual earth pressures as consolidation and pore water pressure dissipation in the core continued. Although it was recognized that the filters may have crushed in the zone against the core it was considered that they would continue to function as an effective filter in the event of leakage through the core.

There was a risk that if there was a significant flood event then it would not be possible to control the rate of impounding with the 330 m³/s capacity bottom outlet and that this would result in uncontrolled reservoir rising to gated spillway sill at EL 350m. With Shahid Abasspour dam upstream the risk of uncontrolled flooding was mitigated somewhat as it was possible to create nearly 400 million m³ of storage in the reservoir, under the operational rule curve to attenuate a flood. This meant that the risk of uncontrolled flooding through the winter/spring impounding was reduced to a 1 in 10 year event.

CONCLUSIONS
It was concluded that whilst the instrumentation at Masjed-e-Soleiman was consistent with current international practice the number of instrument failures were significantly higher than would be normally expected. In particular the failure of the vibrating wire piezometers, earth pressure gauges and settlement inclinometers make thorough analysis of the embankment difficult. This demonstrated the importance of adopting good installation procedures.

The analysis showed that the pore pressures within the core were just within acceptable limits but were dissipating far slower than had originally been envisaged. Because of the possible risk of arching and hydro-fracture of the clay core it was concluded that the impounding of the reservoir should only be made under strictly controlled conditions to prevent excess hydrostatic pressures in the upstream shoulder. The impounding was staged to allow dissipation of pore pressures in the core to ensure that at no stage would the hydrostatic pressure in the upstream shoulder exceed the minimum effective stress in the core.

There remained a slight risk that uncontrolled impounding would occur under a flood event but this was mitigated against by using the upstream reservoir to provide storage.

The instrumentation provided sufficient data to determine the behavior in terms of pore pressure and soil pressure but there was no effective measurement of settlement / consolidation of the fill due to the failure of the settlement inclinometers and the inability to use the hydrostatic settlement cells until after construction was complete. This was because the hydrostatic instruments require that the instrument houses are constructed at close the elevation of the instrument.

In the event the impounding went well and in accordance to the criteria set out. By early 2003 pore pressures were continuing to dissipate slowly and seepages remained negligible.

ACKNOWLEDGEMENTS
The author would like to acknowledge the valuable advice and assistance given by Iran Water and Power Resources Development Corporation (IWPC) site staff, Nippon Koei Ltd design staff and instrumentation monitoring staff with Daelim Industrial Co Ltd.

REFERENCES
US Army Corp of Engineers Engineering Manual EM1110-2-1908 (1995). *Instrumentation of Embankment Dams and Levees.* US Army Corp of Engineers

ICOLD Bulletin 60 (1988). *Dam Monitoring General Considerations.* Commission Internationale des Grands Barrages

Dunnicliff (1988). *Geotechnical instrumentation for monitoring field performance*

6. Incidents and rehabilitation case histories

Challenges and Limits - The Feasibility of Underwater Rehabilitation Work

C. HEITEFUSS, Ruhrverband (Ruhr River Association), Essen, Germany
H.-J. KNY, Ruhrverband (Ruhr River Association), Essen, Germany
U. MOSCHNER, Ruhrverband (Ruhr River Association), Essen, Germany

SYNOPSIS.   The bottom outlet facilities of many dams all over the world will have to undergo rehabilitation within the near future. Not only the early dams, built in the 19th or in the beginning of the 20th century are affected. Even large hydraulic structures, designed and built in the second half of the 20th century are faced with this problem. Due to various reasons a complete draw down of a reservoir for inspection and rehabilitation purposes has to be considered as not feasible in most cases. That requires manned or unmanned underwater inspection and rehabilitation techniques at the submerged structures of a dam. The paper describes the experiences gained during the underwater rehabilitation activities of the Ruhr River Association and how these experiences can be applied to other projects in Europe at water depths between 20 and 120 meters.

INTRODUCTION

The layout of many hydraulic structures all over the world does not make provisions for repair works of the bottom outlet facilities. This is not only the case for the dams of the very early design periods around the beginning of the 20th century. Even many owners of large hydraulic structures, designed and built in the second half of the 20th century have to face this problem.

At the early design periods the complete draw down of a reservoir for repair purposes was usually considered possible. Therefore the original design of that time did not make provisions for underwater inspection and repair work. Nevertheless even some modern dams have design deficiencies related to inspection and rehabilitation as well. The lack of support structures for emergency gates for instance turns out to be a major problem. Safe working conditions for the diving crews can not always been taken for granted.

Nowadays, due to possible restrictions for water supply, hydropower generation, irrigation, leisure activities and due to the risk of severe ecological problems in and around the water body of a reservoir a complete draw down for inspection and rehabilitation purposes has to be considered as impossible in many cases. Therefore rehabilitation works have to be done during full or partial operation respectively maximum or reduced reservoir levels which allow an unrestricted supply of water, depending on the various purposes of the reservoir.

Extensive underwater work has been done at the reservoirs of the Ruhr River Association (in German: Ruhrverband). These reservoirs provide bulk water for the industry and about 5 million people in the Ruhr area and cover a design and construction period from the beginning of the 20th century until 1966. During the life cycles of these reservoirs of up to one century quite a number of structural and operational deficiencies became evident, not to mention the regular ageing processes. Therefore during the last 10-15 years extensive rehabilitation measures were carried out at the reservoirs of the Ruhr River Association, mainly in order to refurbish the bottom outlet facilities and to adapt the existing hydraulic structures to new operational needs at reservoir levels which allowed an unrestricted water supply. In the following the experiences gained during the underwater rehabilitation activities of the Ruhr River Association are described.

The rehabilitation strategies of the bottom outlet structures of every dam were based upon the following ideas:

- to move the new intakes upstream, away from critical and narrow cross-sections at the upstream foot of the dam
- to use the new intakes as support structures for emergency gates
- to replace the old gates
- to install new intake gates respectively guard valves

Every project led to new experiences in underwater rehabilitation technology, which is described in the following.

Some experiences have been shared with other dam owners in Europe, responsible for reservoirs with water depths between 20 and 120 meters.

UNDERWATER REHABILITATION CASE HISTORIES

<u>The Moehne Dam</u>

Figure 1. Aerial View of the Moehne Reservoir

Introduction
The Moehne Dam (Figure 1) was built from 1908 - 1912 as a curved masonry dam with a height of 40 m, a length of 650 m and a maximum storage capacity of 134.5 million m³. The Moehne Dam was considered as one of the largest dams in Europe of that time. It became known world-wide, when during World War II the dam was severely damaged by an allied bomb attack. The dam was destroyed to a height of 23 m and over a length of 77 m. The following flood wave of about 110 million m³ of water and a height of 6 – 7 m killed more than 1.200 people and devastated the Moehne Valley.

The concept for the rehabilitation of the Moehne Dam
The rehabilitation of the bottom outlets of Moehne Dam can be considered as the first milestone in underwater rehabilitation of the dams of the Ruhr River Association. The work started in 1992 and was finished in 2002. It has been described in (Heitefuss & Kny 2002) and (Klein, Harder & Klahn 2003). Nevertheless two important aspects of underwater work are worth to be mentioned.

Underwater concrete work

After the installation of the new pipework one of the final steps of the construction of new support structures for the emergency gates is usually the underwater concreting. It took various test pourings, until the optimal recipe for the underwater concrete was found. What has been an enormous problem during the Moehne project, turned out to be much easier at the next underwater rehabilitation projects, since there has been a remarkable advance in underwater concrete technology during the last ten years.

Use of a diving platform

When the rehabilitation work started at the Moehne Dam in 1992, this was also the start for the use of a special diving platform (Figure 2) which has been developed by the diving contractor in co-operation with the Ruhr River Association.

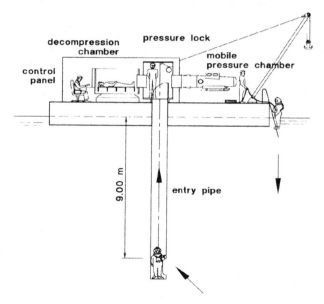

Figure 2. The diving platform with pressure chamber and entry pipe

This diving platform has proven to be a valuable tool for safe diving works at the construction sites of the Ruhr River Association for more than ten years. No diving accident worth mentioning occurred during this time.

The diving platform is equipped with two coupled decompression chambers, which are connected directly to a vertical entry pipe. This pipe reaches 9 m under the water surface and can be filled with compressed air. Thus, the diver can enter the pipe after his diving mission. He remains under pressure, disconnects the umbilical and can take off his diving mask or helmet. With an elevator he is lifted to a pressure lock, where he is undressed by an assistant. Then he enters the decompression chamber for the reduction of pressure. This equipment allows the controlled decompression under dry and warm conditions, which both improves the decomposition of nitrogen in the divers body and prevents, that the diver catches a cold. This improves the safety for the diving crew and increases the effectiveness, since health problems of the divers are getting reduced.

Another technique to improve the effectiveness of the diving activities is the use of oxygen breathing masks in the decompression chamber. If the decompression has reached 0.6 atmospheres and less, the diver breathes pure oxygen in order to accelerate the nitrogen decomposition. According to the new German Safety Codes for Diving, the additional use of oxygen breathing gas extends the so-called ground time for instance at water depths between 30 and 40 meters up to 50 or even 60 % in comparison to the use of a regular breathing gas mixture inside the decompression chamber. Thus, at a water depth of 33 m the diver has a ground time of 80 minutes and a decompression time of 65 min.

The platform is also equipped with a mobile pressure chamber, which can be connected quickly to the pressure lock. In case of a diving accident there are two options. The diver is either placed in the mobile pressure chamber and brought to a hospital via truck or airlift within 30 minutes or medical assistance can be brought in via the second pressure chamber, which in this case is used as a pressure lock.

The Verse Dam

The Verse Dam is a multi-purpose reservoir with an earthfill dam with a concrete cut off and a height of 52 m.. Based upon the experiences from Moehne Dam the intake structures of Verse Dam underwent rehabilitation from 1995 - 1997. This work has been described in (Heitefuss & Kny 2002 / 1997).

At this project it proved, that a preliminary hydrologic study can be a valuable tool for an economical rehabilitation measure.

Before work began, a minimum reservoir level had to be found, which could guarantee the safety of water supply for the adjacent cities and allow reasonable and economical diving and decompression times for the diving crews. The calculations resulted in a minimum water depth of 38 m during underwater work, compared to a regular depth of more than 50 m. Thus the ground time of each dive was doubled.

The cost of the entire rehabilitation project amounted to nearly 2 million €. The cost of supply of steel (RSt 37-2) for the pipe-work and shield was about 8 € per kilogram. The additional cost for the installation of the pipe-work was about 14 € per kilogram of steel.

Figure 3. Aerial View of Ennepe Dam

The Ennepe Dam

The Ennepe Dam (Figure 3) is a curved masonry dam with a length of 350 m and a height of 50 m. It was built during the 4 years earlier than the Moehne Dam, based upon the same design principles. The concept for the rehabilitation of the entire bottom outlet structure was applying the techniques, which had already proven their feasibility at the Moehne and Verse Dam. In fact there was an additional challenge for the engineers of the Ruhr River Association, because it was necessary to install new intakes for bulk water at different water levels (Heitefuss & Kny 2002 / 2001).

Another major difference to the rehabilitation projects at Moehne and Verse Dam was use of stainless steel. This required special conditions during production and installation of the components in the plant and on the construction site.

The underwater rehabilitation work at Ennepe Dam took 5 years. The estimated cost of the entire bottom outlet rehabilitation project was 4 Mio. EURO. The cost of supply for the stainless steel was between 11 and 14 € per kilogram. The additional cost for the installation of the pipework was between 20 and 30 € per kilogram of stainless steel.

UNDERWATER REHABILITATION PROJECTS

Mornos Dam, Greece

Introduction

The engineers of the Ruhr River Association have been involved in a feasibility study for the rehabilitation of the bottom outlet of Mornos Dam, Greece. With a storage capacity of 780 million m³, a height of 126 m and a crest length of about 800 m Mornos Dam is one of the largest earth dams in Europe and very important for the water supply of Athens. Due to the dimensions of the upstream reach there is no access to the bottom outlet gate with conventional diving techniques. In the following some options for the inspection and rehabilitation, as well as some ROV–techniques, methods of saturated diving and pipe freezing are described.

Description of the Bottom Outlet and the Transfer Device

The bottom outlet of Mornos Dam has a capacity of 400 m³/s. The horizontal reach of the intake tunnel has a length of more than 450 m and a diameter of almost 10 m. It narrows to a square profile of about 3 x 3 m. The bottom outlet of Mornos Dam is equipped with two roller gates.

Right after the commissioning of the Mornos Dam it became evident, that the water losses of the upstream bottom outlet gate (considered as emergency gate) amounted up to about 0.5 m³/s. A so-called transfer device was built in order to make the upstream gate of Mornos Dam revisable. The basic idea of the use of the transfer device (Figure 4) was to pull the upstream valve under balanced pressure into the top of this pressure chamber, in order to be able to operate with the top of the transfer device separately from the water pressure inside the penstock.

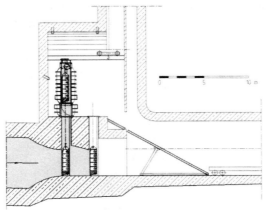

Figure 4. Use of transfer device inside the gate chamber of Mornos Dam

The use of the transfer device is a very complex and difficult operation. A failure has to be avoided. Therefore the transfer device has to go through a process of structural calculations and proofs.

The Use of Underwater Technology - Access with Divers from Upstream

At water depths of more than 60 m conventional diving reaches its technical and economical limits. The maximum water level of Mornos Reservoir requires the technique of saturated diving. The divers stay inside a system of chambers (so-called "habitat") for several weeks and remain on "ground depth". The divers are brought down to their working depth with a mobile diving bell. The supply of the diver out of the bell with breathing gas and heating water is maintained by means of an umbilical. According to the relevant safety regulations the length of the umbilical is limited to 30 m.
Therefore the use of this diving technique at Mornos Dam has to be ruled out due to the very long distance between intake and closure gate. For the same reason autonomous diving techniques is impossible too.

Access with an ROV (Remotely Operated Vehicle) from Upstream

During the last years the so-called ROV's became widely used in underwater technology. The water depth of 120 m is no problem for a modern ROV. The main difficulty for an ROV at Mornos Dam is the enormous horizontal distance. A very powerful ROV has to be used, which is capable of driving 450 m from the intake tower to the upstream gate.

Access to the Bottom Outlet with an ROV from Downstream

It was examined, if the use of an ROV from downstream offers technical and economical advantages. By using the two gates as a lock it is possible to enter the bottom outlet with an ROV (Figure 5). This procedure requires the penetration of the umbilical through the downstream steel plate, which is no major problem. Then the upstream gate can be raised that much, that the ROV can pass safely underneath.

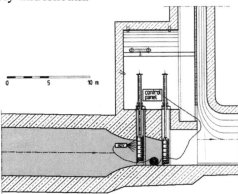

Figure 5. Inspection of Bottom Outlet with ROV from Downstream

Access with Divers from Downstream at full Storage Level
A technique is presented below, which allows diving activities at full reservoir level on the upstream side of the bottom outlet gate. This requires the use of the method of saturated diving. By installing a specific flange in the downstream gate a pressure chamber can be connected to the bottom outlet gate.

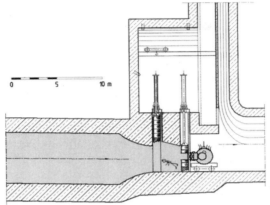

Figure 6. Access with Divers from Downstream

Access with Divers from Downstream at reduced Storage Level
A method of access with divers (Figure 6) at full storage level is described above. At a draw-down of the reservoir to about 80 - 90 m dives with partial saturation get applicable.
At these depths breathing gas mixtures like TRIMIX would be used, but there are still enormous decompression times from a dive with a ground time of 30 minutes:

water depth [m]	decompression time [min]
80 m	175 min
90 m	225 min
100 m	260 min

Use of non-conventional techniques
The basic requirement for the access to the bottom outlet under atmospheric conditions is an emergency gate, which has to be positioned in front of the intake trumpet upstream of the closure gate. Due to the conditions at Mornos Dam there is **no** way to install a conventional temporary emergency gate. Therefore in the following an approach is described, which possibly facilitates the access to the bottom outlet under atmospheric conditions with a non-conventional method.

Rehabilitation of the Bottom Outlet by means of Freezing
In the last 20 years the methods of freezing became widely used in the fields of geotechnical- and offshore-engineering. In the field of pipeline engineering the so-called pipe-freezing - the placing of ice plugs in pressure conduits for inspection and repair purposes - became a common method in the last years. By means of a so-called jacket the coolant is brought onto the pipe from outside. By freezing both the pipe wall and the medium inside an ice-plug is produced, which is connected tightly to the pipe wall.
The installation of an ice-plug of this size inside a bottom outlet has never been done before. Therefore material testing has to be conducted. The creep of ice-samples of this size has to be examined. For the production of an ice-plug of this size the problem of convection has to be solved

Rehabilitation of the Bottom Outlets of Early Embankment Dams
The experiences gained during rehabilitation of the Ruhr River Association's dams respectively their bottom outlets could be applied not only to a very large hydraulic structure like Mornos Dam. Also very early embankment dams with Puddle Clay Core could be refurbished, applying the same or adapted underwater rehabilitation techniques. A typical example is Lower Vartry Dam (Figure 7), near Roundwood, County Wicklow, Ireland.

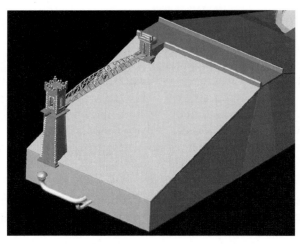

Figure 7. Schematic View of Upstream Side of Lower Vartry Dam

As a follow-up of the 12th Conference of the British Dam Society, *Reservoirs in a Changing World*, held at Trinity College, Dublin, September 2002 some aspects of the rehabilitation of the bottom outlet of Lower Vartry Dam have been discussed. It can be stated, that the basic rehabilitation techniques, which have already proven their feasibility at the Moehne, Verse and Ennepe Dam can be applied at Lower Vartry Dam as well, which might be (in brief):

- removal of sediment in the inner culvert and replacing it with concrete
- installation of guard valves inside the outer culvert behind stop-wall
- installation of new pipework and regulating valves in outer culvert
- use of the existing ball plug as a temporary emergency gate
- use of an additional valve at the intake as permanent additional guard valve respectively emergency gate.

Experiences of the Ruhr River Association show, that this work can be done at full reservoir level under full operation of Vartry Waterworks. Techniques like pipe-freezing to stop the flow during construction can not be used at Lower Vartry , since the pipes are made of cast iron, which has a tendency to fracture at extremely cold temperatures.

CONCLUSION

A number of successful underwater rehabilitation projects carried out by the Ruhr River Association has proven, that down to water depths of 50 – 60 m the rehabilitation techniques have become state of the art. Nevertheless, a number of fatal accidents in professional diving during the last years indicate, that safe working conditions should not be taken for granted. The dam owners have to insist and enforce, that the diving contractors provide the best possible safety features and working conditions. Otherwise fatal accidents are almost inescapable. Well equipped diving platforms with decompression chambers should be a must on every underwater rehabilitation site. It has also be stated, that the inspection and rehabilitation of very large hydraulic structures with water depths of more than 60 – 80 m and extreme dimensions especially at the upstream intake are still a challenge. The basic layout of many of these structures turns out to be rather disadvantageous for inspection and rehabilitation. Apparently minor design deficiencies can prove as extremely costly with regard to rehabilitation. The design of new structures should focus on this problem much more.

Some of the sophisticated inspection and rehabilitation techniques described in this paper (like ROV and saturated diving) can be considered as state of the art, but in combination with unusual features of the hydraulic structures (like extreme dimensions) there is still a lot of practical knowledge to be made. Techniques like pipe-freezing seem to be not ready for the practical use in large hydraulic structures yet.

Therefore the international exchange of experiences in the field of inspection and rehabilitation of large hydraulic structures is vital for the future of our water infrastructure.

ACKNOWLEDGEMENTS

The authors would like to acknowledge the generous permission of Dublin City Council to use some material on Lower Vartry Dam for this paper.

REFERENCES

Fleming, E. (2002). *Some Aspects of early Irish dam construction.* In P. Tedd (ed.), Reservoirs in a changing world, Proc. of the 12th conference of the British Dam Society held at Trinity College, Dublin, Ireland, 4–8 September 2002, Thomas Telford, London, 3-14

Klein, P., Harder, L. & J. Klahn (2003). *Möhnetalsperre - Sanierung der Grundablässe, der Schiebertürme und des luftseitigen Mauerwerks, Moehne Dam – Rehabilitation of the bottom outlets, the gate towers and the downstream masonry face.* Journal "Die Wasserwirtschaft" 93 (2003) 10, Vieweg & Sohn, Wiesbaden, Germany

Mantwill, H. & F. Campen (1993). *Rehabilitation of the bottom outlet works of the Moehne Dam.* Ruhrwassermenge 1994 (Annual Report on Water Resources Management 1994). Ruhrverband (Ruhr River Association) Essen, Germany

Heitefuss, C. & Kny, H.-J. (1997). *Rehabilitation of the intake structures at the Verse Dam, Germany.* In E. Broch, D.K: Lysne, N. Flatabo, E. Helland-Hansen (eds), Hydropower '97, Proc. 3rd intern. conf. on hydropower, Trondheim, Norway, 30 June–2 July 1997. Rotterdam: Balkema: 405-411

Heitefuss, C. & Kny, H.-J. (2001). *Rehabilitation of the bottom outlets of Ennepe Dam / Germany with stainless steel pipes.* In G. Midttomme et al (eds.), Dams in a European Context, Proc. of the ICOLD eur. symp., Geiranger, Norway, 25–27 June 2001. Lisse: Balkema: 429-434

Heitefuss, C. & Karopoulos, C. (2001). *Strategies for the Rehabilitation of the Bottom Outlet of Mornos Dam / Greece.* In G. Midttomme et al (eds.), Dams in a European Context, Proc. of the ICOLD eur. symp., Geiranger, Norway, 25–27 June 2001. Lisse: Balkema: 419-428

Heitefuss, C. & Kny, H.-J. (2002). *Underwater work as a means for the rehabilitation of large hydraulic structures under full operation and unrestricted water supply.* In P. Tedd (ed.), Reservoirs in a changing world, Proc. of the 12th conference of the British Dam Society held at Trinity College, Dublin, Ireland, 4–8 September 2002, Thomas Telford, London, 167-178

Ericht and Dalwhinnie Dam refurbishment and protection works

K J Dempster, Scottish and Southern Energy plc, UK
M Gaskin, Scottish and Southern Energy plc, UK
R M Doake, Faber Maunsell Limited, UK
D Hay-Smith, JBA Consulting (formerly Faber Maunsell Limited), UK

SYNOPSIS.   Following the statutory inspection of Loch Ericht reservoir both Ericht and Dalwhinnie Dams have been recategorised A from category B and assessed in relation to their capacity to safely pass a PMF event combined with wave surcharge allowances.

The paper describes the investigation, identification of the requirement for protection and subsequent design of works to primarily prevent wave surcharge levels overtopping the existing crest levels of both dams. Further refurbishment and protection works were also identified in relation to concerns over the ability of the spillway\corewall interface to resist erosion, poor spillway basin configuration and the potential vulnerability of the scour penstock during spill conditions at Ericht dam.

INTRODUCTION
The investigation and subsequent works carried out at Loch Ericht reservoir were required following the 10 yearly statutory inspection under the Reservoirs Act 1975 (1) which was carried out in June 2000 by Dr A K Hughes. The reservoir was recategorised A (general/minimum) from its previous category of B under the Floods and Reservoir Safety Guide (2). The various structures associated have therefore been assessed in relation to their capacity to safely pass a Probable Maximum Flood (PMF). Concerns were also raised over the vulnerability of the scour penstock and general spillway basin configuration.

DESCRIPTION OF RESERVOIR AND DAMS
Loch Ericht reservoir is situated approximately 75km northwest of Perth and was completed in stages over the period 1928 to 1954. The reservoir is one of the main storage reservoirs within Scottish and Southern Energy plc's (SSE) Tummel valley cascade hydro scheme system and provides long term seasonal storage from a catchment extending to 135.22km^2.

The reservoir is formed by the construction of Ericht and Dalwhinnie dams. Ericht dam at the southwest of the Loch comprises of sections of concrete gravity; concrete corewall with downstream grass covered embankment as support and is approximately 340m long and 14.3m maximum height above ground level. There is also a homogenous earth embankment section with grass covered upstream and downstream faces, approximately 65m long and 2.1m maximum height above ground level. Dalwhinnie dam at the northeast end is an embankment dam with a central concrete corewall supported by both upstream and downstream embankments, the upstream face is protected by concrete slabs and the downstream face is grass covered, approximately 350m long and 4.5m maximum height above ground level. The volume stored within the reservoir is 230 million m^3 with a surface area of 23.27km^2 at spillway level of 359.359mOD and water length of 24.4km. General sections of both dams are shown in Figures 1 and 2 respectively.

Scour from the reservoir is provided via a penstock that was added in 1957 as an extension to the original culvert through the concrete dam section. The penstock is a 2.13m diameter steel plate section extending 15.7m from the toe of the dam within the spillway basin. An anchor block with a 1.83m diameter disperser valve is located at the end of the penstock.

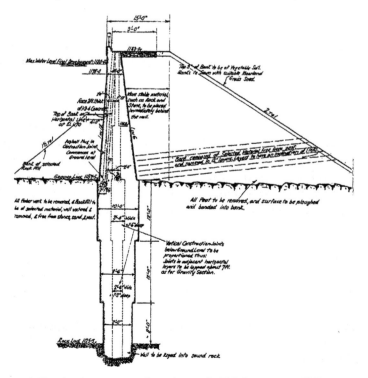

Figure 1 Typical cross section through Ericht corewall Dam

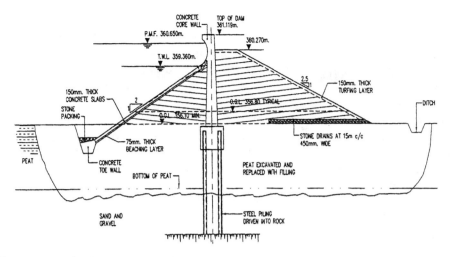

Figure 2 Typical cross section of Dalwhinnie Dam

RESERVOIR RECATEGORISATION

As a key element of the initial review the potential consequences, in particular the incremental consequences between PMF and dam breach were considered.

Ericht Dam failure

The inundation maps prepared and subsequent consequence study fully supported the recategorisation of the dam on the basis that should Ericht Dam breach other cascade failures would be likely in the Tummel valley reservoir system and hence a significant impact on communities beyond the next reservoir in cascade. The mapping also showed the cumulative effect on a number of isolated properties along the shoreline of Loch Rannoch.

Dalwhinnie Dam failure

Under normal conditions and flood events Dalwhinnie Dam prevents flow from the Loch Ericht catchment from passing into the River Truim. Should the dam breach the reservoir inundation mapping and subsequent consequence study clearly demonstrated that there would be an unacceptable level of inundation and significant impact on Dalwhinnie and the downstream communities. Application of Category A is therefore also appropriate for Dalwhinnie Dam.

INVESTIGATION PHASE

As part of the statutory inspection SSE led and implemented a detailed investigation to allow assessment of the impact of recategorisation and to prepare options for subsequent detailed design and implementation.

PMF and wave surcharge assessment

SSE completed flood studies for both PMF and 1 in 10,000 year return period events under worst case conditions of the syphons unprimed, due to some longstanding doubt over their operation and a snowmelt rate of 70mm/day. The following key flood results were obtained.

> 10,000 year (FSR) (3) 360.065mOD
> 10,000 year (FEH) (4) 360.348mOD
> PMF (FSR)   360.575mOD

The wave surcharge levels using normal approaches were estimated for each of the component dams with a straight line fetch of 3.7km adopted for Dalwhinnie Dam rather than the bent fetch of 24.4km. The identified deficiencies are summarised in table 1.

Table 1 Wave surcharge assessment

Element	Wave surcharge	PMF + wave	Crest	Deficiency
Ericht gravity	0.64m	361.215m	360.58m	N/A
Ericht corewall	1.02m	361.595m	360.58m	1.015m
Ericht saddle	1.73m	362.305m	361.19m	1.115m
Dalwhinnie	2.93m	363.505m	361.19m	2.315m

As a result of the above analysis various parts of the dams were considered to be vulnerable and would be effected under extreme flows. Such sections required to be protected or modified in order that wave overtopping would not erode embankment sections, which if allowed too could ultimately lead to a breach of one or more of the dam sections.

Survey and Site Investigation

A full topographic survey was completed at both dams and the surrounding area in order that key dimensions and physical layouts could be confirmed. Site investigation works followed to confirm ground conditions and to provide information for the subsequent design of remedial works. Investigations comprised ten cable percussive boreholes, seven trial pits and 18 Macintosh probe penetration tests at Dalwhinnie Dam with two boreholes and seven trail pits at Ericht Dam. Disturbed samples were taken for subsequent grading analysis, seven falling head permeability tests were carried out and insitu standard penetration tests were made in granular material to assess the relative density. The sulphate contents and pH values of ground water and soil samples were determined. Piezometer standpipes were installed in four boreholes at Dalwhinnie Dam with pressure transducers attached to dataloggers and the water levels monitored and related to reservoir level.

Hydraulic model

In order to examine concerns raised over the vulnerability of the scour penstock and poor spillway basin configuration a 1 in 50 scale physical model of the spillway, scour penstock anchor block and adjacent river channel was constructed and tested by ABPMer. The model was built to provide an understanding of the flow mechanisms existing on the downstream side of the dam and in particular examine hydrodynamic loading and scour on the valve structure and the corewall embankment where it intersects with the spillway section of the dam.

The model confirmed that whilst the existing velocities and differential head across the penstock were not significant at $2ms^{-1}$ and 0.4m respectively the penstock would be submerged and the protruding body of the disperser valve may be vulnerable, in particular during flow build-up. Winter operation is to empty the penstock to avoid freezing, but the penstock was not designed to be submerged under this condition. With the high replacement cost of the valve if damaged by debris, it was considered prudent to encapsulate the penstock and provide a protection wall.

The water levels within the spillway basin were found to be at a level were erosion of the corewall embankment was possible, especially with an eddy between the penstock and the embankment toe. The optimum configuration and top wall level for a spillway basin training wall was developed using the model.

Localised infilling of the unlined spillway floor were also modeled to improve conditions during routine operation and avoid problems with water ponding around the penstock.

When comparing each of the configurations tested, the addition of a slab, penstock protection and a baffle wall did not significantly affect the hydrodynamic environment. The exception was an increase in eddy speed adjacent to the scour valve, however the scheme does provide substantial protection to the valve and penstock against impact of debris in flow from the dominant direction. The addition of the corewall embankment toe protection progressively reduced the strength of the eddy but with a corresponding detrimental effect on water levels and flow speeds in other areas.

Option Study
Following review of the various elements of the investigation measures were considered to prevent the wave overtopping. Comprising of reducing the reservoir operating level to create further freeboard; additional spillway capacity to reduce the flood lift; providing wavewalls to prevent overtopping; providing downstream protection and to create a rougher upstream face to absorb wave energy thus limit run-up.

Reducing the reservoir operating level would place restrictions on the generation output from Rannoch Power Station and would require large elements of the diverted catchment to be turned out during extended periods to maintain freeboard levels.

Limited potential exists for economically adding further spillway provision at Ericht and Dalwhinnie dams due to the nature of the embankment and corewall sections and the excessive overtopping levels that required to be mitigated. The main Inverness to Perth railway line traversing 100m downstream of the dam compounds this at Dalwhinnie.

Wave overtopping prevention by the addition of wavewalls is well proven and could be combined with additional upstream face rip-rap and or slope reprofiling to absorb energy and reduce wave heights. An optimum balance between upstream face rip-rap protection, wave wall and downstream erosion protection was considered the best solution at this stage.

The penstock protection and spillway basin improvement works required a compromise between the construction costs of implementing them and minimisation of the hydraulic forces and scour velocities.

At this stage SSE prepared an option study report to summarise the findings and to provide the basis for detailed design. A subsequent contract was awarded to Faber Maunsell Limited to carry out the detailed design.

DESIGN PHASE
PMF Reassessment
Following the interim guidance for owners and panel engineers issued by DEFRA (5) the PMF was reassessed by bench marking against the FEH 10,000 year rainfall depth for the critical storm duration. The all year PMP was 241 mm against an FEH 10,000 year depth of 285.65mm for an 18.5 hour storm. The PMF hydrograph was generated assuming the modified PMP storm depth (equal to the FEH 10,000 year rainfall) and routed through the reservoir. The view was also taken that the syphons would prime under

such conditions and full account taken of this. This resulted in an inflow of $1684m^3s^{-1}$, outflow $459m^3s^{-1}$ and maximum water level 360.65mOD. An increase of $322m^3s^{-1}$ for inflow, $43m^3s^{-1}$ outflow and 150mm above an equivalent FSR estimate. Also 75mm over SSE's previous assessments which were considered to be conservative by assuming unprimed syphons.

Wave surcharge Reassessment
Wind-wave generation in most reservoirs is governed by fetch limited conditions for wave generation and deepwater conditions for wave propagation. However, these conditions were considered not to prevail for waves approaching Dalwhinnie. A detailed reassessment of wave conditions during the mean annual and the 1 in 200 year wind-wave event was carried out and is reported upon separately (6). The mean annual significant wave height is estimated at 2.12m for the PMF level of 360.65mOD. The 1 in 200 year significant wave height is estimated at 1.47m for the top water level of 359.37m. A significant reduction over previous estimates.

Fetch limited and deep water conditions apply at Ericht Dam, and the wave conditions approaching both the corewall and embankment dam sections was reassessed using the standard Donelon/JONSWAP method, as recommended in Floods and Reservoir Safety (2) and a bent fetch slightly longer than previously adopted. The mean annual significant wave height is estimated at 1.14m for the PMF level. The 1 in 200 year significant wave height is estimated at 1.55m.

The above estimated wave conditions for both Dalwhinnie and Ericht Dams were used together with the maximum flood levels to calculate wave overtopping discharge for freeboard assessment and wave loading for the structural design of wave walls. A methodology for deriving impact loading, occurring when waves break directly on the structure, was developed to provide an improved prediction of impact forces due to concerns over damage and instances of failures of wave walls, reported separately (6).

Value Engineering
A value engineering meeting was held to discuss preliminary design options for the works required. Formal value engineering techniques were used to evaluate options for overtopping protection at Dalwhinnie Dam, while the remaining items were discussed more informally.

The basic options considered for Dalwhinnie Dam were A) placement of open stone asphalt layer on upstream face with additional wave wall; B) placement of rip-rap on upstream face at existing 1:2 slope with additional wave wall and C) placement of rip-rap on upstream face at 1:4 slope with additional wave wall. All options assumed crest protection would be installed. Permutations included infilling the maximum depth section in the foreshore to limit the incident depth limited waves to the average depth condition, installation of downstream protection, use of grouted rip-rap to reduce the stone size, and inclusion of a tandem rock breakwater upstream to reduce incident wave height.

A value tree with importance weightings assigned to each criteria and the options were evaluated in more detail with the aim of identifying the best value alternative to be carried forward to detailed design. A decision matrix was developed from the weighted value criteria identified during the structuring of project objectives. The results of the decision matrix are shown in Table 2.

Table 2 Dalwhinnie Dam decision matrix results

Option	Description	Total rating
A iii)	Open stone asphalt layer	8.6
B iii)	Rip-rap at 1:2	6.9
BG iii)	Bituminous grouted rip-rap at 1:2	6.9
C iv)	Rip-rap 1:4	7.4
C v)+	Rip-rap at 1:4, tandem breakwater	6.5

The matrix analysis showed clearly in favour of option Aiii), placement of open stone asphalt on the upstream face, with wave wall, crest and downstream protection. This was partially due to the significant cost savings of this option; estimated to be approximately £200k cheaper than the next cheapest option considered.

Design Solutions
At Dalwhinnie Dam the upstream face will be overlaid with a 250mm thick layer of open stone asphalt and the low area in front of the dam infilled to the general level of 356.8mOD at the mitres of the dam. The wave wall will be raised by precast concrete unit's approximately 2m height, supported by an insitu concrete beam formed at the base of the existing wave wall on the upstream side, and anchored to the upper part. The crest and downstream face will both be armoured with concrete reinforced grass, to increase the tolerance to wave overtopping discharge. The works are generally shown in Figure 3. Mass concrete corewall extensions are also required at either abutment to prevent floodwater bypassing the dam and eroding the downstream embankment toe.

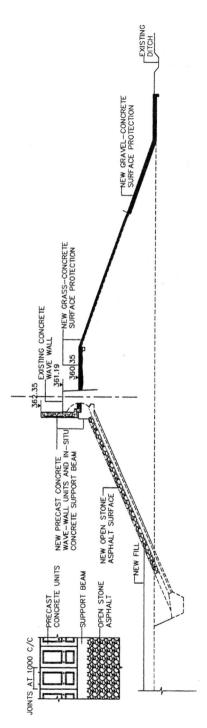

Figure 3 Dalwhinnie Dam

The corewall section of Ericht Dam requires a wave wall 1m in height to be added to limit overtopping to an amount acceptable for an unprotected downstream face. A reinforced concrete wall anchored onto the corewall section will provide this. Rip-rap is to be placed on the upstream face of the embankment dam at a slope of 1V: 2H to reduce wave run-up and overtopping discharge. A low berm will be formed above the crest level, negating the need for a wave wall. The crest of the embankment will be reinforced with grass-concrete blocks, to increase the tolerance to wave overtopping. Both sections are indicated in Figure 4.

In order to provide protection to the exposed penstock concrete encapsulation beyond the toe of the dam will be carried out and a baffle wall added to protect the protruding disperser valve. Permanent access to the interior of the penstock will be provided by 1m diameter flanged branch pipe. Improvement of the spillway training and invert protection will consist of placement of a reinforced concrete slab on the invert of the spillway channel to a maximum level of 347mOD, draining towards the river channel downstream. A reinforced concrete training wall along the interface with the corewall embankment be constructed to a nominal height appropriate for frequent spill events, with the remaining slope to be protected with grass-concrete blocks to prevent erosion during extreme events. The penstock and spillway works are shown in Figure 5.

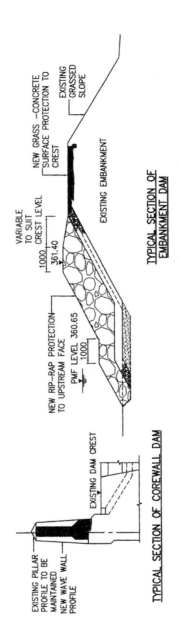

Figure 4 Ericht corewall and embankment dam works

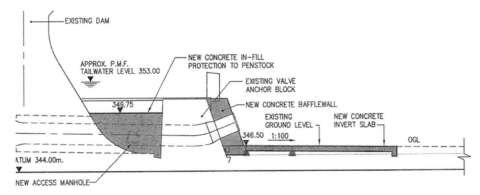

LONG SECTION THROUGH SCOUR OUTLET

Figure 5 Ericht penstock and spillway works

IMPLEMENTATION

Consents

An application to implement the works was made under the Electricity Act 1989 (7) in December 2002. This Section 36 consent remains outstanding one year on for what should have been a minor consent application. No EIA was required and consultation processes were carried out with each local authority, Scottish Natural Heritage and local estates.

Contract Strategy

Tender documents were based on the NEC Engineering and Construction Contract (8) with an activity schedule, all for implementation of the works during 2003 with a reservoir draw down over 18 weeks. However due to consent delays the works have been deferred to 2004 with the subsequent increase in costs. Estimated costs are £700k and £300k at Dalwhinnie and at Ericht respectively.

Water management issues lead to the adoption of sectional completion on the spillway and scour penstock protection works in advance of the main works to allow compensation water to be released downstream of Ericht in the event of plant failure at Rannoch Power Station. This will also provide a further control on the reservoir level should it be required.

Valve and penstock Refurbishment

The original plan was to remove in advance and refurbish the disperser valve to coincide with the sectional completion of the penstock civil works. Shot blasting and repainting of the internal surfaces of the penstock and the addition of an access manhole was included within the civils scope to avoid interface issues during concrete works. Due to the consent delay SSE decided to mitigate this and carry out all of the penstock mechanical works in advance and awarded the works to Isleburn MacKay & MacLeod. The works were completed in December 2003 at a cost of £80,000.

CONCLUSIONS

Both Ericht and Dalwhinnie Dams have been recategorised A, which was fully supported on the basis of inundation mapping and consequence studies. Subsequent investigation demonstrated that certain elements of the structures would be vulnerable under PMF conditions and required to be modified.

PMF re-estimation increased inflow by 24%, outflow by 10% and the resultant flood lift indicated by 14% following the application of the FEH 10,000 year rainfall depth as an estimate of PMP. In this situation the difference in level adopted was relatively small in relation to the overall surcharges being considered, and the economic implications were generally acceptable. It may even provide some degree of insurance against subsequent changes to future methodologies.

Wave surcharge reassessment concluded that depth limited prediction methods reduced the significant wave heights compared to those estimated using the standard wave run-up method. A methodology for estimation of wave impact forces on the wavewall extensions at Ericht Dam has been established.

A value engineering exercise established an open stone asphalt system combined with a wave wall extension at Dalwhinnie Dam as the optimum solution, previously unconsidered in the investigation stage.

Delays to the consent process were partially mitigated by carrying out the penstock mechanical works in advance. Future reservoir projects will be considered closely and where appropriate not be subject to the section 36 consent processes.

REFERENCES
(1)   Reservoirs Act, 1975, HMSO, London 1975

(2)   The Institution of Civil Engineers, (1996), Floods and Reservoir Safety Third Edition, Thomas Telford, London.

(3)   Natural Environmental Research Council, (1975) Flood Studies Report, Natural Environmental Research Council, London.

(4)   Flood Estimation Handbook, 1999, Institute of Hydrology, Wallingford

(5)   DEFRA, interim advice available on website

(6)   Wave assessment on Loch Ericht, Long Term Benefits & Performance of Dams, 2004, Thomas Telford, London

(7)   New Engineering Contract : the engineering and construction contract, 1995, Thomas Telford, London

(8)   Electricity Act, 1989, HMSO, London.

Wave Assessment on Loch Ericht

DEBBIE HAY-SMITH, JBA Consulting, previously FaberMaunsell
RICHARD DOAKE, FaberMaunsell
WILLIAM ALLSOP, H R Wallingford
KIRSTY McCONNELL, H R Wallingford

SYNOPSIS.   This paper uses the example of two dams to illustrate generic problems and solutions to the analysis of waves and wave forces on wave walls.

Dalwhinnie Dam and Ericht Dam impound Loch Ericht, which straddles Perthshire and Highland Regions in Scotland. Inadequate freeboard at both dams means that remedial works are required to increase the wave overtopping protection. In both cases, standard design methodologies had to be extended to achieve a credible design basis for the works.

INTRODUCTION

Loch Ericht is a natural loch drained by the River Ericht, flowing in a southerly direction into Loch Rannoch. The loch level was first raised in 1930/31 by the construction of Ericht Dam at the south western end, and further in 1937 by the raising of Ericht Dam and construction of Dalwhinnie Dam at the north eastern end. The reservoir is now over 24 km long. The reservoir is owned and operated by Scottish and Southern Energy plc, and supplies the Rannoch hydo power station as part of the Tummel Hydro Electric scheme.

Following a statutory inspection in June 2000 under the Reservoirs Act 1975 (1), Loch Ericht was recategorised from Category B to Category A. As such, both Ericht Dam and Dalwhinnie Dam were subject to more severe design flood standards, the result of which was that freeboard was inadequate for both dams. Concerns were also raised over the arrangement of the scour penstock and general spillway basin configuration at Ericht Dam.

FaberMaunsell (FM) was retained by Scottish & Southern Energy plc (SSE) for the detailed design of the works.

Long-term benefits and performance of dams, Thomas Telford, London, 2004, 473–487

The works comprised wave overtopping protection and corewall extension at Dalwhinnie Dam, wave overtopping protection at Ericht Corewall Dam and Ericht Embankment Dam, and penstock protection, spillway channel training and invert protection at Ericht Dam.

This paper focuses on the assessment of wave conditions and wave forces for the design of wave overtopping protection works at Dalwhinnie Dam, and at Ericht Corewall Dam. A full description of the project is reported separately (2).

WAVE ASSESSMENT AT DALWHINNIE DAM

General
The wave surcharge on a dam is a function of the wave height and other wave characteristics. Initial estimates of the wave characteristics at Dalwhinnie Dam, and the corresponding remedial works required, were based on standard methods for waves developed in deep water conditions. It was appreciated that these methods were not necessarily fully appropriate for the conditions at Dalwhinnie Dam because of the extremely long, narrow nature of the reservoir, and shallowness in the approach region to the dam. A full review of the wave assessment methodologies was therefore carried out.

Wave prediction in reservoirs
Most wave prediction methods are based on measurements carried out in oceanic and coastal waters, with fetch lengths and fetch widths very different from those found in most UK reservoirs. The Saville/SMB method was the standard method in UK prior to the production of the 3rd Edition of Floods and Reservoir Safety Guide (3). A full review of available methods such as Saville/SMB, JONSEY and Donelan/JONSWAP can be found in HR Wallingford Report EX1527 (4).

Following concern that the Saville/SMB method did not provide good predictions of waves on long, narrow reservoirs, measurements of wind and waves were made notably at Megget reservoir and Loch Glascarnoch (5). It was concluded that while none of the methods gave particularly good agreement with measured wave heights for all wind speeds and directions, Donelan/JONSWAP gave fairly good agreement for a wide range of wind directions, and any errors in predicted wave heights were likely to be conservative. This simplified Donelan/JONSWAP method was subsequently recommended in the Floods and Reservoir Safety Guide, 3rd edition (3).

For the Loch Ericht study, it was felt that further investigation of the appropriate wave prediction method would yield little without site specific data, and that the simplified Donelan/JONSWAP method should be retained.

Duration limited wave generation

Wind-wave generation in most reservoirs is governed by fetch-limited conditions for wave generation. A duration factor is applied to the wind speed in the method described above to take account of the fact that usually fetch lengths in inland waters are small and waves fully develop within 15 minutes; waves are thus "fetch-limited". However, Loch Ericht is unusually long, with a fetch approaching Dalwhinnie Dam of 24.4 km, shown in . Examination of the wave characteristics of the JONSWAP spectrum reveal that the minimum wind duration required to develop waves of this size is greater than 2.5 hours (refer to Figure 9 of BS 6349 Part 1 (6)). Waves on Loch Ericht are therefore "duration-limited".

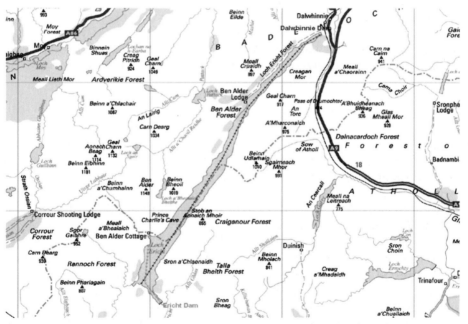

Figure 2: Loch Ericht showing fetch to Dalwhinnie and Ericht Dams

Various duration factors are given in CIRIA 83 Rock Manual (6), and these are applied to the mean annual maximum hourly wind speed to estimate mean annual maximum wind speeds for increased durations. For a duration of 2.5 hours, the appropriate factor is 0.97. Calculation of the design mean annual maximum wind speed and resulting significant wave height is shown in Table 1.

Table 1: Calculation of Dalwhinnie Dam significant wave height – deep water

fetch (m)	24400
fetch direction °N	239
50 year max hourly wind speed U_{50} (m/s)	23.50
return period adjustment f_T	0.79
altitude adjustment f_A	1.36
over water adjustment f_W	1.31
duration adjustment f_D (2.5 hour duration)	0.97
direction adjustment f_N	1.00
mean annual max wind U (m/s)	32.06
deep water significant wave height H_S (m)	2.84

Wave conditions at Dalwhinnie Dam

At Dalwhinnie Dam, shallow water depths extend approximately 1 km into the reservoir from the dam.

The onset of shallow water processes depends on the water depth (d) in relation to the deep water wavelength. The deep water wavelength was estimated assuming small amplitude wave theory:

$$L = \frac{gT^2}{2\pi} \qquad \text{giving } L_{op} = 54.33 \text{ m}$$

For deep water wave conditions,   d/L > 0.5
For shallow water wave conditions,   d/L < 0.05

The ground level at the toe of the dam is approximately 356.8 mAOD. The average depth of water at the toe for a Probable Maximum Flood (PMF) level of 360.65 mAOD is therefore 3.85 m. The lowest part of the toe towards the north end of the dam is 356.113 mAOD, giving a maximum depth of water of 4.535 m.

In this case, d/L = 0.08, in deepest water at the toe. Deep water wave conditions, where wave speed is determined solely by wavelength, therefore do not apply. Asymptotic shallow water wave conditions where wave celerity is determined solely by water depth are also not fulfilled. The speed of waves approaching Dalwhinnie Dam is therefore determined by both wavelength and water depth.

This result implies that shallow water processes will affect the waves approaching the dam.

Estimate of shallow water wave characteristics
Reducing water depths lead to the transformation of incoming deep water waves by refraction, shoaling and eventually breaking. On the assumption that the wave period is constant, these processes affect the wave height and wavelength.

Refraction
Wave refraction is a consequence of the wave moving out of deep water. The portion of the crest in shallower water has its celerity reduced and is progressively turned parallel to the bed contours.

Refraction due to the bathymetry of the foreshore upstream of Dalwhinnie Dam is difficult to assess without more detailed bathymetric data than was available. Budget and contract time constraints meant that numerical modeling could not be accommodated. It was therefore assumed that incident wave fronts are parallel to the loch bed contours which are themselves parallel to the dam face and no refraction takes place.

Shoaling
Shoaling is the increase in wave height and decrease in wavelength and wave celerity caused as waves propagate in reducing water depths. Using linear wave theory, this effect can be expressed as a shoaling coefficient K_S; the equation for is given in CIRIA 83 Rock Manual (6) as follows:

$$K_S = 1/\{[1+2kd/\sinh(2kd)]\tanh(kd)\}^{0.5}$$

Where k = wave number = $\dfrac{2\pi}{L}$ d = water depth

At the toe of the dam, the wavelength can be estimated using first order wave theory as follows:

$$L = \frac{gT^2}{2\pi} \tanh(\frac{2\pi d}{L})$$

For average water depth, $L = 33.55$ m
For maximum water depth, $L = 35.89$ m

This wavelength is used to estimate the shoaling coefficient to give a new estimate of significant wave height at the toe of the dam shown in Table 2:

Table 2: Shoaling wave heights

	Average d = 3.85 m	Max d = 4.535 m
Wavelength L (m)	33.55	35.89
Shoaling coeff K_S	1.128	1.063
Wave height H_S (m)	3.20	3.02

Wave breaking

As waves approach a shoreline, and the water becomes shallower, they may become unstable and break, either through steepness induced breaking, or depth induced breaking. In shoaling water, breaking is usually caused by the latter, but both should be considered.

Steepness induced breaking occurs when $H/L \leq (H/L)_{max} = 0.14\tanh(2\pi d/L)$
Depth induced breaking occurs when $H/d \leq (H/d)_{max} = \gamma_{br}$ (breaker index)

For regular waves, theoretical γ_{br} is 0.78, but in practice, for irregular waves, γ_{br} is found to be 0.5-0.6. A summary of the wave breaking calculations are shown in Table 3.

Table 3: Summary of wave breaking at Dalwhinnie

	Steepness induced breaking		Depth induced breaking		Comment
	H/L	H/L$_{max}$	H/d	H/d$_{max}$	
Average depth d = 3.85 m	0.096	0.086	0.83	0.6	steepness and depth induced breaking
Max depth d = 4.535 m	0.084	0.092	0.67	0.6	depth induced breaking

Depth induced breaking will be the most critical event in this situation. The theoretical breaker index of 0.78 occurs in 4.0 m water depth. The reported figure of 0.5 for irregular waves would cause waves to break in about 5.0 m of water, and a breaker index of 0.6 would cause breaking in 5.5 m of water. Depth induced breaking could therefore occur anywhere from 200 m from the dam, up to the dam face itself, assuming a foreshore slope of 1 in 120.

CIRIA 83 Rock Manual (6), figure 121 gives design graphs for shoaling on uniform slopes with breaking. For a foreshore slope of 0.01 or shallower (1 in 100), and relative water depth (d/L$_{op}$) of 0.07, the resulting significant wave height was found to be 1.86 m.

Owen (8) also suggested a simple method to provide an estimate of the upper limit to the significant wave height in any depth of water. The method describes simple empirical equations for varying foreshore slopes. For a slope of 1 in 100, the equation is as follows:

$$\frac{H_S}{d} = 0.58 - \frac{2d}{gT_m^2}$$

giving $H_S = 2.12$ m   for average   d = 3.855 m
        $H_S = 2.47$ m   for max   d = 4.535 m

These two methods are the most up-to-date empirical methods available. They result in a fairly wide range of wave height and in the absence of any refining detail, we must adopt the more conservative figures from Owen's method.

The transformed wave characteristics at the toe of the dam for the mean annual wind wave therefore become:

Table 4: Shallow water wave characteristics – Mean annual wind

	Average depth d = 3.85 m	Max depth d = 4.535 m
significant wave height H_S (m)	2.12 m	2.47 m
peak wave period T_P (s)	5.90 s	5.90 s
mean wave period T_M (s)	5.13 s	5.13 s

These results indicate that waves incident on Dalwhinnie Dam are affected by the depth of water during a PMF event and break before or on the dam face. It should be noted that if wave breaking had not been indicated, the shoaling wave height is significantly higher than the deep water wave height, as shown in Table 2, and the resulting design wave height would have exceeded the deep water wave height.

Change in wave height distribution
Wind generated waves are irregular, having varying wave heights and wavelengths. The significant wave height used in many design methods represents the mean of the highest one third of the waves, or the wave height which is exceeded by 14% of waves. In deep water, wave heights tend to follow a Rayleigh distribution. However, as shown in Figure 109 in CIRIA 83 Rock Manual (6), the wave height distribution in shallow water is affected by wave breaking. This shows that the proportion of waves higher than H_S reduces due to shoaling and breaking.

Ideally, the entire wave spectrum should be transformed for shallow water effects and a new design wave selected from the transformed spectrum. However, this is not considered necessary as transforming deep water significant wave height through shallow water should produce a conservative result in terms of estimating wave overtopping discharge for wave wall design.

200 year wave height
The Floods and Reservoir Safety Guide, 3rd edition (3), recommends that if the calculated wave surcharge is greater than the flood surcharge, as is the case for Dalwhinnie Dam, then the total surcharge should be calculated again assuming the reservoir at initial reservoir condition plus the wave

surcharge resulting from a 200-year wind speed. The higher surcharge should then be used.

For Dalwhinnie Dam, the water depth is less in this situation, and because the waves are depth limited, the 200 year significant wave height is less than the mean annual maximum significant wave height. The condition of PMF plus concurrent mean annual wind wave event is therefore more severe and this event was used as the design basis for the overtopping protection works. The final design wave height represented a significant reduction compared to previous estimates assuming deep water and fetch-limited conditions, and allowed a more economical solution to be designed, details of which are reported separately (2).

WAVE FORCES AT ERICHT DAM
At Ericht Dam a new wave wall to the corewall section was required to provide sufficient freeboard. The height of the new wave wall would be designed to ensure acceptable overtopping discharges during the more severe of the following conditions:

- Peak pool elevation during the PMF event, plus mean annual maximum wind-wave
- Top Water Level plus 1 in 200 year wind-wave

Initial investigation of the wave conditions approaching the dam indicated that the new parapet wall could be subject to impact wave loading, when waves break directly on the structure. Following evidence of damage to the breakwater at Amlwch on Anglesey, failure of a wave wall at Porthcawl and other instances, HR Wallingford has advised that wave impact forces be included in the analysis of all such walls.

For Ericht, a methodology for deriving impact forces and calculating effective forces for Ericht Corewall wave wall was developed to provide an improved prediction of impact forces, taking into consideration the duration of impact forces and 3-D spatial effects.

Wave loading on structures
Wave loading on vertical or composite structures can be either pulsating loads or impact loads:

Pulsating or quasi-static wave loading arises when a wave impinges directly against the structure, the wave surface rises up and applies a quasi-static pressure difference on the structure.

Impact or dynamic pressures occur when a wave breaks directly on the structure due to the particular combination of foreshore slope, water depth and wave characteristics. Wave impacts are generally of high magnitude, but of short duration, and may be too fast for massive structures, such as the corewall itself to respond to, but may be more critical for smaller structure components, such as the parapet wave wall. They are also spatially limited, so average loads will decrease with increasing section length.

Initial investigations indicated that for the mean annual wind/wave event at PMF level, wave loadings at Ericht Corewall Dam were pulsating. However, impact loadings might act on the proposed parapet wall during the 1 in 200 year wind/wave event occurring at top water level.

Development of wave impact loading methodology
Methods to estimate pulsating wave loads are reasonably well established. Goda's method (6) to estimate pulsating wave forces was used. However, methods to estimate impact loadings are far from comprehensive and, in particular, methods to estimate the wave pressure distribution in order to estimate the effective pressures on the wave wall alone had not been published.

Advice was sought from HR Wallingford, who has published the most recent design methodologies for estimating wave impact loadings. These methodologies were still incomplete with regard to a few aspects of wave impact loadings that were necessary to complete the calculations at Ericht.

The aim was to produce a methodology to provide an improved prediction of impact forces for Ericht Corewall wave wall, taking into consideration the duration of impact force and 3-D spatial effects.

The resulting methodology has been developed from published design methods (9), (9), (10), (11), supplemented with data held by HR Wallingford.

Procedure to derive effective wave forces on Ericht corewall parapet wall
Four sections of the west corewall dam were considered to investigate wave loading. Sections 1 and 2 were treated as composite structures, having a berm in front of the corewall, as shown in Figure 2; while sections 3 and 4 were treated as vertical walls. In the absence of survey data, a foreshore slope of 1 in 20 was assumed as this gave the highest design wave heights. The relevant structural and hydraulic data are given in Table 5 and Table 6.

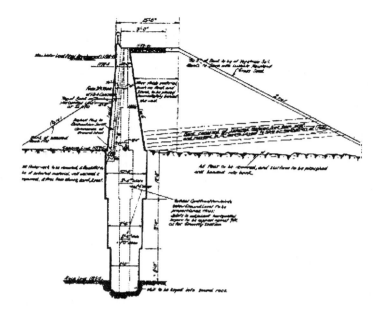

Figure 2: Typical section through Ericht Corewall (sections 1 and 2)

Table 5: Corewall structural data

Section	1	2	3	4
Top of wall (mAOD)	361.57	361.57	361.57	361.57
Wall base (mAOD)	360.58	360.58	360.58	360.58
Top of berm (mAOD)	356.62	356.62	N/A	N/A
Bed level (mAOD)	353.42	354.64	356.77	358.3
Foreshore slope	0.05	0.05	0.05	0.05

Table 6: Hydraulic data for Ericht Corewall Dam

Water level (TWL mAOD)	359.4
1 in 200 year wave height H_s (m)	1.55
1 in 200 year wave period T_s (s)	3.48
Deep water wavelength L_{op} (m)	24.9

The developed methodology is as follows:

1. Pulsating wave forces

As a first estimate of wave induced loads on the wall, Goda's method (6) was used to estimate pulsating wave forces, taking into consideration wave shoaling and shallow water depth-limiting effects. The results are shown in Table 7.

Table 7: Pulsating wave loads

Section	1	2	3	4
Total force (kN/m)	90	87	68	21
Total force on wave wall (kN/m)	14	13	10	3
Pressure at top of wall (kN/m^2)	10.84	10.43	12.16	0
Pressure at wall base (kN/m^2)	16.21	15.74	6.85	5.15

2. Identify likely loading conditions from parameter map
In order to assess whether wave impact forces are likely to occur, the geometric and wave parameters were checked against the parameter map given in (11). For the wave conditions considered, impacting waves are possible at all four sections.

3. Predict wave impact force
Impact forces were calculated using Allsop & Vincinanza's equation (12):
$$F_{imp,1/250} = 15(H_{si}/d)^{3.134} \rho_w g d^2$$

4. Impact force duration and dynamic effects
Impact force durations were then estimated from HR Wallingford test data. Using the outer envelope, the impact rise time was estimated at 0.4 seconds. Data within the average band suggested a rise time of 0.04 seconds. These impact durations were then compared to the natural frequency of the wall, estimated using a standard cantilever formula to lie between 0.014 and 0.028 seconds. In both cases, the impact rise time exceeds the natural frequency of the wave wall. It is therefore reasonable to assume that impact wave loads will not be significantly damped, and the wall will effectively experience this load as quasi-static.

5. Pressure distribution
The prediction methods above give total wave forces over the full active depth, not just the parapet wall. In order to establish the proportion of load acting on the parapet wall section, the pressure distribution over the height of dam must be known.

For pulsating wave forces, Goda's method assumes a trapezoidal pressure distribution. Although this was not intended by Goda to predict pressure distributions, the general level of wave forces predicted is well validated, and no better guidance is available. The proportion of wave force acting over the parapet wall must therefore be estimated using this distribution.

For impact forces, Hull (13) analysed measurements of wave impact pressures by McKenna (10) and others from which Hull developed a method for estimating pressure distributions as a function of the pressure at still water level (p_{max}) given in Table 8.

Table 8: Impact force pressure distribution

Pressure (kN/m^2)	Elevation (m relative to SWL)
$p_f = 0$	$1.2\ h_s^2/h_b$
$p_e = 0.08\ p_{max}$	$0.4\ h_s^2/h_b$
$p_d = 0.4\ p_{max}$	$0.17\ h_s^2/h_b$
$p_c = p_{max}$	0
$p_b = 0.4\ p_{max}$	$-0.25\ h_s^2/h_b$
$p_a = 0$	$-0.9\ h_s^2/h_b$

Where h_s = depth of water at toe of berm
h_b = height of top of berm above bed level

Note that this step could only be carried out for Sections 1 and 2 as the method is only applicable for composite structures. Figure 3 shows the pressure distribution plots for Section 1.

The effective pressures acting on the wave wall alone were estimated. Assuming a linear distribution between the pressure at the top of the wall and the base of the wall, the total force acting on the wall was then calculated.

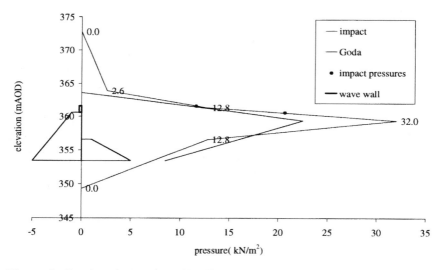

Figure 3: Section 1 pressure distribution

6.   *Three dimensional effects*
Wave impact forces are spatially limited. Increasing unit length leads to a greater reduction in force, but unit length is more likely to be driven by construction methodology. For Ericht parapet wall, 1.5 m was considered to be the longest practicable pre-cast or in-situ cast unit. Guidance is given in (11) for methods to calculate the reduction in wave impact forces.

Assuming a wall unit length of 1.5 m resulted in a reduction in impact force of 9%.

7.   *Resulting design wave forces*

Following the methodology set out above, Section 1 was found to give the highest force and pressure estimates for both pulsating and impacting loads. The results are given in Table 9.

Table 9: Wave loads on wave wall section

Loading	Impacting	Pulsating
Total force on wave wall (kN/m)	16	14
Pressure at top of wall (kN/m^2)	11.63	10.84
Pressure at wall base (kN/m^2)	20.66	16.21

The reductions in impact loading to account for 3-D spatial effects, and the pressure distribution of that loading have resulted in a total impact force on the critical Section 1 that is only marginally larger than the pulsating force. The impact on the design of the wave wall due to wave impact loading is therefore minimal. However, it is recommended that for cases where impact forces appear to be smaller than the Goda pulsating forces, the Goda force should always be used for design.

CONCLUSIONS

The Ericht project provided an exceptional opportunity to explore the limitations of standard methods to assess wave conditions and wave forces generated on reservoirs, and to extend those methods in a practical context.

The wave assessment at Dalwhinnie Dam has shown that not all reservoirs conform to the assumptions implicit in the standard method to estimate significant wave height using the simplified Donelan/JONSWAP, described in Floods and Reservoir Safety, 3rd edition (3), namely that:

* Fetch limited waves are generated on reservoirs
* Deep water conditions apply for waves approaching any dam

Reservoir engineers should be aware of these implicit assumptions and check whether they apply in any given location. Shallow water processes, in particular, can have a substantial effect on the wave height incident on the upstream face of a dam, and may offer significant savings. However, if approach characteristics are such wave breaking does not occur, design wave heights higher than deep water conditions may result.

Initial analysis of wave loads at Ericht Corewall Dam to be used in the design of a wave wall indicated that available methods to estimate impact loadings are far from comprehensive. In particular, methods to estimate the wave pressure distribution to allow an estimate of the effective pressures on the wave wall alone had not been published.

A methodology was developed to provide an improved prediction of impact forces for Ericht Corewall parapet wall, taking into consideration the duration of impact force and 3-D spatial effects. Impact forces can be substantially higher than pulsating forces. However, in this case, because of the position of the parapet wall relative to the still water level where the maximum impact pressure occurs, the impact force did not greatly exceed the estimated pulsating force.

REFERENCES

(1)   Reservoirs Act (1975), HMSO, London
(2)   Dempster KJ, Gaskin M, Doake RM, Hay-Smith D (2004), *Ericht and Dalwhinnie Dam refurbishment and protection works*, Long Term Benefits & Performance of Dams, Thomas Telford, London
(3)   Institution of Civil Engineers (1996), *Floods and Reservoir Safety Third Edition*, Thomas Telford, London
(4)   Owen MW (1987), *Wave prediction in reservoirs: a literature review*, Report EX1527, Hydraulics Research, Wallingford
(5)   Owen MW & Steele AAJ (1988), *Wave prediction in reservoirs – comparison of available methods*, Report EX 1809, Hydraulics Research, Wallingford
(6)   British Standards Institution (2000), *British Standard Code of Practice for Maritime Structures, Part 1, General* Criteria, BS 6349: Part 1, BSI, London
(7)   CIRIA/CUR (1991), *Manual on the use of rock in coastal and shoreline engineering*, Simm JD (Ed) Special Publication, CIRIA, London
(8)   Owen MW (1980), *Design of sea walls allowing for wave overtopping*, Report EX924, Hydraulics Research, Wallingford
(9)   Allsop NWH & Kortenhaus A (2001), *Hydraulic aspects*, Chapter 2 of *Probabilistic Design Tools for Vertical Breakwaters*, pp61-156, ISBN 90 580 248 8, Balkema, Rotterdam
(10)   McKenna JE (1997), *Wave forces on caissons and breakwater crown walls*, PhD thesis, Queen's University of Belfast, September 1997, Belfast
(11)   Allsop NWH (2000), *Wave forces on vertical and composite walls*, Chapter 4 in *Handbook of Coastal Engineering*, Editor J Herbich, McGraw-Hill, New York

(12) Allsop NWH & Vincinanza D (1996), *Wave impact loadings on vertical breakwaters: development of new prediction formulae*, Proc 11th International Harbour Congress, Antwerpen, Belgium
(13) Hull P (2001), *Wave impact loading and its effects on blockwork structures*, PhD thesis, Queen's University of Belfast, September 2001, Belfast

An Incident at Ogston Reservoir

A.K. HUGHES, KBR, UK
P. KELHAM, KBR, UK
D.S. LITTLEMORE, KBR, UK
S.D.R. HARWOOD, Severn Trent Water, UK

SYNOPSIS.   In the autumn of 2001 an incident occurred at Ogston
Reservoir which led to the catastrophic failure of the pipework in the draw-
off shaft.   An uncontrolled release of water commenced which was only
prevented by the quick actions of two operatives in the shaft. This paper
describes the investigations carried out to establish the cause of failure, the
remedial works which were carried out and the lessons learnt.

INTRODUCTION

Ogston Reservoir is owned and operated by Severn Trent Water.
Completed in late 1959, the reservoir is situated about 6 km north west of
Alfreton, Derbyshire.   The treatment works, situated immediately
downstream of the reservoir, supplies water to areas in North East
Derbyshire, Chesterfield and Sheffield.

The dam, which is an earthfill embankment with central puddle clay core,
has a height of 19.8 metres and is 213 metres long. It impounds a maximum
storage of 6,180,000 cubic metres of water.

The forebay tunnel, overflow shaft, valve tower, and combined overflow
and draw-off tunnel are situated in the centre of the embankment and
constructed of mass concrete. The complex arrangement of these structures
is shown in Figures 1 and 2, with the draw-off tower forming a single
structure with the overflow shaft. The overflow tunnel and draw-off tunnel
are also formed as one structure.

There are three levels of draw-off comprising 24″ (600mm) diameter cast
iron pipework and in-line guard and duty gate valves, feeding into a
common 30″ (760mm) diameter draw-off stack. Water passes vertically
downwards in the stack to join into a similar diameter cast iron draw-off

Long-term benefits and performance of dams, Thomas Telford, London, 2004, 488–502

main which passes along the discharge tunnel under the embankment to Ogston Water Treatment Works.

The scour facility prior to the incident comprised 30″(760mm) diameter cast iron pipework with a 30″(760mm) guard gate valve known as G4, and a 700mm duty butterfly valve known as S1.

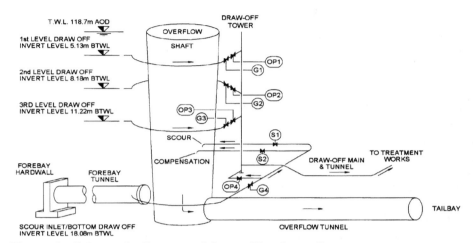

Figure 1: Schematic diagram of draw off and overflow arrangement

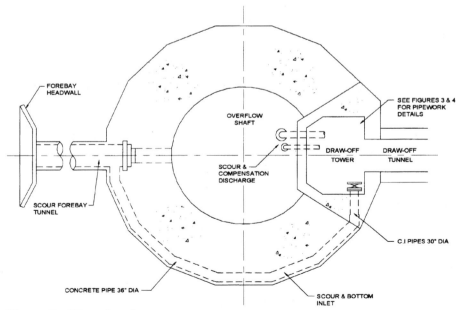

Figure 2: Plan showing scour arrangement

Discharge of scour water is via a pipe outlet through the wall of the draw-off tower into the base of the overflow shaft. A 9" (225mm) branch connection from the scour pipework incorporated a Larner Johnson streamline valve, known as S2, for the release of compensation water. The butterfly valve (S1) was a recent replacement for the original 30"-24"-30" (760-600-760mm) diameter Larner Johnson streamline valve which had been found to be in a poor condition, difficult to operate and requiring rehabilitation or replacement. The original and modified layouts of the scour valves in the draw-off tower are shown in figures 3 & 4.

INCIDENT

As part of the refurbishment process to return the Larner Johnson to a serviceable condition alternative valve options were considered due to the extent of the refurbishment work that would be required on the original valve. A value engineering exercise was carried out and a 700 mm diameter butterfly valve was chosen to replace the Larner Johnson. A Panel Engineer was not involved in the value engineering exercise, however subsequently one was consulted on the proposal to install a butterfly valve. The Panel Engineer, having carried out some calculations, commented that the velocities appeared to be high and recommended that confirmation be sought from the manufacturer as to the valve's fitness for purpose with respect to the maximum expected velocity and its location within the pipework arrangement. This confirmation was provided and the valve was obtained and installed.

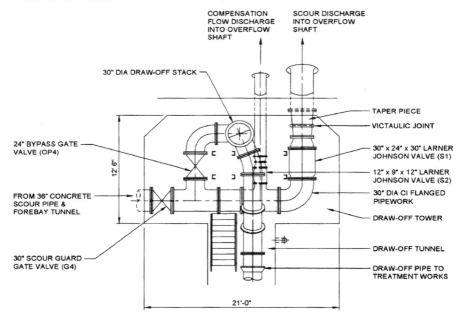

Figure 3: Original scour valve layout in draw-off tower

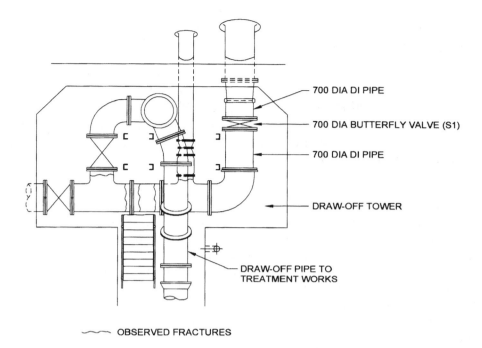

OBSERVED FRACTURES

Figure 4: Modified scour valve layout in draw-off tower prior to incident

Following the installation of butterfly valve S1 some difficulties were experienced. Initially the valve was found to be very stiff to operate and a number of modifications were made including increasing the diameter of the operating hand wheel and increasing the capacity of the gearbox. During the commissioning tests on the butterfly valve, the pipework immediately upstream, including the compensation water branch, suffered catastrophic failure resulting in the sudden uncontrolled release of water from the scour pipe into the base of the draw-off tower where there were two men trying to operate the valve. This discharge quickly started to fill the draw-off tunnel until it blew the doors open at the downstream end allowing water to discharge back to the downstream tail-bay area.

Guard valve G4 was subsequently shut to isolate the discharge by the men going back through the discharging waters.

ADVICE

Dr Hughes, who was the appointed Inspecting Engineer at that time, having recently carried out a routine inspection of the reservoir in accordance with the Reservoirs Act 1975, was informed of the incident. Details and recommendations arising out of his subsequent site visit were included in his report.

Technical advice was provided to the owner throughout the project by Dr Hughes, and by the Review Panel, the owner's retained experts, headed by Mr R E Coxon.

Kellogg, Brown and Root (KBR) were appointed to design and supervise the construction and installation of all temporary, enabling and permanent works involved in the restoration of the scour and compensation facilities, taking due account of all personnel health and safety and reservoir safety considerations.

During the initial site inspections it became apparent that the draw-off tower and draw-off tunnel were not safe places to work, since catastrophic failure of the pipework had occurred in a number of places and had left the scour guard valve G4 unsupported and unrestrained in the base of the draw-off tower. Fortunately the flanges on the scour pipework either side of valve G4 appeared to have survived the surge pressures generated by the failure of butterfly valve S1. However, because of the uncertainty regarding the condition of the valve G4 and the adjacent flanged puddle pipe it was considered unsafe to operate the guard valve. Therefore there were now no means of effecting scour draw-off from the reservoir should the need arise and so 'emergency' remedial works were recommended by Dr Hughes 'in the interests of safety'.

Plate 1: Fractured 30" scour pipe

Plate 2: Fractured compensation pipework

Plate 3: Damaged draw-off tunnel doors

The following permanent works were deemed to be necessary:-

- Reinstatement of the scour and compensation pipework and valves.
- Reinstatement of associated accesses and floor stagings where necessary.
- Reinstatement of tunnel access doors.

It was immediately apparent that:-

- The full extent of damage was unknown.
- That the working areas were very restricted.
- The reservoir would have to remain partially full during the works in order to protect the fisheries and the shoreline nesting bird population, as agreed with English Nature (the shoreline was a designated SSSI).
- The compensation flow of 6 Ml/d would have to be maintained.
- No record drawings existed – although construction drawings were available.

It was also recommended that:-

- An additional 600mm diameter washout facility be provided in the raw water supply pipeline to provide scour facilities and greater control of the reservoir water levels in the short term.
- A temporary bulkhead should be installed on the scour forebay tunnel headwall to enable safe access into the draw-off tower to facilitate the investigation and repair works. In order to assess the feasibility of the bulkhead installation an underwater survey of the scour forebay headwall would be required.

PROPOSED APPROACH

The approach proposed was to work closely with the owner to ensure the safety of the reservoir whilst undertaking the necessary investigations and surveys required to formulate a strategy for the method of repair.

Therefore, in order to achieve a successful outcome to the project, an 'Operational Plan' was written with the owner to:

- Provide the owner with sufficient information to operate the reservoir so as to meet water supply requirements and the needs of the scour valve repair project. In the case of the latter it was arranged to reduce water levels over an agreed timescale to meet the start date for the repair contract.

- Identify the steps necessary to ensure the safety of the reservoir, company project personnel and the public.
- Identify key contacts and responsible persons.
- Ensure that all statutory and legal requirements were met.
- Provide a framework for liaison with all interested parties.
- Ensure that all environmental issues were fully recognised and managed.

The Operational Plan was considered to be a 'live' document subject to continual review and update as the project proceeded. New operating control curves were drawn up and a number of draw-down and refill scenarios as well as 'unusual events' modelled to assist the operators and contractors engaged to undertake the surveys, investigations and permanent works.

It was not considered likely that the embankment would fail in the event of the scour pipework failing upstream of guard valve G4, however the uncontrolled release of water which would take place through the draw-off tunnel and the eventual draining of the reservoir had to be considered.

PROJECT PROGRAMME AND ENABLING WORKS
The Operational Plan included a very detailed programme and methodology covering all activities necessary to control the reservoir level over the winter period and achieve a managed draw-down to the lowest draw-off level in the spring of 2002 to facilitate the installation of the replacement scour pipework and valves in the draw-off tower.

An additional 600mm gate valve washout facility was provided, via a 600mm branch off the 24″ raw water supply main to the treatment works. This was installed and commissioned before the onset of winter.

An underwater survey was undertaken by divers. The objectives of the survey were to:-

- Determine the silt levels and accessibility of the scour forebay headwall.
- Undertake a survey of the forebay tunnel headwall.
- Assess the feasibility of installing a temporary watertight bulkhead. The bulkhead would be used to enable dewatering of the forebay tunnel and scour pipework around the overflow shaft, and allow examination of the embedded puddle pipe (immediately upstream of valve G4) in the draw-off tower wall. Information obtained would be used in the design of the subsequent valve replacement works.

PERMANENT WORKS

It was immediately evident from the inspection following the failure of the butterfly valve S1 that the surge pressures generated by the incident caused fracture and complete failure of the scour pipework. What was not known was whether the surge pressure had caused overstressing of the valve bodies and other fittings which did not show any visible signs of failure. It was possible that the surge pressures had caused damage to:-

- 30" scour guard valve G4 and the associated puddle flanged pipe set into the wall of the draw-off tower.
- 36" concrete scour pipe that ran around the base of the overflow shaft.
- 24" bypass valve OP4, 90 degree bend and connection with the draw-off stack.
- 9" Larner Johnson streamline valve S2.

In addition, the movement of the scour pipework may have caused damage to the 30" and 9" compression couplings downstream of the Larner Johnson streamline valves on the scour and the compensation pipework respectively.

The temporary and permanent works were designed, therefore, to replace the majority of the above pipework and valves by the construction of a temporary bulkhead on the scour forebay tunnel headwall, so as to provide a safe environment inside the draw-off tower for construction operatives and supervisory staff.

The temporary and permanent works involved:

- Contractor designed watertight bulkhead on scour forebay tunnel headwall with a facility to dewater the tunnel by pumping.
- Removal and replacement of all damaged and suspect pipework and valves.
- Construction of new thrust blocks.
- Carrying out in situ non-destructive testing of all built – in pipework, all couplings and any pipework likely to be left in position.
- Modifications to platforms, ladders and stairs as required.
- Replacement of damaged doors at entrance to access tunnel.

Following an assessment of the options for replacing the scour valves, KBR recommended that the butterfly valve should be replaced by a Larner Johnson valve. Fortunately it was possible to track down the original Larner Johnson valve which had been removed and have it refurbished for subsequent installation by the appointed contractor.

Because of the dangers associated with entering the draw-off tower and also the lack of detailed drawings there were a number of uncertainties and concerns at the time of tendering regarding the feasibility of using a bulkhead to facilitate the de-watering of the scour forebay tunnel and pipework. These uncertainties included:-

- achieving an adequate seal between the bulkhead and scour forebay tunnel headwall;
- the structural capacity of the headwall to support the bulkhead;
- the quantity of leakage into the forebay tunnel and scour pipework around the overflow shaft;
- the structural capacity of the scour forebay tunnel to withstand the proposed de-watering;
- the feasibility of manoeuvring the Larner Johnson valve along the draw-off tunnel;
- whether the Victaulic joints could be refurbished or replacements found.

CONSTRUCTION PHASE

Prior to awarding the contract detailed interviews with tenderers were held to ensure that their proposed methodology, risk assessments and strategies for dealing with the project uncertainties detailed earlier in this paper had been properly considered and evaluated. Following this process Norwest Holst Construction Ltd was appointed as Principal Contractor.

The contractor successfully completed the safe refurbishment of the scour facility in October 2002 by following the basic order of procedure detailed in the Operational Plan. The principal activities were:-

- Installation of a temporary bulkhead on the scour forebay headwall to allow dewatering of the forebay tunnel and the safe removal of the damaged scour pipework and butterfly valve.
- Carrying out a detailed survey of valve shaft pipework.
- Removal of scour guard valve G4 and testing of the embedded puddle pipework.
- Installation of anchor frame on puddle pipe flange and new 30" scour guard valve. It is worth noting that outline pipework and valve designs were carried out by KBR at an early stage to facilitate the early procurement of the valves. The Contractor was given the detailed design of the pipework and valve arrangements following an accurate survey of the existing pipework in the tower. This survey could only be undertaken once the bulkhead had been fitted.
- Installation of new scour and compensation pipework and refurbished Larner Johnson valves. The Contractor elected not to

dismantle the Larner Johnson scour valve once it had been factory refurbished. Following delivery to the site a specially designed trolley enabled the valve to be moved along the tunnel and then positioned in the base of the shaft.
- Commissioning of valves and pipework.
- Removal of temporary bulkhead.

Plate 4: Draw-off main within the draw-off tunnel

Plate 5: Transporting the Larner Johnson Valve

INVESTIGATION OF VALVE FAILURE

Even though it was clear from the initial visit that the butterfly valve had been installed in a far from ideal position almost immediately downstream of a bend and discharging into almost free air with zero downstream pressure it was essential to find the reasons for the catastrophic failure witnessed at Ogston. Therefore an investigation was devised to determine the physical condition of the damaged pipework including:

- remaining wall thickness
- degree of corrosion
- evidence of welding
- flange rating
- strength

and to investigate the cause of the failure of the butterfly valve by:

- establishing that the valve had been constructed to the manufacturer's specification
- identifying the point and mode of failure
- performing strength tests on the failed components
- determining whether the valve had any locking device

Plate 6: Failed butterfly valve in position and failed gearbox component

Following removal from the draw-off tower the butterfly valve and several pieces of pipework were taken to an independent testing laboratory for detailed examination and testing. A visual inspection of the valve identified no external damage, however, an internal investigation of the valve and gearbox made some interesting findings. The principal findings of the investigation were:

- The gearbox fitted to the valve was undersized for the application. The connection between the valve and the gearbox failed as a result of the excessive torque required to operate the butterfly valve beyond 50% open. This was due to the calculated presence of full cavitation and uneven flow profile, caused by the close proximity of the bend and the positioning of the valve.

- The capability of the gearbox was considerably reduced by one of four screws used for coupling the gearbox to the valve drive sleeve being missing.
- When the connection between the valve and gearbox failed there was nothing to prevent the valve slamming shut.
- There was little external corrosion of the pipes; however internal corrosion within the structure of the metal had reduced its tensile strength to some degree.
- The estimated surge pressures generated by the instantaneous closure of the butterfly valve would have resulted in the failure of new pipe to the same specification as that installed.

In summary, the failure of the pipework was due to the torque required to operate the butterfly valve being underestimated by its manufacturer and the gearbox being too small for purpose. This problem was exacerbated as one of the screws was not fitted into the gearbox drive sleeve and the remaining screws were not able to take the applied load. They subsequently failed allowing the valve disc to be rotated by the water flow, slamming closed and bringing the water flow to a sudden halt. The resulting change in momentum caused a pressure surge estimated to be in excess of 55 bars. This surge caused several sections of the pipework to fracture releasing a considerable quantity of water.

CONCLUSIONS

This paper describes an incident which put operatives at risk and resulted in the sudden uncontrolled release of water following the catastrophic failure of scour pipework. The failure of the pipework was caused by the fitting of an inappropriate valve and gearbox for the required duty and system configuration. The lessons to be learnt include:-

1. The specification for the design of a valve should take account of, inter alia, its purpose, location, fixings, hydraulic loading, adequacy and configuration of existing pipework and valves, thrust/tension resistance, intended operating procedures, frequency of use, accessibility and ease of operation, maintenance and facility for subsequent removal.
2. The valve manufacturer should design the valve, together with gearbox, actuator, etc, to meet the specification and should certify compliance by providing supporting calculations and details of works tests.
3. Due regard should be taken of such phenomena as cavitation and spiralling flow.

4. Engineers should understand how various valves work and consider their possible modes of failure. In the case of butterfly valves it needs to be recognised that a failure of the connection between the gate and the gearbox will result in the gate slamming shut instantaneously.

5. Expert advice both in terms of mechanical plant and reservoir safety should be sought when considering the replacement or refurbishment of valves in existing scour and draw-off arrangements.

ACKNOWLEDGEMENTS
The authors would like to express their thanks to Messrs Neil Williams (Principal Engineer Reservoirs) and Ian Elliott (Director of Engineering) of Severn Trent Water for permission to publish this paper.

Marmarik Dam Investigations and Remedial Works

L. SPASIC - GRIL, Jacobs, Reading, UK
J. R. SAWYER, Jacobs, Reading, UK

SYNOPSIS.   Marmarik dam is a multipurpose embankment dam on the Sevan – Hrazdan cascade in Armenia. The dam is situated in one of the most seismically active regions in Armenia and in the vicinity of the reservoir numerous landslides could be seen. The dam was commissioned in January 1975 and twenty days later significant subsidence of the clay core occurred causing 14m settlement of the dam crest. The dam has never been rehabilitated and the reservoir has never been impounded.

The dam was investigated by JacobsGIBB ltd as part of the World Bank funded 'Technical Investigation of 60 Dams'. In addition to the slope failure, further issues include high regional seismicity, landslides adjacent to the dam and reservoir, inadequate spillway capacity, rehabilitation of the derelict outlet works and concerns regarding the foundation cut-off.

INTRODUCTION

Marmarik dam, situated in Kotayk Marz in Armenia, was originally designed to provide water for the future aluminium mining industry, a cement factory, two thermal power plants, irrigation of 2,000ha and flood water regulation. However, as the aluminium mining industry was never developed the dam changed ownership and the new owner became the Ministry of Water Resources. Marmarik dam is a part of the Sevan – Hrazdan cascade which significantly contributes in overall regional energy balance and provides water for irrigation systems and six power plants.

The dam was commissioned in January 1975 and twenty days after the commissioning significant instability of the embankment occurred, causing a 14m settlement of the dam crest over half of the dam crest length. Immediately after the subsidence a local company was commissioned in 1975 to investigate causes of the dam failure. It was found that the failure occurred as a result of high pore pressure in the clay core that was placed with a high moisture content. The dam has never been rehabilitated and therefore the reservoir has never been impounded. The river is diverted in a tunnel through the left abutment. This uncontrolled diversion has been left in operation since the construction period.

DESCRIPTION OF THE DAM

Embankment
Marmarik dam, Figure 1, was originally designed as a 64m high embankment with a clay core and compacted fill shoulders. The original dam crest was at 1914masl. The design cross section is shown in Figure 2. The shoulders were originally designed to be built of gravel from a borrow area some 5km downstream of the dam. However, only the bottom 5m of the embankment was constructed from gravel as the further use of the gravel borrow area was not permitted. Thereafter the embankment shoulders were constructed from a compacted sandy silt from borrow areas within the reservoir, but to the original design slopes and with no filters.

The central part of the dam is founded mainly on granular river alluvium and the abutments are founded on a thick layer of cohesive colluvial deposits. The designed foundation anti – seepage measures comprise a cut – off bored secant pile wall constructed up to 30m deep through the central part of the alluvial foundation and a grout curtain through the colluvial foundation at the abutments.

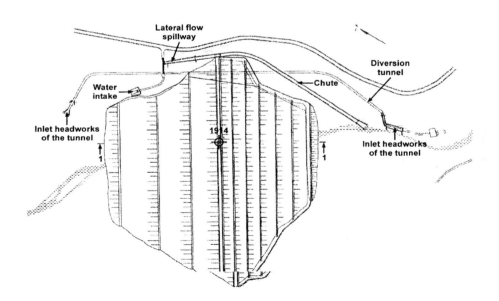

Figure 1. Marmarik reservoir- plan

Diversion tunnel

At its present state the river is diverted into a 3.2m diameter, D shaped diversion tunnel designed for a temporary condition, for a flood of 50m³/s with a return period of 20 years. Since its completion, there have been three occasions on which the incoming flood exceeded the designed value but the flood was absorbed in the reservoir storage volume without risk of overtopping the dam. There is a side weir at the outlet end, which permanently maintains a minimum water level of 1.4-1.5m in the lower section of the tunnel. For that reason the complete inspection of the tunnel has never been carried out in the past.

Spillway

The spillway, situated on the left abutment comprises:
-       A 60m long side-channel inlet weir
-       A culvert under the dam crest
-       A 6m wide discharge chute with a variable gradient

Outlet Works

The outlet works consist of a tunnel leading to a short connecting shaft to the diversion tunnel which is 11m below. The connecting shaft contains two 1.0m diameter outlet pipes which are cast into mass concrete which fills the shaft. A 50m deep, 6m diameter gate shaft is located at 5m offset from the connecting shaft. The gate shaft contains an emergency closure gate and a maintenance gate for each pipe.

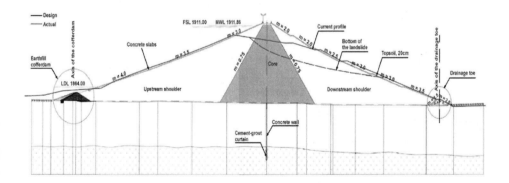

Figure 2. Cross Section 1-1

INVESTIGATIONS CARRIED OUT

Immediately after the subsidence of the dam, a local Armenian company was commissioned to investigate causes of the dam failure. The investigations were carried out between 1975 – 1978.

Under the 'Technical Investigation of 60 Dams' the following investigations and surveys were undertaken during 2002 - 2003:

-   Topographic survey of the dam and the diversion tunnel
-   supplementary ground investigations of the dam
-   microseismic survey to establish site specific seismic parameters
-   landslides hazard assessment and landslide ground investigation
-   investigations of the diversion tunnel
-   investigation of the efficiency of the foundation cut-off

The investigations undertaken during 2002-2003 are described below in more detail. Based on the results of the investigations and the findings of the investigations carried out during 1975-1978, geological, geomorphological and seismic conditions at the dam site were assessed as well as the status of the dam, foundation anti- seepage measures and the diversion tunnel.

Supplementary ground investigation

Supplementary ground investigation carried out to validate previous investigations included 480m of drilling through the dam, trial pitting, in – situ permeability testing and laboratory testing.

Microseismic survey

Microseismic survey comprised the following works:

-   Seismic Refraction- carried out at 48 measuring points in the reservoir and 24 measuring points on the dam
-   Measurements of ground micro - vibrations by using SMACH –SM and OMNILIGHT instruments - p-wave velocities were recorded in the surface deposits and in the bedrock, as well as the peak horizontal accelerations, vertical geomagnetic field and the distribution of predominant frequency spectra

Landslides hazard assessment and landslide ground investigation

Landslides of a seismogenic origin are widespread along the whole length of the southern (right) bank of the Marmarik River canyon. Four potentially hazardous seismogenic landslides were identified within the Marmarik reservoir area that may influence the dam safety, namely landslides N1 to N4. The landslides are shown in Figure 3. Landslide hazard assessment was carried out based on the analyses of satellite images and aerial photos that were taken in 1948, 1976 and 1986 as well as the field surveys carried out in 1975-78 and 2002.

Landslide N1 is located some 2 km to the S-SE of the Marmarik Dam, in the upper reaches of the Kiarkhana River, which is a lateral inflow of the Marmarik River and as such it poses a low hazard to the dam. The landslide was not investigated further.

Landslide N2 is located some 250m to the south of the dam. The landslide is situated close to the confluence of Kiarkhana with the Marmarik river. During 1969-1974, the material from the toe of landslide was excavated for construction of the Marmarik dam. The excavation destabilised the landslide leading to a development of presently active secondary landslides. The landslide was investigated during 1975-1978 site investigation. Thickness of the landslide varies from 30 to 80m, total volume is about $94 \times 10^6 \mathrm{m}^3$.

Landslide N3 is 1.6km upstream of the dam and it was reactivated a number of times in the past, most recently during dam construction when the soil from the toe of landslide was excavated and used for the fill material. The landslide was investigated during the 1975-1978 site investigation. Thickness of the landslide varies from 40 to 60m, the total volume is about $16 \times 10^6 \mathrm{m}^3$.

Landslide N4 is 5.2km upstream of the dam and at its toe, it branches into two landslides separated by some 700m. This landslide is the most distant from the dam, but it is the largest in volume. If it is triggered it could dam the Marmarik River and create a lake which, if the natural dam is breached, could induce a flood inflow into the reservoir. This landslide was investigated during 2003. It was found that landslide comprises a layer of rock debris with a soil matrix up to 50m thick, over a thin slip surface that overlays the in - situ rock.

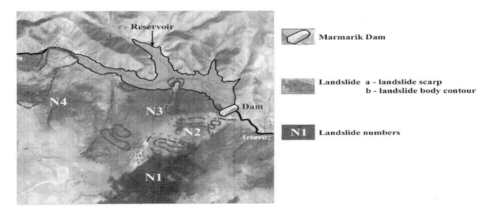

Figure 3. Landslide hazard map for the Marmarik reservoir

Investigations of the diversion tunnel
Investigation of the diversion tunnel comprised the following:
- Initial walk through and visual inspection
- Intrusive drilling through the tunnel lining
- Non destructive testing using a calibrated Schmidt hammer and a hand held ultrasonic meter to determine quality of the concrete.

Investigation of the efficiency of the foundation cut-off
As the reservoir has never been impounded, there is a significant uncertainty about the efficiency of the foundation anti-seepage measures, especially the ones through the alluvial foundation. The investigation of the effectiveness of the cut-off through the alluvial foundation was therefore accomplished by carrying out water pressure tests and indicator tests in the test section located in the centre of the river channel. The test section comprised three holes located 15m u/s of centreline (Hole1), 5m u/s of centreline (Hole 2) and 5m d/s of centreline and drilled down to the bedrock.

The water pressure tests were used to measure the difference in response of piezometers (placed in the foundation material upstream (Hole1) and downstream (Hole 3) of the cut – off) to water pressure applied in Hole2 drilled upstream of the cut – off. In Holes 1 and 3 piezometers were installed 5m below the fill/alluvium interface and 5m above the alluvium/bedrock interface. The difference of the response in Holes 1 and 3 is a measure of the permeability of the cut –off.

The water pressure test method was supplemented by the introduction of an indicator (salt solution) into the borehole and a comparison of the concentrations of the indicator throughout the borehole.

GEOLOGICAL CONDITIONS
Geology of the dam site comprises deeply weathered and fractured granodiorites and metamorphic complex of the Oligocene age. The dam site is located in fault-controlled river valley following the trend of a major northeast to southwest trending regional fault.

The central part of the embankment is founded on alluvial deposits (coarse sandy gravel) which fill the entire river valley and which are underlain by weathered granodiorite. The alluvial deposits vary in thickness between 10m and 30m and have a hydraulic conductivity of 10^{-4} to 10^{-5} m/s.

On the abutments the embankment is founded on colluvial materials, mostly silty clays. The colluvium covers the valley sides to a thickness of up to 20m and is derived from the weathering of granodiorites, with landslides in some areas. The hydraulic conductivity of the colluvium is 10^{-6} m/s.

Seismological conditions

Regional seismicity
Marmarik dam is located in a highly seismic area of Armenia. Some 13.7km north of the dam site runs the largest and most active Pambak – Sevan Fault. This is the main regional fault, 490km long, which in the past generated earthquakes of magnitudes up to 7.4. Also, very close to the site (5.2 km away), to the west of the dam, is the Garni fault, 198km long, which in the past generated earthquakes with magnitudes up to 7.0. The dam is directly situated on Marmarik fault, 30km long which joints the Garni fault. However, as no tectonic activity has been registered along the Marmarik fault in the Holocene, the fault is regarded to be seismically inactive.

Seismic design parameters
Seismic design parameters have been assessed based on the methodology given in Reference 1 as well as the site specific seismic hazard assessment.

The method in Reference 1 gave the following design accelerations (return period of 475 year):
- Ground acceleration: $a_{pk}= 0.144g$
- Acceleration at the dam crest: $a_{pk}= 0.555g$

Site Specific Seismic Hazard Assessment was carried out using Deterministic Seismic Hazard Assessment (DSHA) and Probabilistic Seismic Hazard Assessment (PSHA). For site specific response the results of the microseismic survey were used.

The DSHA was used for assessing the maximum credible earthquake (MCE). Two past earthquakes were analysed; Mmax=7.5 along Pambak – Sevan fault and Mmax=7.1 along Garni fault. These earthquakes produced a peak horizontal acceleration of 0.44g and 0.82g at the dam's base and the crest respectively.

The PSHA produced the following peak horizontal acceleration:
a = 0.32g at the base and 0.6g at the crest (Return period of 100years)
a= 0.43g at the base and 0.81g at the crest (Return period of 250years- magnitude saturation occurs after 250years)

Based on the above analyses, the following design peak horizontal accelerations were recommended for checking stability of the dam:
- OBE= 0.32g at the base and 0.6g at the crest (Return period of 100years)
- MCE= 0.44g at the base and 0.82g at the crest (return period of large number of years)

Liquefaction analysis
Liquefaction assessment of the fill material was carried out based on the particle size distribution that did or did not liquefy during past earthquakes, Reference 2, and also the methodology given in Reference 3. According to the Japanese Seismic standard, the liquefaction potential is evaluated by calculating the liquefaction resistance factor, F_L. A soil layer having the liquefaction factor $F_L < 1.0$ is susceptible to liquefaction. An $F_L < 0.6$ was obtained for the fill in the top 10m of the dam (slipped mass) for an average seismic acceleration of 0.5g. Therefore some 60% of reduction in shear strength properties for that zone could be expected to occur during a strong earthquake.

STUDIES CARRIED OUT

Hydrological and Flood Routing
Two methods were used to analyse the flood inflows into the reservoir. The first, the SNIP method (Reference 4), is based on standard Russian techniques and is in general use in Armenia. The second is a statistical method that uses all annual maxima flow data recorded in the region and is derived from the approach developed during investigation of floods in the British Isles (Reference 5). The following results have been obtained for the 1:10,000 year peak flow:
- Regional Method: 147 m³/s
- SNIP: 138 m³/s

In addition to the 1:10,000 year flood, the flood that would result from breaching of the landslide dam due to reactivation of the Landslide N4 (see above) was also considered. The estimated peak inflow for this scenario was 1920m³/s, with a volume of 2.4 million m³.

The flood routing was carried out for the event of a 1:10,000 year flood as well as the event of a failure of a dam created by the N4 landslide. The flood routing was done for the existing condition (empty reservoir), for the design condition with the dam at its full height (FSL at 1911masl) and for an intermediate condition (partial impoundment).

Foundation seepage
Seepage through the dam foundation was analysed for two typical sections, namely for the deepest section with the piled cut-off and the abutment section with a grout curtain only. For a conservative assumption that the anti-seepage measures are ineffective, the total leakage through the dam foundation was assessed at about 100l/s.

Stability analyses

Existing condition of the embankment
Stability analysis of the upstream and downstream slope of the dam at its present condition was carried out using the parameters obtained from the investigations. The analyses demonstrated that the dam was stable with the reservoir empty. However, if impounded, the dam would be unsafe during rapid draw down (u/s slope) and steady seepage (d/s slope).

Stability after the remedial works to the embankment are implemented
Stability analyses was also carried out for three options for the remedial works. The embankment remedial works were developed so that minimum required factors of safety were satisfied for all loading conditions.

Stability of the landslides
The analyses carried out for the Landslide N2 showed that for sliding occurring along the predefined slip plane, factors of safety obtained were lower than unity even in the aseismic conditions. For possible new slip surfaces occurring within the landslide material, factors of safety obtained in aseismic conditions were higher than unity. However, in the case of an earthquake, slippage would occur. The slippage would most likely occur in a direction perpendicular to the ground contours, towards the Kiarkhana river and away from the dam and therefore would not directly affect the dam safety.

It was shown that the stability of the Landslide N3 is largely influenced by the reservoir water level. If the reservoir is filled the landslide would be re-triggered. The volume of the unstable mass was estimate to be 200,000m^3. It was shown that this mass would immediately raise the reservoir level by some 20cm. In addition a wave of 1.5m height would be induced. Such a wave, with its run up of some 2.8m would therefore need a minimum freeboard of 3m in order to prevent the dam from overtopping if the reservoir was full.

The volume of a potentially unstable mass for the Landslide N4 was assessed by stability calculations to be 2400m^3/m of the landslide length. That volume could create a 21m high natural dam which could impound a 2.4 million m³ lake. As the river flow is some 3-4m³/s the volume would be filled within a few days. In a major storm event this could take less than one day. The landslide 'dam' has been considered as an earth embankment and analysed for a dambreak. The analysis indicates a peak flood flow of 1920m^3/s and a flood volume of 2.4 10^6 m^3 (see above).

SUMMARY OF FINDINGS AND THE REMEDIAL WORKS

Embankment

Current crest elevation is approximately at 1900masl for a good part of the dam. Presently the upper part of the failed embankment material forms the top part of the core and the downstream shoulder. There is a very high perched water table within the dam body. Stability analyses demonstrate that the dam is stable in its present condition. However, if impounded to a level 3m below the current crest, the dam would be unsafe during rapid draw down (u/s slope) and steady state seepage (d/s slope). Furthermore, due to the high seismicity of the region, peak ground accelerations at the crest of about 0.6g could be generated. These accelerations are likely to cause liquefaction and strength reduction in the loose landslide material in the top part of the downstream slope and further contribute to the embankment's instability. It is therefore proposed to rehabilitate the embankment to improve its safety. Three options are developed as follows:

- Option 1 - Reinstate the dam to the full height with the crest at 1914 masl; Full storage level at 1911masl, total storage volume 36×10^6 m^3
- Option 2 - Reinstate the dam to elevation of 1905masl; Full storage level at 1902masl, total storage volume 24×10^6 m^3
- Option 3 - Reinstate the dam to elevation of 1889masl; Full storage level at 1886masl, total storage volume 10×10^6 m^3

The earthworks proposed for the above options are shown on Figure 4.

Foundation anti - seepage measures

Foundation anti – seepage measures in the central part of the dam comprise a secant bored pile wall that was constructed through the granular alluvium into the bedrock. The field tests carried out in the deepest section indicated that the cut-off would reduce the overall foundation permeabily and the leakage would not exceed 100 l/s in the worst case scenario. Nevertheless, to reduce the uplift under the downstream shoulder, it is recommended that, for all three options of the embankment remedial works, 20 m deep, 200mm dia toe wells are installed along the downstream perimeter at 10m centres. The wells will comprise a perforated plastic tube wrapped in geotextile and placed inside a hole in a sand surround. Each well will discharge water into a collector trench which runs along the perimeter of the dam.

Landslide hazard

Four potentially hazardous landslides were identified in the vicinity of the dam. The landslides N1 poses a very low hazard to the dam. The landslide N2 is also likely to pose a low hazard, but because of its proximity to the dam it is recommended that monitoring instruments are installed in two monitoring profiles.

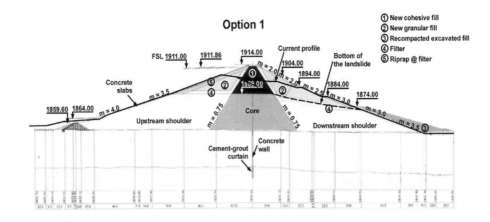

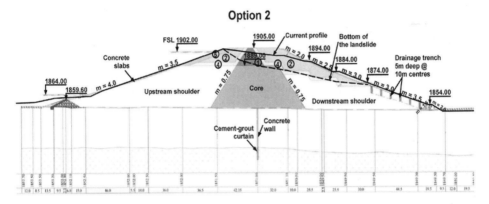

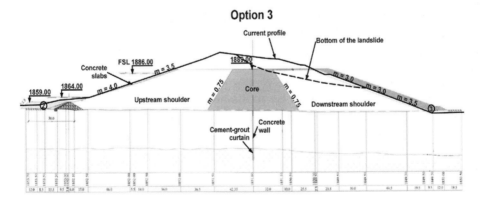

Figure 4. Options for dam rehabilitation

If the Landslide N3 slides into the reservoir it could create a 1.5m high wave and an allowance in the freeboard of 3m is made to accommodate the runup of such a wave. This landslide will also be monitored in two monitoring profiles.

If the landslide N4 collapses, it could block the Marmarik river and create a natural dam which if breached could create a 'dambreak' flood. It is recommended to install 1m high fuse gates (HydroPlus or similar) over the whole length of the spillway crest which could be activated should the landslide occur and the reservoir level needs to be lowered. Alternatively the spillway could remain conventional but the freeboard could be increased to 4m by provision of a 1m high concrete crest wall. It is also proposed to install monitoring instruments on the landslide and monitor the slope movements.

Diversion Tunnel
In its present state the river is diverted into a diversion tunnel designed for a temporary condition. The tunnel was inspected and investigated. The tunnel lining is of a satisfactory strength and the voids between the concrete and the rock are only of a limited extent. The tunnel is therefore considered to be stable in the short term. However the following remedial measures are recommended to enable its operation in the long term:
-   Mass concrete plug upstream of the inlet pipes
-   Consolidation grouting as a circumferential fan to a depth of 15m around tunnel over a 50m length downstream of the plug and backgrouting of tunnel lining in areas of voids.
-   Replacement of tunnel invert downstream of plug
-   Drainholes to be incorporated into invert to minimise hydrostatic loading.

Spillway
The existing spillway is in poor condition and requires substantial remedial works (Option 1). For Options 2 and 3 a new spillway is required at a lower level.

Outlet works
The outlet works require substantial refurbishment.

COST ESTIMATE FOR THE REHABILITATION OPTIONS
Costs for the three rehabilitation options are as follows:
-   Option 1 - $10,5 M
-   Option 2 - $7.5M
-   Option 3 - $5.3M

A cost of decommissioning and breaching of the dam was estimated to be around $5M. It is likely that the client will go ahead with the rehabilitation Option 2.

REFERENCES

1. *Seismic Design Standard of Republic of Armenia II.2.02 – 94.* (1994)
2. Tushida, H. (1970), *Prediction and counter measures against the liquefaction in sand deposits*, pp.3.1 – 3.33, Abstract of the seminar in the Port and Harbour Research Institute
3. *Japanese Seismic Code, Earthquake Resistant Regulations*, (1992) A World List 1992, IAEE
4. *Soviet Design Standard, SNIP 2.01.14 – 83 – Calculation of Design Floods* (1983)
5. *Natural Environment Research council. Flood studies report*, 1975, reprinted 1993.

An update on perfect filters

P R VAUGHAN, Professor Emeritus, Imperial College of Science, Technology and Medicine, London
R C BRIDLE, Rodney Bridle Ltd, Amersham

SYNOPSIS.   Recent work shows that the probability of failure of dams resulting from internal erosion is often higher than that resulting from other threats.   Filters to protect dams against erosion are therefore important. Most of our existing dams are not protected from internal erosion by filters. The 'perfect' filter equation links permeability of filters to the floc size of the soil they will retain. This permeability approach is useful in establishing the vulnerability or otherwise of existing dams to internal erosion because the permeability of fills can be determined by in-situ permeability measurements in boreholes. Floc sizes can also be simply determined using the principles of Stokes' Law in the laboratory.   Some samples display murkiness which obscures the results. Examples of the use of perfect filters are given, including examples of retro-fitting of filters in dams in which they were not originally installed.

GUARDING AGAINST INTERNAL EROSION

It has long been suspected, and recent reservoir safety work for Defra (KBR & BRE, 2002) has demonstrated, that the probability of failure resulting from internal erosion of existing British dams is often greater than from the two other major threats, overtopping and earthquakes. Internal erosion is the process in which soil particles are eroded from the walls of cracks and discontinuities in earth dams by water flowing through them, often at high velocity because of the high hydraulic gradients through dams. Continued erosion leads to enlargement of the discontinuity, often as 'pipes' through the structure, which may erode back from the downstream end initiating a process of slope instability, crest lowering and overtopping that may ultimately cause failure. Internal erosion can be contained by 'filters', non-cohesive soils, usually medium silts to sands, which are sized to retain the soil particles eroded from the soil to be protected (the 'base soil') while allowing water to pass through. This prevents the development of erosion 'pipes' and thereby protects the structure.

How to design filters, particularly filters for existing dams, is likely to become an important dam safety issue in the coming years, and it is timely to update the information available.

CORES, COHESIVE SOILS, CRACKS AND EROSION

In dams, the element most vulnerable to erosion is the waterproofing element, the core, usually of clay. The protection of a dam core is probably the most critical function that a filter must perform. The consequences of failure can be severe damage and even catastrophe.

The vulnerability of cohesive clay cores to erosion arises because cohesive soils are able to sustain open cracks. Cracks or other leakage paths may form through cores during construction, during first filling, because of settlement, arching, hydraulic fracture or other causes. Filters should ensure that the presence of openings does not lead to loss of material from them.

VULNERABILITY OF EXISTING BRITISH DAMS TO INTERNAL EROSION

Most British dams are not equipped with filters and are therefore not equipped to resist internal erosion should it arise. Measures such as puddling, using very wet fill and wetting clay fills to make them softer, were all intended to make these vulnerable soils flexible and able to deform without cracking as the dam deformed in response to foundation settlement, water level variations, earthquakes and other loads.

To further reduce the vulnerability of narrow puddle clay cores in the older British dams, a zone of 'selected fill' was often placed on either side of the core. It was easier for early dam builders to use finer but non-cohesive soil as transition. It was easy to dig and compact and, following the exhaustive discussion at the inquest on the disaster at Dale Dyke dam, which failed in 1864 (Binnie, 1978), the desirability of well rammed fine-grained transition fills was understood and acted upon, more often than not. The 'selected fill' in the transitions may be of a grading that would provide filter protection to the core, as Vaughan (2000a) found at Ladybower, but it may often be cohesive and therefore able to sustain open cracks, making it too vulnerable to erosion and unable to act as a filter.

Fortunately, instances of internal erosion proceeding to serious damage are rare (Charles, 2001). Cohesive cores have considerable resistance to erosion unless they crack or develop concentrated leakage paths for other reasons. Thus satisfactory behaviour in operation may continue for ever. However, erosion may be occurring very slowly and not yet been revealed. Although the general experience is that dams grow safer with time, there is no justification for assuming that because they have not leaked or failed after a given time, they will never leak or fail.

In assessing the risks of internal erosion of dams, one of Vaughan's (2000a) conclusions was that 'usually there is considerable warning, allowing corrective action to be taken'. However, this is not always the case. Catastrophic wash out before remedial action can be taken is the big danger. The risks should be assessed by investigating the dam and, from a knowledge of its properties, evaluating the mechanisms by which internal erosion might develop and the speed at which it might occur. Appropriate defensive measures and surveillance routines can then be put in place.

FILTER DESIGN METHODS
Many methods have been put forward for filter design (e.g. CIRIA/CUR, 1991). Most apply to coarse materials, such as used in coastal protection, but the application of them results in the design of successively coarser layers, each of which is sized so that grains or particles from the adjoining layer will not pass through its neighbour. In an ideal filter, the pore spaces between particles should be just small enough to prevent the passage of the smallest of the protected grains. There is a wide range of sizes in the any granular material and a similar range of pore sizes. Consequently, most filters depend on some of the protected material moving into the filter to make it effective. This is called 'self-filtering'. Most filter rules for non-cohesive soils allow for this.

FILTER DESIGN FOR CLAY CORES
In dams, the element most vulnerable to erosion is the waterproofing element, the core, usually of clay. This poses special problems in filter design because using traditional rules to design filters to protect cohesive soils usually leads to filters of sizes which are themselves likely to be cohesive. These would be capable of keeping cracks open like the core they are intended to protect. Clearly, this offers no effective protection to vulnerable cohesive clay cores and it is generally accepted that different design principles should be applied.

These different principles address the issue of the actual size of the clay particles that filters must retain. Clay particles exist in nature in flocs, groups of individual particles. The floc size is related to the clay type and the pore water chemistry. In some circumstances, such as changes in pore water chemistry brought about by introducing water with differing chemistry, the flocs can be dispersed, partially to form smaller flocs, or completely to be dispersed into individual clay particles. In laboratory particle size distribution tests, the clay portion is artificially dispersed using a dispersant, and the sizes of individual clay particles are determined. Clays are defined as being 2 microns (0.002 mm) or smaller. Clay flocs are larger than this, often around 10 microns (0.01 mm), the medium silt size in the standard particle size distribution.

DISPERSION OF BASE SOILS

Chemical dispersion of the clay in dam cores has a history of causing erosion and washout in arid parts of Australia, Brazil, the U.S. and elsewhere. It is produced by a combination of the chemistry of the clay and the percolating water. Several chemical situations have been identified as causing it (Aitcheson & Wood, 1965; Emmerson, 1967; Stratton & Mitchell, 1976; Perry, 1987). Aicheson & Wood (1965) refer to a dam in Australia which washed out immediately when the water impounded was changed to relatively pure fresh water after several years of successful operation while holding water of a higher salt concentration. They also describe how arid conditions can lead to a ped structure with a much higher permeability than is expected in a clay fill. The large voids in such fill allow the dispersed particles which have been eroded to pass through them.

Dispersive soil is a special case. The authors know of no examples encountered in UK. However, as a precaution, all soils likely to be used in dams should be tested in prospective reservoir waters to demonstrate non-dispersion.

CRITICAL FILTER DESIGN

The most commonly used filter design method is the 'critical filter' approach developed by USDA Soil Conservation Service (1986) and Sherrard & Dunnigan (1989), also given in ICOLD (1994).

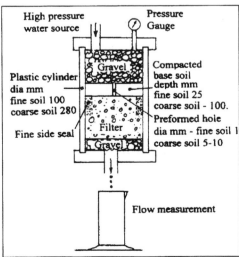

Figure 1 'No erosion' apparatus for the critical filter test

The method was based on an empirical laboratory 'no erosion' test, using the apparatus shown in Figure 1. Samples of base soil and prospective filter were tested by passing water under pressure through a small diameter hole in the base soil into the filter. If water discharged from the filter is clear, it is judged adequate; if water discharged from the filter is not clear, the filter is inadequate. From the results of many tests, the filter gradings that would protect the several groups of core materials were recommended. The groups of core materials are defined using the conventional (i.e. the dispersed, deflocculated) particle size distribution.

'PERFECT' FILTER DESIGN

The alternative design method for filters for clay cores is the 'perfect filter' method. It was devised after sinkholes developed at Balderhead dam on first filling in 1967, as shown on Figure 2 (Vaughan et al 1970; Vaughan & Soares, 1982; Vaughan, 2000b). Segregation was identified in the erosion debris from the clay core found in the damage zone. The sand found in the eroded crack was the remains of the core fill, as the particle size distribution diagram on Figure 2 shows. The sand had been retained by the filter (designed to methods that precede both perfect and critical methods) but finer silt and clay-sized materials had passed through the filter because it was too coarse.

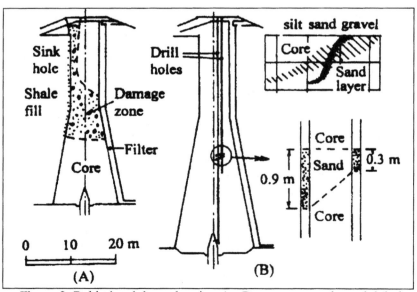

Figure 2 Balderhead dam, showing A - Damage zone where sink hole formed and B – Erosion hole filled with water washed sand

The perfect filter is required to retain the finest material which might be eroded from the walls of a crack in the core. This was taken to be the finest material obtained by mechanical dispersion of the clay in the appropriate water. This was usually clay flocs of around 10 μm (0.01 mm) particle size. Since this was a lower bound approach, no safety factor was required.

The design of a perfect filter involves two steps: first, the determination of the size of particle which must be retained and, second, the filter grading which is required to retain it. The filter grading required in design rules for non-cohesive soils is based on the finer sizes present, usually the 15% size. When the filter design was evolved it was found that the size of particle retained correlated well with filter permeability. The permeability of a filter is determined by the size of the continuous pores through it. Moreover the permeability is likely to vary with particle shape and it will vary with density of packing. While for uniform soils the permeability correlates with such an approach quite well, for well-graded soils the permeability depends on finer sizes and cannot be correlated with a particular percentage size.

The size of particle retained by a given filter was found experimentally by preparing different sizes of particle and passing them in dilute suspension through the filter (Vaughan and Soares, 1982). Either the sediment passed through the filter immediately or it sealed the surface, causing the flow rate to decrease rapidly. There was a small zone where the sediment clogged the surface more slowly. This was counted as retention. The test was more difficult to interpret when it was performed at a larger scale on filters containing gravel-sized particles. The results are summarised on Figure 3:

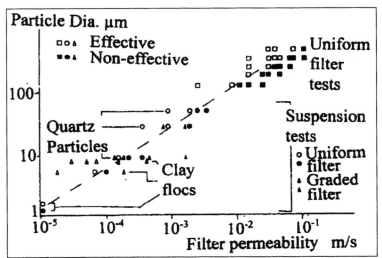

Figure 3 Summary of 'perfect' filter tests determining filter permeability required to retain base soils of various particle (and floc) sizes

Vaughan and Soares (1982) found that the relationship between filter permeability and the size of particles retained could be expressed as:

$$\delta_R = 1.49 * 10^3 \, (k)^{0.658}$$

where: δ_R = size of smallest particle retained in microns (10^{-6} mm)
k = permeability of filter (m/s)

The application of these findings to the erosion at Balderhead is illustrated on Figure 4. The grading of the core is shown, as is the grading of the portion of the core material retained by the 'actual' filter. This is the sand shown on Figure 2 above. The D_{15} range of the 'critical' filter that would have been provided to protect the core is also shown, as is the grading of the 'perfect filter'. It can be seen that the critical filter would have been too coarse to prevent the erosion that occurred through the cracks. Note also the modified core grading showing how it curtails at the minimum floc size, about 7 microns (0.007 mm), medium silt size.

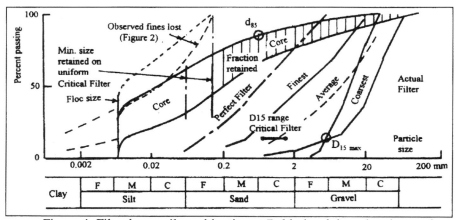

Figure 4 Filter base soil combination at Balderhead dam showing perfect filter, critical filter and observed segregation

COMPARISON OF PERFECT FILTERS WITH CRITICAL FILTERS

It is of interest to compare critical filters with perfect filters. This has been done by Vaughan (2000b) and the results are summarised on Table A below. The results are for filters of the appropriate critical filter base soil groups. No critical filter method 'no erosion' tests have been performed. The comparison has been made in terms of the minimum size of particle retained. This has been deduced from Sherrard & Dunnigan (1989) for the critical filters by first estimating the permeability of the critical filter from the relationship between permeability and the D_{15} size (Vaughan, 2000b).

The size of particle retained is then deduced from the perfect filter relationship.

Table A: Comparison of perfect and critical filters

Dam	Perfect Filter Design		Critical filter details deduced from Sherrard & Dunnigan (1989) Filter Provided			
	Floc Size δ_R (μm)	Permeability k (10^{-5} m/s)	Core Soil Group	D_{15} of filter (μm)	Permeability (10^{-5} m/s)	Size retained δ_R (μm)
Ardingly, UK	10	22	2	700-1500	319 - 1228	34-82
Carsington, UK	8	16	1	180	29	7
Cow Green, UK	6	10	2	700-1500	319-1228	34-82
Dhypotamus, Cyprus	6	10	2	700-1500	319-1228	34-82
Empingham, UK	10	22	1	90	9	3
Evinos, Greece	11	26	2	700-1500	319-1228	34-82
Kalavasos, Cyprus	5	8	2	700-1500	319-1228	34-82
Monasavu, Fiji	20	13	1	70	5	2
Balderhead, UK	7	13	2	700-1500	319-1500	34-82

The Critical Filters are more conservative than Perfect Filters for Group 1 cores (plastic clays) (e.g. 3 microns against 10 microns actual floc size at Empingham) and significantly less conservative for Group 2 cores (well graded sandy clays) (e.g. 34-82 microns against 7 microns at Balderhead). This is despite Group 2 cores giving poorer field performance. For the Group 1 cores the critical filters are more conservative than the perfect filters, despite the latter being able to arrest the smallest particle which may develop during erosion.

DETERMINATION OF FLOC SIZE

To use the perfect filter design method, the floc size of the core soil must be known. It is commonly determined using standard particle size analysis techniques (e.g. hydrometer) on samples slaked in reservoir water only, NOT subjected to the usual chemical dispersion process. Figure 2 above shows the results for the Balderhead core material. Often samples with and without dispersion, and sometimes without dispersion but in distilled, not reservoir, water, are also tested; these are the so-called 'double' and 'triple', respectively, dispersion tests. While the minimum floc size often shows up well in these tests, it is not always clear.

A simpler test (Head, 1992), which normally shows the floc size clearly, is based on Stokes Law, which relates the size of bodies falling through a liquid to their size. In our case, the smallest flocs sink slowest and can be seen as a falling front above which is clear water. The rate of fall of the front can then be used to determine the size of the smallest flocs present by using the version of the Stokes Law formula below:

$$D = 0.005\ 531\ \{(\eta\ H) / (t\ (\rho_S - 1))\}^{0.5}$$

where: D = minimum floc size (mm)

   H = distance floc front falls (mm) in time t (mins)

   t = time (mins) to fall H (mm)

   ρ_S = mass density of soil particles, should be measured, but is commonly in the range 2.6-2.7

The dynamic viscosity of water, η, varies with temperature, as follows:

Temperature (°C)	Dynamic Viscosity, η (mPa-s)
10	1.3037
15	1.1369
20	1.0019
25	0.8909

The rate of fall of flocs of the sizes normally encountered is quite rapid and Stokes law tests can be done quickly. The table below shows the time that flocs of various sizes take to fall 300 mm and gives information on the floc sizes and the filters required to retain them:

Mins to drop 300 mm	Terminal velocity mm/s	Floc size microns	Floc texture	Permeability perfect filter m/s	D_{15} uniform perfect filter mm	Texture D_{15} perfect filter
5	1.00	32.9	Coarse silt	3.04E-03	0.681	Coarse sand
15	0.33	19.0	Medium silt	1.32E-03	0.424	Medium sand
45	0.11	11.0	Medium silt	5.73E-04	0.265	Medium sand
90	0.0556	7.8	Medium silt	3.38E-04	0.196	Fine sand
180	0.0278	5.5	Fine silt	2.00E-04	0.146	Fine sand
360	0.0139	3.9	Fine silt	1.18E-04	0.108	Fine sand
1080	0.0046	2.2	Clay (deflocculated)	5.12E-05	0.067	Fine sand

THE 'MURKINESS' PROBLEM

Sometimes in the Stokes Law test the falling front is not visible. The sediment can be seen to arrive at the base of the measuring cylinder, but the water above remains 'murky' and opaque, so that the falling front cannot be seen. The source of the murk is not known. It usually persists for extremely long periods, longer than even the smallest clay particles would take to settle, and it seems unlikely it comprises dispersed clay flocs. Its source may be the same as the source of 'colour' in treated water, although it is more severe, making the water opaque, not transparent as 'coloured' waters are. It seems prevalent in alluvial soils, perhaps because organic materials are present. It complicates a simple and useful test, easily done in the field, and research into its source and how to overcome the murkiness without affecting the validity of results would be valuable.

PERMEABILITY AND GRADING OF FILTERS

The use of a relationship that relates retained floc size to the permeability of the filter reflects the fact that permeability is related to pore sizes. However, measuring the permeability of a potential filter is less convenient than measuring its grading and the expression below (Vaughan 2000b), which is for uniform filters, is useful to give an early indication of the grading of potentially suitable filters:

$$k = 3 * 10^{-8} (D_{15})^{1.767}$$
where: D_{15} = D_{15} size of uniform filter (in μm, microns)
      k   = permeability of filter (in m/s)

Note that the actual filter, if not uniform (i.e. $D_{60}/D_{10} > 1$), will have a different permeability, and therefore a different filtering capability, and the permeability of candidate filters should be measured before they are used.

The permeabilities of filters retaining clays flocs are low and their drainage capacity is therefore limited. If filters are protecting fills that include permeable layers that may allow substantial quantities of seepage to pass, it may be necessary to provide a coarser drainage filter downstream of them to allow the seepage to escape freely. To pass the quantity, the hydraulic gradient across the low permeability, low capacity filter is high, and the gradient along the high permeability, high capacity drainage filter is low.

FILTER PROPERTIES

Filters should be non-cohesive, at least as placed. The 'sand-castle test' described by Vaughan & Soares (1982) is a convenient and quick means of proving non-cohesiveness at source. Granular soils may bond with age and develop cohesion, although so far as the authors know, no problems have been reported from this cause.

Filters must be internally stable and self-healing. Kenney & Lau (1985) and Lafleur et al (1989) give methods to check the internal stability of non-uniform filters.

A further check on the suitability of filters can be made by passing 'muddy' water made from (reservoir) water containing the base soil through a layer of the filter in a permeameter. Adequate filters retain the 'mud' and clear water passes through. Inadequate filters allow the muddy water to pass.

IN-SITU PERMEABILITY AS A GUIDE TO THE VULNERABILITY OF EXISTING DAMS TO INTERNAL EROSION

The filtering capacity of non-cohesive shoulder fills can be assessed from in-situ permeability measurements. For example, Vaughan (2000a) found non-cohesive silty sandy gravel transition fill at Ladybower to have a maximum permeability of $4*10^{-6}$ m/s. This provides perfect filter protection to the adjoining clay core in which the minimum floc size is about 10 microns. As the fill tested may not be uniform, use of a lower bound to the permeabilities measured may be appropriate. The filter relationship between the transition and the general shoulder fill should also be checked as transition fills may erode into coarse shoulder fills.

The perfect filter equation makes the connection between floc size retained and filter permeability, as follows:

$$\delta_R = 1.49 * 10^3 \, (k)^{0.658}$$

where: δ_R = size of smallest particle retained in microns (10^{-6} mm)
     k = permeability of filter (m/s)

The equation was derived for non-cohesive filters with permeabilities ranging upwards from 1×10^{-5} m/s. Use of the equation to determine the floc size of soils that would be retained by soils with in-situ permeability less than 1×10^{-5} m/s should be cautious. If the soils are cohesive, improbable results emerge (Tedd et al, 1988). In practice, this means that in low permeability fills, the cohesiveness of the soil should be checked, and the floc size of cohesive materials should be determined in the laboratory.

Note that samples taken from boreholes in fills with substantial proportions of granular materials are likely to have lost fines and not be properly representative of the in-situ fill, consequently laboratory permeability tests do not give usable results. In-situ tests are needed, usually from piezometers installed in boreholes. These may also serve for measuring pore pressure in the investigation of old dams

However, as Charles et al (1996) point out, the sand in the sand-pockets in piezometers installed in fill will usually have a permeability up to about $2*10^{-5}$ m/s. If this is less than the fill in which the piezometer is sited, it will appear that the fill will retain smaller flocs than it is capable of retaining, an unsafe situation. A progressive approach to determining the permeability, and hence the filtering capacity, of fills that may be required to protect against internal erosion is therefore recommended, commencing with in-situ permeability tests in boreholes.

DAMS WITH PERFECT FILTERS

There is a growing body of dams with perfect filters, as listed on Table B below:

Table B: Dams with perfect filters

Dam	Perfect Filter Design		Filter Provided			
	Floc Size (μm)	Perm-eability (10^{-5} m/s)	Filter Soil Type*	Perm-eability (10^{-5} m/s)	Size re-tained δ_R (μm)	D_{15} of filter (μm)
Ardingly, UK	10	22	ns	9	3	230
Carsington, UK	8	16	psg	1 to 10	1 to 3	80-170
Cow Green, UK	6	10	ns	2	1	110
Dhypotamus, Cyprus	6	10	sng	1	1	1000
Empingham, UK	10	22	ng	8	3	100
Evinos, Greece	11	26	sng	10	3	220
Kalavasos, Cyprus	5	8	sng	4	2	600
Monasavu, Fiji	20	13	cr	4	2	210
Balderhead, UK	7	13				
Melton Mowbray, UK	4	12	ns	10	3.5	150
Audenshaw, UK	6	23	ns	10	3.5	

* ns = natural sand   psg = processed sand and gravel   ng = natural gravel
cr = crushed rock sng = natural sand and gravel screened to remove coarse sizes

It has always proved possible to find or make perfect filter gradings for the cases listed above, although this was sometimes difficult. Dounias et al (2000) describe how river gravels were used as the core filter at Evinos Dam. Hughes et al (2001) and Bridle (2003) describe the filter investigations at Audenshaw and Melton Mowbray respectively.

It must be emphasised that the perfect filter is only required to protect against erosion by continuous reservoir flow through cracks or other flow paths which are in cohesive soils, and which can sustain such an opening without sealing by collapse. This is typically a core, but where the foundation is of erodible clay, a short length of perfect filter blanket is often added on the foundation downstream of the core, where significant hydraulic gradients exist. The principle of the Perfect Filter for cohesive soil is that erosion through concentrated reservoir flow is prevented. Intrinsically, no cause for such flow is presumed. A relatively thin layer of filter has been considered acceptable.

It is inevitable that the filter provided is less permeable than the Perfect Filter required. This gives a safety factor, although one is not required.

RETRO-FITTING FILTERS
In dams which are found to be unacceptably vulnerable to internal erosion, filters will be required. This presents some challenges. Although perfect filters will protect fills for which they are designed, fills in old dams may be variable. Also, if the filters are incomplete and do not cover the entire exposed fill, erosion may still occur. Protecting against foundation erosion is particularly difficult. Methods of retro-fitting filters to meet these challenges will have to be devised, probably derived from previous experiences, a few examples of which are described here.

At Lower Tamar, a filter layer was placed below a weighting berm on the downstream slope to collect and filter seepage passing through the core (Kennard, 1972). Care should be taken to make sure that arrangements such as this have a sufficient weight to secure against a concentrated leak (Vaughan, 2000a). Bailey (1986) describes the provision of a filter wall to prevent erosion through tension cracks near the top of the core and the installation of a geotextile filter behind a retaining wall at the toe of the downstream slope to filter seepage passing through Upper Litton dam. Talbot & Ralston (1985) give examples of retro-fitting of filters to deal with cracks and potential internal erosion in dams, including flood dams.

Jairaj & Wesley (1995) describe the construction of a filter wall drain using a bio-polymer slurry at Hays Creek dam. The wall drain was excavated using slurry support in the usual way, and the trench filled with filter sand placed by tremie pipe, displacing much of the slurry. Water and sodium hypochlorite was pumped through the sand/slurry in the trench to break down and remove the remaining slurry, leaving the sand as a filter at the required permeability in the trench.

Filter collars can be provided near the downstream ends of culverts and pipes through dams to limit risks of erosion along the interface between these structures and the dam fill. Talbot & Ralston (1985) advocate filter collars, and give information on suitable dimensions and positioning.

CONCLUSION
The perfect filter approach to providing filter protection against internal erosion in dams provides a rigorous means to design safe, effective filters. It can also be conveniently used to assess the vulnerability to internal erosion and the need for filters in existing dams. The aim of this paper is to make the perfect filter approach accessible to European, including British, dam engineers to assist them in keeping their dams safe from damage through internal erosion in the long term.

REFERENCES
Aitchison G D and Wood C C (1965). Some interactions of compaction, permeability and post-construction deflocculation affecting the probability of piping failure in small earth dams. *Proc Int Conf Soil Mech & Foundn Engng, Montreal.* Vol. 2, pp 442 - 446.

Bailey M C (1986). Design and construction of new spillways and other remedial works to the Litton reservoirs near Bristol. *Journal of the Institution of Water Engineers and Scientists*, Vol 40, No 1, pp 37-53.

Binnie G M (1978). The collapse of the Dale Dyke dam in retrospect. *Quarterly Journal of Engineering Geology*, Vol 11, No 4, pp 305-324.

Bridle R C (2003). Melton Mowbray flood storage dam. *Dams & Reservoirs, Journal of the British Dam Society*, Vol 13 No 1.

Charles J A (2001). Internal erosion in European embankment dams. Dams in a European Context. *ICOLD European Symposium, Keynote Lecture, reports from European Working Groups and late papers.* Norwegian National Committee on Large Dams, Oslo.

Charles J A, Tedd P, Hughes A K and Lovenbury H T (1996). *Investigating embankment dams – a guide to the identification and repair of defects.* Building Research Establishment Report. Construction Research Communications Ltd, Watford, 81 pp.

CIRIA/CUR (1991). *Manual on the use of rock in coastal and shoreline engineering.* Construction Industry Research & Information Association, Special Publication 83, Centre for Civil Engineering Research & Codes Report 154. CIRIA, London. ISBN 0 86017 326 7

Dounias G T, Dede V and Vaughan P R (2000).. Use of river gravel for the core filter of Evinos dam. *Filters and drainage in geotechnical and environmental engineering, proceedings of the third international conference - Geofilters 2000, Wolski W & Mlynarek J, eds.* Balkema Rotterdam. ISBN 90 5809 146 5

Emmerson W W (1967). A classification of soil aggregates based on their coherence in water. *Australian Jour. Soil Research.* Vol. 5,47 - 57.

Head K H (1992). *Manual of soil laboratory testing.* 2 vols. Pentech Press, London. ISBN 0 7273 1318 5.

Hughes A K, Lovenbury H and Owen E (2001). Audenshaw Reservoir. *Dams & Reservoirs, Journal of the British Dam Society*, Vol 11 No 1.

KBR & BRE (2002). *Reservoir safety - floods and reservoir safety integration.* Report XU0168 for Dept for Environment, Food and Rural Affairs, www.defra.gov.uk/environment/water/rs/02/index.htm, London

International Commission on Large Dams (1994). *Embankment dams - granular filters and drains.* Bulletin 95. ICOLD, Paris.

Jairaj V and Wesley L D (1995). Construction of a chimney drain using bio-polymer slurry at Hays Creek dam. *International Water Power & Dam Construction,* February, pp 30-33.

Kennard M F (1972). Examples of the internal conditions of some old earth dams. *Journal of the Institution of Water Engineers and Scientists,* Vol 26, No 3, pp 135-154.

Kenney T C and Lau D (1985). Internal stability of granular filters. *Canadian Geotechnical Journal.* Vol 22, No 2, pp 215-225.

Lafleur K, Milynarek J and Rollin A L (1989). Filtration of broadly graded cohesionless soils. *ASCE Journal of Geotechnical Engineering Division,* January, Vol 115 Dec.

Perry E B (1987). Dispersive clay erosion at Grenada Dam, Mississippi. *Engineering Aspects of Soil Erosion, Dispersive Clays and Loess. ASCE Geotech. Spec. Publicn. No 10,* pp 30 - 45.

Sherrard J L and Dunnigan L P (1989). Critical filters for impervious soils. *ASCE Journal of Geotechnical Engineering Division,* January, Vol 115 (GT7) pp 927-947.

Sherrard J L, Dunnigan L P and Decker R S (1976). Identification and nature of dispersive soils. *Jour. Geotech. Engng. Div. ASCE* Vol. 102, No GT4, pp 287-301

Stratton C T and Mitchell J K (1976). Influence of eroding solution composition on dispersive behaviour of a compacted clay shale. *Dispersive Clays, Related Piping and Erosion in Geotechnical Projects.* American Soc. Testing Materials Spec. Tech. Publicn. 623, pp 398 - 407.

Talbot J R and Ralston D C (1985). Earth dam seepage control, SCS experience. In *Seepage and leakage from dams and impoundments. Volpe R L and Kelly W E eds, Proceedings of a symposium sponsored by the Geotechnical Division in conjunction with the ASCE National Convention, Denver, Colorado, May 5 1985.* American Society of Civil Engineers, New York. ISBN 0-87262-448-X.

Tedd P, Claydon J R and Charles J A (1988). Detection and investigation of problems at Gorpley and Ramsden dams. *Proceedings of BNCOLD Conference,* Manchester, paper 5.1.

USDA Soil Conservation Service (1986). *Guide for determining the gradation of sand and gravel filters.* Soil Mechanics Note No 1 210-VI. United States Dept of Agriculture, Soil Conservation Service, Engineering Division, Washington

Vaughan P R (2000a).   Internal erosion of dams – assessment of risks. *Filters and drainage in geotechnical and environmental engineering, proceedings of the third international conference - Geofilters 2000, Wolski W & Mlynarek J, eds.* Balkema Rotterdam. ISBN 90 5809 146 5

Vaughan P R (2000b).   Filter design for dam cores of clay, a retrospect. *Filters and drainage in geotechnical and environmental engineering, proceedings of the third international conference - Geofilters 2000, Wolski W & Mlynarek J, eds.* Balkema Rotterdam. ISBN 90 5809 146 5

Vaughan P R, Kluth D J, Leonard M W and Pradoura H H M (1970). Cracking and erosion of the rolled clay core of Balderhead dam and the remedial works adopted for its repair. *Transactions of 10th International Congress on Large Dams, Montreal,* Vol 1, pp 73-93.

Vaughan P R & Soares H F (1982).   Design of filters for clay cores of dams. *ASCE Journal of Geotechnical Engineering Division*, January, Vol 108 (GT1) pp 17-31.

Remedial drainage to Laggan and Blackwater gravity dams

R P WALLIS, Civil Engineer and Reservoir Supervising Engineer, Alcan Primary Metals (Europe) Ltd, Fort William, UK.
A C MORISON, AR Panel Engineer, Halcrow Group, Swindon, UK.
R GUNSTENSEN, Engineer, Halcrow Group, Inverness, UK.

SYNOPSIS. Alcan's 48m high Laggan Dam and 26m high Blackwater Dam have both been reassessed for extreme floods and seismic loading, and stability at both was found to fall short of modern guidelines. In the case of Laggan Dam the critical load case was the PMF, which would overtop the substantial masonry walls of the spillway bridge. Blackwater Dam stability was found to be marginal under normal conditions and unsatisfactory under both extreme load cases because of its slender section and serious doubts that it could carry tensile stresses, particularly at the foundation contact. An assessment of alternatives found that remedial drainage provided the cheapest satisfactory solution at both dams. At Laggan Dam this involved drilling from the dam crest to intersect the gallery and from the gallery into the foundation. At Blackwater Dam, which has no gallery, the solution involves inclined holes from the downstream face to intersect the dam/foundation interface.

The paper sets out the studies, investigations and design of the remedial works, and implementation of the work at Laggan Dam. Implementation is still to take place at Blackwater.

ALUMINIUM IN THE SCOTTISH HIGHLANDS.
The British Aluminium Company was formed in 1894 with the aim of developing the new process of electrolytic reduction, which depends on the availability of cheaply produced electricity for the commercial production of aluminium. The company's first hydroelectric power station was built at Foyers on Loch Ness in 1896. It produced 3MW of power and was capable of satisfying one tenth of the world demand for aluminium, which at that time stood at two thousand tons per annum; this compares with over 18 million tons today.

Long-term benefits and performance of dams, Thomas Telford, London, 2004, 532–543

The Company expanded their facilities in the Highlands with an aluminium smelter in Kinlochleven in 1907 and Lochaber Smelter in Fort William in 1929. Both of these are powered by their own hydroelectric schemes with Kinlochleven generating 20MW and the Lochaber Scheme 65MW. These schemes were each massive engineering undertakings in their day with a combined catchment area of 940sq km.

The Foyers Smelter was closed in 1967, although the hydropower scheme was taken over and redeveloped by Scottish Hydro Electric. In 1981 the British Aluminium Company was taken over by Alcan. The smelter at Kinlochleven was closed in 2000. However its hydropower scheme was retained and refurbished to produce supplementary power for the Lochaber Smelter. Lochaber smelter and hydropower scheme are still in operation. It is impressive that the original hydropower developments have stood the test of time, both being largely still in operation in their original form.

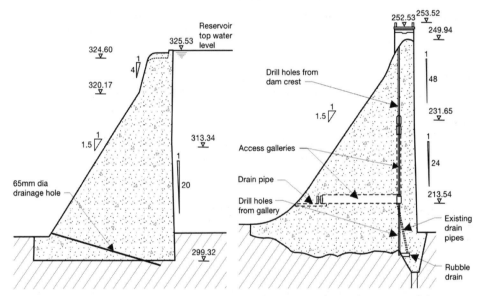

Figure 1   Sections of Blackwater and Laggan dams (not to scale)

BLACKWATER DAM

Background

Blackwater Dam impounds the main storage reservoir for the Kinlochleven Hydro Electric Scheme. The dam is situated some 70 miles north of Glasgow on the west side of Rannoch Moor. Access to the dam is via a 5 mile long rough track suitable only for small 4WD vehicles. The reservoir supplies water to the power station in Kinlochleven, 8km from the dam, via a covered free-flow channel and surface penstocks. Francis turbines and AC generators were installed in the power station following the closure of the aluminium smelter in 2000, replacing the original 1908 DC Pelton units.

The 26 meters high concrete gravity dam was completed in 1909. It is 948 metres long and 503m of the crest, ie more than half the length of the dam, is overflow spillway at an elevation of 325m aOD. The dam section is shown on Figure 1. There is a central valve tower on the upstream face from which water is drawn-off through pipework in the base of the dam. The reservoir is 12 kilometres long and holds 111 million cubic metres at top Water Level. At the time of construction the reservoir was thought to be the largest in Europe. Reservoir levels often reach spill level during the winter months and are typically drawn down twelve to fifteen metres during the summer. The dam has been in continuous service since it's construction.

The dam is composed of a mass concrete matrix in which are embedded heavy granite displacers. Hearting concrete in the dam was a 1:5 mix with up to 50mm aggregate and a slightly richer 1:4 mix with 19mm aggregate used for the facings. Rock was quarried locally to the site but much of the sands and gravels were imported from a gravel bank in the tidal loch near Kinlochleven. The rock foundation was excellent with almost no fissures being found in the foundation area. The maximum depth of excavation to sound rock was only 4.6m. Cement mortar 50mm thick was laid on the rock foundation with a further 25mm thick layer placed prior to the placing of the concrete by derrick cranes. Large granite displacers weighing up to 10 tons were embedded in the concrete with many of them bridging the lift joints. Further details are given in the 1911 ICE construction paper [7].

The section of Blackwater dam is slender [Figure 1] and stability has always been recognized as marginal. The narrow section was commented on in the discussion following presentation of the construction paper in 1911 [7] in relation to other dams of the period. This appears, with hindsight, to be because uplift was not properly taken into account in the design. Moreover, at some time after this, the top water level was raised by about 0.9m by infilling the lower parts of the crest in the original stepped spillway.

The vertical cracks found in the dam shortly after completion were of great interest to the civil engineers of the time and may well have been influential in the inclusion of vertical contraction joints in subsequent dams. Interestingly a small water supply dam built shortly afterwards in the valley below has vertical joints. There were seven main cracks which opened up to 2.4mm wide. Water wept through the cracks and attempts to seal them using silicate solution and fine grout were largely unsuccessful. Subsequent treatment using peat introduced into the water on the upstream face of the dam were more successful. Practically no water was leaking through the dam by 1910, the cracks having apparently having been sealed either by sediments and peat in the water or by leached free lime from the cement. This built up as a hard white deposit on the downstream face dam and still builds up today, particularly below any small leaks and weeps.

The first inspections under the 1930 Act in 1933 and 1943 reported that the leakage through the contraction cracks was small, but stability was investigated in 1935, which included taking cores from the dam.

In 1963 the deterioration of the upstream side of the horizontal construction joints was considered to be a problem. Significant effort was put into sealing the upstream face of these joints and the vertical cracks by breaking out unsound mortar and concrete and refilling with mortar overlain with bitumen reinforced with a fiberglass mat. This significantly reduced the amount of leakage through the dam. Periodic maintenance of the upstream face sealing has been carried out since 1963.

In 1979 the stability of the dam was reassessed using information gathered from new core holes and piezometers, following which the frequency of inspections increased to five and then ten years. This work was recorded in an 1982 ICOLD paper [9].

Leakage readings are taken every two months from six fixed points downstream of the dam and also from the base of the draw off tower.
Foundation piezometers are read twice a year at high and low reservoir levels. In recent years some have tended to show a rising trend, and their condition has deteriorated, making readings less reliable. Tower plumb bobs are monitored annually. Movement stations are surveyed every four years.

Recent studies and investigations
Blackwater Reservoir lies above the town of Kinlochleven and is Category A to the standards in Floods and Reservoir Safety [2]. Following the 1993 periodic inspection, a hydrological assessment was carried out in 1995 [10] based on the Flood Studies Report [1]. This calculated a peak PMF outflow of $1043m^3/s$, compared with the original design flood of about $377m^3/s$. Calculated maximum flow depth over the 503m long spillway in its present configuration was 1.09m. This is within the height of the wave wall, although deficient in the recommended wave freeboard. A subsequent check has shown that the water level produced by a 1:10,000 year flood based on the Flood Estimation Handbook [5] is less than this.

A Flood Stability Assessment [11] of the dam was carried out in 1996 for the revised PMF design flood level. No provision for foundation or dam body drainage was included in the design, but uplift pressures are measured at the six piezometers installed in the 1970s. The gravity method stability analysis concluded that Blackwater Dam relies on tensile strength on the lift joints to provide the recommended factors of safety under both normal and PMF conditions. If tensile strength is ignored and cracks exist on lift joints in the upstream face, the factor of safety against overturning is about 1.15, as against a recommended value of 1.5 in the Engineering Guide to the Safety of Concrete and Masonry Dam Structures in the UK [4], and could drop to as low as 1.03 under PMF conditions. Both were considered unacceptably low. The main recommendations of the report were:

- that an investigation of the vertical tensile strength of the lift joints in the dam be carried out; and
- that the performance of the dam under seismic loading be included in the review.

These recommendations were reviewed and endorsed in 1997 in a Section 10 Inspection report, which also recommended that any unacceptable deficiency in stability should be remedied.

Historical research revealed that formation of the lift joints is described in the construction paper [7] as follows:

"Before concreting a new layer the surface of the old one below it was thoroughly cleansed, roughened, and covered with 1 inch of cement mortar, upon which, while still fresh and soft, the new concrete was deposited. This was done to ensure a sound and watertight seam between layers.

In this connection mention may be made of a particular characteristic of the rotary cement. A fine, brown scum formed on the surface of the concrete, and, if left, set hard with a skin like glass, this making it difficult to obtain a sound joint with the next layer. It became necessary, therefore, to destroy this skin by brushing the surface when partly set and thereby leaving it rough. Care was taken to have this done always."

The hearting of the dam was constructed incorporating large granite displacer blocks, weighing up to 10 tons. The extent of these is clear from the construction photographs [Figure 2]. Particular care was taken to include these across lift joints, and they have a significant influence on the tensile strength at the dam body. Inspection of the joints at the upstream and downstream faces, however, suggested that the concrete material in the joints is significantly weaker than the rest of the dam concrete [Figure 3].

Records of borehole investigations of the dam in 1935 and 1978 were examined, as were details of the 1978 investigation given in the 1979 Stability Report [8] and ICOLD paper [9]. However neither of these considered the tensile strength of the lift joints.

Figure 2   Displacer blocks on lift joints at Blackwater Dam

Figure 3   Raked-out joint in upstream face at Blackwater Dam

Alcan awarded a site investigation contract to Exploration Associates in July 1998 for drilling horizontal cores at joints and vertical cores through the dam and subsequent laboratory testing. Work commenced on site in August

1998 and was completed in October 1998. The factual report covering the drilling and testing was submitted in December 1998 [12]. Few of the lift joints were recovered intact and significant areas of honeycombing and poor joint bonding were found in the cores.

A further Stability Review [13] taking the collected data into account, concluded that, while concrete tensile strength could not be expected at lift joints, the rock displacers provided an acceptable degree of tensile strength across lift joints in the dam body. However unacceptable cracking could still occur at the dam foundation/rock interface. The cheapest option to remedy this was found to involve drilling inclined holes from the downstream face to intersect the foundation contact and so relieve uplift pressures.

An initial pseudostatic seismic analysis based on the UK Seismic Guide [3] had indicated tensile stresses at the dam heel likely to produce excessive cracking, leaving the post-seismic cracked section only marginally stable. However a lower 1:10,000 peak horizontal ground acceleration of 0.2g was derived from the Application Note to the Seismic Guide [6] and confirmed against site-specific seismic accelerations calculated for Scottish and Southern Energy dams in the area [15]. A more detailed 2-dimensional dynamic analysis using EAGD-SLIDE [14] subsequently demonstrated that, with the drainage works required for flood stability in place, direct seismic failure is most unlikely, although cracking may still occur, and that post-seismic stability is sufficient that any damage caused can be assessed and, if necessary, can safely be dealt with after the event.

Implementation
Design drawings and Tender documents for the recommended remedial drainage works at Blackwater Dam were prepared in 2003 and the works are planned for construction in 2004 and 2005.

LAGGAN DAM

Background
Laggan Dam impounds 40 million cubic meters for the Lochaber Hydro Electric Scheme. The dam is situated some 100 miles north of Glasgow at an elevation of 250m in Glen Spean, west of Spean Bridge. Access to the dam is via a short road immediately off the A86 trunk road. Now owned by British Alcan the reservoir supplies water via Loch Treig to the power station at the aluminium smelter in Fort William.

Completed in 1934, the dam is a conventional mass-concrete gravity dam some 48m high between general foundation and spillway crest level. It is slightly curved upstream in plan, but was designed as a purely gravity structure. The whole crest of the dam is a free-overflow spillway except for a central block housing siphons and gate control equipment. The spillway crest is broken into 29 bays by piers, supporting bridge arches. The upstream and downstream faces of the bridge consist of massive masonry wave walls. Six siphon pipes embedded in the dam concrete supplement the crest spillway discharge. The siphons make and break automatically at pre-set reservoir levels using a system of air valves.

The dam was built in 7 blocks with both copper strip and hot poured asphalt water stops in the joints. The bulk of the dam body is constructed of mass concrete with a nominal 15N/mm^2 characteristic design strength. However recent testing of cores from the dam has shown average strengths of 28.5N/mm^2, and a peak strength of 36N/mm^2. Higher strength concrete was used in the external faces. The dam concrete contains about 5% of granite displacers. The dam foundation rock is fresh or slightly weathered granite. After the dam was completed, gunite was applied to the upstream face to reduce leakage through any contraction cracks that appeared. The gunite was applied in two layers onto a wire mesh fixed to the face of the concrete.

The dam is generally in excellent condition with practically no leakage or signs of movement. The original gunite facing has not been a success and can be seen lifting away from the face of the dam at low reservoir levels. Until the remedial drainage works, completed in 2001, no works of any significance had been required on the dam since construction.

The total leakage from the foundation drains into the gallery is monitored. Leakages are small and increase with increasing reservoir level. The foundation drainage system is tested annually by forcing water by means of a packer into each drain pipe in turn and noting connections to adjacent pipes. This monitors the integrity and porosity of the rubble drain system.

Recent Studies and Investigations

Laggan Dam is situated upstream of Roy Bridge and Spean Bridge and is considered as Category A by the standards of Floods and Reservoir Safety [2]. Following the 1993 periodic inspection, a Flood Hydrology report [10] was prepared in 1995 using the Flood Studies Report [1]. This calculated the routed PMF outflow through the siphons and over the free spillway on the dam crest as 2073m^3/s with a corresponding flood rise of at least 3.07m. This compares with the original design flood of 396m^3/s and corresponding flood rise of 0.87m above spillway crest given in the 1937 construction paper [16]. This assumes that the spillway discharge is unrestricted. Taking the effect of the arches into account, the peak PMF outflow is restricted to 1620 m^3/s but the peak flood level rises to 4.09m above spillway crest level, or 0.54m above the top of the masonry wave wall on the dam crest bridge. A check has shown that the water level produced by a 1:10,000 year flood based on the Flood Estimation Handbook [5] is less than this.

Halcrow carried out a Flood Stability Assessment of the dam [17] in 1996 for the revised flood water level. The highest sections of Laggan Dam contain a foundation gallery and copper drain pipes are provided between the gallery and a rubble drain on the foundation at the back of the cut-off trench, but the dam body above the gallery is undrained, and drainage of the section below the gallery into the copper pipe drains is doubtful. The construction paper [16] reports that only 50% uplift pressure was allowed for in the original design of the dam. This was justified by the inclusion of the gunite layer.

The stability analysis concluded that the dam body stability relies on tensile strength at the lift joints to provide the recommended factors of safety under both normal and PMF conditions. Under normal conditions the dam structure requires a tensile strength of 0.3MPa at undrained lift joints to meet recommended factors of safety against tensile failure but would remain stable should cracking occur. Under the PMF the vertical strength at the lift joints to provide the recommended factor of safety of 2 against tensile cracking would need to be 0.7MPa. Should the upstream face fail in tension and a crack develop, the dam would become unstable against overturning under PMF conditions, although this assessment ignores the moderate arch shape of the dam. The main recommendations of the report were:
- that an investigation of the dam concrete and in particular of the vertical tensile strength of the lift joints in the dam be carried out; and
- that the performance of the dam under seismic loading also be reviewed.

These recommendations were endorsed in a Section 10 Reservoirs Act report in 1997, which also recommended that appropriate measures be taken to ensure adequate stability of the dam.

Alcan awarded a site investigation contract to Exploration Associates in July 1998 for drilling vertical cores from the dam and subsequent laboratory testing. The report was submitted in December 1998 [18]. Over 70% of the lift joints in the cores were found to be intact, and some of those broken were fresh, having been broken in the drilling process. Average tensile strength at intact joints tested was 0.7MPa.

A further Stability Review [19] was undertaken in 1999 using the data on the dam concrete from the drilling investigation. Higher than previously expected concrete density of 2450kg/m^3 improved the results, but not to an acceptable extent. Stability at PMF was acceptable in the upper third of the dam, at the drained foundation and where the section was drained by the gallery, but was unacceptable in the lower two thirds of the dam where the section was undrained. Pseudostatic seismic analysis showed that the seismic load case was less critical than the PMF, and that, while the dam section could be expected to crack under extreme seismic loads, post-seismic stability was acceptable, particularly when subjectively taking into account the curvature of the dam in plan.

Consideration of remedial options concentrated on providing adequate drainage to the dam section, and it was concluded that this was best done by drilling from the dam crest to intersect the gallery, and from the gallery into the foundation. This included both vertical and inclined holes. The layout of these was complicated by the curvature of the dam in plan, the need to drill holes down through bridge piers and across the spillway openings and the presence of siphon outlets, a bottom outlet and associated equipment embedded in the dam body. A new survey and 3-d AutoCAD model of the dam were prepared by RBJ Surveys Ltd of Glasgow at a cost of £15,000 to confirm the setting out of the holes.

Implementation

The drainage design required drilling: -

- 980m of minimum 65mm diameter vertical and inclined open drain holes from the dam crest up to 40m deep to intersect the gallery.
- 350m of minimum 50mm diameter vertical and inclined open drain holes from within the gallery to intersect the rock foundation.

In total there are 66 holes spaced at nominal 3m centres, but adjusted where necessary for reasons of access and to avoid built in parts such as the siphon pipes. All holes drain into the 0.9m wide by 1.8m high gallery. All of the holes were technically challenging due to the accuracy required to intersect the gallery or the very cramped conditions within the gallery.

The contract was put out to tender to six contractors and four competent tenders were submitted. The successful contractor was Ritchies Ltd of the Edmund Nuttall Group who successfully completed the works within the 16 week contract period in summer 2002. The contract value was £170,000.

The main Health and Safety hazard identified in the risk analysis was working in the gallery where access was via two small vertical shafts. Most importantly a rescue procedure was developed with the aid of the Mines Rescue Service. This was tried out in a full mock rescue. Rope access personnel were trained and retained on site in case a rescue was required. Other systems developed included those for communication, noise and dust control and movement of materials and drilling equipment.

Environmental control adjacent to a major watercourse was of prime importance. Great care was taken to reduce the risk of spills of oils and fuels and emergency procedures were developed. On the dam crest drill stems were shrouded and vented to a dust collection system with the cement and rock dust being disposed of off site. Dust suppression in the gallery was achieved using water mist injection into the compressed air.

All holes were drilled using down the hole percussive hammer techniques. A crawler mounted rig was used for vertical and inclined holes on the dam crest roadway. The contractor elected to drill at 95mm diameter in order to reduce drill string deflection with the top 3metres of each hole being cored and cased to aid directional control through the rubble filled piers and the "air gap" between the underside of the roadway bridge and the curved spillway below. The drilling system drilled well through light reinforcement and rock displacers in the concrete. At the final count only three out of 30 holes drilled from the dam crest just missed the gallery, this was likely caused by the drill head "glancing off" embedded steelwork. However all three were later located using a vibrating tool and intersected from the gallery. A special drill rig was fabricated to work in the confines of the gallery and used successfully to drill 50mm diameter holes inclined and vertically upwards and downwards. However progress was slow and a rock drill with an air leg was used to increase production rates on some of the shorter inclined holes.

Figure 4   Drilling equipment at Laggan Dam crest and gallery

CONCLUSIONS

Both Blackwater and Laggan Dams are historic concrete gravity dams of their period, constructed before modern design criteria were established. Both have served, and continue to serve, the original function for which they were designed. Despite the differences in their designs, remedial drainage to relieve uplift pressures has been adopted as the most appropriate and economic measure to improve stability at both dams to upgrade them to meet modern standards.

REFERENCES

1 Flood Studies Report. Institute of Hydrology. 1975.
2 Floods and Reservoir Safety. Thomas Telford. 3rd Edition, 1996.
3 An Engineering Guide to Seismic Risk to Dams in the United Kingdom. BRE. 1991.
4 Engineering Guide to the Safety of Concrete and Masonry Dam Structures in the UK. CIRIA Report 148. 1996.
5 Flood Estimation Handbook. Institute of Hydrology.1999.
6 An Application note to An Engineering Guide to Seismic Risk to Dams in the United Kingdom. Institution of Civil Engineers. 1998.
7 The Loch Leven Water–Power Works. A H Roberts. ICE Proceedings Volume 187. Nov 1911. Paper No 3923.
8 Report on an Investigation into Factors Affecting the Stability of Blackwater Dam, Argyllshire. Halcrow. Revised reprint. May 1982.
9 Investigations at Blackwater Dam, Argyllshire, Scotland. Martin R. 14th ICOLD Congress, Rio de Janeiro, Vol 1, pp1-9. 1982.
10 Lochaber and Kinlochleven Power Schemes. Flood Hydrology Report. Halcrow. March 1995.
11 Blackwater Dam. Flood Stability Assessment. Halcrow. August 1996.
12 Alcan Dams Coring Contract. Blackwater Dam. Factual Report on Ground Investigation. Exploration Associates. Report No. 138161. December 1998.
13 Blackwater Dam. Stability Review. Halcrow. October 1999.
14 Blackwater Dam. Dynamic Stability Analysis. Halcrow. January 2001
15 Seismic Hazard at Scottish Hydro-Electric Dams. EQE International Ltd. Report No 414-01-R-0. 23 April 1998.
16 The Second Stage Development of the Lochaber Water-Power Scheme A H Naylor. ICE Proceedings 1937. Paper No 5089.
17 Laggan Dam. Flood Stability Assessment. Halcrow. August 1996.
18 Alcan Dams Coring Contract. Laggan Dam. Factual Report on Ground Investigation. Exploration Associates. Report No. 138161. December 1998.
19 Laggan Dam. Stability Review. Halcrow. October 1999.

Papan dam studies and remedies.

C. MAKINSON, JacobsGIBB Ltd, Reading, UK.

SYNOPSIS. As part of a major irrigation refurbishment programme in Kyrgystan, seven major dams built in the Soviet era have been examined and rehabilitation measures designed and costed. Perhaps the most complex of these was Papan Dam near the city of Osh on the Silk Road. The dam is a 100m high gravel embankment with a grouted core, set in a very narrow limestone gorge. High regional seismicity and local fault alignments all increase the risk of failure of the works, and the provincial capital downstream is only part of the consequent hazard. This paper describes the inspection and investigation process, and the difficulties of identifying the seepage pattern in a three dimensional context. The justification of remedial measures currently in progress, and comprising a 70m deep diaphragm wall through the upper half of the dam core, is discussed. A short description of the bottom outlet and spillway and their current condition is given, together with comments on the parallel issue of reservoir operation and flood freeboard.

INTRODUCTION

Papan Dam has a height of 100m, above over 20m of alluvium, but is set in an extremely narrow gorge. The crest level is 1290 masl (metres above sea level) but the adjacent cliffs soar another 300m higher, with site access only by tunnel. The 90m crest length reduces to under 20m gorge width at foundation level, and this inner river channel winds within a doglegged and possibly faulted canyon. The embankment dam was constructed in three height stages, to different standards and concepts, and provided with a combined bottom outlet and spillway comprising an intake tower and tunnel.

The Papan water storage project was inspected under the World Bank funded Kyrgyz Irrigation Rehabilitation Project and a dam examination report (DER) issued in June 1999. This report was based on the historic drawings and records made available to the consultant TemelsuGIBB (Joint Venture). The technical and safety evaluation raised concerns in relation to:

- Earthquake resistance and possible fault break
- Flood routing and discharge reliability
- Seepages through the dam and high phreatic surface within the downstream shell
- Operation of hydromechanical equipment.

Long-term benefits and performance of dams, Thomas Telford, London, 2004, 544–558

The situation of the dam as understood at this time is described by Jackson & Hinks in Dams 2000 (Reference 1). The recommendations of the Panel of Experts included restricting the maximum normal operating level of the reservoir to 1270 masl, until implementation of rehabilitation works permit safe operation to the full supply level of 1282 masl. The owner had already self-imposed such a reservoir level restriction since 1990, in response to seepages observed on the upper downstream slope at higher reservoir levels.

The site investigations were carried out in 2002 and were immediately followed by preliminary design of full depth core and curtain rehabilitation measures. Monitoring of the expanded piezometer layout over the following reservoir operating cycle gave a better indication of the internal seepage regime. Eventually the measured phreatic surface was used to calibrate a two dimensional seepage model, in which alternative cut-off works could be examined. This led to a contract being let for construction of a 70m deep plastic concrete diaphragm wall within the dam core from crest level, at a much reduced cost compared with treating the full 120m depth. However, the three dimensional reality of the seepage pattern is more difficult to define, and the adequacy of the remedial works will only be confirmed by piezometric monitoring, before and after the diaphragm wall construction expected in 2004.

The long investigation, monitoring and rehabilitation path for the embankment has given time for the other defective operation and safety aspects to be studied and resolved. In particular the bottom outlet capacity is limited as much by downstream interests as operational constraints. Rather than construct a second independent spillway, a compromise maximum reservoir operation level has been selected which will store and reduce the design flood peak while maintaining the irrigation benefit of the project. This resolves the apparent spillway discharge deficit in western eyes whilst improving on the typical Russian reliance on flood storage. Many local dam projects have no spillway at all, since they are oversized reservoirs for current yield and may be raised in future to suit water demand.

SITE INVESTIGATION

The layout and key sections for Papan Dam are shown in Figure 1, and construction took place from 1975 to 1985. The construction stages are indicated and the unusual downstream shell zoning that resulted. The few remaining piezometers had indicated a high phreatic surface across the downstream shell, suggesting a defect in the water tightness of the centrally located core/curtain and an internal hydraulic control under the downstream berm. The lower core was known to consist of a nine-line grouted zone constructed within a selected clean gravel fill, supported by sandy gravel shells. Details of the geometry of later stages relied on an obviously super-

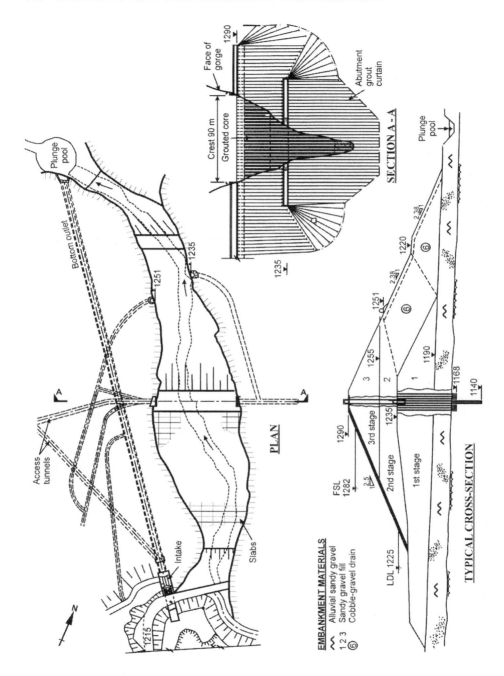

Figure 1. Papan Dam Plan and Sections

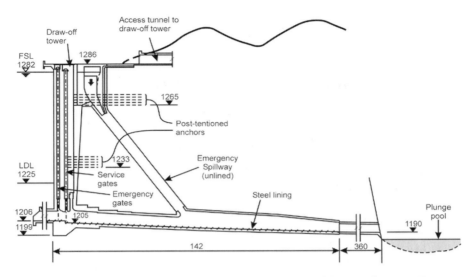

Figure 2. Combined Intake of Bottom Outlet and Spillway of Papan Dam

Fig.3. Investigation on Dam Crest   Fig.4. Downstream Slope from Berm

superceded planning drawing, but it was known that the upper core and shells were placed across the full dam width as a single sandy gravel zone followed by three lines of grouting near the dam centerline. The right abutment grout curtain at the upper level had been re-grouted in 1990 to reduce leakage into the gallery from 120 to 52 l/s.

A suitable layout of boreholes and piezometers was instructed with four percussion boreholes defining the cross-section including a 120m deep hole through the core zone. Fifteen rotary holes with piezometers were added in the downstream shell of the dam body, and four 20m deep trial pits - two from the crest into the core. From the tunnel and upper gallery four sets of two boreholes, inclined upstream and downstream of the curtain, were instructed and five minor holes at low levels within the abutments. The local water well drilling organization, called the Kyrgyz Geological Expedition, carried out the work. They provided a heavy-duty percussion rig (Figure 3) and lorry mounted rotary drills, the shafts being subcontracted to the Kyrgyz hydro institute (Kyrgyzhyprovodhoz) who also supervised day to day and carried out the soil testing (density and gradings).

Limitations & Results

Not surprisingly the $100,000 budget and one-month programme had to be doubled and trebled respectively. The recently independent state drilling enterprise had a learning curve on contractual obligations and did a magnificent job with the available equipment. The first deep hole alone took three months and the congestion on the narrow downstream slope prevented simultaneous drilling of more than two or three holes. Although accustomed to pumping out tests for wells, the drilling team found difficulty with constant head and falling head permeability test procedures within extremely deep boreholes. Providing formulae, instructions, occasional supervision and review of calculations is not enough to obtain accurate data from inexperienced testers, as we shall see below.

The steel casing type piezometers and acoustic (non-electrical) sounding equipment endemic to the region are difficult to reconcile with western practice –but the Kyrgyz in turn do not consider a narrow bore plastic pipe and electrical sounder subject to condensation on the pipe walls as reliable in cobbly-gravel fill, with calcareous deposition in all drains or pipework, and subject to frequent seismic shaking. Crucially the steel casing extends two metres below the response length and is capped at the bottom end. This provides a collector 'bucket' for washing out and removal of drilling mud during commissioning of the piezometer, since the rotary drilling depends on mud (and even lorry loads of loess) for stability of the holes within gravel fill. However, it also guarantees a piezometer reading even though the phreatic surface is far lower.

With hindsight there were also errors in the much discussed piezometer layout. Basically the arrangement of lines of piezometers yielding dam cross sections, and transverse cross-gorge sections was appropriate, but economies and misconceptions combined to frustrate this simple plan. The diamond shaped zone 6, placed in stage two as the downstream part of the downstream shell, proved to be such coarse cobble-gravel that permeability tests were meaningless and drilling was often curtailed by total loss of drilling fluid – all in the area thought to suffer from an exceptionally high phreatic surface. Setting out on a loose gravel surface (Figure 4), disfigured by a temporary zigzag access road, and confined by near-vertical, irregular canyon walls, was based on offsets from the crest and from a downslope steel access ladder. The narrowness and winding geometry of the inner gorge meant that many 'deep' holes simply hit rock prematurely. The layout had in any case been planned for monitoring seepage conditions for lake levels at or above 1270 masl, because of the high-level seepage reported. Eight out of 18 piezometers installed in the dam body are so shallow that they are never going to register a water level until after rehabilitation. Put simply, every deep piezometer counts for interpretation and there is no redundancy in the system.

The exploration in the dam core was particularly useful in indicating that the grouting had been only partially successful. The deep borehole appeared to indicate a large-scale window of high permeability in the lower core. Together with the 20m deep by one metre square, wood-braced, shafts a good impression of the upper core was also obtained. Eye witness accounts of construction had been useful in establishing that the gravel here was placed in 60 cm layers with non-vibrating roller compaction. The dumping and dozing placing sequence resulted in significant segregation, and only the top of each layer was compacted, resulting in effective horizontal stratification with permeable sub-layers and thin aquicludes. Only occasional boulders, wedges or sills of cemented gravel or plain cement grout takes were found in the upper core constructed by grouting this segregated, sandy gravel material. This scenario is consistent with the seepages observed at high levels on the downstream slope whenever the reservoir rises above about 1270 masl.

Geology & Faults

The strong crystalline Carboniferous/Devonian limestone forming the walls of the gorge is extensively jointed, with variable bedding and locally karst chimneys visible. No new rock cores were extracted as the whole area of the gorge had been extensively explored prior to construction, with addits, geophysical survey and deep coring. There appears to be a halo of stress-relieved, open jointed rock in the canyon walls, and a single line grout curtain on the dam core centerline. A major sub-horizontal, open plane of discontinuity is visible at a level around 1280 masl on the upstream right abutment. Prior to construction the groundwater flowed from right to left across the canyon, and the plateau above the gorge also imposes an intermittent flow towards the gorge. The South Katarsky fault runs along the side of the reservoir and across the upstream toe of the dam, and represents a major regional thrust fault. Overlying intact Middle Quaternary terrace deposits show that movement has not continued since that time. Conceivably the gorge location is determined by an associated tear fault, although its presence was discounted by the original (Tashkent) Design Institute, who opportunely searched for any evidence of breccia zones below the river or differential movement of the walls. The bends of the river gorge would require two suites of short en echelon faults to explain the erosion pattern. This subject was much discussed but not effectively proven as an active fault feature capable of past and future movement rather than a simple joint alignment. The definition of an active fault as defined for New Zealand in Reference 4 was found useful: repeated movements in the last 500,000 years or a single movement in the last 50,000 years. Detailed procedures to determine previous movements were proposed but not rigorously applied, due to the featureless massive limestone and lack of significant terrace remnants within the narrow gorge. This type of dam could in any case withstand minor fault movement without failure.

PRELIMINARY DESIGN

An interpretative report of the site investigation was prepared by GIBB just before the delayed end of the drilling contract and was sufficient to define the various material zones, sub-zones and their characteristics. This left the important piezometric monitoring during the reservoir operating cycle to a later date. The initial phreatic surface, determined on the dam cross-section for a 1248 masl reservoir level on 11/10/2000, confirmed that the inner downstream lower shell of first stage construction was saturated. However it also indicated a near horizontal water level just above the top of foundation alluvium running from the midpoint of the downstream shell to the dam toe. (30m lower than the high levels previously recorded on a single piezometer). Water was also appearing in the right bank lower gallery slightly in advance of and above the water level in the inner downstream shell. At this juncture all parts of the core curtain system were considered as possible culprits for the various leakage phenomena.

Accordingly, whilst awaiting more significant monitoring data, a design report was commissioned covering full depth rehabilitation of the dam core/curtain. This involved the feasibility of diaphragm walling or grouting to 120m depth in gravel, technical methodology and cost estimates. The assistance of Mr Gabriel Jorge, ex-S.American manager for Soletanche was obtained and four detailed projects drawn up:

- Full depth bentonite cement and silica gel grouting using tube-a-manchette on 9 lines. (Project 1)
- Full depth plastic concrete diaphragm wall construction with hydrofraise equipment. (Project 2)
- A hybrid project combining an 85m deep diaphragm wall with lower core grouting by angled holes from the ends of the lower gallery. (Project 3)
- Full scale re-grouting of the abutment grout curtain, or part thereof, to supplement the 1m embedment and 10m contact grouting halo included in the other three rehabilitation alternatives. (Project 4)

The estimated construction costs for the three alternatives core rehabilitation projects, with associated investigation and control monitoring, were $US 12.2 Million for core grouting Project 1, $US 7.5 Million for full depth diaphragm wall Project 2, $US 7.5 Million for the Hybrid Project 3. These prices were based on worldwide rates and included a 20% contingency for the isolated location and over half a million $US for mobilization. Each project was programmed to be completed in a single year in view of the snowbound winter conditions. The Hybrid Project 3 was capable of being subdivided into two phases with $US 4.8 Million allocated to the 85m deep diaphragm wall and $US 3.5 Million to lower core grouting from the ends of the existing lower galleries. Project 4 to extend the contact grouting halo to full abutment grout curtain rehabilitation was estimated at $US 3.6 Million, including the same 20% contingency and $US 0.25 Million for mobilization. The possibility of treating only part of the abutments on a pro-rata cost was mooted. The Interim Design Report A covering these matters was issued in March 2001, shortly followed in April 2001 by an initial Monitoring Review Report B from which a decision on the appropriate project to adopt or adapt was expected to emerge.

PIEZOMETRIC MONITORING

The gradual reservoir rise in the winter of 2000/2001 was from a base level of 1230 masl in June up to a peak of 1263 masl in mid-March as the snow-melt season progressed. Thereafter the irrigation releases exceeded inflow, but data up to August 2001 was subsequently analysed and added to the report graphs and figures. Although this reservoir range is rather limited compared with the full range of 1225 to 1282 masl, and has not been amplified in the subsequent years, it was sufficient to derive some surprising conclusions and to illustrate the limitations of the piezometer layout. The data was plotted against time in comparison with reservoir level and onto an idealized dam cross-section plus three gorge sections. These gorge sections were located downstream of the core, at midslope and through the downstream berm, conveniently breaking the data into manageable parcels and focusing on areas of interest. In addition the situation in the left and right abutments were separately analysed. The initial finding of a low-level, near horizontal water table across the downstream half of the downstream shell was confirmed for the range of reservoir levels as a sort of internal stable tailwater level, at around 1195 masl with a crossfall of under 1.5m (downstream gradient less than 0.02 below the berm). This prompted an enquiry into the historical record of high piezometric levels recorded at intervals over many years in the downstream slope. This was traced to a single standpipe piezometer (n8') that, although indicated on the drawing as extending down into the alluvium, in fact terminated at a high level –similar to the ghost readings it had been producing. Pouring ten metres of water into the piezometer resulted in a swift return to the residual level. Therefore all the initial stability checks for the 1999 DER, finding the dam just stable with some crest subsidence under earthquake MDE acceleration of 0.72g, were very conservative, since they were based on a false premise with the downstream phreatic surface 30m higher than reality.

The downstream berm feature is not some temporary cofferdam feature obstructing flow, but a drainage zone. A simple calculation of the flow through the narrow inner river channel alluvium, using measured insitu permeability at the measured low hydraulic gradient, indicates that the discharge is insufficient to maintain equilibrium so the rock walls of the inner gorge are also carrying seepage flows. Looking at the upstream gorge section the water levels in the inner downstream shell zone (first stage construction) are close to 1228 masl and conceivably fed from higher levels on the abutments. The left abutment plots are all based on dry readings, and this whole abutment is considered to be a groundwater sink. On the right abutment the readings are near the bottom of the inclined piezometers from the upper gallery down to 1250 masl (possibly in mud or trapped end sections), and water overflows from the lower gallery piezometers at 1235 masl level. The upstream piezometers faithfully reflect the reservoir level and appear to indicate a head drop of 3 to 8m across the grout curtain.

As part of the selected rehabilitation measures these raking piezometers from the high level will be extended downwards to overlap with the lower gallery and thus give a reliable reading in the critical range for intermediate reservoir levels. Additional rock drainage measures will probably convert them to permanently dry piezometers, unless right abutment seepage is a major source for maintaining the water level in the adjacent shell zone. The midslope gorge section shows a water level in the shell at 1201 masl, just 7m above the internal tailwater level and giving the overall downstream gradient as 0.05. There is thus an internal waterfall 27m high close to the downstream slope of the first stage construction. This was picked up during drilling as a high permeability (75m/d) sub-zone, and may simply represent the leading edge of the second-stage, diamond-shaped, internal drainage zone 6 of cobble-sized gravel or a rock boulder layer of rejects or riprap on the then downstream face. By accident or intention an inclined internal chimney drain has resulted downstream of which the internal phreatic surface is insensitive to reservoir level fluctuations.

Finally it was discovered that the line of existing piezometers immediately downstream of the 25m wide crest, again indicated on the drawing provided, was actually just one defunct piezometer. The reading on piezometer n2', thought to be on this line, actually corresponds to the gorge section 40m downstream of the dam core axis on which the line of new piezometers were installed (supposedly to fill the gap). There is thus still a gap in the critical area just downstream of the core where piezometric readings might indicate whether the seepage is passing through upper, middle or lower core. This may be indicated as a concave or convex phreatic surface joining reservoir level at the single upstream shell piezometer to the established upstream row of piezometer readings. Again this defect in the piezometer layout will be remedied with a new row of instruments installed to control the diaphragm wall performance. The seepage contribution of the much grouted right abutment and of the alluvium below the core (also grouted on just three lines) also remain unknown factors. Before taking a decision on rehabilitation measures and priority areas to be treated, a computer seepage model was commissioned in an attempt to resolve these issues.

SEEPAGE STUDIES

The two-dimensional seepage model was prepared using the program SEEP/W and the assistance of Clare Glenton. We started with the zone boundaries and permeabilities from the Site Investigation interpretation, and this created a phreatic surface as for a homogenous section, exiting onto the downstream slope near the top of berm level 1220 masl. This was plainly due to the supposed window in the lower core. Of course the water test results from one borehole towards the back of the core does not imply the window cuts through the whole 9-line grout curtain. Taking a probability view on the mass permeability of the lower core the window was closed down to 10% of its potential flow. Thereafter the model required 8 further iterations in order to adjust the phreatic surface to fit the monitored version. Each time the permeability of one or other key zone was modified, and in some cases the vertical/horizontal permeability ratios. This calibration process intrinsically accepted the overall seepage pattern of the piezometric data in preference to measured in situ point values of permeability. Even supposing that the insitu tests were accurate, they need not be representative of complex zones subjected to years of high gradients and particle migration (suffosion). Of course the zone interpretation in terms of geometry, permeability relative to adjacent zones, material type and placing method were respected as far as possible.

To avoid potential errors at the beginning, peak or end of the monitoring series, three calibration curves of phreatic surface were used corresponding to reservoir levels:

- 1262 masl – either side of the peak 1263 masl.
- 1250 masl – descending reservoir stage, and when it became available
- 1242 masl.-close to the August 2001 minimum of 1241.5 masl.

The model based on the first two of these phreatic surfaces, and incorporating significant modifications of initial permeability values, was then able to predict the third. The model replicates in two dimensions what is in truth a three-dimensional seepage pattern, with probable transfers from the right abutment and to the left abutment rock. It is thus superior to a simple zone/permeability model lacking calibration and a reasonable predictive tool, which was named the basic revised seepage model. However, it is not a unique model since the permeability of the ultimately controlling upper core zone is not affecting the calibration against the observed phreatic surfaces. This point was difficult to communicate since the original hypothesis of a downstream hydraulic control below the berm had permitted simple extrapolation of phreatic surface data to the highest flood levels. Nevertheless the model was able to simulate the observed seepages on the upper downstream slope at high reservoir levels, which was the primary concern of the owner. This was only possible because the particular program indicates flows above the zero pressure line. Hydraulic gradient contours were produced for the 1286 masl reservoir flood level, and values of 6.5 occur in the second stage shell adjacent to the top of the drainage zone, compromising the filter relationship between sandy gravel and rockfill. Even higher gradients are registered at the adjacent external slope around 1250 masl, indicating sloughing may be expected where seepage losses had indeed been observed at high reservoir levels. It also estimated unit width seepage flows at chosen sections, but care has to be used in deriving three-dimensional flows in the gorge situation.

By this stage the Panel of Experts including Professor Raymond Lafitte, Jonathan Hinks and Professor Bektur Chukin had co-opted a grouting specialist from Moscow, with records of the original lower core grouting. Based on his evidence they inclined to suspect that seepage loss flows were more likely to be crossing the core/curtain alignment through the upper core than elsewhere. The model was used to determine the effect of a diaphragm wall cutting off the upper core zone (first phase of the hybrid Project 3). Distinct depths of penetration of 85, 75, 70 and 60m were modeled and hydraulic gradients and seepage flows derived. All the plots were similar. Gradients behind the toe of the wall and within the downstream shell do not exceed 2.5, which should avoid suffosion effects, or at any rate avoid significantly increased particle migration over the current situation. Total leakage predictions are subjective, but unit width flows decrease only gradually with depth of cut-off wall and by up to 12% in comparison with the no cut-off version. The real benefit is thus only to cut off the horizontal seepage on privileged paths through the segregated sandy-gravel fill in the upper shoulder of the dam.

BOTTOM OUTLET & SPILLWAY

The combined spillway and bottom outlet system were described in Reference 1 and are shown here on Figure 2. TemelsuGIBB calculations for flood routing in the DER 1999 were based on western concepts such as passing the full PMF over a spillway, ignoring the bottom outlet. Dambreak would not only destroy the provincial capital Osh but also pass through the main agricultural zone of the Ferganah Valley, and cross several national borders as a flood wave on the Syrdarya River (Reference 2). There are numerous local precedents for designing dam projects with massive freeboard so as to absorb flood peaks as reservoir storage. In the unusual case of Orto Tokoi dam the spillway capacity had to be severely restricted to increase the live storage and water yield. In most other Kyrgyz dams additional spillway provision was recommended, and in the case of Papan this implied an expensive separate tunnel spillway with the possible addition of a mini-hydro station to augment the two 9m long weirs at the intake tower (Figures 2 and 5).

Understandably the client stalled on this issue until the necessary embankment rehabilitation was decided and permissible dam operating level was established. The possibility of glacial lake bursts within the far upstream catchment was mooted (Reference 1), but dismissed after examination of possible natural dams originating from landslides, fault movements or moraine deposits. Reservoir routing of PMF and 1:10,000 year floods had been carried out for the temporary maximum reservoir level of 1270 masl and for the normal maximum of 1282 masl (spillway crest level). The flood hydrographs comprise a narrow 2-day peak of 652 and 464 m^3/s respectively superimposed on snowmelt base flows of around 150 m^3/s. For the 1282 masl reservoir starting level, dam crest level of 1290 masl and the intake platform at 1286 masl were compromised by the extreme cases even with maximum gate discharge of 260 m^3/s. The two huge wagon gates would be difficult to adjust in a flood situation and only a 1:1,000 year flood could be accommodated at the irrigation setting of 20 m^3/s, although in reality the combined tunnel becomes the limiting factor with a 345 m^3/s maximum open channel flow. The inclined spillway shaft linking to the tunnel was said to be a textbook design, but no calculations or model tests were provided. The condition and presence of the shaft lining have not been confirmed (no access), the steel tunnel lining downstream of the gates has previously been repaired, the gates have vibration and aeration problems, and the tunnel outlet and plunge pool are sub-standard.

The reasons for this situation being tolerated only gradually became clear. First the maximum gate discharge is only 160 m³/s as permitted by the gate manufacturer not 260 m³/s as calculated for two fully opened gates. There is a continuous minimum discharge of 20 m³/s compensation water for Osh city potable water. Fundamentally there is no benefit in filling the reservoir and even after rehabilitation a 1275 masl normal top water level will be applied. Flood routing then becomes much simplified with 45Mm3 extra flood storage, doubling that for the 1282-88 masl flood range. The gate vibrations occur at small openings and have now been measured and found to be within acceptable limits of the Soviet code (SNIP). The obstruction to the aeration shaft has been lifted and an aeration slot will be created, and minor welded insitu reinforcement of gates and linings added, rather than proposed major off-site gate modification. The tunnel outlet rock-support wall and plunge pool amplification at the inaccessible dam toe will be carried out, after diverting the compensation flow by a temporary adit linked to the tunnel.

DIAPHRAGM WALL CONSTRUCTION
TemelsuGIBB first proposed a no-action strategy based on maintaining a normal top water level of 1270 masl. The only example of diaphragm wall construction through a high dam core in a gorge of this magnitude appears to be Mud Mountain Dam in the USA, where problems due to arching of the core occurred (Reference 3). A separate Kyrgyzhyprovodhoz study had shown that the irrigation benefit of the extra 12m depth to 1282 masl (75Mm3 extra reservoir volume) was zero for the agricultural areas developed, due to unreliability of supply in most years. The Panel of Experts however did not accept no-action, since flood rises will temporarily affect the upper dam and long-term storage levels might not respect present agreements as the responsible personnel retire or change. The owner agreed that the opportunity to introduce a partial cut-off at limited expense from current funding should not be missed. The options of 60 or 70m depth were further explored, with consideration of later phases should lower core or right abutment grouting subsequently prove necessary. A key factor was that the Hybrid Project 3, with drillholes angled sidewards and downwards fom the confined and plugged ends of the lower galleries to achieve a nine-line curtain was simply impracticable. At this stage, after 20 years of consolidation of embankment gravel fill, the gallery ends could be simply joined by excavating across the partially grouted core – permitting sub-vertical groutholes and minimal overlap with the diaphragm wall. A diaphragm wall depth of 70m (down to 1220 masl) was eventually selected to give a generous 15m penetration into the lower core.

Contract documents were prepared based on the Engineer's scheme, but asking the Contractor to carry out final design to suit his capacity and plant for excavation and production, delivery and placing of plastic concrete mixes. The Iranian firm JTMA Co. was awarded the diaphragm wall construction contract early in 2003 at a sum close to the estimate. Competitors had priced the works at up to twice this sum, anticipating difficulties with the site location and inhospitability. The Contractor has reduced the typical panel length to 3.0m and volume to 210 m3, from the French based 8.8m and 616 m3. This gives a more manageable task in the congested area of the short dam crest with tunnel access only. The other advantage is that to achieve a nominal 1.0m depth key into rock, the hydrofraise has to cut a square panel base jutting into abutment rock as a triangle. Given the extremely high rock strength (estimated at 200 MN/m^2) the advantage of smaller triangles of the hard, siliceous, crystalline limestone to grind away is evident. The joint detail has also been simplified, cutting into the preceding panel and inserting a groutpipe rather than drilling separately.

Maintaining a one-metre wall thickness for the reduced wall depth assists the joint overlap being guaranteed against lateral deviation during excavation. The alignment of the wall has been moved a few metres downstream, to the back of the lower core, for fear of the hydrofraise being severely damaged by bumping into old steel grouthole casings left in the fill. Not surprisingly given the logistics of reaching Osh, of fabricating batching plant in Tehran and upgrading the hydrofraise excavation rigs in Italy, progress this year has been slow and limited to preparatory and auxillary works. It is now anticipated that the diaphragm wall construction will proceed in the spring of 2004. Contractual niceties come second to physical possibilities in remote locations and extreme winter weather.

CONCLUSIONS
The Final Dam Design Report was compiled to record this four-year study and issued in October 2002. While every effort was made to speed the process, the assumptions made to facilitate design packages were often proved unsatisfactory in the long term. The difficulty in obtaining and then verifying information and even drawings should not be underestimated. Piezometer layout sections were extremely misleading, and 10 years of hand plotted phreatic surfaces in full reservoir conditions simply erroneous. As for dinosaur research, one must first assemble an awful lot of fossil bones, and then ask the right questions in order to correctly interpret them functionally. In these days when safety inspections of 60 dams at a time are required, with two days input allotted for each and only limited access to and around the site, the inspector must explore the 'facts' and try to understand the mindset of the owner/operators. The painstaking

investigation, monitoring and design review process applied at Papan dam were a rare opportunity to focus on the individuality and idiosyncrasies of a standard Soviet dam design applied to a narrow canyon. There were no short cuts, and only perseverance by all parties concerned will lead to satisfactory rehabilitation works. The very active role of the client, Project Implementation Unit for Kyrgyz Irrigation Rehabilitation (Ministry of Agriculture, Water Resources and Processing Industry) and invaluable assistance of Mr Fedotov (Kyrgyzhyprovodhoz) in pursuing these technical issues, providing information in difficult circumstances, and obtaining value for money solutions is acknowledged.

Figure 5. Papan dam – Reservoir and Intake Tower

REFERENCES
1. Jackson, E. A. and Hinks, J. L. (2000). Dam safety in Kyrgyz Republic. *11ᵗʰ Conf. British Dam Society,* Bath, pp 262-271.
2. GIBB Ltd. Tashkent, December 2002, *Aral Sea Basin Program,* Dam Safety and Reservoir Management Report.
3. Graybill, K. and Levallois, J. (1991). Construction of a cut off wall with the hydrofraise through the core of Mud Mountain dam. *17ᵗʰ ICOLD,* Vienna, Q.66, R.49, pp 880-900.
4. Hatton, J.W, Foster, P.F. and Thomson, R. (1991). The Influence of foundation conditions on the design of Clyde Dam. *17ᵗʰ ICOLD,* Vienna, Q.66, R.10, pp 157-178.

The Washburn Valley Reservoirs – spillway improvements

J. R. CLAYDON, Yorkshire Water
D. L. KNOTT, TEAM
I. C. CARTER, TEAM

SYNOPSIS. The Washburn Valley reservoirs comprise a cascade of four impounding reservoirs situated about 12 km to the west of Harrogate in North Yorkshire. The three lower reservoirs, Lindley Wood, Swinsty and Fewston, were formed between 1875 and 1879, by the construction of earth embankment dams with puddle clay cores. The upper reservoir, Thruscross, is a mass concrete gravity dam constructed in 1966. The upper three reservoirs supply water to Leeds while the lowest, Lindley Wood, provides compensation flows to the River Washburn.

The three lower dams would be overtopped during the Probable Maximum Flood (PMF). The situation was complicated by the publication of the Flood Estimation Handbook (FEH), which led to a review of the conceptual design and a lengthy delay, which was recovered by carrying out works at two dams in one season, instead of one per year as originally planned.

The rehabilitation works consisted of crest raising and spillway modifications at the three embankment dams.

- Lindley Wood: a 3 m high earth embankment was built downstream of the existing crest road, which will be inundated during extreme floods.
- Swinsty: the crest was raised 1.2 m and the multi-span masonry arched bridge replaced by a clear span.
- Fewston: the crest was raised 0.9 m and the multi span masonry arched bridge replaced by a clear span.

At £6.5 M this is one of the largest reservoir safety rehabilitation schemes undertaken by *Yorkshire Water* (YW). It was successfully completed in 2003 on time and within budget by team working.

Long-term benefits and performance of dams, Thomas Telford, London, 2004, 559–568

BACKGROUND

The Washburn Valley reservoirs comprise a cascade of four impounding reservoirs. The three lower reservoirs, namely Fewston, Swinsty and Lindley Wood were formed between 1875 and 1879, by the formation of embankment dams with puddle clay cores. The upper reservoir, Thruscross, was completed in 1966 with the construction of a mass concrete gravity dam. The upper three reservoirs supply water to Leeds whereas the lowest reservoir, Lindley Wood, provides compensation to the River Washburn.

FEASIBILITY STUDY

All four reservoirs are 'large raised reservoirs' in accordance with the Reservoirs Act 1975. The Statutory Inspection Reports published in July 1997, re-designated all of the dams under Category A, as defined in Floods and Reservoir Safety [1]. Previously the dams had been classified as Category B and the spillway capacity determined as satisfactory. *Mott MacDonald* carried out a study of the options for dealing with the increased design floods. This study considered all four dams in cascade and included physical hydraulic models of the spillways for the three lower dams, built at *Hydraulics Research Wallingford*. The upper dam, Thruscross, was found capable of passing the design flood and therefore is not discussed further.

There are considerable benefits from being able to undertake iterative hydrology, hydraulic calculations and model testing at the same time. Changes to any one component can affect the others. For example, removal of spillway restrictions at Fewston and Swinsty changed the hydraulic characteristics, reduced the flood attenuation, and increased the required freeboard at Lindley Wood.

The five arch masonry bridges at Fewston and Swinsty significantly reduced spillway capacity at moderate discharges. Various schemes to preserve their appearance were considered but none were found to be practical and it was therefore decided to seek planning consent to remove them. The outline designs to pass the PMF through the cascade included:

- Fewston   A 2 span bridge, raised crest road and wave wall.
- Swinsty   A 2 span bridge, raised crest road and wave wall.
- Lindley Wood   Crest raised by 3 m.

Considerable out of channel flow was predicted at all three dams. The model testing provided excellent information on depths and velocities. YW adopted its normal practice of extending the spillway rating curves to flows 10% higher than anticipated and recording the test performance on video.

FLOOD REVIEW AND IMPACT OF FEH

In October 2000 *TEAM*, a working agreement between *E C Harris, Arup* and *MWH*, were appointed to carry out a feasibility review, detailed design and project management for implementation of the scheme. This was soon after the publication of the Flood Estimation Handbook (FEH) [2], which complicated the situation significantly.

The flood assessment techniques contained within the FEH are on a different basis to the Flood Studies Report (FSR) [3], and were publicised as being the "the replacement for the Flood Studies Report".   Interim guidelines on their application were published by DoE, summarised as:

- If the overflow capacity is adequate to present standards (i.e. Reference 1), then do nothing.

    The overflow capacity had been found to be inadequate – hence 'do nothing' was unacceptable.

- If new or improved spillways are required, then follow one of the following three options:

    1. If practicable, then postpone work on spillways until new guidance is available.

        This was impracticable, since the recommendations were "in the interests of safety" and therefore mandatory, furthermore it was not known when new guidelines might become available.

2.   If (1) is not practical, adopt a 2-stage improvement, if this is technically, financially and environmentally acceptable.   The first stage is to increase the spillway capacity using FSR.   The second stage is increasing the capacity further, if subsequent higher standards are recommended.

   This approach was adopted, with the works designed to allow future crest and wall raising.

3.   If (1) and (2) are not practicable, increase the capacity using FEH rainfall or worst case PMF.

Revised PMF

The FEH methodology was claimed to include latest thinking on catchment characteristics, which updated the Flood Studies Report.   The new procedures were incorporated into a review of the flood hydrology, which resulted in an increased PMF from this "hybrid" approach.   The following table compares previous and new PMF values and existing and proposed flood defence levels for the three reservoirs.

It was decided to adopt a precautionary approach and design to the higher flows and levels.

	Estimated PMF outflow (m^3/s)		Flood Defence Level (m OD)	
Reservoir	FSR	FEH "hybrid"	Existing	Required
Fewston	405	442	156.58	157.55
Swinsty	454	498	140.20	141.28
Lindley Wood	504	536	93.22	96.09

PROGRAMME OF WORK

It had originally been intended to improve spillway capacity sequentially, working upstream from Lindley Wood in 2001 and finishing with Fewston in 2003.   The flood review had effectively lost a year from the programme.

It was decided to carry out the remedial works on the lower two reservoirs, (Swinsty and Lindley Wood) during 2002 under a single contract.   This presented parallel difficulties of maintaining compensation discharges and water supplies, which were overcome by careful control of reservoir level.

The works at Fewston followed under a separate contract in 2003, recovering the time lost.

The contracts were let by competitive tendering to a select list using NEC ECC Option A contract conditions. Both contracts were won by *Morrison Construction*, who were able to transfer staff, cabins and 'lessons learnt' from Swinsty to Fewston.

WORK CARRIED OUT

The main components of work at each dam are outlined below:

Lindley Wood

Lindley Wood dam is 330 metres long and was a maximum 21 m high with a capacity of 2920 Ml. Remedial works included:

•   Raising of flood defence levels by about 3 m. This was achieved by construction of a new embankment above the existing one, thanks to the unusually wide crest. The new embankment comprised granular fill with side slopes of 1:2. An HDPE membrane was laid over the upstream face, terminating within the existing clay core at the bottom and rising above peak still water level at the top. The design of the crest raising was unusual in that the existing wide crest allowed the construction of the new embankment downstream of the existing access track. In extreme conditions both the track and existing valve towers will flood. Rather than opting for a scheme with higher capital costs that would ensure the track and valve tower did not flood, YW accepted this arrangement as a 'business risk' since it would not pose a threat to reservoir safety. There is no wave wall, however one could be built on top of the new embankment in future.

•   Increase of spillway capacity by the demolition of existing footbridge, the construction of a new reinforced concrete headwall structure and by making provision for out-of-channel flow by creating reinforced grass revetments utilising proprietary pre-cast concrete blocks.

Swinsty

Swinsty dam is 460 m long and was a maximum of 20 m high with a capacity of 4655 Ml. Remedial works included:

•   Raising flood defence levels by about 1.2 m, which was achieved by the construction of a new 2.25 m high reinforced concrete wave wall to replace the existing and raising the crest road level by approximately 1.2 m, in granular fill. A sheet pile cut off embedded into the existing puddle clay core and extending into the wall base ensures a continuous water barrier to above peak still water level. The wall can be raised by 0.5 m.

• Increase of spillway capacity by demolition of the existing five arch bridge and replacement with a new single span bridge with the soffit level set above the PMF level. Provision for out-of-channel flow by construction of additional bunding and provision of reinforced grass revetments utilising proprietary pre-cast concrete blocks. The replacement of the bridge at Swinsty was undertaken as a 'design & build' element within the contract, and designed to be lifted 0.5 m in the future. The main beams for the bridge were prefabricated and delivered to site as single 30 m long units;

Fewston dam wave wall and new road bridge.

Fewston

Fewston Dam is 430 m long and a maximum of 21 m high with a capacity of 3814 Ml. Remedial works include:

• Raising of flood defence levels by about 0.9 m by the construction of a new reinforced concrete wave wall to replace the existing. The wall is typically 2.8 m – 3.2 m high. Crest road levels have been raised by approximately 0.9 m. A sheet pile cut off embedded into the existing puddle clay core and extending into the wall base ensures a continuous water barrier to above peak still water level. The wall can be raised by 0.5 m.

- The crest road at Fewston is a public highway and as such these works are subject to the approval procedures of North Yorkshire County Council and the wave wall has been designed to provide vehicular impact containment to P2 level in accordance with BD 52/93 "The Design of Highway Bridge Parapets".

- Increase of spillway capacity by demolition of the existing five arch bridge and replacement with a new single span bridge with the soffit level set above the PMF level. The bridge is similar to Swinsty.

- Provision for out of channel flow by construction of additional bunding where necessary and provision of reinforced grass revetments utilising proprietary erosion control geotextile;

DESIGN ISSUES

Although many of the elements of the three designs were common to each, a number of issues required special consideration:

Revetment Protection System

Revetment protection systems were designed on the guidance of CIRIA Report 116 – Design of Reinforced Grass waterways. Maximum anticipated out of channel flow velocities for the three spillways are as indicated in the table below:

Reservoir	Maximum estimated out-of-channel velocity
Fewston	6.0 m/s
Swinsty	7.1 m/s
Lindley Wood	9.7 m/s

Flow velocities at Fewston and Swinsty resulted in geotextile erosion control matting and interlocking precast concrete blocks respectively to be chosen as the preferred method of protection. The peak velocities at Lindley Wood were anticipated to be in excess of those velocities covered by the

CIRIA guidance (8 m/s maximum). However, one of the authors of that report confirmed that the interlocking pre-cast concrete block system could withstand sustained flows at velocities up to 10 m/s, if installed with sufficient attention to detail, hence this system was adopted.

<u>Environmental Issues</u> The Washburn Valley constitutes part of the Nidderdale Area of Outstanding Natural Beauty and planning restraints have required that as far as possible the existing landscape be preserved or enhanced. Detailed Planning Consent was sought for all three dams in a single application in order to reduce the chances of delays and permission was obtained with acceptable conditions. All new structures are required to be fully clad in natural stone work, including the bridges, and measures such as ecological surveys, archaeological studies and tree preservation strategies were employed in order to minimise the impact of the works.

<u>Lindley Wood Cottage</u>

This disused dwelling was originally intended for demolition as it was considered an obstruction to the dam raising works. However plans were altered when two bat colonies were discovered within the roof void. Bats are protected species and a mitigation strategy needed to be agreed with DEFRA in order that permission to remove the habitat could be given.

Lindley Wood Cottage

The most straightforward mitigation was to build another bat roost nearby, carefully replicating the conditions in the hope that the bats would move, however this would have meant delaying the work by at least one year and possibly longer. Alternatively, the raising could have been done by a complicated realignment of the crest around the house, in order that the structure might be left intact. The solution adopted in order to facilitate both the crest raising and the maintenance of the bat habitat was to build the cottage into the raised dam embankment. The ground floor was filled with lightweight concrete and the existing first floor became an electrical plant room. Landscaping around the house was designed to maintain flight paths and bat tiles were built into the roof to maintain access for the bats. The bats returned to breed in 2003, helpfully discharging the planning condition.

Recreation

The area is popular with ramblers and YW has promoted circular walks, which pass through the construction sites. Temporary footpaths were erected and maintained to segregate pedestrians from traffic.

All the reservoirs are active fisheries, which were able to continue in use during the work. New permanent tracks were built to allow access to the drawn down waterline in order to enable Fewston Reservoir to be restocked with rainbow trout.

Water Control Measures

YWS undertook to maintain water levels in the reservoirs within a pre-determined range below existing overflow weir levels so as to ensure that construction could not only proceed safely but also so that water supplies to Leeds could be maintained. The criterion for the upper limit was based on a 1% chance that the level would be exceeded during the critical construction period when work is undertaken on the dam or spillway. A procedure was formulated by *YWS*, *TEAM* and *Morrison Construction* whereby water levels would be monitored and contingency plans brought into action in the event of the reservoirs rising above various threshold levels.

The contingency plan was called into operation on one occasion during the works at Lindley Wood and it worked well. The contractor mobilised plant and materials to protect the open excavation over a weekend, scour discharge to the river was maximised and the bags of stone and clay were removed without getting wet the following week. The client accepted the financial risk of invoking the emergency measures and the incident was covered under the cost component schedule of the contract, including costs to accelerate the works back on to programme.

CONCLUSION

By the time the work at Fewston dam was nearing completion the project team of consultant, client and contractor were working so well together that they wanted to move straight on to the next dam upstream. Regrettably, all good jobs come to an end and this one was finished on time and below budget.

ACKNOWLEDGEMENTS

Thanks are due to Ian Farmery of TEAM, for permission to use some text previously published in Water Projects UK.

REFERENCES

1. Institution of Civil Engineers (1996) *Floods and Reservoir Safety,* 3rd Edition.
2. CEH Institute of Hydrology (1999) *Flood Estimation Handbook.*
3. NERC (1975) Flood Studies Report.

Rehabilitation design of Acciano rockfill dam after the September 1997 earthquake

R. MENGA, Enel.Hydro S.p.A. – ISMES Division, Seriate , Italy.
M. EUSEBIO, Enel.Hydro S.p.A. – ISMES Division, Seriate, Italy.
R. PELLEGRINI, Enel.Hydro S.p.A. – ISMES Division, Seriate, Italy.
R. PATACCA, Umbra Acque S.p.A. – Ponte S. Giovanni, Italy.

SYNOPSIS.   The Acciano rockfill dam was originally designed without taking seismic action into account. The area where it is located is now classified in the 2nd seismic zone, according to the current Italian regulations. On September 26th 1997, an earthquake of magnitude $M_w=5.5$, one of the largest seismic events of the last 20 years in Italy, occurred in that area and caused some visible damage to the dam. Subsequent investigation programmes and structural assessments were carried out to evaluate the residual safety margins of the dam in order to identify possible rehabilitation provisions to comply with the Italian standards for seismic design.
This paper describes the evaluation of the post – earthquake condition of the dam and outlines the assessments to validate the rationale of the rehabilitation project.

THE DAM

The Acciano rockfill dam is located in the centre of Italy (Perugia Province); it was built between 1976 and 1980 to impound water for agricultural use during period of deficient supply. The reservoir capacity is 1.7 million m^3.

The Acciano dam has a zoned embankment with a curvilinear axis; an internal central impervious core and external rockfill shoulders. The embankment is characterised by three berms, at different elevations: two are located at the downstream side and one at the upstream. The structure reaches a maximum height of 28.5 m, and it is 182 m long along the crest, at elevation 531.5 m a.s.l. The faces have a slope (Fig. 2) equal to 1:1.4 from the crest to the first berm (el.513 m), and 1:2.5 in the bottom part. The shoulders were built by dry compaction of two different materials: the zone above el. 513 m with rockfill and the part below with a gravelly sand. The core is of silty-clay.

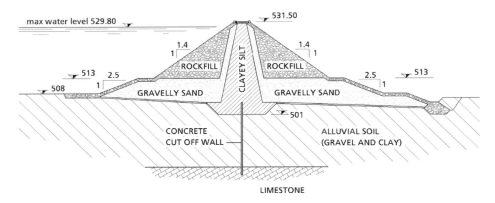

Figure 1. The main cross section

The embankment is founded over an alluvial soil, which is below the main section, about 20 m thick. Between the main section and the shoulders, the thickness of this alluvial layer decreases to zero and the dam is directly founded on a marly limestone rock. To reduce seepage, a concrete diaphragm wall 0.6 m thick was built below the core to bedrock. Cement grouting was carried out to enhance the hydraulic performance of the rock foundation, in particular at the abutments.

At the right abutment the rock is fractured to locally highly fractured: a grout shield 60 m long from the crest elevation and 30-40 m deep into the abutment rock mass was therefore added.

The dam has two outlet works, both located within the rock on the left abutment: a bottom outlet (el. 506.9 m a.s.l) and an overflow spillway (el. 528.50 m a.s.l.) which can release, at the maximum water level, a discharge outflow of 38.8 m³/s and 86.2 m³/s respectively, which overall corresponds to the 1000 year flood. The bottom outlet is a tunnel of precast reinforced concrete that, in its central part, lies within the dam body. In this part the discharge gallery is closed by two sliding gates, which are operated from the control tower located in the reservoir near to the left abutment.

The monitoring system comprises: a topographic collimation to observe planimetric displacement evolution and settlement of the crest and at the downstream berms; eight Casagrande piezometers placed downstream respect to the dam body and into the left and right foundation rock. The monitoring records did not show any anomalous response before the earthquake took place. The dam had been operational for 11 years and the reservoir reached 522 m a.s.l. water level before the earthquake struck.

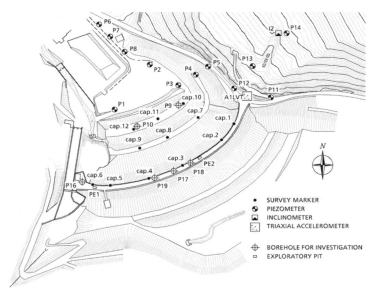

Figure 2. The instrumentation network (partial) and boreholes for investigation

OBSERVED EFFECTS OF THE EARTHQUAKE TO THE DAM

On September 26th 1997, an earthquake of magnitude Mw=5.5 occurred in the area of Nocera Umbra, one of the largest seismic events of the last 20 years in Italy, causing some visible damage to the crest structure.

Shortly after the event some wide cracks appeared in the asphalt paving of the crest: two longitudinal cracks, close to the crest edges, and two transverse ones, near the extreme ends of the dam, were observed (Fig.4). The upstream crack appeared the most significant and spanned the entire crest. Settlements up to 15 cm of the rigid reinforced concrete edges of the crest road were measured in the main cross section, where also a lateral spreading of about 10 cm was also found. The downstream lower berm settled about 2 cm. In the following month, settlements of the dam increased by only a small percentage, mainly due to the dissipation of the excess pore pressure in the foundation soil (a 0.05 MPa increase was measured in the alluvium shortly after the event). The following Figure 3 shows the chronological measurements of displacements on the crest and on the two berms.

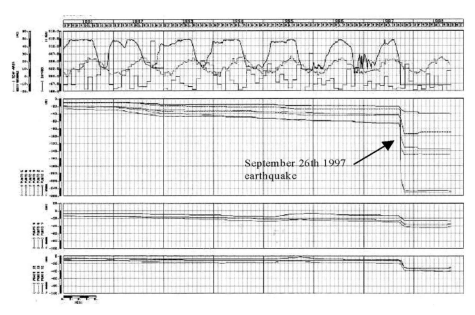

Figure 3. The chronological diagrams of displacements on the crest and berms (Midas® - Ismes Software)

In the following table, settlements of the crest and the two berms before and after the earthquake are reported.

Table 1. The measured vertical displacement values [mm]

Marker		from jan 1986 to sept,26 1997	After Earthquake	%
Crest	P1	-25	-16	64
	P2	-51	-84	165
	P3	-77	-171	222
	P4	-74	-177	239
	P5	-52	-97	187
	P6	-32	-58	181
Berm 1	P7	-18	-9	50
	P8	-32	-10	31
	P9	-26	-9	35
Berm 2	P10	-13	-18	138
	P11	-22	-17	77
	P12	-15	-18	120

Two square exploration pits 3m wide (PE in Figure 2), dug one year later, did not show any evidence of deepening of the two transverse cracks. No extension of the observed crack path and of complementary damage has been observed.

Figure 4. The damage on the crest

CALCULATED EFFECTS OF THE EARTHQUAKE GROUND MOTION

A large number of stations of the national accelerometer network were operating in the Umbra-Marche region. The nearest to the Acciano dam was that of Nocera Umbra, 11 kmaway. From that record, a site spectrum has been derived as input at the dam foundation for use in structural assessments. Processing of the Nocera Umbra records was made adopting the Sabetta-Pugliese attenuation laws. Site amplification data, based on local measures of micro tremors, gave evidence that no local amplification need to be incorporated. The resulting response spectra (horizontal and vertical component) are depicted in Fig. 5 together with the corresponding accelerograms.

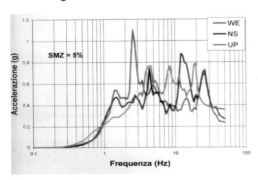
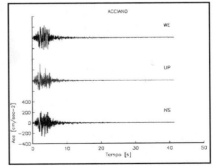

Figure 5. Calculated spectrum and time history below the dam

It may be appreciated that the highest accelerations concentrate within the interval 1.5-10 Hz, where the natural frequencies of the dam also fall. The dam crest could suffer accelerations up to 1.1 g.

A Finite Element (FE) model (Fig. 6), representing the dam body and a portion of the foundation rock, has been set up to determine: the actual distribution of accelerations within the dam body; the principal dynamic properties which are vibration modes and the associated frequencies. The material model is elastic with assigned decay law of material properties (damping and shear modulus). The parameters incorporate the stiffening contribution of the interstitial water and of the short duration of loading.

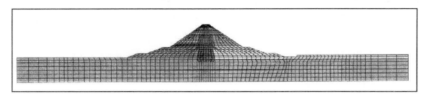

Figure 6. The F.E. mesh for dynamic analyses

The calculated accelerograms were applied at the base of the model. The earthquake (40 s duration) can reduce significantly the stiffness of dam materials: straining reaches the order of magnitude of about 0,01%, a threshold for a significant reduction in stiffness for many materials. This threshold was confirmed by resonant column tests run on specimens taken from the core material.

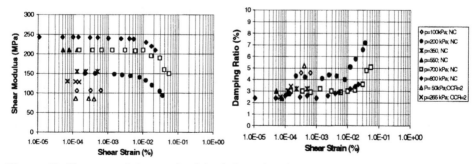

Figure 7. Shear modulus and critical damping by resonant column tests

As a further evidence of the decay, the modal analysis calculated 3.67 Hz for the first natural frequency before the seismic input is applied and 2.60 Hz at the end.

SEISMIC STABILITY EVALUATION
The evaluation of the safety margins after the shock were made by applying the Newmark method, which determines, within a limit equilibrium

approach, the residual sliding displacement of a given portion of the dam body suffering a given acceleration record. In this case the accelerograms obtained by processing the actual records have been applied.

The most critical surfaces were determined again by the limit equilibrium method by Bishop (simplified with use of circular potential sliding surfaces), where the Italian regulatory seismic input was applied as a pseudo-static inertial contribution, based on a constant acceleration of 0.07 g. The approach is consistent to that used by the designer, who took into consideration static loads only.

In both assessments the same physical and mechanical material properties were used for the materials in the dam body and in the foundation. They are given in Table 2 and result from the design phase as well as from tests.

Table 2. Design physical and mechanical parameters

Material	γ_d [kN/m³]	c' [kPa]	φ' [°]
Clayey silt	16.7	30	25
Rockfill shoulders	19.6	0	40
Gravelly sand	21.6	0	35
Alluvial soil (gravel and clay)	17.7	0	30

The water level was taken at the maximum operating (529.8 m a.s.l.) for conventional checks and to the much lower one (514.0 m a.s.l.) present when the earthquake took place.

The assessments confirm that the dam for most critical surfaces complies with the Italian regulation (1.4 and 1.2 for static and seismic conditions respectively). Critical surfaces located in the upper portion of the embankment have a reduced safety margin for static loads (1.17 compared to 1,4) and near to 1 for the seismic condition.

A thorough visual inspection of the embankment slopes did not show even local which could be attributed to sudden unstable conditions. It may be deduced that a higher shear resistance is available at the surface, where confinement due to overburden is a minimum (see Fig 8).

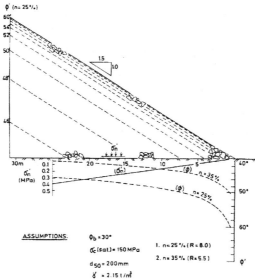

Figure 8. Available shear strength, as friction angle in a rockfill embankment approaching to the surface. σ'_n is the confining normal stress. Source NIT report

The effects of the actual earthquake have been evaluated by the modified Newton method (by Makadisi-Seed) The Newmark method allows the evaluation of the permanent displacement of a slope subjected to an earthquake, assuming that the motion occurs along arcs or planes as in the usual static analysis of stability. Direct integration has been used to compute the magnitude of the dynamic motions produced by the earthquake. The fundamental parameter in the analysis is the critical acceleration Kc, i.e. the pseudostatic acceleration corresponding to a unit safety factor against sliding in the limit equilibrium analysis.

The following steps were taken:
• For each critical surface the critical acceleration was determined at the centre of mass of the given portion of the dam body, defined as that bringing to unit the safety coefficient against sliding. This is the conventional value for unstable response to occur under the form of progressive displacements.
• By checking the critical value with the actual horizontal acceleration record of the given earthquake instant where the critical acceleration is exceeded are determined and the corresponding displacements cumulated.

The maximum critical acceleration value is calculated for critical surfaces originated nearby the crest and ending at the upstream portion of the embankment, and among them for those having the toe below the water table. In any case all the surfaces examined could withstand pseudostatic horizontal accelerations in excess of 015g.

The residual displacement reaches 15 cm, for the most vulnerable surface (Fig. 9) while, for most cases, such displacement is lower than 6 cm.

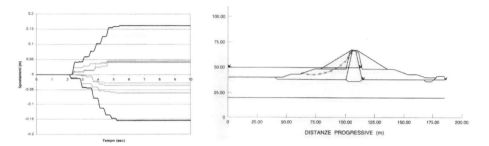

Figure 9. Residual displacements and a potential failure surface

The values obtained are modest if compared with threshold suggested in the literature (Lambe and Whitman for earth dams, NIT report), which are metric.

It may be concluded that the cumulative damage is moderate, and concentrated on the crest area. Considering that the actual earthquake can be associated to a SSE (rare), with a 475 year return period, the overall performance of the dam confirms the high safety margins incorporated in the Italian regulation even for static loads.By processing safety condition in terms of allowable acceleration, Paoliani concluded that millimetric to centimetric displacements can be associated with a OBE to SSE earthquake, which is characterised by PGA's well in excess (0.20-0.28g respectively) of that required by the Italian standards for dams (0.07g at the Acciano dam site). It was concluded that there were grounds for rehabilitating the dam.

EXPERIMENTAL ASSESSMENTS
Evidence of the state of critical materials in the dam body and foundations were acquired by an extensive site and laboratory investigation in support of the rehabilitation design. Emphasis was given to the core material, in order to ascertain its strength and stiffness properties. Self-healing and self-sealing capabilities, which mainly relate to the amount and the mineralogical composition of the clay fraction in the core material were of specific interest. They can counteract the possible formation of damage under the form of shear bands/microcracks, affecting therefore the fundamental barrier

function of the core. Tests were also performed on the foundation materials, the alluvial soils and the rock itself.

Some *in situ* standard penetrometer and permeability tests (Lefranc and Lugeon) provided a framework for data obtained in laboratory on specimens. Test boreholes have been drilled from the crest to the foundation through the core, and some from the downstream berm (see Fig.2).

The laboratory test programme for specimens from the core consisted of: triaxial consolidated drained and undrained tests to derive frictional properties and undrained cohesion; direct permeability tests on undisturbed specimen to evaluate a possible anisotropy; load controlled oedometer tests to derive stiffness and permeability, to assess overconsolidation, and, finally, resonant column tests, for further assessment of dynamic properties to determine effects on the dam induced by the earthquake.

A few unconsolidated undrained tests were run on specimens of the alluvial foundation layer resting under the dam body. Uniaxial load tests were run on the foundation limestone rock to derive its mean strength properties.

RESULTS
The core is of a silty-sandy clay with gravel. The clay fraction increases, reaching 50% at increasing depths into the core. The upper material is much more dominated by the sandy and gravel fraction (Fig. 10).

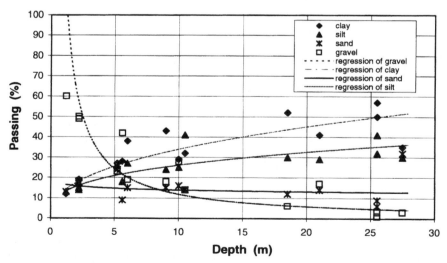

Figure 10. Material composition of the core by grain-size determination.

The clay is inorganic, of plasticity ranging from medium to high, CL (top portion of the core), CH (deeper core) according to Casagrande classification. The consistency index, 0.96, reveals a solid-plastic clay.

According to pocket penetrometers and S.P.T. tests, the core material is of good consistency, stiff in the upper part to very stiff in the deeper one, corresponding to a uniaxial compressive strength (UCS) ranging from 200 to 300 kPa.

The core material was found to be nearly saturated. The average water content is 24% and the wet density of the deeper core material is 19.3 kN/m^3, reaching 20-21 kN/m^3 in the upper, more gravelly portion. The core shows, from oedometer tests, some light overconsolidation. The estimated preconsolidation stress ranges from 400 to 500 kPa. Strength properties from drained and undrained consolidated triaxial tests run at several confining stresses in normally consolidated conditions can be described, according to the Mohr Coulomb criterion, as c' = 45 kPa and φ' = 23°. The same tests run on core upper material suggest a more marked frictional response, c' = 7 kPa and φ' = 34°, in Figure 12.

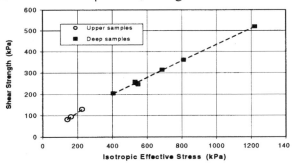

Figure 11. Shear strength vs. isotropic effective stress

Tests run on slightly overconsolidated specimens (OCR=2 corresponding to in situ conditions at the given elevation) show some, peak resistance. All considered, the overall strength envelope for the core can be defined by c'=15-45 kPa and φ'=27-23°, values adopted in the design stage are still represented.

Overconsolidation is weak and the general stress-strain response is quasi-ductile. Specimens collapse in triaxial tests at shear strains from 3.5% to 5% for overconsolidated specimens, and in excess for normally consolidated.

The undrained cohesion (c_u) has been found remarkably high, about 250 kPa. Some overestimation respect to values obtainable by other tests is due to the test bias. Upper specimens, which had to be reconstructed and consolidated to separate the gravelly fraction from the cohesive one gave lower, but still significant values (80-150 kPa).

The oedometric modulus is about M=30-15 MPa, the latter in the overconsolidated range within the prevailing stress state (up to 600 kPa). The consolidation coefficient is estimated as between 1×10^{-8} and 5×10^{-8} m^2/s. The shear modulus varies from 80 MPa a 250 MPa with increasing confinement stress (Fig. 12).

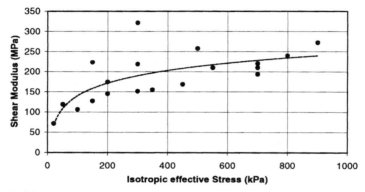

Figure 12. Shear modulus vs. isotropic effective stress by different tests.

Pin Hole test results indicate that the core material is not dispersive. The test has been run with distilled water; determination of chemical species in water taken at the dam site does not reveal any potential adverse effect with the clay minerals stability. Hydraulic conductivity from triaxial test run at several confining stresses and, indirectly, from oedometers, varies, within the stress range of interest (up to 600 kPa) from 1×10^{-10} (extrapolated) to 17×10^{-11} m/s (Fig. 13). No significant anisotropy of hydraulic conductivity has been observed. Lefranc tests indicate values from 5.0×10^{-9} m/s to 5.0×10^{-8} m/s for the core material.

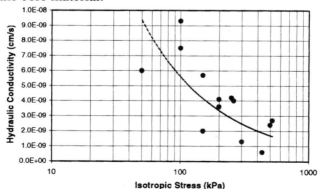

Figure 13. Hydraulic conductivity vs. the isotropic stress

Two principal soils have been identified in the foundation, one with a granular character, the other more cohesive. The latter has displayed undrained cohesion values of 70-80 kPa, and hydraulic conductivity (by Lefranc tests), similar to the average one of the core, 1.0×10^{-9} m/s.

DESIGN CONCEPT FOR REHABILITATION

The assessments of the impact of the Marche-Umbria 1997 earthquake and the properties of the core and cohesive foundation materials investigated by *ad hoc* laboratory tests and by *in situ* determinations revealed that the core retains satisfactory properties after the earthquake shock, which are very near to those adopted in the design phase. It is therefore justified to proceed with a seismic rehabilitation essentially based on providing additional confinement to the core, and higher margins of local safety against sliding to the slopes, by reshaping the embankment slopes. The above objectives can only be achieved by the addition of rockfill material.

The freeboard has been increased to 0.75 m, and the slopes shaped to reach 1:2 upstream and 1:1.8 and 1:2.2 downstream. It is proposed to rebuild a small portion of the top of the dam body (the first two metres) . Some overburden is provided, at the downstream toe, to increase the factor of safety against piping.

Grouting is proposed to enhance the performance of the foundation materials, soils and rock, against seepage.

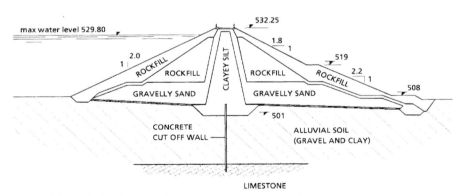

Figure 14. Rehabilitation design: main cross section

Data obtained from the tests allowed all the necessary analyses in support of the remedial works. These were basically:
- Stability checks of the dam body and foundation to comply with Italian standards.
- Seepage evaluation and checks for piping.

- Stability of structures within the dam body, such as the outlets, etc., and seismic checks on the structure of the gate tower.
- Check of punching of the concrete diaphragm into the clay core, during the earthquake motion and possibly reactivated by the consolidation effect of the alluvial soils due to the weight of the new rockfill material (more than 55000 m^3).

The design has proved compliant with regulations with regards to the above effects.

The assessments confirm that the rehabilitation project significantly improves the static and seismic safety margin with respect to the original configuration, varying from 40% for the surfaces located in the upper portion of the shoulders to 25% for deeper surfaces.

The evaluation of settlement of the dam, in the short and long-term conditions respectively, showed a maximum value of 6 cm and 11 cm. The maximum shear strains induced in the material core is 0.3% low compared with deformability of core material as observed in triaxial tests. The core material is able to withstand the overburden without displaying global or local damage effects. The check for local punching, before and after the seismic event, indicates that maximum shear stress (75-80 kPa) keep well below the undrained cohesion of the core.

CONCLUSION

The Acciano dam has provided evidence that factors of safety for seismic design incorporated in the Italian dam design code can effectively provide a significant seismic resistance capacity. The damage is confined to the crest of the dam and the condition of the core has remained suitable to allow remedial works to be implemented.

Such conclusions could only be made following a comprehensive testing and modelling programme outlining the critical role that such methods can play in assessing the current safety condition of existing dams.

REFERENCES

Paoliani, P. (2001) *The behaviour of Acciano earth dam during the Umbria-Marche earthquake of September 1997*. Rivista Italian di Geotecnica 2/2001.

Makadisi, F., Seed, H.B. (1978). *Simplified procedure for estimating dam and embankment earthquake-induced deformations*. J.Geot Eng. Div. ASCE, 104, pp.849-867.

Newmark, N.M. (1965). *Effects of earthquakes on dams and embankments*, Geotechniques, 15, n.2, pp.139-160

Sabetta, F., Pugliese, A. (1987). *Attenuation of peak horizontal acceleration and velocity of Italian strong-motion records 198.* Bulletin of seismological society of America.

Seed, H.B. (1966). *A method for earthquakes resistant design of earth dams,* J. of the Soil Mechanics and Foundations Division Vol.92, SM1, pp.13-41.

ICOLD (1991). *First Benchmark Workshop on Numerical Analysis of Dam.* Bergamo (Italy) May 28-29 1991 – Organized by the "Ad hoc Committee on Computational Aspects of Dam Analysis and Design of CIGB-ICOLD in cooperation with ISMES and Dam Engineering.

Lambe, T.W., Whitman, R.W (1979). *Soil Mechanics.* John Wiley & Sons

Hydropower Development (1992). *Rockfill Dams.* Norwegian Institute of Technology Report n.10

Prakash, S. (1981). *Soil dynamics.* Mc Graw Hill.

Bowles, J.E. (1988). *Foundation analysis and design.* McGraw Hill.

Lessons from a dam incident

R C BRIDLE, All Reservoirs Panel Engineer, Amersham, UK

SYNOPSIS.   Panel Engineers learn much when they are called out to deal
with dam incidents. This paper attempts to share the lessons from an
incident at an anonymous dam. Fifteen lessons are identified, which if
followed, will lead to a greater understanding of the properties of dams and
their behaviour if failure threatens. It is recommended that this knowledge
be used to compile 'emergency handbooks' to equip those handling
emergencies to take previously planned measures to minimise the risks to
lives and property downstream and to release water quickly from threatened
dams.

LEARNING FROM EXPERIENCE

Panel Engineers learn much when they are called out to deal with dam
incidents. Our dams will present less of a threat if all concerned in dam
safety learn from and react to these lessons. Taking a cue from the ICOLD
(1974) publication 'Lessons from Dam Incidents', this paper attempts to
share the lessons from a dam incident. The lessons are mostly not new.
Some are already statutory, many appear in the embankment (Johnston et al,
1999) and concrete (Kennard et al, 1996) dam guides, and others were
mentioned in the guidance on preparation of section 10 reports (Dams &
Reservoirs, 2001). However, putting them together in the context of a dam
incident makes their relevance and usefulness all the more obvious, and
will, I hope, encourage all owners to prepare 'emergency handbooks' on
their reservoirs to assist those charged with handling any emergencies to
deal with them promptly and effectively, without making an already
difficult situation worse.

ANONYMOUS RESERVOIR

The dam in question will remain anonymous, the lessons are not site
specific and naming the reservoir serves no purpose. I would ask readers to
try to live through the experience as I did and add it to their own experience,
perhaps equipping themselves to deal with future incidents all the more
competently. Of course, I recognise that my approach was far from perfect
and many of the lessons that I learnt will not be new to all readers!

Long-term benefits and performance of dams, Thomas Telford, London, 2004, 584–596

ANONYMOUS ACKNOWLEDGEMENT
I would like to say that the owners of the anonymous reservoir responded magnificently to the demands of the incident, maintaining close liaison with the police and emergency services, and providing, without hesitation, all the extra people and equipment needed to deal with it.

THE SYMPTOMS AND WHAT CAUSED THEM
The first lesson related to establishing the cause of the problem. I was called out because a hole had appeared in the downstream slope of a typical British dam. It was about one metre deep and about 800 mm in diameter, and looked to me like the surface expression of piping. We probed down and seemed to reach a solid bottom and we jumped on the bed of the hole and it didn't collapse, but I thought that perhaps the movements that had led to the formation of the hole had also formed some kind of temporary arch across the erosion pipe. No other reason for the formation of the hole seemed obvious. This and the fact that a rush of water in the culvert at the toe of the dam had been reported at about the time the hole had been spotted, led me to think that internal erosion was probably the cause. Had something triggered it, or had there been slow erosion for years that had finally manifested itself as the hole? BGS at Edinburgh reported no seismic activity and water level in the reservoir had been constant for months, and long-term slow erosion looked like the culprit.

I felt it was important to identify the cause of the problem because the incident could then be handled accordingly. But after pondering, experimenting and investigating and finding no other symptoms to convincingly confirm this diagnosis, I realised that it was not going to be easy to identify the cause. Worse, not knowing the cause, I would have to recommend further action after making a judgement on whether the hole had resulted from a single event that would not re-occur, or whether it was the result of some event that would re-occur or continue, possibly at an accelerated rate, and lead to escalating damage. Not a satisfactory situation for a supposed expert to be in, but a useful first lesson - there will be many unanswerable questions during the course of an incident, the first being what has caused the symptom triggering the incident, but even without knowledge, you must recommend appropriate action, usually erring on the side of caution.

LESSON 1 – It is rarely possible to quickly identify the cause of the symptoms that lead to incidents, consequently initial actions cannot respond to the cause, they need to be generally cautious

GETTING TO KNOW THE DAM

In an effort to see where the leakage causing the erosion was, the water level was kept high, better monitoring arrangements were put in place, and I studied drawings and read a paper about repairs done at the dam many years before. Here was lesson two - write down your experiences in a reputable journal. I found being able to 'talk' to my predecessors invaluable and eliminated much of the conjecture that inevitably arises during incidents.

LESSON 2 – Record your experiences

IMPACTS OF IMPROVEMENT WORKS AT OLD DAMS

The paper revealed that there was much stony fill in the dam. The hole may have resulted from settlement of this fill, because it had been wetted, perhaps for the first time, by water released after a huge thunderstorm from perforated pipes recently laid along the toe to drain the mitre. Lesson three – think carefully about the impact of supposed improvement works at old dams. You probably know relatively little about them, especially the fill in their shoulders, and therefore you know little about the impacts your safety works might have. Don't enter into even simple improvement works lightly. Incidentally, the perforated toe drains were replaced by unperforated ones, which reduced flows reaching the culverts during rain, as Figure 1 shows.

LESSON 3 – Think carefully about the impact of works to improve old dams

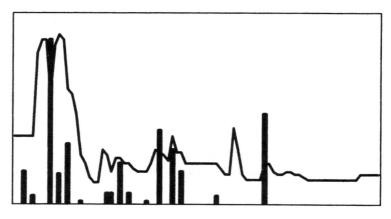

Figure 1 Rainfall and culvert flow over time

SAFETY FIRST

Then came lesson four. I realised that my forensic investigations had distracted me from the alarming fact that the real cause, whatever it was, might cause failure of the dam, releasing the water from the still full reservoir on to the unsuspecting public downstream. I ordered the reservoir to be lowered forthwith.

LESSON 4 – Take precautions first, don't take chances.

RAPID DRAWDOWN

Lowering water level inevitably leads to the questions of how low and how quickly? On how quickly, I was very cautious because the paper had told the story of repairs to the upstream slope, and I didn't want to add to our woes by allowing water to escape from the reservoir through a slip of the upstream slope caused by too rapid a drawdown of the water level. The uncertainty could have been reduced if a rapid drawdown analysis had been done – lesson five. The guidance – 300 mm a day – is conservative in well-drained slopes (Reinius, 1948), but not conservative in clay slopes if the water level is drawn down a long way (Morgenstern, 1963). Modern numerical methods to analyse safe drawdown rates are also available (e.g. Dounias et al, 1996). All require knowledge of the fill in the upstream shoulder. In many British dams the shoulder fill is stony and well drained, consequently rapid drawdown rates will not cause failure. Some dams have clay upstream blankets as well as cores. Lowering the water level in these situations could leave a high water level between and the blanket may be ruptured.

LESSON 5 – Investigate fill in upstream shoulder and do rapid drawdown analyses

EMPTYING CAPACITY

The outlet pipework at the reservoir had an enormous capacity, though there was no rating curve. Lesson six – work out the opening v discharge relationship for the scour and other outlet valves. The maxima should be entered in the Prescribed Form of Record, Part 8, but a rating curve or table would be more helpful in an emergency situation.

LESSON 6 – Know how much water the emptying pipes can release

DOWNSTREAM RIVER CAPACITY

If I had not been constrained by drawdown failure worries, we could have lowered water level at a terrific rate, very re-assuring when there is a problem. However, the discharge might have gone out of bank and flooded properties downstream. Lesson seven – identify pinch points and low-lying properties near the river downstream, estimate likely in-bank flow capacity, send scouts out to pinch points when releasing water from the reservoir.

LESSON 7 – Know the capacity of the river downstream

CRITICAL RESERVOIR CAPACITY

The next question that arose was how far should we lower water level? Having postulated internal erosion, there was a possibility of a low-level erosion pipe working back towards the reservoir, the symptomatic hole being a vertical tributary from it, and this could release the whole reservoir downstream. I thought that the consequences of this would decrease as the water level retained became lower; perhaps at some critical level there would be no significant damage downstream. The dambreak analyses (they had been done previously, otherwise another lesson would have been learnt) were dusted off and trials done with progressively smaller volumes of water escaping from the reservoir. The result? Even at 50% full, the number of properties at risk, while less than half of those when the reservoir was full, was still large, many up to 30 kilometres downstream. This confirms what we know (or it could be another lesson?), that the extent of dambreak impacts relates more to the river slope downstream of the dam, and therefore the flow velocity, than to the volume of water released. However, it would have been good to know about this before the emergency. At reservoirs with gentle downstream river slopes there might be a critical reservoir level at which no damage would occur downstream in a dambreak. Lesson eight – do dambreak studies with differing retained reservoir volumes to give guidance on how far to lower water levels in an emergency. LESSON 8 – Know how much water needs to be released to reduce the threat to acceptable levels.

CONSTANT SURVEILLANCE

The emergency authorities had some direct questions. They needed to know when we would be able to tell them to evacuate people downstream. How would we know when the dam had started to fail? We expected that there would be accelerating movements, sounds of running water in the culvert, outbursts of dirty water into the culvert, subsidence of the crest and other manifestations of failure by internal erosion that we know about from the literature. The police noted that all these could occur at any moment. The order to evacuate depended on direct visual evidence. They instructed us to provide constant surveillance, with shelter and phones, at the dam. Lesson nine – faulty reservoirs should be under surveillance 24 hours a day, seven days a week. Provide experienced people, with a cabin and ready access to phones, they would be the ones who would trigger evacuation if failure seemed inevitable.
LESSON 9 – Faulty reservoirs should be under constant surveillance

DEFORMATIONS ON THE RUN UP TO FAILURE

The police questions about how would movements accelerate before failure occurred was a tough one. I have seen finite element analyses of movements prior to stability failures (e.g. Vaughan et al, 1989), but if internal erosion was occurring and the eroding water was somehow escaping without coming into view, what deformations could be expected and would there be any change in the rate at which they occurred. Also would they be

visible to the naked eye, or would they only be discernible by survey methods? Lesson ten – we should know more about deformations before failure, and what means would be required to monitor them.
LESSON 10 – How do dams deform on the run-up to failure?

SURVEY MOVEMENT MONITORING

We had set up monitoring pegs, read and co-ordinated in position and level daily by surveyors. As pre-failure deformation rates seemed (to me at least) something of an unknown, it was agreed that if the rate of deformation increased to be 25 mm or more between successive days, I would be called out to judge if failure was imminent. I relied on my innate knowledge of dams and earthworks to be able to make the judgement! I admit that by this time, when asked the question how likely did I think it was that failure would occur, I felt confident enough to say it was less that 50-50, mainly because there seemed no evidence of any serious signs of the causes of the damage, or of any changes in the dam's profiles.

The fact that we had simple pegs monitoring movement was re-assuring, and I recommend this is done in such situations, assuming that they don't develop rapidly, because it does provide evidence of malfunction or no malfunction, the latter being important in providing evidence to justify de-mobilising the emergency arrangements, when as often occurs, 'incidents' turn out to be non-incidents. Whether pegs should be permanently installed could be considered, though incident specific additions would probably be needed also. Lesson eleven – provide simple survey movement monitoring arrangements to monitor behaviour of dams during incidents.
LESSON 11 – Provide simple survey movement monitoring devices

FILL TYPES AND FAILURE MODES

While all these precautions were being dealt with, the matter of the cause of the incident, the 'hole', remained in question. If it was the top of an internal erosion pipe, the pipe below might be a sizeable cavern, and I felt that this precluded excavating into it. If a large hole was exposed, the excavator might fall into it. How would we stop an enormous hole forming, exposing the downstream side of the core, leading to its collapse and release of water, a terrible scenario? Could large quantities of filter be assembled and quickly shoved into the void if such a disaster struck? Such considerations had only one answer; don't dig into the hole until more is known about it. I have to confess that boreholes were to be done for other reasons at the dam, and this may have made such a decision easier.

However, if internal erosion were the cause, a knowledge of the fills, in the core and in the upstream and downstream shoulders, and the nature of the rocks in the foundation, would assist in assessing how vulnerable the dam would be to internal erosion. I knew something of the upstream fill because of the paper, and the core was puddle clay, but I knew nothing of the downstream fill, except that some, placed as part of the early repairs, was stony. The foundation, as seen in the culvert floor and in exposures near the dam, was open jointed sandstone.

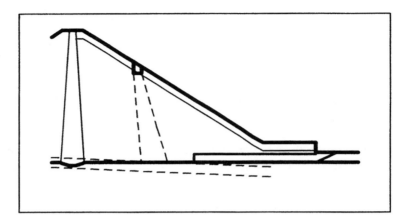

Figure 2 Erosion at interface between open-jointed foundation and fill; eroded materials carried downstream in joints in foundation rock, not visible in culvert. Subsidiary pipe leads to the hole on downstream slope.

As Figure 2 indicates, I postulated that water flowing in the open-jointed foundation rock might have slowly eroded through the base of the fill, including the core. Sediment in suspension was not visible in the culvert flows, but 'dirty' water containing eroded materials may have drained away below the foundation/fill interface and not been visible. An erosion pipe may have extended through the core, and the 'hole' may have been the top of a subsidiary pipe. This was worrying, but there were re-assuring signs, including no sign of local crest settlement (also being monitored by the surveyors) and no whirlpools in the reservoir (although they would not be formed if the upstream end of the erosion pipe broke out of the upstream shoulder at depth.).

Many of the anxieties about excavating into the dam and the potential for development of erosion pipes would have been dispelled if the properties of the fills were known. In our case the downstream fill was predominantly stone and gravel, not likely to be eroded, or to sustain open erosion pipes!

The property of the fills that can be most readily used is its in-situ permeability, easily measured in boreholes. Permeability values can be used as a qualitative indicator of the texture of the dam fills.   However, the permeability values can be used quantitatively in drawdown stability analyses of the upstream slopes to establish the safe rate of drawdown.   In the downstream slope the permeability values can be used in the 'perfect' filter equation (Vaughan & Bridle, 2004) to determine the filtering potential of adjoining fills, the shoulder fill against the core fill, for example, thereby indicating whether the dam is capable of self-filtering or is vulnerable to internal erosion.

LESSON 12 – Know the properties of the fills in the dam, core and shoulders, particularly their permeabilities.

WARNING THE PUBLIC AND ASSISTANCE WITH EVACUATION

When we considered what to do should the dam fail, we expected that police and owners' staff would enter vulnerable areas and advise residents to evacuate, assisting them as necessary.   However, we learnt that modern 'duty of care' obligations would preclude sending staff, including police, into a situation where their lives may be at risk.   The concern was that the flood wave velocity, and therefore the period during which police and owners' staff could safely assist evacuees, was very uncertain.   This made it impossible to devise any system of tracking and warnings that would be sure to get people clear without failing to meet reasonable 'duty of care' obligations.   There may have been time to help those living far from the dam; those nearer would have to be assisted by other means - a 'sky shout' helicopter!   People would be told to evacuate by a voice from the sky. Many wouldn't be expecting trouble and might not believe their ears.   How would they know which way to go to be safe?   Rehearsals might have helped, but were not advised in the context of a possibly imminent failure, they may have caused panic.   Leaflets were precluded for the same reason.

An unsatisfactory situation, but it leads to a most important lesson, lesson thirteen – there is not time to safely evacuate those living close to a suspect dam after the flood wave is released, i.e. before failure has definitely commenced.   They may have to be evacuated before failure has commenced.   Some dam failures have been telegraphed by clear signs, (e.g. Baldwin Hills, ICOLD 1974) giving more time for evacuation, but this is not always the case.

Some of the advice and recommendations that the police gave may have been coloured by my earlier less than 50-50 decision.   I certainly felt confident enough not to insist on any pre-failure evacuation - but this is a decision that we will be called on to make in future incidents.   It would be more effective if it were pre-planned and those living in the floodway were made aware of it

LESSON 13 – The time available to evacuate those in the floodway will be limited, plan to start evacuation before failure has definitely commenced.

PREPARING FOR DAM DISASTERS

Although dambreak analysis had been done at the dam, the incident showed that the contingency plan to deal with such a situation was generic and not site specific. This leads to lesson fourteen – that it is necessary to make full preparations to deal with evacuations to avoid dambreak floods. The provisions of the Water Act (2003) empower the Secretary of State to call for 'flood plans' to be prepared at reservoirs and it seems likely that they will be required for all high hazard reservoirs, where large numbers would be at risk should a dambreak occur. The statutory requirement for flood plans should lead to effective plans being put in place for high hazard reservoirs. Owners will still need to make plans, appropriate to the hazard posed, for all their reservoirs. Clear plans are needed and the emergency services and the public should know the plans, and what actions they will need to take if a dambreak occurs. They may need to be made aware of them by rehearsals, signposts, talks, and whatever other means seem appropriate, and there may be a need for routine refreshers.

LESSON 14 – Make clear contingency plans to deal with dambreak risks, alert the public to them and train them to evacuate effectively

BEING PREPARED AND 'EMERGENCY HANDBOOKS'

And that leads to the final lesson – owners, advised by dam professionals, can do a great deal to reduce the impacts of dambreaks. We have worked together under the Act and the good practice that has been developed around it, to make dams safe. We cannot eliminate all risk but we can prepare to deal effectively with the residual risk.

We tend to think that dealing with emergencies comes down to mobilising the contingency plan and evacuating people at risk. But a great deal can be done by analysis and at the reservoir, completely within the owners' control, to be prepared for emergencies. This improves the possibility of being able to control situations effectively, reducing the numbers of lives at risk by prompt evacuation and perhaps completely averting dam failure. An 'emergency handbook' about the reservoir would do much to equip Panel Engineers and owners' staff handling emergencies to deal effectively with a crisis. They would be much more in command of the situation; they would not be experimenting. They would know the consequences of opening up valves and other actions, and could balance the risks.

The fundamental objective is to save lives by prompt evacuation and by emptying, or lowering the water level in, the reservoir as quickly as possible without exacerbating an already difficult situation.

A most difficult decision is ordering evacuation, because to be effective at most reservoirs it needs to commence before it is certain that the dam will fail. Present knowledge does not equip us to be able to quickly detect how close to failure a faulty dam is. Our trained instincts and the performance of the suspect dam probably help us on this. A few years of focussed research would likely lead to better guidance. All the other technical information needed for emergency handbooks could be assessed from published information, some of it dating from many years ago.

If we knew, as we entered an emergency situation, who should be evacuated promptly, how many turns of the scour valve would lead to out of bank flow at critical points downstream; how many turns of the scour valve would precipitate rapid drawdown failure of the upstream slope and how much water should be released to reduce the risk to an acceptable level; we would be in control. We could give the emergency authorities more specific advice. If our observations of the performance of the stricken dam warranted it, we could go into the 'risk zone' and take a chance on local downstream flooding and upstream slope failure if it would bring the water level down more quickly and thereby reduce the number of lives at risk.

Also, if we knew more about the fill in our dams, particularly its vulnerabilities to drawdown failure and internal erosion, we could provide permanent or emergency stand-by measures to deal with them, thereby further reducing the probability of occurrence of emergencies or failure.
LESSON 15 – Be prepared, by finding out more about our dams and assembling emergency handbooks to equip us to deal with emergencies as effectively as knowledge allows.

INCIDENT OR DISASTER?
The incident had a happy ending, the dam did not fail, the reservoir was refilled and has performed satisfactorily since. As indicated on Figures 3 and 4 below, I concluded that the hole had formed following deformation brought on by a combination of circumstances associated with new drains and wet weather. The heavy rainfall discharged from the (temporarily) slotted toe drain and wetted the stony fill, which had been kept dry by the deep layer of clayey fill on the surface of the downstream slope. The wetting caused some movement of the fill as points were wetted and collapsed. The clayey surface fill had been cut through at top and bottom of the slope for new drains, and may have stretched a little. There was some evidence in a trial pit of soil movement into voids in the stony fill, local internal erosion.

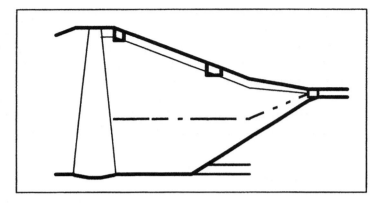

Figure 3 Cross section

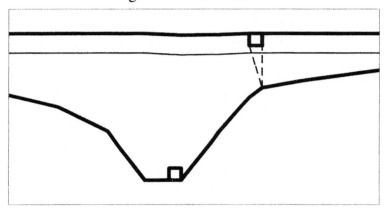

Figure 4 Longitudinal section

Figures 3 & 4 Conjectured mode of hole formation. Figure 3, cross-section, toe drain released water into stony fill, which may have settled slightly. Toe drain and crest drain cut through clayey surface fill, made slope less stable, allowed water in, surface fill moved. Figure 4, longitudinal section, foundation profile at hole position created differential settlement and adjustments in stony and clayey fill increased stress on clayey surface fill.

THE LIMITS OF OUR KNOWLEDGE AND USING WHAT WE KNOW
My final remarks relate to Panel Engineers and the expectations put on them. It would be easy to think that if you are appointed as a Panel Engineer, you must know enough about dams to respond expertly to any emergency. But as you have read, dealing with an emergency certainly revealed shortcomings in my knowledge, an admission shared, I imagine, by all but those entirely lacking in humility! While we will never know

everything, it is disappointing that we do not routinely require safe drawdown rates and susceptibility to internal erosion to be checked using published techniques. Pre-knowledge on these issues would improve the safety of our dams and our effectiveness in dealing with emergencies. However, a proper sense of humility should not lessen our effectiveness in dealing with incidents. Those working with you will want clear and positive instructions as to what they are to do. I hope the lessons I've listed, when put into practice and assembled in emergency handbooks, will make that easier to achieve.

SUMMARY OF LESSONS

No	Lesson
1	It is rarely possible to quickly identify the cause of the symptoms that lead to incidents, consequently initial actions cannot respond to the cause, they need to be generally cautious
2	Record your experiences in a reputable journal, being able to 'talk' to predecessors is invaluable and eliminates much of the conjecture that inevitably arises during incidents
3	Think carefully about the impact of works to improve old dams
4	Take precautions first, don't take chances
5	Investigate fill in upstream shoulder and evaluate safe drawdown rate
6	Work out the opening v discharge relationship for the scour and other outlet valves
7	Identify pinch points and low-lying properties near the river downstream, estimate likely in-bank flow capacity, send scouts out to them when releasing water from the reservoir.
8	Do dambreak studies with differing retained reservoir volumes to give guidance on how much to lower water levels by in an emergency.
9	Faulty reservoirs should be under surveillance 24 hours a day, seven days a week. Provide experienced people, with a cabin or other shelter and ready access to phones.
10	We should know more about deformations before failure, and what means would be required to monitor them.
11	Provide simple survey movement monitoring arrangements to monitor deformations during incidents.
12	Know the properties of the fills in the dam, core and shoulders, particularly their permeabilities.
13	The time available to evacuate those in the floodway will be limited, plan to start evacuation before failure has definitely commenced.
14	Make clear contingency plans to deal with dambreak risks, alert the public to them and train them to evacuate effectively.
15	Be prepared, by finding out more about our dams and assembling emergency handbooks to equip us to deal with emergencies as effectively as knowledge allows.

REFERENCES

Dams & Reservoirs (2001). Guidance on preparation of section 10 inspection reports. *Dams & Reservoirs, Journal of the British Dam Society*, Vol 11 No 1, May.

Dounias G T, Potts D M and Vaughan P R (1996). Analysis of progressive failure and cracking in old British dams. *Geotechnique*, Vol 46, No 4, pp 621-640

International Commission on Large Dams (1974). *Lessons from Dam Incidents*. ICOLD, Paris.

Johnston T A, Millmore J P, Charles J A and Tedd P (1999). *An engineering guide to the safety of embankment dams in the United Kingdom*. Building Research Establishment Report. Construction Research Communications Ltd, Watford, 102 pp.

Kennard M F, Reader R A and Owens C L (1996). *Engineering guide to the safety of concrete and masonry dam structures in the UK*. Construction Industry Research & Information Association Report 148. CIRIA, London.

Morgenstern N (1963). Stability charts for earth slopes during rapid draw-down. *Geotechnique*, Vol 13, No 4, pp 129-150.

Reinius E (1948). *The stability of the upstream slope of earth dams*. Mederlanden Statens Kommitte for Byggnadsforskning, Stockholm, No 12.

Vaughan P R and Bridle R C (2004). An update on perfect filters. To be published, this conference.

Vaughan P R, Dounias G T and Potts D M (1989). Advances in analytical techniques and the influence of core geometry on behaviour. *Clay Barriers for Embankment Dams. Proceedings of conference held in London*, October, pp 87-108.Thomas Telford, London, 1990.

Comments on failures of small dams in the Czech Republic during historical flood events

J. RIHA, Brno University of Technology, Zizkova 17, 66237 Brno, CZ.

SYNOPSIS.   During the catastrophic floods, namely those in July 1997 and August 2002 and also during local flood events, which occur almost every year in the Czech Republic, more than 100 failures of small dams were identified during last decade. After careful analysis of typical small dam failures, the reasons for dam collapses were found and assessed. During the flood events the most frequent failure of small dams was by breaching due to dam overtopping. The majority of small dam spillways suffer from insufficient capacity, inconvenient structure and arrangement. At some places spillways were blocked by broken gates, clogged jammed racks or floating debris. Moreover, in some cases the bottom outlets were not maintained and out of order. In the paper, several examples of unsuitable spillways and other dam appurtenances are shown.

INTRODUCTION

During the August 2002 flood, which affected the middle and western Europe, about 70 breached small dams were identified in the Czech Republic. This paper deals with the Blatna region in the south of Bohemia (see Figure 1), where 10 small dams were breached. The reasons for failures of two small dams in the area were analysed in more detail (dams of the Metelesky and Melin ponds). The failures caused disastrous damages in the villages of Metly and Predmir located downstream of the ponds, the breach outflow of two dams mentioned caused overtopping and failure of five small earth dams downstream of the ponds and finally flooded the town of Blatna (see Figure 2). In the analysis, the hydrological conditions in April 2002 were assessed in context with the capacity of bottom outlets and emergency spillways of both small dams. Finally the breaching mechanism was reproduced and the peak flood discharge was estimated based on comprehensive field data on the failure process during the night of 12th to 13th August 2002.

Long-term benefits and performance of dams, Thomas Telford, London, 2004, 597–608

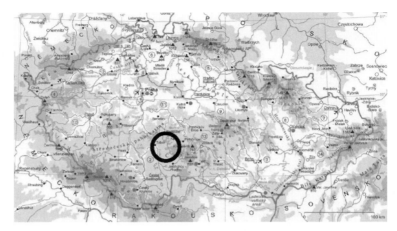

Figure 1: The map of the Czech Republic with the area of interest

Figure 2: Town of Blatna during the August 2002 flood

BASIC DATA

August 2002 flood

On August, 5th 2002 a cyclone developed above the Western Mediterranean, which proceeded northeast and reached the eastern Alps during August 6th. Heavy rainfall occurred over southern Bohemia with local showers of high intensity, which temporarily ceased on the morning of August 8th. After this cyclone the second one followed coming from British Isles to the southeast. On Saturday, August 10th the cyclone regenerated above Italy and continued to the north. During the 11th and 12th August the cyclone reached the Czech Republic, where the long lasting precipitation struck almost all the country. The most intensive rainfalls occurred in the mountainous regions to the southwest and northwest of Bohemia, and in the area of interest the three-day total was about 160 mm. On August 13th the rainfall intensity reduced and on 14th it completely ceased.

As the catchment soaked completely during the first precipitation event, the runoff percentage (runoff coefficient) during the second precipitation event was considerable. Due to the relatively long duration of rainfall an extreme flood was generated throughout the Vltava river catchment. Moreover, at some places local showers of considerable intensity caused runoff concentration at smaller streams. At bigger streams on downstream reaches the discharge exceeded the 500 year flood, and at smaller streams, especially at upper catchment portions it was estimated to be a 1000 year flood. This was the reason for the breaching of a great number of small dams with insufficient spillway capacity. Details of the dams breached during the August flood in the vicinity of Blatna are given in Table 1.

Table1: List of small dams breached in the Blatna region

Name of the pond	Dam height in metres	Total reservoir volume in thousand m^3	Reservoir area in hectares	Reason for failure
Belcicky	6.7	788	39.4	Overtopping
Buzicky	2.7	900	60.0	Overtopping of side dam
Dolejsi	2.6	334	30.0	Overtopping
Horejsi	4.0	232	22.4	Overtopping
Luh	3.8	48	6.0	Overtopping and improper outlet location
Melin	6.2	250	11.4	Slide of the downstream slope
Metelsky	8.5	1037	51.4	Overtopping at two places
Mlynsky	2.6	160	12.7	Overtopping
Podhajsky	2.9	225	15.0	Overtopping
Pusty	3.5	65	5.5	Overtopping

Details of the ponds studied

In this paper, the results of the analyses of only two ponds, namely Melin and Metelsky, are given. Both ponds are situated at the Metelsky brook about 12 km to the North of the town Blatna just upstream of the village Metly (see Figure 3). The catchment area of the pond Metelsky is about 15.5 km^2 and is covered by agricultural land (30%) and forests (70%).

The *Melin* dam is about 6.2 m high, the dam body is homogeneous, made of sandy clay with estimated hydraulic conductivity between 1.5×10^{-8} to 4.5×10^{-8} m/s. The upstream and downstream slopes are 1:1.5. The dam crest is uneven with 0.60 m differences in the crest level, and the lowest part of the dam crest is close to the bottom outlets. The dam crest is overgrown by trees and bushes. The root system of the vegetation disturbed the upper portion of the dam body, which is much more permeable than the lower part. The upstream slope of the dam is faced by stone pavement, and the downstream slope is grassed. The pond was equipped by one wooden bottom outlet with the maintenance and service shaft located at its upstream end (Figure 4). The dam is provided with two emergency spillways, one at the left bank abutment, the other at the right one. The total spillway capacity is 10.5 m³/s for the water level at the minimum dam crest.

The *Metelsky* dam is about 8.5 m high with upstream and downstream slopes 1:2. The dam body is heterogeneous, created by upstream clayey blanket and sandy downstream shoulder. At the upper portion close to the dam crest the clayey sealing is missing or is degenerated by the root system of the plants grown on the dam crest (Figure 5). The upstream slope of the dam is faced by stone pavement, and the downstream slope is grassed. The pond is equipped with two wooden bottom outlets in a bad condition due to ruptures permitting seepage and rinsing of the sand from the dam to the pipes. At the left abutment the dam is equipped by an emergency spillway with a capacity of about 9.5 m³/s. An auxiliary spillway (capacity 2.5 m³/s) is formed by the local right bank road.

Hydrological data
Both ponds are constructed and operated as through-flow. The Melin pond is fed by three streams, the pond Metelsky is fed by two tributaries with total catchment area 15.83 km² with the peak level 712 m above sea water level (SWL). The N - year discharges at the dam sites are given in Table 2.

Table 2: N-years discharges Q_N in [m³/s] at the dam sites

N	1	2	5	10	20	50	100
Q_N - Metelsky	5	7	10	12	15	19	23
Q_N - Melin	3.3	-	6.8	8.9	11.0	15.0	18.0

The flood hydrograph corresponding to the "natural" August 2002 flood at the dam profiles was derived using a rainfall – runoff model (Figure 6). The results of the modelling were compared with results obtained from the calibrated hydrodynamic model. The flood routing model calibration was based on the traces of the flood at the site. The flood routing in the area downstream of both ponds was considerably influenced by their collapse.

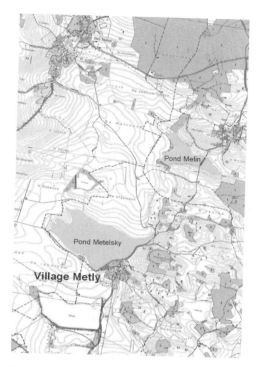

Figure 3: The detailed map of the ponds and Metly village

Figure 4: Melin – dam breach with remaining service shaft of the bottom outlet

Figure 5: Dam Metelsky - the cross section at the right breach

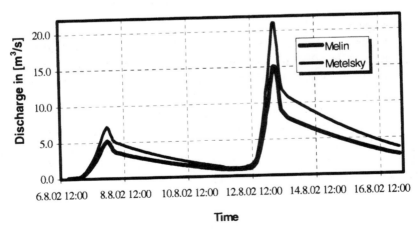

Figure 6: Derived flood hydrograph in August 2002 at the dam sites

ANALYSIS OF THE FAILURES OF THE SMALL DAMS

General comments
When comparing previous data about spillway capacities with flood hydrographs, it is obvious that the main reason for the dam failures was insufficient spillway capacity. Nevertheless, the purpose of the study was to provide a complex analysis of the event. Therefore, the following effects and their combination were assumed:
- dam erosion due to overtopping;
- loss of the dam body stability due to slide of the downstream face;
- internal erosion of the dam body.

The detailed analysis showed that the failures of both dams were caused by the combination of effects mentioned above. The analysis was carried out in following steps:
1. The reconstruction of the event using witness testimony provided by criminal police, local inhabitants and by the traces of water level at banks and upstream face of the dams.
2. The setting up of a numerical model consisting of rainfall-runoff, dam break and flood routing models. During this work, bottom outlet and spillway capacity rating curves were derived carefully.
3. The model calibration was based on the knowledge obtained in step 1. The calibration scenario resulted in the real flood and dam break discharges and possible reasons of the failures.
4. Finally, several aditional scenarios dealing with possible manipulation with bottom outlets combined with the temporary side spillway 'on-site' installation were solved. The main goal of these scenarios was to prove that no measures were capable of averting dam failures.

The upper pond – Melin
Due to very low hydraulic conductivity of the homogeneous dam body, the seepage through the dam material was assumed to be very low. Anyway, the site investigation showed that the upper portion of the dam body of the thickness 0.3 to 0.5 m is composed of weathered grained humus material, the structure of which is disturbed by the root system of the vegetation (Figure 4). This material is of a significantly higher permeability. After the water level reached the higher position close the dam crest, more intensive seepage through the weathered layer probably caused the instability of relatively steep downstream face slope (1:1.5).

Detailed assessment of seepage conditions does not indicate suffosion trends in general. Nevertheless, the old wood pipe was found at the place of the breach, the rest of the pipe having been flushed down and dispersed downstream up to a distance 300 m from the dam site. As the wood pipe was quite old and damaged by cracks, it was concluded that the sandy material of downstream shoulder was flushed off by the pipe and caused the subsidence of the dam crest. This probably contributed to the dam failure.

The results of hydrological and runoff modelling indicated a rapid increase in the inflow discharge to the pond on the evening of 12 August 2002 with the peak discharge approximately 15 m^3/s (Fig. 7), which corresponds with 50 year flood (Table 2). The flood routing through Melin reservoir showed a transformed flood peak discharge 13.5 m^3/s and shift of the peak by 2.5 hours. At the same time, the spillway capacity was about 10 m^3/s at the water level at the dam crest.

The height of the wind driven waves was estimated to be between 0.55 and 0.60 m. The dam was locally overtopped by the wind waves for the period of 3 hours at the place with the lowest dam crest, i.e. at the location of the dam breach.

The Melin dam failure was caused by combined action of leakage through the upper portion of the dam below the dam crest and the wind waves overtopping the dam crest. These factors caused the dam failure in the section of the lowest crest, where its subsidence was probably caused by suffosion of the sandy material into the damaged wooden pipe. The process of breaching was accelerated by slides of the relatively steep downstream face of the dam. The dam break peak discharge at the dam site was estimated to be 150 m^3/s and this was verified by the calibrated flood routing model in the valley downstream of the pond. The resulting dam breach opening was of almost rectangular shape with the 5 m depth and 15 m width.

Dam at Metelsky pond
In case of the Metelsky dam, an overtopping was the primary reason of the failure. During the natural flood, the retention capacity of flood surcharge was exhausted due to malfunction of bottom outlets and unsufficient spillway capacity. At the same time the reservoir inflow increased considerably due to the breach of Melin dam located approximately 2 km upstream from Metelsky pond. Melin dam break peak at the inflow to the Metelsky reservoir was transformed to "only" 130 m^3/s, the flood wave volume corresponding to the Melin reservoir volume was about 600,000 m^3.

The detailed modelling of the event showed that the water level during the flood event was about 0.6 m above the spillway crest. At that time the Melin dam break wave entered the Metelsky pond and caused dam overtopping in two places. The resulting peak discharge was about 550 m^3/s, and the total volume of the flood wave was estimated to be 2.3 million m^3. The widths of two breach openings were 35 m and 27 m, the breach depth was about

7.7 m. Final dam break flood hydrographs compared with the hydrological flood are shown in Figures 7 and 8.

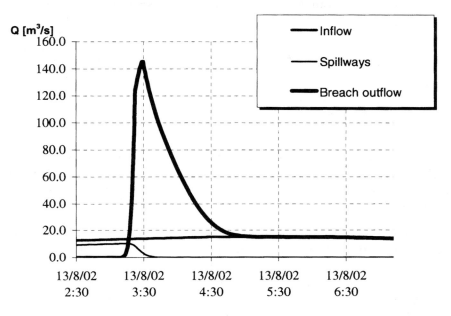

Figure 7: Melin reservoir - inflow and outflow

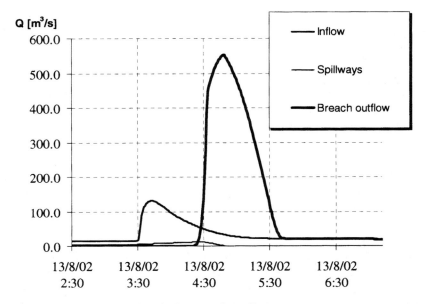

Figure 8: Metelsky reservoir - inflow and outflow

Several additional phenomena contributed to the dam failure and accelerated dam breaching. The 4 hours' action of wind waves of height 0.75 to 0.80 m and local overtopping caused local slides and disturbance of grass cover at the downstream face prior to the dam overtopping and this accelerated the destruction process of overflowing water. Moreover, the upper portion of the dam body was disturbed by the root system of the vegetation and by the action of animals. The left breach opening was located at the place of the old abandoned wooden bottom outlet pipe. The site investigation proved internal erosion and flushing of finer particles to the disturbed pipe (Figure 9) and consequent settlement of the dam body at this location.

Figure 9: The rest of cracked wood bottom outlet

The facts mentioned were not the primary cause of the failure, but accelerated the dam collapse and contributed to earlier overtopping. The important circumstance was inadequate technical safety surveillance of the dam and poor maintenance of dam body and equipment.

CONCLUSIONS

The causes of the failures of two small dams assessed can be summarised in the following statements:

- In case of Melin dam the failure was primarily caused by insufficient capacity of both spillways corresponding to 5years flood discharge (Q_5), while the peak flood was estimated as Q_{50} to

Q_{100}. Additional factors contributing to the failure were local slides of downstream slope, extreme seepage through upper loose portion of the dam, action of wind driven waves, potential privileged seepage paths and suffosion along the wood pipe outlet and malfunction of bottom outlet due to its improper structure.

- In case of Metelsky dam the failure was caused by an extreme hydrological situation combined with the breaching of the upstream dam Melin. The partial factors were practically same as in case of Melin dam.

The following conclusions and recommendations were put forward for further remedial and reconstruction activities at the sites of interest:

- Before the reconstruction of the dams, the revision and carefull surveillance of the state of dam bodies should be carried out. The restoration of the ponds cannot proceed in their present state. Careful assessment must be focused also on the design parameters of dam equipment, namely bottom outlets and emergency spillways.
- The reassessment of present safety classification of small dams should be done with respect to potential danger from insufficiently equipped small dams and based on the new dam safety standards and actual hydrological data.
- The manipulation regulations should reflect the optimal function of the entire reservoir system, which consists of approximately 20 small reservoirs in the Blatna region.
- The potential flood prone area specification should contain the inundation due to potential dam failures.

It is true that during local extreme flood events, on average from two to four small embankment dams (height less than 15 m) are overtopped and breached every year in the Czech Republic. During the extreme regional floods in 1997 and 2002 more than 100 small embankment dams failed and about 50 levees breached in the Czech Republic.

We recognize that the deficiencies mentioned in the structure, arrangement, parameters, operation and maintenance of small dams and their appurtenance are not rare phenomena in the Czech Republic (or in other countries). Remediating the present situation does, however, require time and money and it is also a difficult problem in relation to property and land ownership. Private dam owners (e.g. angling clubs) usually are not able to finance the remedial measures that are required. The state financial support is not systematic and is not steadily anchored in the present legislation, which in many cases is still not prepared for the private ownership of small waterworks.

REFERENCES

Riha, J. et al. (2002) *The Assessment of the Failures of Small Dams in the Blatna Region.* Research Report. Brno Technical University, 30 p., 35 appendices.

Acknowledgement: The paper was prepared as the part of the solution of the grant project 103/02/0018 of the Grant Agency of the Czech Republic.

Detailed investigation of an old masonry dam

Dr.-Ing. ACHIM JAUP, Lahmeyer International GmbH, Bad Vilbel, Germany

SYNOPSIS.   The gravity dam, described in this paper, is now over 100 years old. After such a long period of operation, a rehabilitation of the upstream face of the dam is necessary. During the last rehabilitation between 1965 and 1967 a protective shotcrete layer was applied on the upstream face. In the present rehabilitation the shotcrete layer will be maintained and a drained geomembrane will be installed. The geomembrane will be fastened by a system, which consists of inner and outer stainless steel profiles. These profiles are anchored onto the shotcrete layer.

Because the strata for fixing the anchors (the shotcrete layer) is now over 30 years old – and requires maintenance – major testing(e.g. georadar) of the shotcrete layer was carried out to give a guarantee of a sufficient bearing capacity of the anchor. The testing of the shotcrete is important to establish a basis for the design of the anchorage system.

This paper will describe the construction of the geomembrane lining with special consideration of the fastening of the steel profiles and the testing of the shotcrete layer.

INTRODUCTION

The dam was built by Prof. Intze between 1898 and 1900, and was one of the first Dams in Germany. The dam is used for the drinking water supply. The original dam had a height of 34 m and was designed as a gravity dam and is curved in plan (r = 176 m). In 1934 the dam was heightened to 38 m. The base width is 23.6 m and its width decreases to 4.5 m at the dam crest. The crest has a length of 215 m.

The dam consists of masonry with a volume of 47,000 m³. The masonry consists of quartzite greywacke and a lime mortar with river sand and fly ash. Revetted masonry formed the upstream and downstream faces. The reservoir has a storage capacity of 2,855,000 m³.

PREVIOUS REHABILITATIONS

There were two main rehabilitation activities in the past. The first was between 1950 and 1952, the second between 1965 and 1967.

Rehabilitation from 1950 to 1952

Investigations of the dam foundation showed that there was a high permeability on joints on the silty clayey slate, which were mainly rectangular to the dam axis. This explains a high water pressure in the dam foundation. Piezometer readings showed values up to 75 % of the hydrostatic pressure of the reservoir level. So after 50 years of operation the dam was rehabilitated by cement grouting, to seal the dam body and the dam foundation. The objectives were:

- Sealing the masonry and dam foundation
- Reduction of seepage water
- Avoiding the ongoing deterioration of the bond between the masonry and the lime mortar

Rehabilitation in 1965 to 1967

In the previous rehabilitation the sealing of the masonry was not satisfactorily achieved.

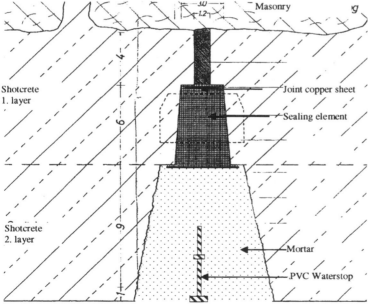

Figure 1. Cross section through joints of shotcrete layer

The main aspect of this rehabilitation was to seal the upstream face of the dam. The rehabilitation consisted of a reinforced shotcrete lining in panels of 5.0 x 5.0 m. Figure 1 and 2 show the shotcrete lining and the joints as a plan and as a photograph.

Each panel was anchored into the dam by 16 anchors with a diameter of 20 mm and a length of 2.50 m. The shotcrete was applied directly onto the masonry in two layers, each 10 cm deep. The joints between these shotcrete panels were formed with a sealing element and mortar, see figure 1.

Figure 2: photo of existing shotcrete lining

Despite these complex rehabilitations the sintering process deterioration was only slowed down slightly. As the joints were not really flexible joints the joints were disturbed due to thermal movement in the dam. Hence the water could leak through these joints into the dam.

CURRENT REHABILITATION

After over 30 years of operation since the last rehabilitation works in 1967, it was necessary to rehabilitate the upstream face of the dam. It was the wish of the owner to maintain the shotcrete layer. So the rehabilitation plan will be implemented by the following methods:

- Breaking out a gallery
- Installing drilled drainage holes
- Sealing the upstream face of the dam by application of a watertight geomembrane

A gallery in combination with drilled drainage holes is a very common method of reducing the subsoil water pressure and is thus not described in this paper.

Sealing concrete faces with a PVC geomembrane has been carried out now for over 40 years, especially on the upstream faces of dams, mainly either RCC dams or masonry gravity dams, e.g. the Brändbach Dam in Germany Veyhle/Jaup (2002) or Scuero/Vascetti (1996).

The geomembrane concept is in general quite simple: the synthetic impermeable liner extends over the whole area, which is to be sealed, not only joints or cracks. It is conceptually equivalent to a single waterstop, which covers the whole area. The geomembrane is attached to the structure as a separate element. The geomembrane is mechanically fastened by means of steel-profiles. The elasticity of the system allows it to bridge cracks that may develop due to external loads or changes in temperature.

The geomembrane consists of a flexible polyvinyl chloride (PVC) membrane, extruded in homogeneous mass from a flat die and heat bonded during manufacturing to a non-woven, needle punched geotextile (100% Polyester). The purpose of the geomembrane layer is to provide watertightness, while the purpose of the geotextile is to help to protect the geomembrane from puncturing and to give dimensional stability. The geomembrane is situated on the geogrid. This geogrid consists of a polymer–plastic, which serves as a drainage layer between the geomembrane and the dam. This construction enables a discharge by gravity of any drainage water that could infiltrate between the waterproofing liner and the dam body.

The Carpi PVC geocomposite is anchored by tensioning profiles on the shotcrete. The tensioning profiles belong to the patented Carpi-System, see figure 3. They are made of two parts, the inner and outer profile. The inner profile is anchored to the concrete structure by anchors. The outer profile is connected to the inner profile by means of a special adjustable threaded device. The connection of the two profiles creates a clamping effect to the geocomposite. Pre-tensioning gives the geocomposite best adhesion to the existing concrete surface.

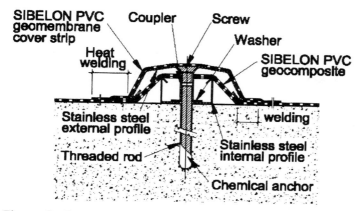

Figure 3: Cross section through pretensioning profiles.

INVESTIGATIONS OF THE DAM

As described in the previous section , the geomembrane is fastened onto the shotcrete lining by anchors. Hence the lining had to be tested in detail, as described in the following sections, to ensure a safe construction for the current rehabilitation.

Visual investigation

The whole surface of the upstream face of the dam was tapped with mechanical devices. A suspended platform was installed to allow access to the whole area of shotcrete. This first stage had the following objectives:

- Finding of areas where bond between the two shotcrete layers was poor
- Removal of loose shotcrete
- Marking on site of all repair areas

Findings

The visual inspection showed that the shotcrete itself was in good condition. Just one panel of the shotcrete layer was bad. In half of the panel there was no bond between the first and the second layer. Debris from the shotcrete was falling down during the tapping procedure. It was decided to remove the panel and to replace it completely.

In contrast to the shotcrete layer the panel joints were in a worse condition. This was because the joints were not flexible joints, see figure 1. Thermal movements of the dam caused high compressive stress in the joint material, which often resulted in cracks. More than two thirds of the joint material had to be replaced. The geomembrane needs a smooth subsurface with no protrusions, so these joints had to be reconstructed.

Shotcrete testing

For stability analysis of the anchoring system detailed knowledge of the shotcrete was necessary. Therefore five concrete cores with a diameter of 80 mm were drilled and analysed. The joint material was also examined.

Findings

- The concrete was classified as a concrete C20/25.
- All cores showed a steady microstructure. Also the transition zones between masonry and shotcrete and in between the shotcrete layers showed no voids.
- The geometry of the shotcrete layer could be confirmed: thickness 20 cm and two layers of reinforcement.
- In both the shotcrete and the joint material ettringite was found. Where ettringite fills the pores completely, an internal pressure arises and the microstructure is disturbed. As described above, the shotcrete is in good condition (no hollows, good microstructure), so the shotcrete will not be damaged by the ettringite. Figure 4 shows a microscopic view of the shotcrete. A pore with slight ettringite can be seen. Generally these very small pores are found in the shadow of the reinforcement. Further information on ettringite can be found in Jungermann (2000). In the joint material the pores were filled with the ettringite and the microstructure was destroyed. All the factors which lead to damage of the microstructure can also promote the formation off ettringite. The existence of ettringite crystals in concrete cracks is, as a rule, only a consequence and rarely the cause of the cracks, Stark/Bollmann (2000). This matches with the visual investigation, that the joint material was in a bad condition.

Survey

The entire dam was surveyed with a 3-D laserscanner. This scanner combines values from distances, angles and inclination to calculate 3D coordinates of each point of the dam. This method allowed a detailed design of the rehabilitation work considering the exact geometric condition of the dam.

Findings

This survey confirmed the general assumed dimension of the dam.

Figure 4: Microscopic view of shotcrete

Georadar

As mentioned above the geomembrane is fastened on to the shotcrete with anchors in vertical lines. To obtain detailed information of the shotcrete at each vertical line of a profile the shotcrete was surveyed by georadar, which had the following objectives:

- Recording of anchor plates of the shotcrete layer
- Recording of reinforcement (missing or overlapping reinforcement)
- Recording the structure of the concrete
- Recording the thickness of the shotcrete

In the Georadar technique electromagnetic impulses are radiated into the structure. Signals are reflected at interfaces between two materials (e.g. shotcrete - masonry) or at objects (e.g. reinforcement) and are registered at the surface. On the basis of the magnitude and the form of these reflections information on the structure can be obtained. The recording of the data is digital. First results can obtained during the measurement and a detailed evaluation including graphical presentation of the results will be made later. Figure 5 shows an example of the evaluation of the Georadar investigation

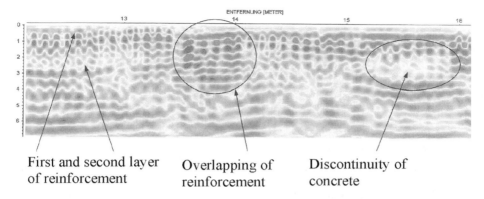

First and second layer of reinforcement   Overlapping of reinforcement   Discontinuity of concrete

Figure 5: Example of evaluation of Georadar investigation

Findings
With the georadar survey the good condition of the shotcrete lining could be confirmed. There is a high consistency with the core drilling results, e.g. the thickness of 20 cm of the lining and the existing 2 layers of reinforcement. Discontinuities, e.g. the right side of figure 5, were in all cases minor.

CONCLUSION
As a result of all the investigations the main statements were:
- The shotcrete was classified as a reinforced concrete C20/25
- The lining has a thickness of 20 cm
- No big discontinuities were found
- The joints of the shotcrete lining had to be reconstructed

The results of all the investigations were satisfactory to decide that the geomembrane will be a safe and long-lasting construction. Parameters for anchor fixing were confirmed.

REFERENCES
Jungermann, B. (2000). Zersetzungserscheinungen von Mörteln im dauerdurchfeuchteten, erdberührten Bereich, 2. Kolloquium "Bauen in Boden und Fels", TAE Esslingen
Scuero, A.M., Vascetti, G.L. (1996). Geomembranes for masonry and concrete dams: State of the art, Proceedings, EuroGeo I, p. 153-159
Stark, J., Bollmann, K. (2000). Ettringitbildung im erhärteten Beton – ein oder kein Problem? Beton- und Stahlbetonbau 95, Heft 2, Verlag Ernst & Sohn
Veyhle, D., Jaup, A. (2002): Brändbach: Germany´s first dam rehabilitation using a geomembrane, Hydropower & Dams, Issue 6, p. 84-87

The long term performance and remediation of a colloidal concrete dam.

K J DEMPSTER, Scottish and Southern Energy plc, UK.
J W FINDLAY, Babtie Group, UK

SYNOPSIS. Within Scottish and Southern Energy plc's (SSE) stock of concrete dams a small but nevertheless interesting subset is colloidal dams. SSE own three dams that are part formed using this technique. A brief history of colloidal dams is given followed by detailed information on particular problems at Loch Dubh Dam. Babtie Group (BG) is currently conducting studies leading to remediation proposals.

INTRODUCTION

Colloidal concrete or Colcrete depends on the production of a colloidal grout that is stable but highly fluid and can be injected into prelaid aggregate. The aggregate, from which all material below 1.5 inch must be excluded, is placed in position independently, and the grout is either poured over it and allowed to penetrate downwards, or introduced near the bottom through grouting pipes or channels and allowed to fill upwards. If correctly adopted the method ensures that the aggregate has point contact in all directions. The voidage is therefore less, and proportionately less grout, and therefore less cement, is required to fill it; thermal stresses are reduced, and cumulative contraction is prevented. Cement shortages were a significant issue during the early hydro development period and any reduction in use was sought.

This is a rare form of construction with known shortcomings in terms of performance when compared to conventional mass concrete, mainly due to the high water cement ratio required for placing.

LOCH DUBH DAM

Loch Dubh Dam is situated approximately 10 km north east of Ullapool and was completed in 1956. The dam is a concrete gravity structure of conventional profile, as shown in Figure 1 and set in a small steep-sided valley. It can be described as a medium sized dam in UK terms, due principally to its height of 20m (from lowest foundation). Much of the

Figure 1: General View of Loch Dubh Dam

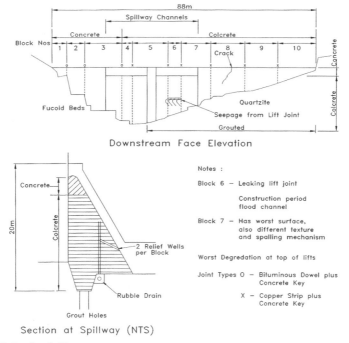

Figure 2: Principal Features

structure acts as the spillweir. The most remarkable feature of the dam is the method of construction using colloidal concrete. An associated feature that is equally unusual is the use of conventional mass concrete for a small number of monoliths on the right abutment. The reason for this decision is not clear although there is reference to a change in foundation rock type as a contributory factor (Refer Figure 2).

Where the dam is founded on quartzite, it is formed in Colcrete with a grout mix of 2 parts of sand to one part of cement by volume and aggregate in the range 37 mm to 225 mm. The overflow sill (upper 1.67 m) is formed in conventional structural concrete (19mm coarse aggregate).

Where the dam is founded on fucoid beds, the main body is formed in mass concrete (100mm coarse aggregate). The upstream and downstream faces have structural grade concrete 0.61 m thick, cast concurrently with the core.

OTHER COLLOIDAL DAMS

During construction of Mullardoch, which is a mass concrete gravity dam, changes in the economics of the day led to the design top water level being lowered and then raised again. The impact of this required a rather unique problem to be solved to thicken that portion of the dam already completed and to restore the 6.1 m removed from the original height. The method adopted for the thickening stage involved casting a slab over the downstream face of the dam, resting on precast concrete ribs on the sloping face but separated from it by a 0.91 m slot so that no bond could form between the two masses. After most of the contraction in the new concrete had taken place, the slot was filled with Colcrete, which was chosen as the material best suited to meet the requirements of minimum shrinkage coupled with good bonding qualities and, to a lesser extent, of high strength and high quality.

Difficulties were not surprisingly experienced during construction and rather than the more traditional form of Colcrete construction the coarse aggregate and grout were placed by "shooting" it into the slot through elephant trunking. Pre-mixing prevented the stones from breaking up and minimised the wear on the trunking. The water cement ratio of the grout was maintained at 0.9 with grout of 3:1 sand/cement mix and the coarse aggregate was crushed rock graded 89 to 63.5 mm. Test results on 57 cores taken demonstrated equivalent average cube strengths of 21.6 Nmm^{-2} and a density of 2387 kgm^{-3}. Current experience would suggest that although there are isolated areas of seepage and deterioration they are no more than those seen at other more traditional concrete gravity structures.

Tummel Bridge Aqueduct was substantially extended during 1957-1959 with the addition of a smolt bay using conventional concrete gravity construction and extension to the existing spillway portion of the aqueduct by buttressing. Colcrete techniques were adopted to form these sections which were formed directly on top of existing concrete gravity wall sections. Subsequently additional drainage holes have been drilled in an attempt to relieve water build up within the structure and to drain the interface between the concrete types. Deterioration of external surfaces due to freeze thaw action and spalling has progressed to an extent where localised concrete repairs should be carried out. The extent of this is perhaps greater than seen on conventional concrete gravity structures.

HISTORY & PERFORMANCE OF LOCH DUBH DAM

Over the years the structure of the dam has exhibited a range of defects. Some of these were not unusual on gravity dams while others were less common. Although disfiguring the external appearance none of the defects were judged immediately critical to safety, although regular monitoring was stipulated by Inspecting Engineers over the period.

The visual defects were surface breakdown and joint leakage while instrumentation revealed high pore pressures. Almost immediately these defects were attributed to the use of colloidal concrete.

The underlying process was believed to be a combination of high seepage through the mass concrete, poor drainage at the downstream face and the north facing aspect of this surface in an area prone to freeze thaw conditions.

These problems were obvious early in the dam's life, if only superficially, by the loss of surface finish. By 1964 the rate of surface erosion was causing concern and measures were put in hand to monitor the areal extent and depth of concrete degradation. These concerns increased during the 1970's and a limit of 75mm loss was set as a trigger for re-facing but by 1984 there was evidence that the process was slowing and to date the maximum loss is only locally at that limit.

Piezometers were first installed in 1968 (18No) as part of a research project into uplift pressures and revealed high levels in the downstream zone that indicated that enhanced drainage was required. The piezometers are simple standpipes. Many of them have been lost over the years and there are now only 6 in service. Although some isolated information is available from the early years the most consistent period of readings comes from the 1990's.

The trend of piezometer readings is not conclusive but shows signs of stabilising after increasing effective drainage in 1971.

Broadly speaking, the drainage gives an immediate drop in uplift of about 5m at full reservoir head (30%) with further dissipation towards the toe. The original series of vertical and horizontal holes have been reamed out where possible and supplemented with inclined connections. There is a possible link between improved drainage and decreased degradation of the surface although this is only a perception over time rather than a measured outcome.

Leakage from a lift joint, low in one of the central blocks, has been observed from the beginning. Accurate measurement of flow is not practical as it sits just above the drowned stilling basin and extends the full length of the joint. The rate of leakage is reported to be increasing.

There is a reasonable photographic record of the appearance of the dam from the 1960's to the present time (1961, 64, 66, 71, 72, 75, 81, 84, 2000 and 02) and this represents one of the best indicators of change in the degradation of the structure. Figure 3 shows a picture of a typical section of degraded face.

Figure 3: Varying face degradation, blocks 3, 4 and 5

The structure is routinely surveyed for settlement and lateral movement. There are no unexpected trends and the dam performs as would be expected of a gravity structure of this size.

The upstream face of the dam has been historically coated with a bitumastic paint although recently this has been allowed to degrade (in the expectation of an enhanced sealing coat).

While the possibility of making good the degradation of the downstream face by applying a finite surface thickness has been mooted it has always been accompanied by a concern that it might worsen the drainage/pore pressure regime even if a suitable application method could ensure adhesion and finish.

The current study has been triggered by the need to set a clear direction for the future maintenance of the asset.

DISCUSSION OF INDICATORS/INFLUENCES

The foregoing section on history describes the noted defects in general terms but as part of the current study the individual drivers were considered and analysed for likely impact. The principal indicators are as follows:

- Leakage
- Surface spalling
- Internal pore pressures
- Cracking

These physical outcomes are driven by a wide range of influences related to the design, construction and performance of this particular dam, namely:

- Colloidal concrete
- Acidic water
- Freeze thaw
- North orientation
- Construction sequence/standards/varying aggregate
- Porosity
- Face saturation
- Movement

The use of colloidal concrete is the first and most obvious deviation from standard gravity dam construction and is thus the prime suspect for poor performance. Although used on several dams in the UK and in the US during the 1950's the use of this technique was not sustained and it is now recognised as having poor strength and impermeability, factors that are likely to affect durability, particularly in an impounding function within a testing climatic environment. Thus the use of colloidal concrete might be expected to contribute directly to leakage and internal pore pressures. A secondary combination of saturation and weakness could also be contributing to the surface spalling. However, cracking is not likely to be a feature of this technique as it was recognised as reducing shrinkage effects.

Being an upland reservoir, acidic water from peaty soils is an influence with a known deleterious effect on cementitious matrix, the more so on weaker mixes. This might be expected to have an impact on the surface finish, particularly etching of the upstream surface but might also be impacting on conditions along discontinuities e.g. open lift joints.

The exposed surfaces of gravity dams in the Scottish highlands are prone to spalling due to frost action, especially if constructed during winter conditions. Additionally, saturated concrete is vulnerable to surface degradation during freeze/thaw cycles, the more so if weak and porous. Although not at a particularly high elevation (190m) the North facing aspect of this dam suggests that it might face the most severe freezing conditions. This might be supported by the difference in condition between the downstream face and the south facing, and submerged, upstream face.

Consideration of the construction details and sequence also reveals potential influences on performance and durability. In general there appear to be a number of variables in the way in which the work was carried out. This included considerable variation in aggregate size, aggregate type and lift height from pour to pour. Present condition also suggests variation in matrix texture (either cement, fine aggregate or additive). Sloping shutters are always an area of construction problems and may be more so with an injected cement matrix. This could explain differences between upstream and downstream face performance. Recognising the way in which colloidal concrete was formed it is perhaps not surprising to find the top edge of lifts showing a preference for early spalling. In some cases lift heights seem to be rather shallow for the technique. The combination of these influences might account for the most severe areas of degradation appearing on the downstream face. More particularly, in common with most gravity dams, a section of the structure was left low to pass construction period floods. A check on the records reveals that this corresponds to the badly leaking lift joint on block 6. While this surface may have suffered from exposure it appears also to have been compounded by the use of a very low lift in the pour immediately below the surface.

Although there is uncertainty over the reasons for the use of conventional concrete on the right abutment monoliths it appears to be related in some undefined way to the change in foundation rock type. However, it seems improbable that there was significant variation in loading response across the foundation thus cracking or movement due to uneven loading seems unlikely. In relation to the one notable crack (crest of block 8) it should be noted that the upper section of the dam (3.3m below crest) was constructed of conventional mass concrete throughout the length of the dam. As this was one of the wider monoliths it may be that the relatively slender block of conventional concrete suffered excessive shrinkage.

Thus it can be seen that there are a number of recognisable reasons for each of the defects indicated at Loch Dubh Dam. The key questions at this time relate to the significance of these processes and their likely progression with time.

Of the defects and processes identified the loss of mass on the downstream face was potentially the most serious in terms of safety of the dam and certainly the one that had been causing most concern over the life of the structure. The other defects (discrete leakage, pore pressures and cracking) although undesirable were not unusual and could be addressed by normal remedial works. However, all of the processes had consequences for the operational life of the structure, some with immediate impact on perceptions of care and others on long term maintenance cost.

Consideration of the above led to a programme of assessments and additional testing to determine the present impact of these processes and the likely consequences of further development. The defining indicators of current impact and future performance loss were identified as follows:

- Stability checks for assumptions of face loss
- Chemical change in seepage water
- Time dependant or event specific degradation of the surface concrete

DISCUSSION OF POTENTIAL OUTCOMES
SSE require a positive direction for future asset maintenance. This programme needs to be underpinned by a comprehensive review of the information available and a clear audit trail from adverse indicator to solution.

The key issue here was whether the defects and processes noted were largely superficial or whether there was a risk of long term erosion of safety margins or functionality. While safety is paramount, the operational function of a dam structure is also a serious issue for a commercial organisation.

Although possibly only cosmetic, the appearance of the degraded surface is an important issue of public perception for SSE. The impression given to senior company management was also relevant in that it reflects the level of investment and care applied to the dam stock of the company.

Upon site inspection in October 2002 and initial review of the records, an immediate impression was that the degradation process was in decline. This was very much a snap judgement and it was necessary to develop a realistic means of checking the validity. This impression did however seem to agree with the most recent periodical reviews carried out by Inspecting Engineers. Central to this was deciding whether the degradation process was time limited or time dependant.

While some information is available on concrete parameters there is nothing definitive within an advancing timeframe. Concrete cube strengths are available from construction in 1955 and there are a few core samples from the piezometer installation in 1968 that were tested for compressive strength. Together with recent tests it can be said only that strength is more than adequate for mass concrete but that there are significant variations across and through the structure (min approx. $15N/mm^2$)

TESTING PROGRAMME AND FURTHER ASSESSMENT OF HISTORICAL INFORMATION

The testing programme was designed to be as simple as possible and manageable with hand-held tools.

While overall compressive strength of the mass of concrete was of interest (for comparison with construction period tests) and would serve to indicate seepage related deterioration, the more important characteristic was the durability of the surface concrete, in particular, whether surface weakness was a one off construction feature rather than a function of continuous exposure.

It was decided that tests could be made for surface hardness and repeated after treating the surface in a defined way. It was also considered possible to core into the body of the dam from the downstream face sufficient to penetrate surface effects. In addition samples were taken of leakage flow for comparison with reservoir water.

Three test panels were exposed, each 1-2 m^2 in area. The approach chosen was to use a Schmidt hammer as a coarse indicator of surface hardness and to use this as a proxy for durability (it is acknowledged that this is using the Schmidt test outside its validated operating methodology). For each panel a matrix of hammer tests was carried out, firstly with the surface as found, followed immediately by a similar set after scabbling back to sound concrete. The tests were then repeated on this cleaned surface after a winter exposure.

Briefly summarising this exercise it was found that the exposed downstream panel was most degraded, followed by the upstream panel with the unexposed downstream panel least degraded. The follow-up tests on the cleaned surface revealed an immediate improvement with little deterioration between second and last tests.

The core strength tests revealed that there were variations in concrete strength but that these were not necessarily a function of seepage rather a reflection of general variability in concrete quality.

A review of stability revealed that the design section had a reasonable tolerance to loss of face material (no unacceptable reduction in FOS up to 150mm loss of material – current maximum of 75mm).

POSSIBLE / LIKELY REMEDIAL ACTIONS

There are broadly two approaches; firstly a universal upgrading of the structure that might be regarded as anticipating future problems. Alternatively, specific solutions to individual defects assuming that the extent of the problem has fully revealed itself.

The long-standing proposal has been to line the upstream face with a membrane. This would have the benefit of dealing not only with point leakage but with general porosity, the associated saturation and its link to face degradation. It would however require a full draw-down of the reservoir (although this would not be difficult as Loch Dubh does not feed direct to generating plant) and more significantly removal of sediment from the toe of the upstream face. This solution assumes that the degradation is essentially porosity/seepage driven and that loss of face material would continue without a global improvement in this characteristic.

The alternative on the upstream face is to deal with the one serious leak along the lift joint and to continue the practice of painting the exposed concrete surface. Depending on the level of the leaking joint in relation to sediment deposits the leak repair work could be carried out by diver without dewatering (using a mechanical clamping system with a strip membrane). Although a replacement bitumastic paint system is envisaged along with some limited concrete repairs to the upstream face, support for this solution depends on evidence that the degradation process is in decline.

In both cases the present level of degradation on the downstream face either has to be accepted or is treated in some way to reduce its visual impact.

As with many reservoir maintenance problems a definitive analysis is not possible and the chosen approach has to rely on an element of judgement.

CONCLUSION
Loch Dubh Dam has a number of visible defects that detract from its appearance and give an unjustified impression of neglect. Most of these features have been in evidence for the majority of the 48 year life of the structure. Despite concerns throughout its life that some of these defects were progressive the current assessment leads to the view that the processes are either in decline in the case of the face degradation, or can be addressed by relatively simple actions in the case of the severe joint leakage.

To be confident of the decision reached the situation has to be regarded from a number of perspectives: reservoir safety, asset performance, asset life and appearance.

In terms of reservoir safety the rate of degradation does not pose a threat to stability and the leakage is not unusual nor is it significant in uplift terms. The crack is not in a critical location and appears dormant.

As with all assets the residual life is of prime importance if the enterprise is to be sustainable. In the case of Loch Dubh Dam there is no reason why the structure should not remain effective for another 50 years even if current face degradation is doubled in extent. Routine maintenance and some remedial works are necessary to sustain that situation.

The appearance of the dam is one of perception and unless studied at close quarters does not raise any great concern. However, the philosophy of SSE is that all their structures should give an appearance of robustness and reliability even if not directly in the public eye. On this basis the current appearance of the dam does not meet the owner's criteria and some action is necessary to improve the situation.

It is implicit that the critical indicators identified here are kept under review, not only in the routine surveillance of the structure but at times of statutory inspections such that the current assumptions are regularly reconfirmed or alternatively that a change of conditions requiring action is identified.

At the time of writing the decision has been made to proceed with the specific actions necessary to reduce the principal defects and specifications will now be prepared for leak sealing, sealing coat and downstream face stabilisation.

Some problems at small dams in the United Kingdom

J.L.HINKS, Halcrow Group Ltd.
P.J.WILLIAMS, Halcrow Group Ltd.

SYNOPSIS.   The UK has a large number of small dams with diverse problems. Some have capacities of less than 25,000 m^3 and therefore fall outside the ambit of the Reservoirs Act, 1975. Many were constructed in the eighteenth and nineteenth centuries to impound ornamental lakes for stately homes. Whilst these reservoirs often require rehabilitation to meet modern safety standards the available funds are frequently tightly constrained.

This paper describes a number of recent case histories chosen to illustrate the breadth of issues which tend to arise, the nature of the solutions adopted and the lessons to be learnt. Some general principles are presented with regard to the rehabilitation of such structures.

INTRODUCTION

The average age of dams for which a construction date is given in the Building Research Establishment Register of British Dams is 100 years (BRE, 1994). 38% have a capacity of less than 100,000 m^3. Of these many are in private ownership and rarely generate sufficient income to pay for inspections under Sections 10 and 12 of the Reservoirs Act, 1975 or for improvements and remedial works.

In February 1986 the Department of the Environment wrote to Panel Engineers urging them to "keep expenditure to a scale justified by the risk" and stressing the importance of amenity, recreation and wildlife conservation. Inspecting Engineers therefore have to steer a careful path between permitting reservoirs to remain in an unsafe condition and imposing demands so onerous (and expensive) that the owners have no choice but to take the reservoir out of service. Of course there is sometimes the option to reduce the capacity of the reservoir to less than 25,000 m^3 and then discontinue it under Section 13. This is almost always to be preferred to abandonment under Section 14 because it dispenses with the need for a Supervising Engineer and 10 yearly inspections.

The following case histories illustrate some of the issues that arise:

Case History No. 1 – Upper Hartleton Farm Reservoir

Upper Hartleton Farm reservoir has a capacity of 59,000 m^3 and a catchment area of 11.5 km^2. It was constructed in 1972 at the same time as Lower Hartleton Farm reservoir immediately downstream. The dams were built of silty clay and performed satisfactorily except for the regular appearance of cavities and internal erosion behind the spillway walls. The cavities appeared at the Lower reservoir in 1979 and 1998 and at the Upper reservoir in 1978, 1995, 2000 and 2003. Since the dams have shown no problems along most of their length it is thought that poor compaction of material adjacent to the spillway walls was the cause of the difficulties.

Figure 1. Axial wall on left side of spillway at Upper Hartleton Farm – leakage was taking place beneath the bottom of the wall.

Following the appearance of the most recent cavities at the Upper reservoir the Supervising Engineer recommended that the Section 10 Inspection due in 2004 be brought forward to 2003. The Inspecting Engineer was concerned about the repeated appearance of cavities over the years and about possible serious problems if heavy leakage were to coincide with the passage of a flood since the river passes through a town in a small culvert a few kilometres downstream of the dams. Consequently he recommended that the fill behind both spillway walls be dug out and replaced with well compacted clay. This was done in October 2003; during the work leakage channels were found in the excavation. At the time of writing there has been no further leakage.

This case history illustrates the importance of good compaction adjacent to structures and of keeping careful records of the behaviour of dams over a long period (in this case 30 years). It also illustrates the desirability of maintaining continuity of Inspecting and Supervising Engineers.

Case History No 2 – Weldon Lagoon, Corby

Weldon Lagoon was built as a flood alleviation reservoir by the Corby Development Corporation twenty five years ago. It had 1 No 1050 mm diameter pipe, 2 No 900 mm diameter pipes and three smaller pipes entering the Lagoon from an urban catchment of about 0.92 km^2 but only 1 No 600 mm diameter pipe controlled by a 225 mm x 225 mm penstock leading out. There was no spillway and with the water up to the crest of the dam the capacity of the reservoir was 30,415 m^3.

The reservoir filled almost to the crest in November 2000 and this prompted the Undertaker to seek an opinion from an All Reservoirs Panel Engineer. The inspection was made on 2 February 2001. The Inspecting Engineer expressed the opinion that the reservoir was a large raised reservoir and that a spillway was needed to pass the PMF (since there were houses downstream). He also said that the spillway should be designed and built before the autumn of 2001.

As well as a new spillway a clay filled cut-off trench was proposed for a length of 50 m along the axis of the dam to a depth of 1.2 m to cut off leakage. Because a spillway would increase the discharges at short return periods the Environment Agency withheld permission for the new spillway under Section 23 of the Land Drainage Act, 1991. However it was pointed out that under Section 23(6) permission is not required for works being carried out in pursuance of another Act of Parliament or any order having the force of an Act. However, everything possible was done to address Environment Agency concerns and, with this in mind, a labyrinth spillway was constructed so that the spillway sill could be set as high as possible.

Figure 2. New Labyrinth Spillway at Weldon Lagoon

There was uncertainty at the time of design regarding the validity of estimates for extreme events obtained using the Flood Estimation Handbook (MacDonald and Scott, 2000). Consequently the 10,000 year outflow, at which there was to be 400 mm wave freeboard, was calculated as 16.5 m^3/sec assuming a rainfall depth, in 83 minutes, half way between that in the Flood Studies Report (125 mm) and that in the Flood Estimation Handbook (213 mm). The 10,000 year rainfall depth obtained from the Flood Estimation Handbook was 73 % greater than ¼ world maximum (123 mm) and was thought to be excessive - this view subsequently gained support from another paper by Messrs MacDonald and Scott (MacDonald and Scott, 2001).   The PMF outflow, at which there was to be nominal wave freeboard, was calculated as 25.8 m^3/sec.

The new spillway was completed in October 2001. On completion of the work a certificate of discontinuance was issued because the capacity of the reservoir was reduced to 17,650 m^3 (ie. less than 25,000 m^3).

The case history illustrates how discrepancies were dealt with between estimates made using FEH and FSR. In addition it shows that there may sometimes be conflicts between the requirements of reservoir safety and those of good practice in flood mitigation. It also illustrates the usefulness of labyrinth type spillways in those situations.

Case History No 3 – Shardeloes Reservoir

Shardeloes reservoir has a capacity of only 50,000 m^3 but a catchment area of 49.8 km^2 . Only 1 mm of runoff would suffice to fill the reservoir from empty. The reservoir is remarkable for having a spillway capacity of only 1 m^3/sec. Because the catchment is largely chalk this has been sufficient to ensure the survival of the reservoir since it was built in the early eighteenth century although it is thought that the dam must have been overtopped in the floods of March 1774 when boats could be rowed along the streets of the town downstream.

Because of the town downstream the dam is assigned to Category A as defined in the Institution of Civil Engineers booklet "Floods and Reservoir Safety" (ICE, 1996). The PMF is calculated at 186 m^3/sec for a saturated catchment. Strengthening the dam to withstand the passage of the PMF would however have been expensive and detrimental to amenity, recreation and wildlife conservation.

However the capacity of the reservoir is less than 1% of the volume of the PMF (5.65 Mm^3). This being so the failure of the dam in a major flood would not make a significant difference to flood levels downstream. It was therefore decided to apply American methodology as described in the article in Dams and Reservoirs on 'Small Reservoirs on Large Catchments' (Hinks, 2003).

Mathematical modelling was first carried out to determine flood levels downstream with and without dam break. A sunny day dam breach was expected to release water at a peak rate of about 11 m^3/sec. Coming on top of a flood of 3.5 m^3/sec the incremental flood depth in the town was calculated as 300 mm. This is considerably less than the figure of 600 mm permitted by American methodology and was therefore judged acceptable.

The new spillway is now being designed with a capacity of 3.5 m^3/sec.

The case history illustrates the relevance of American methodology for small reservoirs on large catchments.

Figure 3. Shardeloes Reservoir

Case History No 4 – Braydon Pond

Braydon Pond is a privately owned Category B reservoir impounding about 100,000m^3. It has a minor road along the crest and two spillways. The central spillway is carried beneath the crest road in twin concrete pipes which were installed by the Highway Authority in 1976. Unfortunately the pipes were surrounded with granular fill so there was considerable leakage downstream when reservoir levels were high. The pipes also became cracked and distorted under the weight of traffic. Eventually the pipes had to be dug out and replaced with new ones surrounded by clay.

The spillway at the right abutment was lowered by 300mm in 2000 to provide greater spillway capacity.

The most recent problem is a major slip in the crest road caused by heavy lorries using the minor road as a short cut. It remains to be seen whether the necessary repairs will be paid for by the owner of the dam or by the highway authority who own the road.

Case History No. 5 – Faringdon House Lake

Faringdon House Lake is an ornamental lake built in the grounds of Faringdon House in Oxfordshire in approximately 1770. The reservoir has a capacity of 33,000m^3 and a catchment area of 0.27 km^2. A particular feature of the reservoir is a spring fed fountain which discharges into the head of the reservoir under a head of 2m. The reservoir has been assessed as category D given that the failure of the reservoir would cause only minor inundation.

The reservoir was one of those that had slipped through the net of the Reservoirs (Safety Provisions) Act, 1930 and was not picked up until 1989 when a Section 10 inspection was instigated and supervising engineer appointed. The reservoir had been somewhat neglected up until this time. The principal defects related to a number of large trees, which had been allowed to grow unchecked on the embankment and the total lack of a spillway. The only outflow from the reservoir was via a 100mm pipe overflow at approximately 0.30m below the crest level. This passed through the 6m high embankment to feed an ornamental cascade on the downstream face. The owner was keen to maintain the essential character of the lake and the Victorian water garden at the dam toe and the remedial options were developed to take this into account. Relatively severe tree surgery removed much of the top weight from the larger trees and reduced the risk of toppling whilst a grass-lined spillway was constructed down the right abutment to carry flow in excess of the capacity of the 100mm pipe. The spillway was designed to pass the 150 year flood of 0.74 m^3/s, with a nominal freeboard and in most years the spillway will operate two or three times a year.

This case history illustrates the need to ensure that reservoirs are entered onto the register. If left unchecked this reservoir may well have failed during a 1,000 year rainfall event which occurred on the catchment in 1998. In addition it has been found that a sympathetic approach with a private owner will generally bear fruit in encouraging implementation of works in the interest of safety.

Case History No 6 – Marston Pond

Marston Pond is believed to date from about 1780. It now has a surface area of about 8 hectares and a capacity of 80,000 cubic metres. The dam is about 500 metres long with a maximum height of about 3 metres.

The dam was quite regularly overtopped and leaks develop fairly frequently.

The problem was to bring the dam up to modern safety standards at reasonable cost without spoiling the fishing and duck shooting in the reservoir.

The dam was classified as Category D on the grounds that there were no houses between the dam and the confluence with a larger river some distance downstream. The low height of the dam was also taken into consideration as was the extensive siltation which meant that the reservoir was generally quite shallow close to the dam.

Figure 4. Glory Hole Spillway at Marston Pond prior to lowering by 450 mm.

In order to pass the 150 year flood of 5.25 m^3/sec without overtopping of the dam the higher of the two spillways was lowered by 450 mm so that it was the same level as the other. After some debate it was decided to allow the owner to install stoplogs between 1 April and 30 September each year. This concession, which is subject to the instructions of the Supervising Engineer, will reduce the capacity of the spillway from 5.25 m^3/sec to 3.3. m^3/sec during the summer months but is expected to have a very beneficial effect on amenity, recreation and wildlife conservation in line with the Department of the Environment letter of 26 February 1986.

This case history illustrates the need to make compromises in order to achieve an appropriate balance between the demands of reservoir safety and those of amenity, recreation and wildlife conservation.

Case History No. 7 – Fawsley Estate Lakes

The Fawsley Lakes were constructed as a series of three ornamental ponds adjacent to Fawsley House in Northamptonshire in approximately 1850. Only one of the three reservoirs comes under the Reservoirs Act with a volume of 120,000 m^3 whilst the other two lakes are immediately upstream, on two separate tributaries, and have volumes of 22,000 m^3 and 23,000 m^3. All three reservoirs have suffered from deterioration over the years and in a statutory inspection some 5 years ago a significant number of items were recommended in the interests of safety. Because the two non-statutory reservoirs posed a risk to the statutory reservoir, items of major maintenance at these reservoirs were included in the recommendation in the interests of safety.

This statutory category B reservoir was deemed to have insufficient freeboard to pass a 10,000 year flood without overtopping of the embankment and there were concerns about seepages through the dam and alongside the walls of the cascade spillway. The spillways of the two upper reservoirs are both largely collapsed and there is significant erosion of the adjacent embankment fill and lack of freeboard. Works are now in hand to rectify these defects but are complicated by the fact that there are multiple undertakers with both the owner of the estate and the local fishing club undertaking work on an ad hoc basis.

This case history illustrates the need to consider non-statutory reservoirs or other constructions that may influence the safety of the statutory reservoir. In addition it demonstrates the potential problems of multiple undertakers in implementing recommendations in the interests of safety.

ACKNOWLEDGEMENTS

The authors would like to express their appreciation to the owners of the various dams for their permission to publish the details given in this paper.

REFERENCES

Building Research Establishment, 1994, *"Register of British Dams"*.
Hinks, J.L. 2003, "Small Reservoirs on Large Catchments" *Dams & Reservoirs*, Vol 13, No 2, June

Institution of Civil Engineers, 1996, *"Floods and Reservoir Safety"*, 3$^{rd.}$ Edition.

MacDonald D.E. and Scott C.W, 2000, "Revised Design Storm Rainfall Estimates obtained from the Flood Estimation Handbook", *Dams and Reservoirs,* Vol 10, No. 2, July.

MacDonald D.E and Scott C.W, 2001, " FEH vs FSR Rainfall Estimates: An Explanation for the Discrepancies identified for very rare events", *Dams and Reservoirs,* Vol 11, No 2, October.

The Discontinuance of Devils Dingle Ash Lagoon

A.K. HUGHES, KBR, Leatherhead, UK
D.S. LITTLEMORE, KBR, Leatherhead, UK

SYNOPSIS. Devils Dingle Ash Lagoon is the principal means of ash disposal for Ironbridge Power Station. The lagoon is impounded by an embankment constructed largely of PFA with an upstream clay core. Filling of the lagoon is entering the final stages and plans for the restoration of the site are currently being formalised.

This paper describes the proposed decommissioning of Devils Dingle Ash Lagoon and the measures taken to ensure that the reservoir will have its storage capacity reduced to less than 25,000m^3 and therefore fall outside the ambit of the Reservoirs Act 1975. The methods used to complete the filling and landscaping of the lagoon whilst maintaining and enhancing the important wildlife habitat that have established around the site are also described.

INTRODUCTION

The Devils Dingle Ash Lagoon has been the main means of ash disposal for the 1000 MW Ironbridge 'B' Power Station since it commenced operation in 1968. It comprises an embankment, constructed mainly of pulverised fuel ash (PFA) impounding a lagoon in a small tributary valley of the River Severn above the village of Buildwas, Shropshire. The embankment straddles the confluence of two small streams flowing down the valley.

The embankment was raised in stages ahead of the ash disposal requirement to a maximum height of 66m. The crest of the embankment has an approximate length of 570m. Approximately 3 million tonnes of ash were used to construct the embankment and another 2 million tonnes were used to fill the lagoon. A compacted clay embankment with stone drainage layers and an upstream rockfill berm was constructed for the initial impounding prior to the availability of the conditioned ash. The main body of the dam was then constructed from compacted PFA with a rockfill berm at the downstream toe. The upstream face of the dam was then sealed with a 3.5m thick clay blanket which is protected from wave erosion by a layer of coarse gravel and rockfill. A vertical wall drain was constructed downstream of the final crest line which connects with a horizontal drainage blanket located beneath the downstream shoulder of the embankment. A cross section of the embankment is shown in Figure 1.

PFA has been delivered to the lagoon in two ways. Conditioned ash with a moisture content of about 23% was delivered to the site by truck between 1967 and 1983. The remainder of the ash was slurried and pumped to the lagoon by pipeline. However, in December 2000 the pipeline delivering slurried ash was ruptured by slope movements along the valley between the power station and the ash lagoon. As a result the ash required to complete the filling of the lagoon and provide landscaping features is being delivered by road.

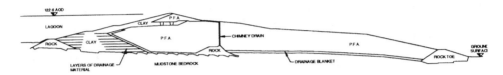

Figure 1: Cross section of the embankment

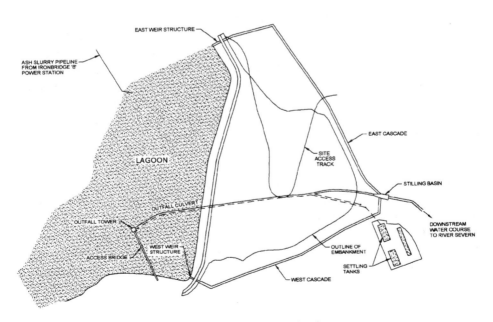

Figure 2: Plan of the embankment at Devils Dingle

OVERFLOW ARRANGEMENTS AND FLOOD CONTROL

During operation of the lagoon supernatant water is discharged over dam boards set in a slot in the side of the 4.6m diameter outfall tower. The water level in the lagoon can be varied by adding or removing these dam boards. Recently water within the reservoir has been held at 122.34mOD although the level can be raised to a maximum of 123.0mOD which coincides with the weir of the outfall tower. Access to the top of the tower is gained via a raised steel platform and walkway from the western (right hand) bank of the reservoir.

Decanted water drops down the tower to the base where a retained pool of water is used to dissipate the energy of the falling water. At the base of the tower a 600mm, reducing to 450mm, diameter pipe is set below in the main weir of the pool to discharge 'normal' flows to stilling ponds downstream of the dam. This pipe runs along the outfall culvert connecting the base of the tower to the downstream toe of the dam. Towards the end of the outfall culvert the pipe is diverted from the main culvert into a smaller secondary culvert that leads into the settling ponds situated just off the toe of the embankment.

During flood events water is initially discharged over the dam boards until the water level in the lagoon reaches 122.6mOD when flows also pass over two cascades located at either end of the embankment. The cascades are constructed of reinforced concrete and form trapezoidal channels with baffle blocks at regular intervals along their length. Each has been designed to discharge a flow of approximately $2m^3/s$ and both discharge to the stilling basin at the toe of the embankment. A weir and venturi flume at the head of each cascade ensures that the design flow is not exceeded even under extreme PMF conditions. As part of the original design the cascade structures were model tested to confirm the arrangements.

As the water level in the reservoir continues to rise overtopping of the outfall tower weir set at 123.0mOD occurs. When the capacity of the 600mm pipe at the base of the tower is exceeded water discharges over the weir directly into the outfall culvert. The flood waters pass down the outfall culvert into the stilling basin before flowing back into the stream leading to the River Severn. The outfall culvert and stilling basin are designed for a maximum flow of $22m^3/s$. The levels of the principal structural elements are summarised in Table 1 below:-

Table 1: Levels of the principal structural elements within the ash lagoon

Structure	Level (mOD)
Embankment Crest	124.30
PMF Flood Level	123.53
Outfall Tower Weir Level	123.00
Cascade Weir Level	122.60
Damboards (Typical weir level)	122.34

A recent hydrological assessment of the site undertaken by KBR reported the catchment area of the ash lagoon to be approximately 1.38 square kilometres with an average annual rainfall (SAAR) of 736mm. The peak inflows for the different flood events within the Devils Dingle catchment area are detailed in Table 2 below:-

Table 2: Peak inflows for the Devils Dingle catchment area during various flood events

Flood Return Period	Peak inflow
Mean annual flood	$0.7 \text{ m}^3/\text{s}$
1,000 year flood	$4.7 \text{ m}^3/\text{s}$
10,000 year flood	$9.7 \text{ m}^3/\text{s}$
Probable Maximum Flood	$19.1 \text{ m}^3/\text{s}$

Prior to the recent period of infilling with the ash lagoon, the attenuation within the lagoon results in the PMF peak outflows being approximately $16\text{m}^3/\text{s}$, with $3.5\text{m}^3/\text{s}$ flowing down the two side cascades and $12.5 \text{ m}^3/\text{s}$ flowing into the outfall tower and along outfall culvert.

REQUIREMENTS OF THE RESERVOIRS ACT
Discontinuance of a reservoir can only be certified if a Panel AR Engineer is satisfied that the impounded volume of a reservoir, excluding any flood storage, has been permanently reduced to less than $25,000\text{m}^3$. However, in situations such as ash lagoons this volume should include any silt or ash deposits that would flow in the event of an embankment breach or failure. Therefore, the volume of 'escapable contents' should be considered in this case.

Ash lagoons such as Devils Dingle are usually operated under a number of interim certificates as the lagoon is being filled to its final level. When filling of the lagoon is completed the final certificate is issued and is immediately followed by a certificate of discontinuance as the lagoon would no longer have any storage available. However, in this particular case the owners of the site were keen that the restoration plan included at least one body of water in order that the wildlife habitat that had established around the lagoon could be retained.

Plate 1: The reservoir at the Devils Dingle Ash Lagoon with draw-off tower access walkway in the background

Given that the current surface area of the lagoon is of the order of 125,000m^2, a single pond with a volume restricted to less than 25,000m^3 would have an average depth of less than 200mm for discontinuance to be possible. In addition any underlying layer of fluid ash would also need to be considered in the calculation of 'escapable contents' and would further reduce the volume of stored water allowed in the final scheme. In order that the final restoration of the site could incorporate some form of stored water feature it was hoped that the ash at the lower levels had consolidated with time, encouraged by under drainage and through drainage into surrounding lower water table. Significant depths of 'fluid ash' would make it not possible to have any form of large ponds within the restoration plan.

Given the large surface area of the current reservoir and the likelihood that a layer of 'fluid' ash exists below the retained water level it was envisaged that 3 or 4 smaller separate water bodies each with escapable contents of less than 25,000m^3 of water and ash would have to be formed rather than a single pond. However, the construction of multiple ponds would have undertaken in such a way as not create a situation where the capacity of each lagoon was considered to be part of the sum of all the lagoons and therefore have a capacity in excess of 25,000m^3. The final design must therefore include lagoons, each one considered to be fully independent of its neighbours and with little likelihood, under any situation including instability, overtopping or piping, of failure of the dividing bunds.

It was considered that the dividing bunds must therefore be designed as engineered structures on a suitable foundation. However there would be no requirement to construct the bunds as linear features, or with uniform cross sections and so it is envisaged that the dividing embankments will be constructed to give the lagoon area as natural an appearance as possible. It is considered that the separating bunds would have to be constructed with typical crest width in the region of 30m and maximum slope gradients of 1V:6H in order to ensure that the dividing embankments remain stable and the lagoons remain independent features.

INVESTIGATIONS AND SURVEYS

In addition to a detailed topographic survey of the site, a bathometric survey of the lagoon was undertaken to determine the levels of the ash within the reservoir. A three dimensional computer model was then developed to determine the remaining void space and to establish a number of discontinuance options using varying quantities of PFA. This design flexibility was required as the actual volume of PFA available for disposal and landscaping is uncertain and largely depends on the operational life of Ironbridge Power Station and the requirement of PFA for other uses. The number of ponds created in the reservoir, the height of the controlling weirs and the height and topography of the ash bunds within the lagoon were all varied to establish the minimum volume of ash required to achieve the discontinuance of this reservoir.

A geotechnical site investigation was carried out to establish the condition of the previously deposited ash and to determine the depth of ash that could be considered to be fluid. Experience from other sites suggested that the low water table around the site and under-drainage may have caused the lower levels of the ash to have partially drained and consolidated. However, the upper two or three metres were likely to be unconsolidated with a high moisture content.

The method of investigation was determined by the soft nature of the ash deposits both in terms of the sampling methods proposed and the ability to move around within the lagoon. Some elevated areas in the lagoon close to the outlets had been 'dry' for many years and as a result the upper layer of ash had become relatively firm and vegetated. However, the level of ash in other areas closer to the embankment was considerably deeper and had been under water for significant periods of time. Due to the positioning of the slurried ash pipeline outlets around the lagoon the surface levels of the ash deposits varied by up to 5 metres. As a result it was decided to use a CPT

(Cone Penetration Test) rig fitted with a Piezocone and mounted on a floating pontoon within the lagoon. The water level in the lagoon would be raised to the level of the two cascades (122.6mOD) by inserting dam boards in the outfall tower and this would enable the pontoon to access as large an area as possible including some of previously 'dry ' areas. Immediately after the completion of the investigation the water level would be reduced to the lowest level possible in order to dry out as much of the ash surface as possible.

In April 2002 the first phase of ground investigation was carried out consisting of forty nine cone penetration tests positioned on a grid at 50m spacings. Each hole was continued to a maximum depth of between 10m and 16m or until 'solid' ash was encountered. In addition six continuous piston sampling holes were carried out to assist in the interpretation of the CPT holes and to enable the geotechnical characteristics of the deposited ash to be determined.

RESULTS OF THE INVESTIGATION

The results of the investigation enabled a depth profile of the ash to be plotted. The results indicated that the majority of the ash deposit had consolidated and drained and that there had been some cementing of the deposits. The results indicated that the ash composition and properties were relatively uniform across the lagoon. A relatively thin layer (<1m) of very soft ash was encountered at the surface of the deposits during the investigation. However, in the event of a breach in the embankment, the PFA deposits would be relatively stable and no significant flow of ash would be expected.

TRIAL FILLING AREA

As part of the preliminary design and prior to the construction of any permanent separating embankments, a trial filling area was established. The trial would not only allow an area of previously submerged ash deposits to be exposed and the proposed foundation to be examined but would also allow a 'constructability' trial to be completed. This would assist the contractor in choosing appropriate plant and methods for completing the remaining filling and the construction of the separating embankments. A site was chosen near to the eastern end of the embankment where the topography of the existing ash surface was suitable and where the trial could be undertaken safely in a position away from the outfall tower.

The trial showed that pushing conditioned ash into the upper layer of ash displaced the majority of the very soft ash deposits present and that the dry ash became founded on a suitable foundation layer. The trial also demonstrated that the method of placing the fill over the previously deposited ash fill was suitable and that a suitable founding layer could be established for the remaining fill and proposed separating embankments.

Following completion of the trial filling area two survey positions were constructed above the areas that had received the greatest depth of fill. The levels of these two survey stations have been recorded on a monthly basis to determine the amount of settlement taking place in the foundation and newly placed fill. The results to date indicate that no noticeable settlement in either the foundation or recent fill is taking place. Therefore, it is likely, given the granular nature of these ash deposits, that the majority of the settlement has occurred during the construction of the trial filling area.

PHASE TWO SITE INVESTIGATION
A second investigation was commissioned in November 2003 after approximately 18 months of filling the lagoon with PFA. To enable comparison with the first phase CPTs, eleven new CPTs were carried out in locations coincident with CPTs from the phase one investigation. These CPTs were carried out using a truck mounted rig and were taken to a maximum depth of 10m. Not all areas of the lagoon were accessible to this truck mounted rig as some areas remain under water.

The Phase 2 investigation was undertaken to determine the condition of the newly placed fill, the changes within the previously deposited ash fill and to establish the presence, or otherwise, of the soft layer previously identified at the surface of these deposits.

The results show that the soft layer was no longer present probably resulting from the method of filling, consolidation as more ash was placed above and the re-distribution of pore water pressures. Also the ash placed above that tested in Phase 1 had improved density and stiffness properties. Therefore, it is considered unlikely that the ash would flow if the embankment were breached.

POND LAYOUT
Based on the results and interpretation of the various investigation phases a preliminary restoration plan was developed. Three ponds are proposed, two of which are to be located close to the embankment adjacent to each of the cascade structures. A third pond is proposed towards the western edge of the lagoon adjacent to and north of the location of the existing elevated walkway to the outfall tower opening. The two ponds located close to the embankment are to have water levels of 122.6mOD controlled by the existing cascade weirs. The third pond will have a slightly higher water level controlled by inlet and outlet structures on the stream entering this pond.

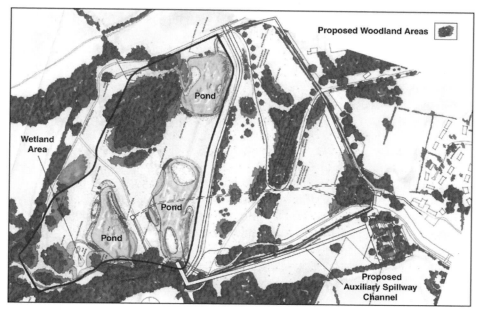

Figure 3: Plan of proposed restoration scheme

Each of the proposed ponds will have a specifically designed profile in order to try and establish a number of different aquatic habitats within the lagoons. Low lying areas close to the incoming streams will also be used to create new habitats such as wetland marginally areas.

Ecologists and landscape architects formed part of the project team that formulated the preliminary restoration plan for the site. Areas of young woodland and other vegetation that has become established within the lagoon area will be preserved where possible and new areas of both 'dry' and 'wet' woodland will be created around the proposed ponds.

PFA will also be used to construct additional landscaping areas on the downstream face of the embankment create a more natural landform and to mask the concrete features on the embankment that for hydraulic reasons tend to follow straight lines. Planting of selected shrubs and trees on the downstream face will also help to disguise the embankment.

MODIFICATIONS TO EXISTING STRUCTURES
In order to return the site to as natural an appearance as possible it will be necessary to carry out modifications to the existing structures associated with the lagoon. Sequencing of the necessary modifications to the overflow tower, cascades and stilling basins must be programmed such that no works are undertaken on these structures prior to the satisfactory discontinuance of

the reservoir. As further ash is deposited in the existing lagoon both the volume of retained water and surface area of the reservoir are reduced. Although the further filling reduces the volume of the reservoir, the benefit of the flood attenuation provided by the lagoon is also reduced. A detailed programme of ash deposition, construction and modifications was therefore developed to ensure that a 'less safe' condition is not created during this process.

During the early planning and preliminary design stages of this scheme the issue of how the site will respond to flood events during and on completion of the restoration plan have had to be addressed. The original proposal for discontinuance includes the decommissioning of the outfall tower and culvert by sealing both ends and filling the void with a PFA/cement grout as this will reduce the future maintenance requirements of the site. The decommissioning of the outfall tower will reduce the discharge capacity of the site to the combined capacity of the two remaining cascade structure approximately 4m³/s which equates to a 1000 year flood event.

Although the reservoir will no longer be subject to the Reservoirs Act 1975 and have a requirement to safely pass the PMF event, the owners were keen that the restoration plan should include measures to protect the embankment against overtopping and possible erosion from storms greater than the 1,000 year event. This would particularly important when maintenance and inspection regimes would be stepped down and in the long term when the site may possibly be sold. The potential for blockage of the existing cascades will also be more likely given the large number of trees and other vegetation to be planted around the proposed ponds

Therefore, it was decided to construct a reinforced grass auxiliary spillway down the right mitre of the embankment, adjacent to the western cascade channel, to give additional spillway capacity. The weir of this structure will be designed in such a way that the discharge capacity of the combined spillways will again be able to pass a PMF event safely and therefore the embankment will be protected from overtopping. Flood flows will pass via a reinforced grass channel into a newly constructed stilling basin at the toe of the embankment where the existing settling tanks are located. The construction of the auxiliary spillway is planned early in the programme prior to discontinuance of the reservoir in order that the adequate discharge capacity is always available during the works. This also will provide greater flexibility in the timing of the remaining ash placement, pond formation and modification of existing structures.

Outfall Tower and Access Walkway

The outfall tower and access walkway will become redundant in the proposed scheme. Removal of the raised walkway will be achieved by the construction of a large ash bund adjacent to the walkway from the western bank of the lagoon out towards the outfall tower. This bund will provide a working platform from which the access walkway will be dismantled and the supporting piers broken down. The bund will also allow access to the top of the outfall tower. It is proposed that the downstream end of the outfall culvert is sealed with a concrete bulkhead and the entire outfall tower and culvert be filled from the above using a PFA cement grout. This will ensure that there will be no long term maintenance issues associated with the outfall tower or outfall culvert.

Plate 2: Eastern cascade channel on the left mitre of the embankment

Spillway Cascades

Following the completion of the restoration plan the two cascades located at either end of the embankment will be in almost continuous use as these structures will control the level of the ponds. The structures are likely to be largely unchanged, however, some screening of the cascades using various types of vegetation will be undertaken in order to reduce the visual impact of these linear features.

Stilling Basin

The main stilling basin will still be required following discontinuance of the reservoir as flows from the east and west cascade channels will enter either side of the stilling basin. As the stilling basin will no longer receive flood flows from the outfall culvert some minor works are proposed to mask the sealed entrance of the outfall culvert and reduce the visual impact this feature.

Main Embankment

The placing of additional ash and topsoil on the downstream face and selected planting is proposed to create a more natural appearance and to create more rounded features and break up the straight lines of site that exist.

Settling Lagoons

The area currently occupied by the settling lagoons will be modified and will be used as a stilling facility for the reinforced grass auxiliary spillway. Measures will be taken to obscure the view of both the auxiliary spillway channel and stilling basin from the village of Buildwas located close to the toe of the embankment.

Pipeline

A 2.5km long pipeline exists between the power station and the lagoon through which slurried ash was pumped up to the lagoon. The pipe varies in depth considerably over its length being some 10 to 12m deep in places. Small land movements adjacent to the pipeline are thought to have caused cracking in the pipe leading to release of water and ash into and onto the surrounding ground on a number of occasions. This release of fluid may have lubricated the surrounding ground to encourage larger slips. As a result of these problems the pipe has not been used for a number of years and all PFA is now transported to the site by road.

It is anticipated that the pipe will be decommissioned as part of the reservoir discontinuance by grouting the pipe with a cement PFA grout. The pressure the pumped grout will need to be adjusted where there are fractures to ensure leakage from the pipe will be minimised.

PROPOSED PROGRAMME

The restoration scheme has already started with the deposition of ash in selected places within the lagoon in line with the final proposals. The construction of the auxiliary spillway is due to commence in the spring of 2004.   Further filling of the lagoon, construction of the ponds and decommissioning of various structures is planned for 2005 and 2006 together with the final landscaping of the site.

ACKNOWLEDGEMENTS

The authors would like to thank the management of Powergen for their permission to publish this paper and in particular Mr Graeme Smith and Mr Colin Pratt at Ironbridge Power Station for his assistance with this project.

Bewl Water spillway remedial works

I DAVISON, MWH
K SHAVE, Babtie

SYNOPSIS.   Bewl Water is a major impounding reservoir in South East England, UK. In the mid-1990's an external inspection confirmed that the crest of the spillway shaft was suffering from severe cracking. A paper, by one of the authors, presented at the British Dam Society Conference in Bangor in 1998 described the investigation that was carried out to determine the cause of the cracking. This paper describes the remedial works contract that was undertaken to replace the pre-cast concrete crest blocks that were suffering from Alkali Silica Reaction, and describes the extensive temporary works over the top of the spillway shaft, the methods used to remove the existing pre-cast units and the design of the new units. The new crest units have been designed to allow the crest to be raised by 350 mm without the need for further major temporary works over the top of the shaft.

BACKGROUND

General
Bewl Bridge Reservoir is a large, raw water reservoir situated approximately 10 km south east of Royal Tunbridge Wells in Kent. The reservoir is primarily filled by the Yalding Pumping Station with additional water from the original pumping station on the River Teise and the relatively small natural catchment of 1.9 km^2.

A 30.5 m high earthfill embankment with a central rolled clay core retains the reservoir. A bellmouth spillway shaft located within the reservoir was designed to discharge floods up to a Catastrophic Flood of 115.5 m^3/s. Water is abstracted from the reservoir via pipework within a 36 m high reinforced concrete draw-off tower adjacent to the spillway tower. Impounding of the reservoir started in 1976 and was full by mid 1978.

Spillway Shaft
The overflow structure, as shown in Figure 1, consists of a vertical shaft from the reservoir bed up to the full supply level of the reservoir. The lower

section of the shaft is 3.5 m in diameter internally and has 500 mm thick reinforced concrete walls. Over its upper 7.75 m, the shaft flares out to give a weir crest diameter of 10.8 m. The crest is formed by a series of 32 pre-cast concrete blocks that form the lip of the weir which is divided into quadrants by anti-vortex piers that prevent a vortex forming in the shaft when it becomes submerged. A 1500 mm deep reinforced concrete beam boat fender surrounds the weir reducing the size of waves impinging the weir and preventing boats from getting too close to the overflow crest. The boat fender is located 2 m away from the weir crest and is supported by four radial beams spanning out from the rear of the anti-vortex piers.

At the base of the tower the shaft turns through a 90° bend into the discharge tunnel which passes under the dam before discharging into the river downstream. The tunnel is a horseshoe in section and is approximately 3.3 m in diameter.

THE PROBLEM

The Supervising Engineer carried out a detailed inspection by boat of the cracking and spalling of the pre-cast concrete blocks around the spillway crest in 1995. In February 1996, McDowells Consulting Engineers carried out a preliminary investigation of the cracking, and recommended further investigation of the boat fender and upper section of the shaft, including removal of core samples from the crest blocks and *in situ* concrete. The cause and extent of the cracking were unknown and it was unclear whether the condition was deteriorating. The Inspecting Engineer considered that ASR was a possible cause and recommended repair or replacement.

THE INVESTIGATION

A detailed investigation was carried out using roped access techniques during the summer of 1997. This investigation included carbonation testing, concrete sampling, a covermeter survey and ultrasonic pulse velocity testing. In addition, a visual crack survey and a photographic record of the inside and outside of the tower were carried out. The concrete samples taken from the *in situ* and pre-cast concrete around the crest of the shaft were tested in a laboratory. Reference 1 gives details of this investigation.

CAUSE OF CRACKING

The investigation concluded that the cause of the cracking was due to Alkali Silica Reaction (ASR) in the pre-cast weir units around the top of the shaft. The units had been secured by dowels during construction, and as the units swelled, due to the ASR, the radial geometry of the units caused them to be forced outwards. This in turn caused the *in situ* concrete to which they were secured also to move outwards and, as a result, tension cracks appeared on the flared section of the shaft and on the anti-vortex piers.

REMEDIAL WORKS DESIGN

Three remedial works options were considered:
1. "Do Nothing"
2. Replace all the crest units thereby removing the problem and preventing further expansion.
3. Selective replacement of units and provision of expansion joints which would allow further expansion of the blocks without causing further distress to the *in situ* concrete.

Monitoring of the crack widths was carried out throughout the winter of 1997 and 1998, by measuring tell-tales with a vernier gauge, to try and determine if the expansion of the concrete was still taking place. A more sophisticated system of monitoring had been considered that would have allowed smaller amounts of expansion to be measured but the estimate of the costs of installing and maintaining the system was not significantly less than the cost of the remedial works.

No signs of movement were detected, although this was possibly due to either the movements being smaller than the accuracy of the vernier gauge or the temporary suspension of the expansion. ASR requires moist conditions to occur and during the monitoring period the reservoir level was below the level of the pre-cast units.

Bewl Water is one of Southern Water's most important assets and since under the worst case scenario deterioration of the spillway could lead to loss of storage in the reservoir, the "Do Nothing" option was discounted early on.

The main portion of the cost of the remedial works was the temporary works required by safety considerations to operate over the water on the outside of the shaft and over the 36 m drop on the inside of the shaft. The difference in costs between the selective and complete replacement of the units was not significant and therefore a decision was made to replace all of the units.

Although replacement of the pre-cast units would prevent further expansion of the top of the shaft, the *in situ* concrete directly under the units had already been cracked, and it was considered that ingress of the reservoir water into the cracks could lead to corrosion of the reinforcement. It was therefore decided to apply a waterproof membrane onto the outside of the shaft using a Flexcrete cementitious coating.

During the inspection several areas of exposed reinforcement were identified on the upper sections of the anti vortex piers. It was decided to carry out a detailed cover meter survey and determine the steel condition by removing the concrete from areas with very low cover. The concrete would then be reinstated using a cementitious compound and the whole anti vortex pier would be then grit blasted and coated with an anti carbonation coating, to reduce the risk of further deterioration of the reinforcement due to lack of cover.

FULL SUPPLY LEVEL RAISING

As a separate project but at the same time as the remedial works design was being carried out, MWH was asked by Southern Water to consider the potential for raising the top water level of Bewl Water by reviewing the amount of wave run-up. During this study a potential for raising the top water level by 350 mm was identified, without raising the embankment crest level or carrying out work on the draw-off tower. This would increase the reservoir storage by 7%, a valuable addition to Southern Water's assets.

It was proposed to carry out the raising by increasing the height of the pre-cast crest units on the spillway shaft. However, full investigation of the effect of the raising and approval of the scheme by the Environment Agency were not possible within the time scale of the remedial works contract, so the shape of the new crest units were designed so that they could be easily raised in the future.

As discussed previously, the major portion of the cost of any works on the spillway is temporary works due to safety considerations. To reduce the risk of working over the centre of the shaft the units were designed so that the future units could be lowered into place and secured from a floating barge without the need for accessing the inside of the shaft.

The blocks to be installed under the remedial works contract were designed with a 200 mm step on the upstream side and a socket in the upstand. The pre-cast section that could be added later to raise the top water level has a corresponding step and socket so that the two blocks can be bolted together and then grouted in place. Sealing of the downstream joints could be carried out by roped access techniques without the need for erecting scaffolding on the shaft.

REMEDIAL WORKS CONSTRUCTION

General

The contract was awarded to Brent Construction following a competitive tender procedure and work began at the beginning of the summer in 1999. Bewl Water is not only a major water supply resource shared by Southern Water and Mid Kent Water but it is also a major amenity used by thousands of visitors each year. The reservoir and its surrounding footpaths are used for fishing, sailing, walking and riding. It was therefore essential that extra care was taken to ensure that no interruption or disturbance was caused to the other users of the reservoir. An excellent working relationship was developed between the Southern Water staff, especially the local rangers, and the contractor. At the end of the contract Brent Construction were presented with a Customer Care Merit award by Southern Water. The contract was complete on programme and under budget.

Temporary Works
Access to the shaft for materials and plant was by barge equipped with Hi-ab crane. Daily access for personnel was by smaller boats.

During the investigation stage of the project, a scaffold walkway had been erected around top of the boat fender. This platform was maintained throughout the remedial works contract, and, in addition, a scaffold platform was erected over the void of the shaft with another hanging walkway suspended around the outside of the weir crest to allow access to both sides of the working area. The central platform effectively blocked the only spillway facility at Bewl Water so the works were programmed for the end of the summer when the water levels are normally low and falling and sufficient storage existed in the reservoir to contain low return period floods without interfering with the works.

Removal of Existing Weir Units
During the design stage consideration was given to the method of removal of the existing pre-cast units. Due to the restrained expansion of the pre-cast units it was considered that a sudden release of any in built stresses could cause problems. Therefore slots were required to be cut through the units prior to their removal. This was carried out by diamond drilling a hole through the base of one of the units nearest an anti-vortex pier in each quadrant. A diamond rope saw was then threaded through the hole and a slot cut upwards through the block (see Figure 2). No sudden release of stresses was noticed.

Further holes were then drilled in the base of each block at the intersection with the *in situ* concrete. A hydraulic jack was then inserted into the hole and pressure applied. The joints on all three sides fractured cleanly and the units were all able to be lifted off in one piece by the crane on the barge.

This operation left a good clean and smooth surface that required little preparation before the new units could be installed. With the units removed it was possible to see that some of the cracking in the *in situ* concrete had penetrated more than half way through the thickness of the shaft walls validating the decision to apply an external waterproof coating to prevent further deterioration.

Installation of New Weir Units

The new pre-cast units were installed using the crane on the barge. The units were placed on thin metal shims and all units were located and levelled prior to any grouting of the vertical or horizontal joints (see Figure 3). Two vertical movement joints were installed in each quadrant, consisting of a 25 mm thick compressible joint filler surrounded by a joint sealant. All other joints were filled with a non-shrink cementitious grout and then pointed with an SBR mortar.

To prevent sliding of the pre-cast units on the smooth surface of the *in situ* concrete, vertical dowels were inserted through holes in the units into the underlying concrete and then grouted in place.

PRESENT SITUATION

The new units have now been in place for four years (see Figure 4) and recent observations made from the balcony of the adjacent draw-off tower indicate that the surface of the replacement units are without deterioration. During this period the reservoir has spilled and this confirmed the quality of the workmanship as spilling was uniform around the perimeter.

The protective coating on the anti-vortex piers is free of defects, maintaining adhesion and without noticeable cracking.

The low water levels in 2003, exposed the coating to the external surface of the spillway shaft, and again, this remains free from defects. No signs of seepage have been noted on the inside of the shaft at high water levels and it appears that the application of the coating has been successful.

ACKNOWLEDGEMENTS

The authors are grateful to Southern Water for permission to present this paper on the remedial works carried out at Bewl Water.

REFERENCES

1.   Davison I (1998). Bewl Water Spillway Investigation, 10th Conf. British Dam Society. Bangor, pp140-150. Thomas Telford, London.

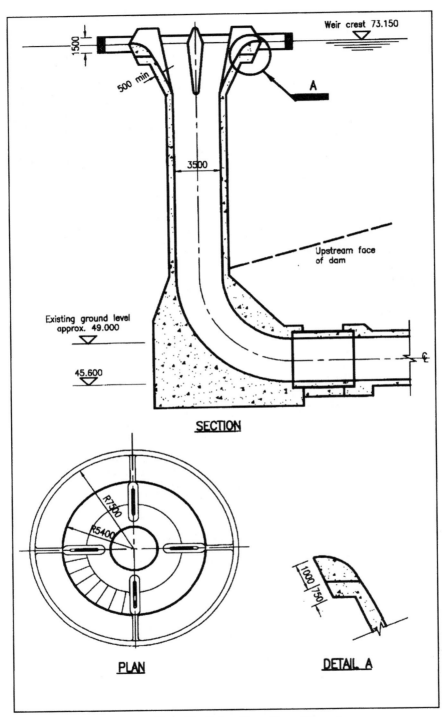

Figure 1 - Detail of the Overflow Structure

Figure 2 - Saw cutting between the existing blocks

Figure 3 - New blocks in place

Figure 4 - Completed spillway crest

Walthamstow Reservoirs No. 4 & No. 5 embankment protection

C.B. PECK, Thames Water RWE, UK

SYNOPSIS.
Walthamstow reservoirs Nos. 2, 3, 4 and 5 are situated in the Lee Valley, north-east London. Reservoirs Nos. 4 and 5 fall within the provisions of the Reservoirs Act 1975, have a common top water level and share a common embankment with Reservoir Nos. 2 and 3 which lie at a lower level. No. 2 and 3 reservoirs are used as settlement lagoons for wash water from the nearby Coppermills water treatment works and were in danger of becoming "silt" bound. The reservoirs are also within a Site of Special Scientific Interest and support fish, birds and wildfowl, including migratory species.

An inspection of the common embankments in 1998, revealed a general lack of protection, including evidence of wave action undercutting the lower toe. Recommendations were made to provide protection to the whole length of the embankment, namely the shore of Reservoir Nos. 2 and 3.

The project involved sinking a chain of timber stakes 3m from the lower toe. A geo-mesh lining was then secured to contain "silt" dredged from Reservoir No. 3. Reeds were then planted in the "silt" to consolidate the protection and enhance the environment. Timber platforms were provided for anglers.

BACKGROUND

Walthamstow reservoirs Nos. 2, 3, 4 and 5 are a chain of reservoirs situated in the Lee Valley, north-east London (Fig. 1). Reservoirs Nos. 2 and 3 were constructed in 1863 and Nos. 4 and 5 in 1866, under the powers of the East London Act of 1853. The two sets of reservoirs share a common embankment.

Long-term benefits and performance of dams, Thomas Telford, London, 2004, 661–674

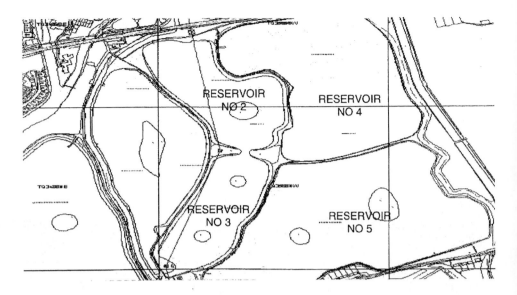

Fig. 1 Location plan of Reservoirs Nos. 2,3,4 & 5
Reproduced from Ordnance Survey mapping on behalf of The Controller of
Her Majesty's Stationery Office © Crown Copyright 100042062

Reservoirs Nos. 4 and 5 are statutory reservoirs, falling within the
provisions of the Reservoirs Act 1975, and have a common top water level
and share a common embankment with Reservoirs Nos. 2 and 3, which lie at
a lower level (Fig. 2). Reservoirs Nos. 4 and 5 are operated as raw water
storage reservoirs and provide a key supply route for stored water to
Coppermills water treatment works, as the final two reservoirs in the gravity
chain. Reservoirs Nos. 2 and 3 are used as settlement lagoons for washwater
from Coppermills water treatment works and are not classified as statutory
reservoirs due to the volumes they hold. These two lagoons were in danger
of becoming "silt" bound.

All the Walthamstow reservoirs form part of a designated site of special
scientific interest (SSSI), which has also been designated as a special
protection area under the EU Birds Directive. This SSSI supports a wide
variety of fish, birds and waterfowl, including migratory species. In
particular they provide a habitat for a colony of herons, which have bred at
the reservoirs since 1928. The environmental management and development
of the site is work in conjunction with English Nature, who act as guardians
of the environmental legalisation. The reservoirs are also used by anglers

and birdwatchers as part of the recreational facilities managed by Thames Water Utilities.

Fig. 2 General view of Reservoir No. 2 next to Reservoir No. 4

Over the past 15 years, three separate incidents have occurred at Walthamstow Reservoirs Nos. 4 and 5, which have affected reservoir safety. The incidents were a downstream embankment slip in 1988, crest settlement between 1986 to 1992 and seepage through the embankment in 1996. Remedial works have been carried out to solve the problems caused by the incidents.

Hydrographic surveys carried out in 1994 and 1998 on Reservoirs Nos. 2 and 3 showed them to be heavily "silted, with a significant increase in "silt" *1 levels between the surveys, as a result of washwater discharge from a newly constructed granulated activated carbon (GAC) / sand separation plant. At the time of the project the inlet to Reservoir/Lagoon No 3 was almost completely blocked with sand and "silt" (Fig. 3). Reprofiling of these reservoirs was identified as being required in the immediate future to maintain their effective use. Issues concerning contamination within the "silt", drying out, transportation and special landfill requirements, ruled out the option of removing the "silt" from site.

*1 the term "silt" referred to in the paper is a general term covering the sediment found in the reservoirs

Fig. 3 Silted inlet of Reservoir (Lagoon) No. 3

During the statutory inspection dated 20th November 1998 of reservoirs Nos. 4 and 5, recommendations were made to inspect the outer embankments. The subsequent inspection, carried out at water level, revealed a general lack of suitable protection including evidence of wave action undercutting the lower toe. The final inspection report recommended that all areas where erosion had taken place, were to be reinstated and erosion protection provided to the whole length of the external bank of Reservoirs Nos. 4 and 5. This is the internal bank to Reservoir (Lagoons) Nos. 2 and 3. This protection was required around the top water level in Reservoirs (Lagoons) Nos. 2 and 3, whose water level is usually constant at around 7.81m above ordnance datum Newlyn (AODN). A project was initiated in September 1999 for the design and construction of 830 metres of bank protection works.

SCOPE OF WORK

Three options were considered, two of which addressed embankment protection only, and one of which addressed embankment protection and washwater treatment as a secondary output.

Option 1: Removal of "silt" from Reservoir No. 3, placing and stabilising it along the external banks of Reservoirs Nos. 4 and 5.

The protection to the external banks in this option, would be provided by dredging the "silt" from Reservoir No. 3 and placing it on the banks to form a "silt" shelf, in which reed beds would be planted. When established, the reed beds will help to keep the "silt" in place and will also provide an environmental enhancement to the area. The length of embankment protected by the reeded "silt" shelf can be identified in Figure 1.

Option 2: Installation of a 2m wide layer of crushed rocks along the external banks of Reservoirs Nos. 4 and 5.

Option 3: Installation of precast concrete mats along the external banks of Reservoirs Nos. 4 and 5.

Options 2 and 3 only addressed the matter of protection of the existing reservoir banks and were unlikely to be favourable from an environmental point of view to English Nature, whose approval was required for any works carried out on these reservoirs.

Option 1 was chosen as it was the only option that, as well as meeting the primary objective of providing protection to the external embankments of the statutory reservoirs, also provided other benefits. Dredging the shallowest part of Reservoir No. 3, will help to maintain the effective use of the reservoirs as settling lagoons for the treatment of the washwater from Coppermills advance water treatment plant. Other benefits included not having to import permanent works materials, which would have created an impact of additional traffic on the restricted local roads leading up to the site. Finally the chosen option was more likely to receive the required environmental approval for the works from English Nature, which in due course was attained. The reed beds provide a new facility for birds such as herons, who already use the site, and also attract new species of birds, and they have also provided biodiversity enhancements to the SSSI.

SURVEYS & TESTING

<u>Hydrographic Surveys</u>
Hydrographic surveys were carried out on Reservoirs Nos. 2 and 3 to
determine the levels and volumes of "silt" in the reservoirs. The
hydrographic survey carried out in 1994 covered all of Reservoir No. 3, but
only part of Reservoir No 2 and therefore did not provide a figure for the
volume of "silt" in Reservoir No. 2. The survey revealed that there was
approximately 50,000m3 of "silt" in Reservoir No. 3, which was
equivalent to 73% of its volume.

The later survey in 1998 covered Reservoir No. 2 as well as No. 3 and took
measurements of the top and bottom levels of the "silt", which enabled the
depth of "silt" to be calculated. In Reservoir No. 2 the "silt" depth varied up
to a maximum depth of 1.1m, and in Reservoir No. 3 up to 2.1m.
Topographical/hydrographic CAD drawings were produced by Thames
Water's Survey Group, which were then used to estimate the volumes of
"silt" in the two reservoirs, which are shown in Table 1.

Table 1 Results of Hydrographic Survey in 1998 for Reservoirs Nos. 2 and 3

Reservoir	Reservoir Capacity at Design TWL (m3)	Volume of "Silt" in Reservoir (m3)	Percentage of "Silt" in Reservoir (%)
Walthamstow No 2	77,000	31,020	40
Walthamstow No 3	68,000	59,400	87

The output from the hydrographic survey carried out in 1998, was primarily
to give an indication of the rate and pattern of the build up of "silt" in
Reservoirs Nos. 2 and 3, but it was also used to determine where best to
dredge the "silt", which was used to form the "silt" shelf.

<u>"Silt" Testing</u>
To provide information on the nature of the "silt" at the base of the
reservoirs to the Contractor Land and Water, several disturbed samples were
recovered using 'grab' sampling techniques from eight separate locations in
Reservoirs Nos. 2 and 3. The results from particle size distribution analyses
and Atterberg Limits indicated that the material fell into two distinct groups.
Two samples contained no fines, one being sand and the other gravel. The
remaining six samples had a fines content varying from between 95 and
99% and liquid limits varying between 135 and 285%. All the samples, bar
one, had a liquid limit of 262% or greater and organic contents around 20%.
By a combination of test results and visual description these samples were
classified as organic "silt"/clay of extremely high plasticity.

The results from the hydrographic surveys and the "silt" testing were used by the Contractor Land and Water and Thames Water, to determine where best to dredge the "silt", that was used to form the "silt" shelf.

DESIGN & CONSTRUCTION

Access to Reservoir Embankments
As there have been minor slips along the external embankments of Reservoirs Nos. 2 and 3 in the past, there was a need to avoid moving any heavy plant along the top or on the embankments, to reduce the risk of causing any further slips. The Contractor was able to carry out all the works from the water using floating craft/machinery (see Fig. 4), with only Land Rovers and small vans being used along the top of the embankments when carrying out the planting of the reed beds.

Fig. 4 Placing of "silt" shelf using machinery on floating craft

Dredging of "Silt"
The Contractor carried out pre and post dredging hydrographic surveys. This information was used to estimate the amount of "silt" dredged and where in Reservoir No. 3 it was dredged from.

The dredging of the "silt" from the reservoirs was carried out using a hydraulic excavator floated on a barge. 3750m3 of "silt" was dredged from Reservoir No. 3, firstly from the area around the outlet from the culvert that brings the washwater into Reservoir No. 3, as this is the location where the larger heavier sand/gravel particles settle out first and had formed banks that were visible above top water level (see Fig. 3). Removing the material from around the outlet helped clear a path for the washwater, creating a more distributed settlement pattern through Reservoirs Nos. 2 and 3. These more coarse particles were a better material for forming the protection shelf being built up all along the 800m length being protected. When the material around the outlet had been exhausted, further "silt" material was dredged from the deepest areas of "silt", identified from the hydrographic survey of Reservoir No. 3. The chemistry of silt samples from Reservoir No. 3 indicated the expected organic rich conditions and elevated sulphide, ammonia, zinc and copper.

There was some existing concrete "rip-rap" embankments protection around the water level on the internal embankments of Reservoirs Nos. 2 and 3. This existing protection was left in place and the new protection shelf formed over the top of the "rip-rap".

"Silt" Shelf Retaining System
A "nicospan" revetment system supplied by MMG, retained the "silt", which formed the 3m wide dredged "silt" planting shelf for the reed beds on the reservoir embankments. The retaining system utilised a geo-mesh lining retained by sinking a chain of timber stakes 3m from the lower toe. "Nicospan" is a prefabricated, double weave revetment fabric made from strong UV stabilized monofilament yarns that are heat sealed to form a series of open pockets each having a width of 220mm, so that posts can be placed into them. The geo-mesh was selected to allow water to pass through but retain the "silt" (Fig. 5).

The posts for the "nicospan" revetment were driven in using a small piling hammer, converted for 100mm posts, mounted on an excavator. The posts were driven, at 500mm centres, into individual pockets of the "nicospan" to progressively "tighten" the revetment. The line of the revetment was agreed with Thames Water's site staff to offer maximum toe restraint to the embankment, but also offer the most ecological benefit. Anchor poles were driven to the rear of the "nicospan" at 100mm centres and, wired to the "nicospan" using galvanized fending wire. The excavator, used to install the retaining system, was secured to a floating pontoon (Fig. 5).

Fig. 5 View of revetment system being placed

The Contractor designed the form and retaining system for the "silt" shelf (see Fig. 6) and Thames Water's Geotechnics Group checked whether the new shelf would affect the stability of the reservoir embankments. The analyses were undertaken for slope angles of 1 in 2 and 1 in 3 using the methodology suggested by Morgenstern and Price (1965) and conservative soil parameters (c' = 0kPa and phi = 37 degrees for the gravels and c' = 3kPa and phi = 20 degrees for the London Clay). The results for failure surfaces within the gravels and the London Clay both with and without the silt shelf are summarised in Table 2

Table 2 FOS Results for failure surfaces within gravels and London Clay with and without the silt shelf

1 in 3 Slope				1 in 2 Slope			
Failure in London Clay		Failure in Gravels		Failure in London Clay		Failure in Gravels	
A	B	A	B	A	B	A	B
1.92	1.80	2.22	2.08	1.36	1.27	1.53	1.40

(Case A FOS without silt shelf and Case B FOS with silt shelf)

The conclusion was that the effect of the "silt" shelf on FOS was minimal and that even with the most onerous combination of a steeper slope and with the failure surface entirely in the London Clay, an acceptable FOS was obtained.

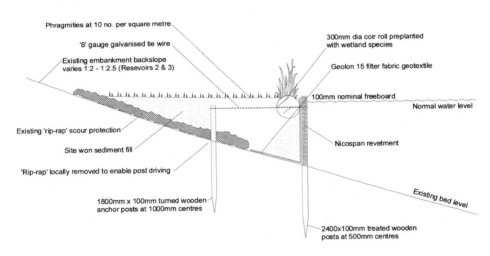

Fig. 6 – Section of "silt" shelf

At the start of the construction period the Contractor formed a short section of the proposed "silt" shelf, which demonstrated the effectiveness of the design, before progressing with the rest of the required 800m length.

After the "silt" shelf was completed (see Fig. 7), reed beds were planted during late spring 2000, which was the best time of the year for their establishment. The reed beds were planted in the "silt" to consolidate the protection and enhance the environment. Timber platforms were constructed at intervals to provide "swims" for the anglers that use the reservoirs.

Fig. 7 View of placed "silt" shelf

Reed Beds

The depth that the reed beds sit in the water was important to their surviving and maturing, and the Contractor formed the "silt" shelf to a level of 8.00m AOD, which allowed for 0.20m settlement of the "silt" shelf. This level ensured that the roots of the plants were always submerged. The finished level of the "nicospan" revetment was 50mm above the top water level. The density of the reed beds planted was ten plants per square metre. The reed beds were planted during May, which was the best time of year for their establishment (Fig. 8). Also the Contractor's design included the use of pre-planted reed "coir" rolls and mattresses, which minimised the chances of die back or natural waste of the reeds.

As these rolls were placed at the front of the shelf and the reeds in the rolls were established, they prevented erosion of the "silt" shelf whilst it consolidated and the planted reeds behind established themselves.

Fig. 8 Planting of reed beds

There is a thriving bird life on the Walthamstow Reservoirs, and the newly planted reeds would be susceptible to damage by the birds, therefore to minimise the damage, netting was placed over the reeds as protection. This type of netting is proven to deter wildfowl interest.

Fishing Platforms ("Swims")
There were ten wooden platforms built out into Reservoirs Nos. 2 and 3 along the 800m of the "silt" shelf and they were approximately 2m long by 3m wide. The swims provided a new safer access to the waterside, and the reed beds either side helped to conceal the outline of the fishermen to the fish. A plan and section of a platform is shown in Fig. 9.

ENVIRONMENTAL ISSUES WITH CONSTRUCTION
There were a few environmental issues identified at the early stages of the project, which were dealt with by the Contractor in a responsible way. There was a need for the Contractor, whilst dredging, to prevent disturbed suspended solids from passing further downstream and into the River Lea. This was done by erecting a geofabric boom sediment curtain at the outlet

from Reservoir No. 2, for the extent of the construction period. This boom curtain was designed by the Contractor.

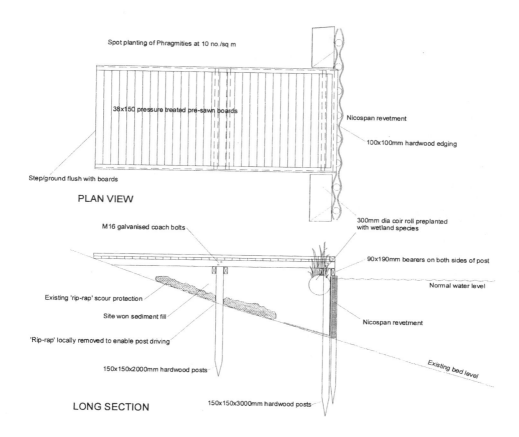

Fig. 9 Section of fishing platform

There are fish and other aquatic life within the reservoirs on the site and the Contractor monitored the dissolved oxygen and ammonia levels in Reservoirs Nos. 2 and 3 at least twice a day during the dredging. If the levels fell dramatically this would be likely to affect the fish and so the Contractor had on site emergency aeration equipment that could be immediately deployed to improve the water quality. The aeration was from a blower feeding a 1m diameter diffuser ring, but this equipment was not actually required to be used.

CONCLUSION

The "silt" shelf and reed beds were completed in May 2000, taking six months to complete, and apart from some minor secondary planting of reeds in early 2001, the reed beds are fully established and along with the "silt" shelf are fulfilling their function of protecting the embankments of the statutory reservoirs and providing biodiversity enhancements to the SSSI.

ACKNOWLEDMENTS

The support of John Harris and Jon Green of Thames Water and James Maclean of Land and Water in preparing this paper is gratefully acknowledged.

REFERENCES

Green J - Walthamstow No.4 and No.5 Reservoirs – A Recent History

Institution of Civil Engineers (2000). A Guide to the Reservoirs Act 1975

Morgenstern, N.R., and Price, V.E., (1965). "The analysis of the stability of general slip surfaces", Géotechnique 15, pp.79-93

Author Index